T0292296

Cosmetic Formulation

Principles and Practice

Cosmetic Formulation

Principles and Practice

Edited by

Heather A.E. Benson
Curtin University, School of Pharmacy and Biomedical Sciences

Michael S. Roberts
University of Queensland

Vânia Rodrigues Leite-Silva
Federal University of São Paulo

Kenneth A. Walters
An-eX Analytical Services Ltd

CRC Press
Taylor & Francis Group
Boca Raton London New York

CRC Press is an imprint of the
Taylor & Francis Group, an **informa** business

Cover artwork designed by Bruna Dallabrida and Laura Dallabrida.

CRC Press
Taylor & Francis Group
6000 Broken Sound Parkway NW, Suite 300
Boca Raton, FL 33487-2742

First issued in paperback 2021

ISBN 13: 978-1-03-209307-9 (pbk)
ISBN 13: 978-1-4822-3539-5 (hbk)

Library of Congress Cataloging-in-Publication Data

Names: Benson, Heather A. E., editor.
Title: Cosmetic formulation : principles and practice / edited by Heather
A.E. Benson (Curtin University, School of Pharmacy) [and three others].
Description: Boca Raton, Florida : CRC Press, 2019. | Includes
bibliographical references and index.
Identifiers: LCCN 2018054576| ISBN 9781482235395 (hardback : alk. paper) |
ISBN 9780429190674 (ebook)
Subjects: LCSH: Chemistry, Technical. | Cosmetics.
Classification: LCC TP151 .C67 2019 | DDC 668/.55--dc23
LC record available at https://lccn.loc.gov/2018054576

Visit the Taylor & Francis Web site at
http://www.taylorandfrancis.com

and the CRC Press Web site at
http://www.crcpress.com

To the inspirational women in my life: my daughter Victoria, my sister Elaine, and my mum Anne.

Heather A.E. Benson

A special thanks to our parents, family, colleagues, staff, students and those who have gone before us. Each of them has helped us in a different but special way.

Michael S. Roberts

I would like to dedicate this book to my mother Celina and father Dan, for all their love and for always and strongly believing in me. They always say: "From our small city Miracatu to the world."

To my husband Nelson and my kids Laura, Bruna and David: You are my inspiration, my love, my everything.

Vânia Rodrigues Leite-Silva

For my grandchildren, Joe, Abbie, Lily, Elsa and Roscoe. Tomorrow's scientists?

Kenneth A. Walters

Contents

Foreword

The title of this book is self-explanatory; however, its impact was clear to me upon reading the entire manuscript.

What have Doctors Benson, Roberts, Leite-Silva and Walters added to this complex and highly important area to our society?

They have been preceded by many thoughtful editors, over many decades, starting with the many editions of Harry, Baran, Barel, and Paye, as well as the new book from mainly the Asian author, Sakamato.

When the reader finishes the book, they will understand its important benefits: Namely, conciseness, focus, and highly useful for the cosmetic formulator.

Each chapter covers vast areas of both old and new literature, combining them in a highly readable manner. No formulator could take the time to study all of this, refine it, and interpret it, in the way that the authors have done, or they would never be able to get to their formulation bench.

The literature is expanding, both basic and applied, and each author was cognizant of their goal, skillfully bringing this information to practice.

All but the most experienced formulators will benefit from their diligent work.

We hope that the editors will continue to update this important book, to the benefit of the international users of such formulation information.

Howard I. Maibach, MD
University of California San Francisco Medical School
San Francisco, California

Preface

> *The nature of beauty is one of the most enduring and controversial themes in Western phi-losophy, and is—with the nature of art—one of the two fundamental issues in philosophical aesthetics.*

Crispin Sartwell*

So what is beauty? Is it really in the eye of the beholder (subjective) or is it a fundamental attribute of an object or an act (objective)? Philosophers from Plato and Aristotle to Hume and Kant, together with many other great minds have considered this without resolve. Whatever the outcome of this philosophical argument, there is little doubt that within our species, beauty attracts attention.

> *You're beautiful. You're beautiful.*
> *You're beautiful it's true*
> *I saw your face in a crowded place*
> *And I don't know what to do.*

James Blunt

With this in mind it is, perhaps, not surprising that the importance of and the desire to 'enhance' our image of beauty occupies a significant amount of cost, time and effort during our daily routines. In essence, we all want to look good to others and to ourselves. While in most cases this can be achieved simply by dressing smartly, presenting a clean façade and smiling, there are often occasions when some extra help is needed.

> *Makeup is no different than clothes and accessories—it's embellishments for your face. And it also gives you creative freedom. You get to have that moment in front of the mirror every morning and give yourself self-love. You're making yourself up beautiful, which is essentially self-love.*

Michelle Phan

This was recognised many centuries ago by the ancient Egyptians. The use of scented oils and ointments was a common means of disguising body odour, and various metallic ores were used to alter face colouring. Thus, the use of materials to alter external body characteristics can be traced to around 10,000 BC. There have been a few improving developments over the past 12 millennia, for example, we can now eliminate body odour rather than simply mask it. We can now alter the colour of our hair and teeth using chemicals, which are relatively safe if used correctly. We can now apply various chemicals to our skin, which will reduce the risks associated with radiation damage and insect-borne disease vectors.

Cosmetics and cosmetic science have evolved over many years. In the time of Elizabeth I, it was the fashion to whiten the face and bosom using ceruse, a mixture of white lead and vinegar, to create this appearance. As Hamlet said to Ophelia 'God has given you one face, and you make yourself another' (Hamlet, Act 3 Scene 1, William Shakespeare, c. 1600). The application of ceruse was not only to create the illusion of youth, it also served to disguise the ravages of such diseases as smallpox and the like. Unfortunately, ceruse was quite a poisonous mixture, but that was in a time when an alternative means to achieve the pale complexion was to be bled. Morton's fork? Nobler Elizabethan women could dye their

---

* "Beauty," *The Stanford Encyclopedia of Philosophy*, Winter 2017 Edition, Edward N. Zalta (ed.), https://plato.stanford.edu/archives/win2017/entries/beauty/

hair yellow using a mixture of saffron, cumin seed, celandine and oil, and other dyes could be applied to the cheeks as rouge (e.g. Cochineal) and the eyelashes (e.g. kohl). Such practices are, of course, still carried out today, but the chemicals and formulations currently used have been extensively evaluated and monitored for safety in use.

Worldwide regulatory agencies, such as the Food and Drug Administration (FDA; United States) and the Scientific Committee on Consumer Safety (SCCS; Europe), together with industry bodies such as the Personal Care Products Council and the Cosmetic Ingredients Review Panel in the United States, and Cosmetics Europe Personal Care Association, are collectively charged with ensuring consumer safety during cosmetic product use. Under these watchful eyes, there is no doubt that the proper use of cosmetics is safe, bar the occasional allergic or irritant reaction. Cosmetic science has also made full use of our expanding knowledge of physiology, molecular biology, toxicology and chemical interactions.

In this volume, the editors have collated many of the recent advances in cosmetic science with some of the fundamental aspects of cosmetic product development, and the various chapters have been authored by established leaders in their respective fields. The volume has been designed to provide an understanding of contemporary cosmetic product development. Chapters 2 through 6 cover the anatomy and physiology of the skin, nail and hair to provide an insight into the environment in which cosmetic products act. Then, Chapters 7 to 16 address the main components that comprise cosmetic products, their properties and applications. Chapter 17 focuses on sunscreens and provides an excellent overview of the development of a sunscreen product from concept to consumer. Chapters 18 through 20 focus on advanced materials and their application in skin care products. Chapter 21 provides an insight into the processes involved from formulation design through to production. The packaging of cosmetic products is essential to ensure stability and importantly to identify a brand, exude elegance and facilitate marketing (Chapter 23). The final chapters (24 to 26) focus on testing, an essential element in the development process to ensure the cosmetic product meets its desired outcomes.

As editors, we owe a debt of gratitude to our publisher, Taylor & Francis, especially Hilary Lafoe and Jessica Poile, for their assistance in the preparation of the manuscript and their infinite patience. We are also extremely lucky to have the support, love and endless encouragement from our partners, Tony, Nelson, Carmel and Peggy. Finally, our thanks and appreciation go to our authors who have given us so much of their time and openly shared with us and you, their extensive knowledge of their subjects. Without them, this book would not have been possible.

Heather A.E. Benson
Vânia Rodrigues Leite-Silva
Michael S. Roberts
Kenneth A. Walters

Contributors

Fabricio Almeida de Sousa
AQIA Industrial Chemistry
São Paulo, Brazil

Newton Andréo-Filho
School of Pharmacy and Biomedical Sciences
Curtin Health Innovation Research Institute
Curtin University
Perth, Australia

Catherine Tolomei Fabbron Appas
UNIFESP – Federal University of São Paulo
Diadema, Brazil

Heather A.E. Benson
School of Pharmacy and Biomedical Sciences
Curtin Health Innovation Research Institute
Curtin University
Perth, Australia

Joke A. Bouwstra
Drug Delivery Technology
Leiden Academic Centre for Drug Research
Leiden University
Leiden, The Netherlands

Antônio Celso da Silva
ABC – Brazilian Society of Cosmetology
São Paulo, Brazil

Chia-Ming Chiang
CMC Consulting
Foster City, California

L. Edward Clemens
Riptide Bioscience, Inc.
Vallejo, California

Lorena Rigo Gaspar Cordeiro
USP – University of São Paulo
Ribeirão Preto, Brazil

Regina Lúcia F. de Noronha
RN Sensory Training and Consultancy
São Paulo, Brazil

Ralph W. Eckert
Department of Pharmacy
Institute of Pharmaceutics and Biopharmaceutics
Philipps–Universität Marburg
Marburg, Germany

Chantra Eskes
SeCAM: Services & Consultation for Alternative
   Methods
Agno, Switzerland

Verda Farida
Department of Pharmacy
Institute of Pharmaceutics and Biopharmaceutics
Philipps–Universität Marburg
Marburg, Germany

Thamires Batello Freire
USP – University of São Paulo
São Paulo, Brazil

Ricardo D'Agostino Garcia
Formular Cosmetic Services
São Paulo, Brazil

Dene Godfrey
Independent Preservation Advice Ltd.
Cardiff, United Kingdom

Jeffrey E. Grice
Translational Research Institute
University of Queensland Diamantina Institute
Brisbane, Australia

Steffen F. Hartmann
Department of Pharmacy
Institute of Pharmaceutics and Biopharmaceutics
Philipps–Universität Marburg
Marburg, Germany

Celio Takashi Higuchi
UNIFESP – Federal University of São Paulo
Diadema, Brazil

Jesse Jaynes
Riptide Bioscience, Inc.
Vallejo, California

John Jiménez
Belcorp
Bogotá, Colombia

Cornelia M. Keck
Department of Pharmacy
Institute of Pharmaceutics and Biopharmaceutics
Philipps–Universität Marburg
Marburg, Germany

Daniel Knoth
Department of Pharmacy
Philipps–Universität Marburg
Institute of Pharmaceutics and
  Biopharmaceutics
Marburg, Germany

Avadhesh Kushwaha
Department of Pharmaceutics and Drug Delivery
The University of Mississippi
Oxford, Mississippi

Ariane Dalanda Silva Ladeira
UNIFESP – Federal University of São Paulo
Diadema, Brazil

Jürgen Lademann
Department of Dermatology, Venerology and
  Allergology
Charité – Universitätsmedizin Berlin
Berlin, Germany

and

Berlin Institute of Health
Humboldt–Universität zu Berlin
Berlin, Germany

Majella E. Lane
Department of Pharmaceutics
UCL School of Pharmacy
London, United Kingdom

Vânia Rodrigues Leite-Silva
ABC – Brazilian Society of Cosmetology
São Paulo, Brazil

and

UNIFESP – Federal University of São Paulo
Diadema, Brazil

Gislaine Ricci Leonardi
UNICAMP – University of Campinas
Campinas, Brazil

Patricia Santos Lopes
UNIFESP – Federal University of São Paulo
Diadema, Brazil

Dayane Pifer Luco
UNIFESP – Federal University of São Paulo
Diadema, Brazil

Brett MacFarlane
Australian College of Pharmacy
Faculty of Health
School of Clinical Sciences
Queensland University of Technology
Brisbane, Australia

Bozena Michniak-Kohn
Ernest Mario School of Pharmacy and Center for
  Dermal Research
Rutgers–The State University of New Jersey
Piscataway, New Jersey

and

Laboratory for Drug Delivery (LDD)
Center for Dermal Research (CDR)
Rutgers–The State University of New Jersey
Piscataway, New Jersey

Cliff Milow
BASF Corporation
Florham Park, New Jersey

Enamul Haque Mojumdar
Drug Delivery Technology
Leiden Academic Centre for Drug Research
Leiden University
Leiden, The Netherlands

Jemima Moraes
Laboratory for Drug Delivery (LDD)
Center for Dermal Research (CDR)
Rutgers–The State University of New Jersey
Piscataway, New Jersey

S. Narasimha Murthy
Department of Pharmaceutics and Drug Delivery
The University of Mississippi
Oxford, Mississippi

and

Institute for Drug Delivery and Biomedical Research
Bengaluru, India

Adelino Kaoru Nakano
Global Innovation, Symrise AG
São Paolo, Brazil

Reinhard H.H. Neubert
Institute of Applied Dermatopharmacy
Martin Luther University Halle–Wittenberg
Halle/Saale, Germany

Jadir Nunes
SLINC – Scientific Learning & Innovation
    Consulting
São Paulo, Brazil

and

ABC – Brazilian Society of Cosmetology
São Paulo, Brazil

Antony O'Lenick
Siltech Corporation
Lawrenceville, Georgia

Apoorva Panda
Department of Pharmaceutics and Drug Delivery
The University of Mississippi
Oxford, Mississippi

Alexa Patzelt
Department of Dermatology, Venerology and
    Allergology
Charité – Universitätsmedizin Berlin
Berlin, Germany

and

Berlin Institute of Health
Humboldt–Universität zu Berlin
Berlin, Germany

Ricardo Pedro
Ipel Itibanyl Produtos Especiais Ltd.
Jarinu, Brazil

Olga Pelikh
Department of Pharmacy
Institute of Pharmaceutics and
    Biopharmaceutics
Philipps–Universität Marburg
Marburg, Germany

Vinam Puri
Laboratory for Drug Delivery (LDD)
Center for Dermal Research (CDR)
NJ Center for Biomaterials
Rutgers–The State University of New Jersey
Piscataway, New Jersey

Tannaz Ramezanli
Ernest Mario School of Pharmacy and Center for
    Dermal Research
Rutgers–The State University of New Jersey
Piscataway, New Jersey

Michael S. Roberts
Translational Research Institute
The University of Queensland Diamantina
    Institute
Brisbane, Australia

and

School of Pharmacy and Biomedical Sciences
Curtin Health Innovation Research Institute
Curtin University
Perth, Australia

and

Department of Medicine
University of Queensland
Brisbane, Australia

Frank Romanski
BASF Corporation
Florham Park, New Jersey

Kishore Shah
Polytherapeutics, Inc.
Lakewood, New Jersey

H.N. Shivakumar
Institute for Drug Delivery and Biomedical
    Research
Bengaluru, India

and

K.L.E. University's College of Pharmacy
Bengaluru, India

Margaret Smith
Syndet Works Pty Ltd.
Bayswater North, Australia

Rose Soskind
Laboratory for Drug Delivery (LDD)
Center for Dermal Research (CDR)
Rutgers–The State University of New Jersey
Piscataway, New Jersey

Pascal L. Stahr
Department of Pharmacy
Philipps–Universität Marburg
Institute of Pharmaceutics and
    Biopharmaceutics
Marburg, Germany

Florian Stumpf
Department of Pharmacy
Institute of Pharmaceutics and
    Biopharmaceutics
Philipps–Universität Marburg
Marburg, Germany

Krishna Telaprolu
Faculty of Medicine
Translational Research Institute
The University of Queensland Diamantina Institute
Brisbane, Australia

Philip L. Tong
Department of Dermatology
Royal Prince Alfred Hospital
Sydney, Australia

and

Immune Imaging Program
Centenary Institute
Newtown, Australia

and

The University of Sydney
Sydney, Australia

Sonia Trehan
Laboratory for Drug Delivery (LDD)
Center for Dermal Research (CDR)
Rutgers–The State University of New Jersey
Piscataway, New Jersey

Kathryn W. Woodburn
Riptide Bioscience, Inc
Vallejo, CA

Kenneth A. Walters
An-eX analytical services Ltd.
Cardiff, United Kingdom

Zoe Webster
Skinnovation Laboratories Ltd.
Nottingham, United Kingdom

1

Cosmetic Products: Science and Senses

John Jiménez, Vânia Rodrigues Leite-Silva and Heather A.E. Benson

CONTENTS

Modern cosmetics are associated with beauty and well-being. They are used to highlight and accentuate features and decrease attention to perceived imperfections. Beauty has been appreciated since the beginning of civilization, and the development of cosmetics has progressed with human history, drawing on both scientific and cultural evolution. In modern Western societies, a relationship is established between body beauty, intelligence and high purchasing power.

There is evidence of body painting used as camouflage for hunting and to show ferocity in inter-tribal wars as early as prehistoric times. The Picts, a tribe living in Scotland in 1000 BC, were named for the Latin term *pictus* (painted) by the Romans (Butler, 2000). Aboriginal body painting has been used by Australian aborigines for thousands of years. The colours and patterns vary between regions and tribes. Hours can be spent applying body paint, often by a relative, in preparation for religious and cultural ceremonies.

Egypt's last queen, Cleopatra (51–30 BC), bathed in goat's milk to give her softer, smoother skin. There is extensive evidence of the Egyptians' use and importance of cosmetics; indeed many were placed in tombs along with the most precious artefacts needed in the afterlife. Items recovered by archaeologists have included leaves of henna from which the Egyptians extracted dye to colour the palms of the hands, soles of the feet, nails and hair, red ochre for face colouring, green colouring from malachite (copper carbonate ore) for the eyes, and kohl for eyelashes and eyebrows. Jars containing oils, unguents and ointments used to cleanse, moisturize and decorate the skin, together with incense, equipment for pulverising and preparing cosmetics, razors, and so forth were frequently recovered in tombs dating as far back as 3000 BC (Poucher, 1941). Indeed there is clear evidence of the use and continued development of cosmetics utilizing minerals, plants and animal products throughout the Middle East, India, China and Europe

(Butler, 2000). From the 15th and 16th centuries onwards, the rapid advancement in trade brought new opportunities that have led to the global industry in cosmetics we see today. Helen Butler provides an excellent and thorough history of the development of cosmetics in her chapter 'Cosmetics Through the Ages' (Butler, 2000). In addition, our knowledge of anatomy, physiology, metabolism, toxicology and other related sciences has provided a greater understanding of the skin, hair and body, facilitating the development of more effective and safer cosmetic products that meet the demands and expectations of the consumers of this century.

Many factors have contributed to the multibillion-dollar cosmetic product industry we enjoy today. Over the course of the last century, the role of women changed with increasing participation in the labour market, leading to greater awareness of self-image but also more disposable income to devote to appearance. Marketing of cosmetic products became more sophisticated with the introduction of new technologies from mass-produced print magazines, to television and the Internet with its highly effective social media and sales platforms. This came at the same time as globalization and economic growth began to open more markets and provide billions more potential customers. In recent decades, it has also become acceptable for men to be concerned with self-image and embrace products to enhance their appearance. The cosmetic industry has embraced and helped to drive technological and scientific evolution, so there is a constant emergence of new brands aided by direct marketing and the participation of social networks. Indeed the industry has benefited from perceiving these emerging trends in society and adapting products and practices accordingly. This includes the move towards more natural ingredients and the elimination of animal testing whilst ensuring that cosmetic products are efficacious and elegant, safe, and produced ethically and efficiently to minimize the impact on the environment.

This book aims to provide an understanding of contemporary cosmetic product development. Chapters 2–6 cover the anatomy and physiology of the skin, nail and hair to provide an insight into the environment in which cosmetic products act. We then address the main components that comprise cosmetic products, their properties and applications (Chapters 7–16). Chapter 17 focuses on sunscreens and provides an excellent overview of the development of a sunscreen product from concept to consumer. Chapters 18–20 focus on advanced materials and their application in skin care products. Chapter 21 provides an insight into the processes involved from formulation design through to production. The packaging of cosmetic products is essential to ensure stability and importantly to identify a brand, exude elegance and facilitate marketing (Chapter 23). The final chapters (24–26) focus on testing, an essential element in the development process to ensure the cosmetic product meets its desired outcomes.

Where does our approach to cosmetic products begin? We suggest that it is the convergence between the senses and science, therefore that is the main theme explored in the opening chapter of our book.

Senses and Science: Art, Science and Communication, and Their Role in Cosmetic Products

Whilst the interaction between the five human senses (touch, feel, sight, smell, taste) and the environment has evolved slowly over thousands of years, the way in which cosmetic products are developed to interact with the senses has evolved much more rapidly. Integral to the development of better cosmetic products is the ability to quantify the interaction with the senses when a consumer is faced with a set of stimuli produced by a cosmetic product, which has changed considerably in recent years. We have witnessed the advance of traditional explicit tests such as basic sensory analysis tests (as discussed in detail in Chapter 24), home-use tests and the application of neuroscience techniques to develop a range of implicit evaluation tests leading to the creation of more innovative concepts for product design.

The cosmetic sector is highly competitive and industries related to this area have now, more than ever, the challenge of expanding the market through innovation. In this search of blue oceans (Chan and Mauborgne, 2015), the sensory experience is the great opportunity to generate a differentiation and it is precisely the blend of science and senses what has allowed the development of new methodologies and new strategies to make cosmetic products more attractive and enhance perception by involving more senses.

TABLE 1.1

Tools to Synergize the Sensorial Experience

Tool	Why?
To discover the consumer's insight	To satisfy the consumer's insight with a differential benefit/claim versus the competitor is a great opportunity for differentiation.
Psychology of the colour	It is a powerful tool with many applications to explore and discover in the formulation of products; design of packaging; and creation of concepts, publicity and advertising.
Sensory additivity	• When stimuli are accurately combined, there is synergy during the perception. This is known as the sensory additivity; it means that $1+1=4$. • When stimuli are not correctly combined there is a decrease in the final perception. This is known as a sensory suppression; it means that $2+2=1$.
Innovation in implicit evaluation	Neuroscience allows the opportunity to enhance the packaging design, formula, products, concept and advertising. • The different neuroscientific techniques make it possible to develop a different form of evaluating the product and introducing novelty in claims.
Synesthesia in design	Synesthesia is understood as the perception of the same stimulus by two different senses. The synesthetic design is an opportunity to surprise the consumer.

The sensory experience of a product is the result of many factors such as emotional, physical, chemical, and neural. It is for this reason that consumers can be considered as multisensory individuals. Table 1.1 lists some tools that can be applied to synergize the sensory experience through senses. Each one of them represents a large area where new methodologies and research projects can be developed with the purpose of satisfying and assessing the needs of the consumer in a different, original and innovative way.

To achieve a proper sensory experience, the primary objective is to seduce the consumer with an effective mixture of aromas, tastes, textures, visual stimuli and sounds. The consumer adapts and reacts to different stimuli that are produced in his senses, and in this way, these stimuli influence the final purchase decision (Stein, 2012). For example, in the case of a perfume, the aroma is the stimulus that communicates and gives value to a concept, to a brand and to an experience. In the case of food, smell and sight are what invites to consume it. Hundreds of sensations are perceived by the sense of touch, from something rough to something smooth, and for a cosmetic product distinguishing viscous, humid or sticky, and so on. Physical contact can lead to positive or negative sensations and therefore has an important influence on having a satisfactory product perception (Malfitano, 2007). It is important that the consumer can test a product before buying it, and it is at this point when the differential and creative details of the product design are decisive.

Discovering the Consumer's Insight

Many times we wonder why a consumer is faithful to an olfactory note, why he prefers a perfume among others, why he likes one colour more than another on the product label and why he prefers a type of texture in an antiaging product. What are the true reasons that lead to consuming, to selecting a specific product, texture or makeup colour?

Cristina Quiñones (2015) has an interesting definition of the insights: 'they are the human truths that allow to understand a deep, emotional and symbolic relation between a consumer and a product'. A true insight is surprising and revealing. Table 1.2 shows examples of insights into different types of industries (Keane, 2018; Millenial Marketing, 2018). The insights generate opportunities for innovation, branding and actionable communication for the companies. They are little-hidden secrets among consumers that explain some of their motivations, fears, behaviour and emotions.

It is motivating when the design of a new product and a new concept starts thanks to the discovery of a powerful insight. At the stage of formulation design and development, discovering and activating the insight becomes a powerful strategy because satisfying that insight with a texture, a sensorial profile, a claim and/or a benefit without a doubt will allow the creation of rewarding and deep, emotional experiences.

TABLE 1.2

Examples of Insights in the Market

Insight	Why Did It Facilitate Innovation?
A study conducted by Activia and research partner GlobalWebIndex revealed that 80% of women in the United States aged between 25 and 55 agree that they are their own worst critic.	In February 2017, Activia launched its 'It Starts Inside' rebranding campaign. The idea was to inspire women to overcome their 'internal critic'.
At puberty, 49% of girls feel paralyzed by the fear of failure, leading them to avoid trying new things.	#LikeaGirl: a successful campaign of Always (P&G) was launched in 2013 and continues to have a positive impact.
7% of Internet users now class themselves as 'eco consumers', those who think brands should produce eco-friendly products.	Nespresso's 'The Choices We Make' campaign focused on efforts of global sustainability and addressed the topic of coffee brands that create unnecessary waste.
Only 2% of all women consider themselves beautiful and only 5% consider themselves pretty.	The Dove 'Real Beauty' campaign is one of the most successful in the cosmetic field. Women easily connect with this insight.

Why is it important to discover the insight? Because the cosmetic formulator can focus on developing innovative textures, amazing sensory profiles, or disruptive claims proposals in a different way in order to satisfy the insight in an innovating way. It is also quite important to differentiate the concepts of insight, verbatim and claim.

Collect many verbatims → Discover the insight → Innovate in claims and benefits

Verbatim

Verbatim is the exact reproduction of a sentence, phrase, quote or text without making any changes from what is written in the source or has been expressed by the consumer (Foro Marketing, 2018). A good strategy to find an insight is to gather as many verbatims as possible and use them to analyze the behaviour and necessities of the consumer.

Insight

Insight is revealed based on the consumer's behaviour, beliefs, necessities and/or wishes (Quiñones, 2015) as it is about the deeper motivations that stimulate people's actions. It is important to bear in mind that an insight is not data, neither observations nor verbatim. There are different methodologies developed by companies who are experts in finding powerful insights.

Claim

A claim is what is promised about a product. Claims can be obtained from sensory analysis, efficacy and clinical tests made using bioengineering equipment and techniques, perception tests with the consumer and neuroscientific tests. A company could gain the opportunity to obtain a differentiation strategy for a new product launch by trying to activate the revealed insights with new, different and original claims.

How do we know that we have achieved the identification of a powerful insight? A powerful insight is able to achieve the following reactions (Dalton, 2016):

- Produces a strong emotional connection with the consumer.
- Resets existing beliefs.
- Facilitates the solution of a real problem and captures new clients.
- Inspires and/or creates an objective and a solution that will be well valued by the client/consumer.

Colour Psychology

Colour theory is important to enhance the sensory experience, as the colour of a formula, packaging or advertising must be selected not only because of the aesthetic aspect, but colours unleash specific signals in the central nervous system and in the cerebral cortex. Photoreceptor cells in the retina of the eye absorb certain wavelengths of light reflected by an object's surface. These cells transform the stimuli into electrical impulses that are sent to the brain via the optical nerves to create the colour sensation (Poyectacolor, 2018). Each colour is associated with one or several emotions, sensations and impressions.

Colour and Its Influence over Emotions

In her interesting book about colour psychology, Eva Heller (2011) collated the opinion of 2,000 people in Germany. She asked about the colour they liked the most, the colour they liked the least, the impressions that each colour caused on them and which of these ones were more associated with specific feelings. Different studies have demonstrated that colour association is not made by likes, but by past experiences that were lived since childhood, which is evidenced by means of psychological symbolism. Tables 1.3 and 1.4 show the most and the least popular colours in this study.

TABLE 1.3

The Most Favourite Colours (Heller, Germany)

Blue	45%
Green	15%
Red	12%
Black	10%
Yellow	6%
Violet	3%
Orange	3%
White	2%
Pink	2%
Brown	1%
Gold	1%

TABLE 1.4

The Least Favourite Colours (Heller, Germany)

Brown	20%
Pink	17%
Grey	14%
Violet	10%
Orange	8%
Yellow	7%
Black	7%
Green	7%
Red	4%
Gold	3%
Silver	2%
White	1%
Blue	1%

We conducted a study in Colombia (Jiménez and Sánchez, 2014), where we asked a group of 209 women (aged 19–84) from a range of different professions about their favourite (Table 1.5) and least favourite colours (Table 1.6). It is interesting to see the similarities between the two studies. For example, blue was the most common favourite colour and together with red was in the group of the three most favourite ones, whereas brown was consistently among the least favourite colours. Despite the difference in countries and cultures, the effect of a colour on emotions can be a general pattern in the population of different countries and therefore across a broad range of markets. This is an advantage for global cosmetic brands that are launching products in several countries.

Table 1.7 shows a summary of some of the emotions generated by different colours (Gobé, 2005; Heller, 2011). The three primary colours are yellow, red and blue. Yellow produces the quickest perception on the eye and is the most contradictory colour as it represents optimism and jealousy, as well as entertainment, understanding and treason. Red represents passion, love, hate and has the oldest chromatic denomination in the world. Children associate red with sweets, candies and tomato sauce. Blue is the favourite colour for 46% of men and 44% of women. It is synonymous with empathy and faithfulness.

Secondary colours are green, orange and violet. Green represents the colour of life, fertility and hope. It is considered a quiet colour. Violet is the last colour of the solar spectrum and the third complementary colour that comes from red and blue. It represents power and luxury and is associated with products for seniors. Orange is the colour of entertainment and joy, and although it is not perceived as popular, it is seen everywhere.

Black and white are considered colours without any value as they do not appear in the chromatic scale: white is colourless, neutral and cold, and does not have any radiation; black represents absorption

TABLE 1.5

The Most Favourite Colours

Blue	34%
Red	13%
Grey	12%
Purple	11%
Green	6%
Pink	5%
Brown	4%
White	3%
Fuchsia	2%

Source: Jiménez, J. and D. Sánchez (2014), *Cosmetics & Toiletries Brasil* **26**(4): 62–68.

TABLE 1.6

The Least Favourite Colours

Green	22%
Yellow	21%
Brown	12%
Orange	10%
Red	7%
Black	7%
Grey	6%
White	4%
Pink	3%
Blue	2%

Source: Jiménez, J. and D. Sánchez (2014), *Cosmetics & Toiletries Brasil* **26**(4): 62–68.

TABLE 1.7

Emotions Generated by Some Colours

Yellow	Blue	Red	Green	Orange	Purple	White	Black
Heat	Air	Attraction	Life	Energy	Senility	Colourless	Death
Source of energy	Synonym of success	Passion	Hope	Warm	Dignity	Nobility	Threat
Positive	Calm	Positive	Patience	Delight	Seriousness	Peace	Fear
Glory	Nobility	Intensity	Abundance	Good humour	Regret	Noble	Sadness
Sun	Sky	Hell, fire	Nature	Vitality	Grief	Light	Night
Luminosity	Intelligence	Feelings	Future	Stimulus	Power	Purity	Mourning
Joy	Rest	Wrath	Youth	Youth	Sadness	Neutrality	Luxury
Calm	Moderation	Violence	Fertility	Modernity	Charm	Brightness	Negativity
Kindness	Harmony	Energy	Organic	Innovation	Misery	Naivety	Power

of radiation and results from a blend of all the colours. Black is associated with luxury concepts in cosmetics. Table 1.8 is a summary of the characteristics of some of the colours and their application to the design of cosmetics (Jiménez and Sánchez, 2010).

Importance of the Chromatic Accord

The chromatic accord is a combination of colours that can add certain effects (Heller, 2011). Like the olfactory accord that when properly done can transmit a concept in perfumery (as discussed in Chapter 16), the chromatic accord can have a meaning and psychological association. A specific colour can produce different effects that are sometimes contradictory when the colour is paired with another colour.

TABLE 1.8

Characterisation and Use of Colours in Cosmetics Design

Colour	Characteristics	In Which Cosmetic Concept Can It Be Used?
Yellow	Yellow colour is identified with heat and energy. This is the colour of joy.	Happiness, joy, glory, associated with vitamin A and C, action, brightness, light, positivism, optimism, it helps to heal wounds.
Blue	Blue is the colour of cold and air; it means nobility, calm and rest for the eyes.	Coolness, clarity and transparency, decrease in physical and nervous tension, denotes preventive medicine, water, moisture.
Red	Red colour is identified with attraction. This is the colour with the greatest impact on the retina. It is related to fire and lust.	Passion, anger, love, violence, fear, emotions, courage, challenge, intimacy, against envy.
Green	Green is the colour of life and can be interpreted as spring or autumn. It is related to water and the development of new ideas.	Renewal, protection, calmness, provocation, sedative, hypnotic and tranquillizer.
Orange	Orange colour transmits energy and good mood, joviality, cheerfulness, party.	It increases and stimulates the state of mind, accelerates the pulse, happiness
Purple	Purple is the colour of senility, of royalty and religion. This colour identifies power and charm.	Ideal colour for male colonies. If it is combined with golden colour it means absolute prestige. It can also be used in products for elderly people.
White	White colour means light, purity, goodness and rejection of heat. It can be associated with home sensations.	It transmits cleanliness, hygiene, clearance.
Black	The colour of hidden things, of the night, of impossible things. It serves to recognize danger and to give a response.	Black colour represents sobriety, seriousness, elegance.

The effect might vary depending on the context in which it is seen, either in art, fashion or nature, and the specific combination of colours might evoke positive or negative feelings.

An accord consists of two to five colours, with a specific percentage of each. There is always one colour that predominates and exerts an influence on the result of the accord. One colour can produce different effects depending on the context in which it is perceived, for instance, red can be sensual or violent, green can be healthy or mysterious, yellow can be bright or painful (Jiménez and Sánchez, 2010). The chromatic accords can have an effect on the synergy of the sensory experience when they are used on the formulation, packaging, advertising and marketing of a product, as the particular composition of colours produces a feeling and a specific effect.

It is interesting to study the effect of the chromatic accords on the perception of the sensory performance of a cosmetic formulation. We undertook a study (Jiménez and Sánchez, 2014) to investigate the effect of the chromatic accords of the truth (Figure 1.1) and the lie (Figure 1.2; Heller, 2011) on the sensory performance of a moisturizing lotion (Table 1.9). Eighteen female panelists, aged 20 to 40 years, who were habitual consumers of cosmetic products, participated in the study.

During the test, every participant was given the following products for evaluation:

- A sample of the moisturizing lotion in a white jar, packed in a box that had the chromatic accord of the truth.
- A sample of the moisturizing lotion in a white jar, packed in a box that had the chromatic accord of the lie.

Participants did not know that they were given the same emulsion. They thought they had received two different products. Each participant was asked:

- Without opening the box, which product will give more benefit to your skin?
- After using the product, mark from 1 to 5 (5 being the best) the effects of moisturizing, softness, stickiness and residue.

The results were very interesting. First, 66.7% of the participants said that they most liked the product packed in the box that had the chromatic accord of the truth, whilst 33.3% preferred the product packed in the box that had the chromatic accord of the lie. This allowed us to demonstrate how the theory of

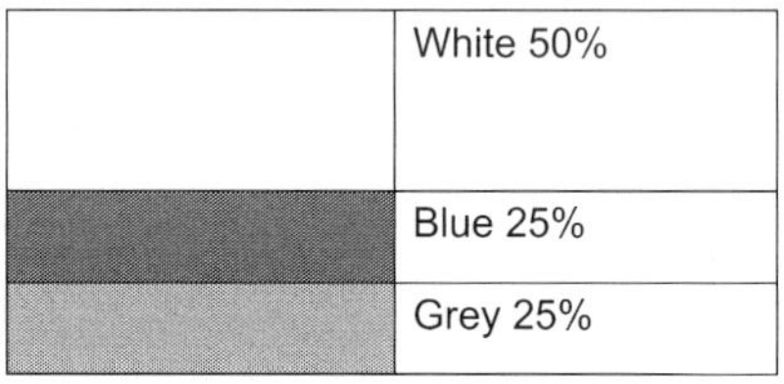

	White 50%
	Blue 25%
	Grey 25%

FIGURE 1.1   Chromatic accord of the truth.

	Yellow 40%
	Purple 15%
	Black 15%
	Green 15%
	Brown 15%

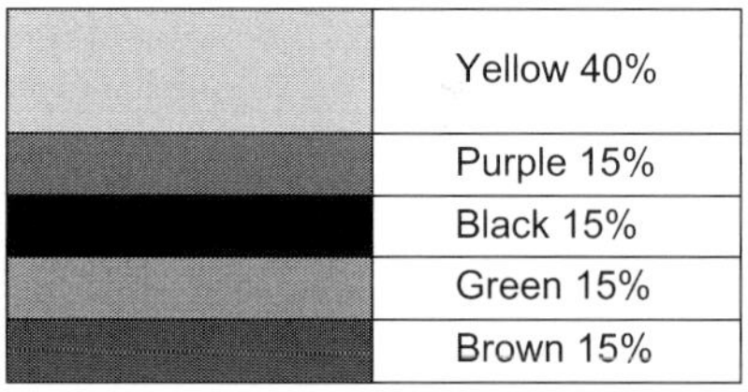

FIGURE 1.2   Chromatic accord of the lie.

TABLE 1.9

Moisturizing Formula Used in the Study by Jiménez and Sánchez (2014)

Ingredient	Percentage (%)
Water	qs
EDTA	0.1
Carbomer	0.2
Propylene glycol	2.0
Dimethicone	2.0
Emollient 1	2.0
Emollient 2	2.5
Cyclomethicone	4
Emulsifier	2
TEA	0.2
Preservative	0.6
Moisturizing active	4.0

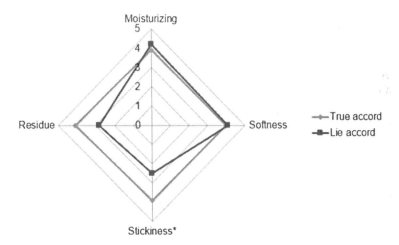

FIGURE 1.3 Sensory analysis for the study of the chromatic accord.

chromatic combinations has an influence on the perception of a product. After the sensory experience, it was the stickiness attribute that was better evaluated in the product packed in the box that had the chromatic accord of the truth (p < 0.05), as shown in Figure 1.3.

Sensory Additivity

The sensory process has several stages (Stein, 2012): stimulation, sensation, perception and answer. Sensory additivity is a very interesting tool to be applied when designing cosmetic products. This tool searches to synergize the final perception of a product based on the right and innovating blend of stimuli. This process is possible because the brain works with different sensory stimuli simultaneously.

Figure 1.4 shows the sequence of the multisensory process of cosmetic products. The product creates a variety of stimuli (aroma, touch, colour, etc.). The consumer transforms these stimuli into neural signs which are retransmitted in a centralized way. Some of these signs are able to converge in individual neurons, generating a multisensory convergence, because different types of stimuli are received by the same membrane (Stein, 2012). This multisensory perception can be synergistic if the combination of the sensory stimuli is suitable (1+1=4), otherwise it is diminished if the combination is not appropriate (2+2=1).

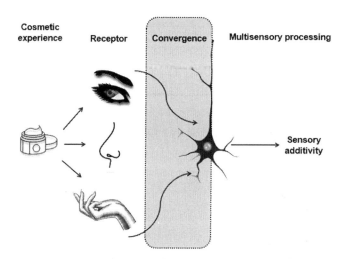

FIGURE 1.4 Multisensory process of a cosmetic product.

This approach within the field of neuroscience is applied by scientists to the design of products and to 'neuromarketing' (Lindstrom, 2010; Pradeep, 2010; Lindstrom, 2011) in different categories such as food, music, packaging and cosmetics. In all studies (Lindstrom, 2010; Pradeep, 2010; Lindstrom, 2011), it was demonstrated that the more the senses are involved in the interaction between a product/brand and the consumer, the more loyalty there will be to the specific brand. Taking advantage of this multisensory process allows the development of more stimulating products and better consumer experiences. One approach that can be developed with the marketing department is to create a product that will surprise the consumer by utilizing an identified insight.

Some examples of the multisensory process in different types of industry are:

- When eating potato chips the sound is essential to perceive good quality. If the chips do not produce this crunchy effect, the consumer concludes that they are not fresh or not good quality.
- Changing the colour in fruit drinks. The consumer expects orange juice to be yellow or orange in colour; any other colour would be strange and unpleasant.
- Playing French music at the supermarket has been shown to increase sales of French wine.
- Some airports have blue light to relieve passengers' tiredness after a long flight, as the brain wakes up more easily under this colour of light.
- When fragrance is added to a formula of liquid soap, a smell of fried food instead of a floral fragrance will drastically diminish the sensory perception of the formula. This is an example of sensory suppression.
- The engine sound of Harley Davidson motorcycles is well recognized. If a customer buys one and the engine sound considerably decreases, the positive perception of the product will also decrease.
- When the iPod was launched to the market, customers considered its most successful feature to be not the sound quality but the sensory experience of touch when interacting with the device.

The Role of Colour in Sensory Additivity

We investigated the theory of additivity and sensory suppression in relation to how colour can affect the perception of an olfactory note (Jiménez and Sánchez, 2014). The following samples were prepared:

Sample 1: Vanilla solution at 2%, in ethanol, with beige colour
Sample 2: Vanilla solution at 2%, in ethanol, with red colour

Sample 3: Mint solution at 2%, in ethanol, with blue colour

Sample 4: Mint solution at 2%, with purple colour

Two groups of 12 women (aged 20 to 40) were given mint sample 3 and vanilla sample 2 (Group 1), or mint sample 4 and vanilla sample 1 (Group 2). Each sample was identified with the corresponding olfactory note name. Each participant was asked to mark her perception on a scale from 1 to 10, where 1 is the lowest value of perception and 10 the highest. The question was 'How much does this sample make you remember mint/vanilla scent (according to each)? Mark the level to which you like each sample.'

The results demonstrated how the colour of the olfactory notes directly influenced how the volunteers remembered the olfactory note and liked the sensation (Figure 1.5). A vanilla sample is not expected to be red and therefore the level of liking and remembering decreases despite the two samples having the same fragrance concentration. The experience is similar to the mint sample. A blue mint sample makes more sense than a purple mint sample, and because of this the level of remembering and liking also decreased for sample 4, although the fragrance concentration was the same in both samples.

We also investigated the effect of the packing box colour on the moisture perception of a moisturizing emulsion (Jiménez and Sánchez, 2014). An emulsion (formula outlined in Table 1.9) was bottled in two identical white jars that were packed in different colour boxes: one white and one blue. Each sample was labeled with a different numeric code. Twelve women (aged between 20 and 40) were asked to evaluate the sensory parameters of moisture, residue, softness and stickiness, and overall which of the two products had a better performance. The participants did not know that they were being given the same formula. It was interesting that 83.4% of participants said that the formula bottled and packed in the blue box had a better performance (Figure 1.6) and that the sensory parameters of moisture and softness were better for the sample packed in the blue box ($p < 0.05$: Figure 1.7).

A sensory perception test for two samples of the same nutritive emulsion but with different olfactory notes was also conducted. The first fragrance belonged to the woody family and the second one to the floral family. Twelve women participants were asked to evaluate the sensory parameters of moisture,

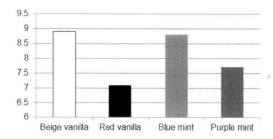

FIGURE 1.5   Sensory additivity/suppression on vanilla and mint samples.

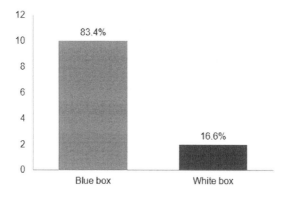

FIGURE 1.6   Preference of the product according to the box colour.

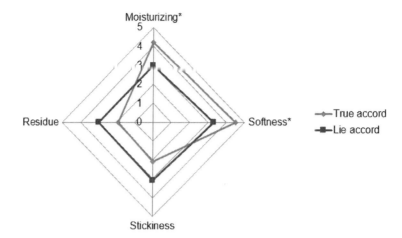

FIGURE 1.7 Sensory parameters evaluated.

softness and residue after application. Again the participants did not know they were evaluating the same product. Residue was evaluated as significantly different for the two samples ($p < 0.05$), showing that the emulsion sensory perception can indeed be affected by the olfactory note. Different olfactory families can have a calming or energizing effect, as has been reported in different bibliographic references of aromacology (outlined in Table 1.10). Table 1.11 shows an evaluation matrix of the sensory additivity/suppression effect that can be used for multisensory evaluation of some types of cosmetic formulas.

Innovation in Implicit Evaluation

Statistics show that approximately 80% of new products fail after a year of being launched, even though these products have been subjected to numerous instrumental and sensory tests during the design and development stage, extensive evaluations in consumer groups, and even clinical tests to support claims. Why would a product that has undergone so many studies and tests fail in the market after the launch?

In a study related to the purchase decision-making process, the neuroeconomist Peter Kenning monitored the areas of the brain of consumers that were activated in response to different brand images, including some that were preferred by panelists and others that they disliked (Barden, 2013). Based on this study Kenning concluded that there was a difference in the brain areas with greater activity when the panelists were shown images of their favourite brand compared to other brands. In this case, the brain showed less reflective thinking activity and more activity in the intuitive decision-making areas. Other

TABLE 1.10

Classification of Some Aromas According to the Calming or Energizing Effect

Effect	Aroma	Effect	Aroma
Calming	Peppermint	Energizing	Vanilla
	Lavender		Chocolate
	Mint		Coffee
	Eucalyptus		Orange
	Tangerine		Orange blossom
	Jasmine		
	Rose		
	Chamomile		
	Geranium		

TABLE 1.11

Evaluation Matrix about the Effect of Sensory Additivity for Some Types of Formulations

Product	Challenge/Benefit	Primary Perception	Secondary Perception	Senses Involved in the Sensory Activity
Facial moisturizing emulsion	Increase the perception of hydration	Hydrated and soft skin touch	Defining which is the shade of blue that the consumer relates to hydration the most. Defining which is the smell of natural source of water that generates the most intensive experience when using a hydrating emulsion.	Touch Sight Smell
Facial firmness cream	Increase the sensation of firmness	Firmer skin touch	Texture that looks like firm rubber with memory effect. Seeing and touching refers to firmness.	Touch Sight
Facial blur	Improve skin imperfections	Visual blurring effect; wrinkles, soft-focus and mattifying effect	Product texture that looks like an eraser.	Sight Touch
Facial filler	Wrinkle-filling effect	Decrease in the visualization of wrinkles	Serum that comes in a pen format and allows filling wrinkles one by one. What type fragrance can boost the filler effect?	Sight Touch Smell
Facial lifting	Tightening effect on the skin	Visual lifting in wrinkle reduction	Sensory effect of rigid touch over the skin	Sight Touch
Exfoliant	Removal of dead cells on the skin	Soft abrasive sensorial beads that are felt when applied	Visual effect of breaking of the rupture of the beads and accumulation of particles that remove skin impurities. Include a sound that increases the perception of exfoliation.	Touch Visual Hearing
Purifying shampoo	Hair cleaning	Soft and non-greasy hair touch	Smell. Determine the fragrance that enhances the sensation of cleanliness to a greater degree.	Touch Smell
Hair perfume	Perfuming the hair	Smell that remains on the hand when the hand passes through the hair	Visual. Pigment that shines in the dark when the hair is perfumed.	Smell Touch Sight
Hair oil	Hair hydration	Soft sensorial and flexible and fresh hair	Paper strip that changes colour when the hair is hydrated.	Touch Sight
Hydrating body lotion	Skin hydration	Water spray texture visual that transmits high hydration	Sensorial with film former that leaves sensory moisture on the skin.	Sight Touch
Deodorant	Odour control	Elimination of unpleasant odour when doing physical activity	Cold sensory release when the active ingredient acts.	Smell Touch
Hand cream	Hydration and softness	Smooth and soft hands sensorial	Fragrance release with hand friction for smoother skin.	Touch Smell

(Continued)

TABLE 1.11 (CONTINUED)

Evaluation Matrix about the Effect of Sensory Additivity for Some Types of Formulations

Product	Challenge/Benefit	Primary Perception	Secondary Perception	Senses Involved in the Sensory Activity
Cellulite treatment	Measurement reduction, cellulite appearance improvement	Visual reduction of adipose tissue accumulation	Equipment that goes through the skin surface and translates the decrease in the accumulation of fat in pleasant music.	Sight Hearing
Fine fragrances	Perfuming	All-day, long-lasting smell	Liberation of aromatic notes that are different from touch perfume or whenever there is an emotion.	Smell Touch
Body splashes	Perfuming	All-day, long-lasting smell	Perception of change in aromatic notes when tasting sweet or acid foods.	Smell Taste
Shaving cream	Facilitate the shaving experience	Sensory smooth skin that allows the blade to slide easily	Cream with a progressive foaming effect that swells and forms a layer of voluminous foam on the skin.	Touch Sight
Aftershave	Skin conditioning after shaving	Cool soft sensorial that decreases skin irritation	Creamy texture that passes over the skin and changes colour slightly after an increase in temperature indicating the presence of sensitive skin.	Touch Sight
Styling gel	Maintenance of styling at high humidity levels	Maintenance and control of the hairstyle, silky and light sensorial	Texture with pigments that glows in the dark when the hair absorbs high humidity indicating the need for reapplication.	Sight Touch

neuroscience studies have also shown that strong brands have an effect on brain activity that leads to rapid decision making (Lindstrom, 2010; Lindstrom, 2011).

We have previously shown that brand experience influences the perception of liking of an anti-aging emulsion (Jiménez and Sánchez, 2010). We acquired a facial product from a well-known and well-positioned luxury brand. We removed the product from the container and used the empty container to package an emulsion that had an unpleasant sensory profile, that is a high waxy, oily residue, low spreadability and high stickiness (product A). The product that was contained in the expensive container was added to another container that was damaged, dirty, scratched and poor labelling condition (product B). Note that the sensorial profile of the luxury product was superior to the sensorial profile of the emulsion packed in the prestigious brand container.

Fifteen women (aged between 20 and 55 years) of average socio-economic level who were regular consumers of facial treatment products participated in the study. They were given the products and asked to rate the following parameters:

- During application: ease of application, level of greasiness, stickiness and smoothness among others.
- After application: feeling of freshness, hydration, stickiness and amount of residue.

It was interesting to note that at the time of the application of product B, two panelists stopped to observe the container in detail and inquired about its condition. The panelists who evaluated product A just glimpsed at the package and immediately proceeded to apply the product. This behaviour indicates there is trust in the brand since none of the women stopped to read the information on the packaging or verify small details. Figure 1.8 depicts the sensory evaluation of the two products after application. Some parameters were better evaluated for the cream with the worst sensorial profile packaged in the luxury brand container (product A) demonstrating how the power of the brand can influence the experience of the product.

Figure 1.9 shows the intuitive purchasing process where the consumer can make a decision in a matter of milliseconds. Knutson et al. (2007) used functional magnetic resonance imaging (fMRI) to show that the nucleus accumbens septi area of the brain is activated once a person perceives a preference for a purchase. This brain area is associated with pleasure, compensation, addiction and the placebo effect. When a purchase is positive, it becomes intuitive, fast and emotional. It can be considered that it is a strategy designed by our brain as a mechanical decision-making process that helps us to better cope with a world saturated with information.

During the initial purchase that can last a few milliseconds the power of the brand is key to influence the decision. However, for the repurchase and subsequent success of the product in the market, the experience of using the formula is essential for adequate sales performance of the product. Thus, one of the objectives of the formulator is to influence the purchasing process by using a sensory profile that increases the neurological response. Neuroscience applied to the research and development formulation process is a great opportunity for the creation of a brand, and the development of products that captivate the hearts and minds of consumers. There is a substantial bibliography associated with the application of neuroscience studies in advertising, branding and marketing, but very little literature on its application in the formulation stage despite the great opportunities available.

Explicit Evaluation vs. Implicit Evaluation

There are two ways to evaluate the impact that a stimulus or product has on the consumer: explicit and implicit evaluation. Explicit evaluation is related to the degree of perception expressed by the consumer in response to a specific stimulus and uses techniques such as focus group sessions, home-use tests and in-depth interviews. Explicit evaluations are a great contribution to the evaluation of products and are widely used in the development process. They provide a lot of information about the performance of the product, the degree of satisfaction and the impact that the product may be generating on the consumer, as shown in Table 1.12. However, it is important to complement these studies with implicit evaluations,

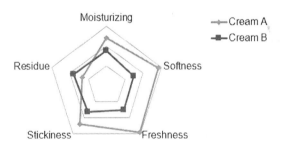

FIGURE 1.8 Sensorial analysis of two emulsions.

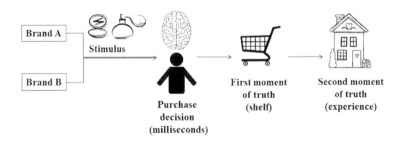

FIGURE 1.9 Purchase decision process.

TABLE 1.12

Types of Motivation in the Consumer

Implicit Motivations		Explicit Motivations	
What to evaluate	*How to measure*	*What to evaluate*	*How to measure*
• Frustration	• EEG	• Colour, odour, flavour,	• Qualitative and quantitative
• Meditation	• fMRI (functional	texture and hearing	tests
• Excitement/like	magnetic resonance	• Efficacy and performance	• Home-use test
• Interest	imaging)	• For instance, 'Soap should	• Focus group
	• Eye tracking	clean and smell good' and	• Efficacy tests with bio
	• Facial expression	'Soap should produce	• Engineering equipment
	• Electrodermal activity	enough foam'	

which seek to assess the internal motivations that lead to a decision, through the evaluation of emotion, attention and memory process (Jiménez et al., 2016). Neuromarketing is the application of neuroscience principles to obtain information about how a person makes consumption-based decisions using these three parameters. One of the great opportunities is to develop applied neuroscience studies during the product formulation stage. The objective is that implicit evaluations complement explicit evaluations, rather than substituting them.

Neuroscience Techniques Applied in the Formulation Stage

Neuroscience is the study of the human nervous system, including the brain, its anatomy and its functions. Neuroscience is essential to understand the different physiological reactions that occur when there is exposure to brands, products and advertising. Neuroscience techniques include physiological or biometric measurements that are captured from body signals such as:

Facial expressions: The human face registers a wide variety of emotional states (Figure 1.10), including changes that are easily observable such as smiling, and small changes that are not easily observable such as muscle movements.

Eye tracking: This tool evaluates saccades (eye movements used to rapidly fixate from one object to another) and pupil dilation in response to different stimuli. Figure 1.11 depicts an example of an eye-tracking reading in front of two images.

Electrodermal activity: Skin transpiration is a parameter evaluated as a response to a specific stimulus. It can be evaluated on the skin of the fingers as shown in Figure 1.12.

Functional magnetic resonance imaging (fMRI): This is a very accurate tool that provides precise images of different levels of brain areas that are active in response to a stimulus (Figure 1.13). However, it is very expensive and not routinely accessible.

Electroencephalography (EEG): This technique measures electrical activity in the brain and is being increasingly used to evaluate formulation-based stimuli. It offers a less expensive alternative to fMRI. An example is shown in Figure 1.14, where the behaviour of different waves as a function of time is observed.

Examples of Eye-Tracking Application

The eye-tracking method provides information about the areas and the sequence where the panelist is paying more attention. This technique can be used during the design stage of formulations to allow the development of new methodologies, differentiated claims and reduce time for decision making. Figure 1.15 shows the application of six different lipstick tonalities in a volunteer and Figure 1.16 shows an example of the reading being made. It is possible to determine which lip colour can have more impact and draw more attention from consumers (Jiménez, 2017).

This methodology also allows us to obtain comparative answers between diverse groups of panelists. Figure 1.17 presents a heat map for a group of 10 men and Figure 1.18 presents a consolidated heat map of 10 women. It is striking because we can observe which colours are more attractive for the diverse groups,

FIGURE 1.10 Example of evaluation of facial expressions.

FIGURE 1.11 Example of eye-tracking reading.

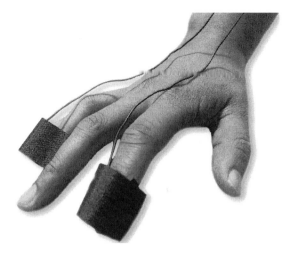

FIGURE 1.12   Example of electrodermal activity.

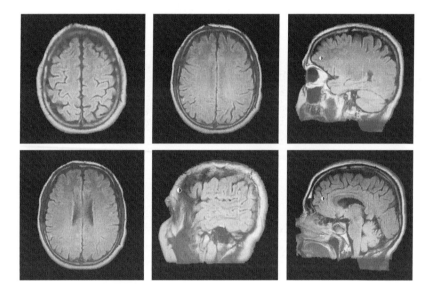

FIGURE 1.13   Example of the study of fRMI.

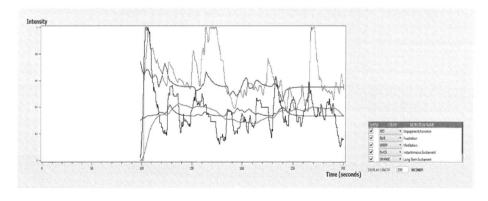

FIGURE 1.14   Example of EEG response curve.

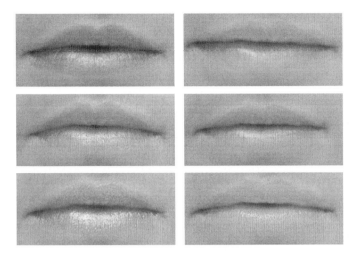

FIGURE 1.15 Different tonalities of labials for evaluation by eye tracking.

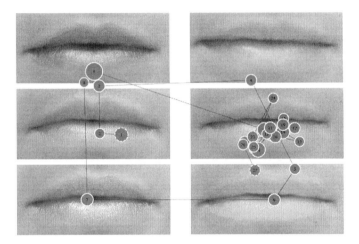

FIGURE 1.16 Example of readings and heat map for a panelist.

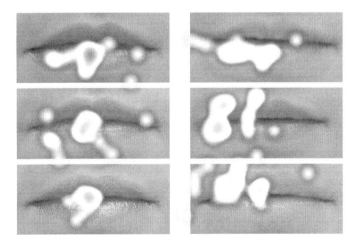

FIGURE 1.17 Heat map for the group of male panelists.

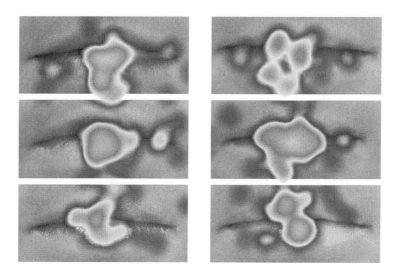

FIGURE 1.18   Heat map for the group of female panelists.

allowing us to accelerate decision-making processes in the design area and also to evaluate different parameters in the concept and advertising of the product.

We developed a new methodology to evaluate the performance of fragrances that we used to evaluate the relationship between fragrance, product name and image (Jiménez, 2015). With this type of methodology, it is important to carry out a prior calibration process to know the performance of saccades movements in a given context. Figure 1.19 presents an example of a reading when there is coherence between the name and the image. The word 'tree' (*árbol* in Spanish) corresponds to the image. Figure 1.20 shows an example of a reading when there is no coherence, since the word 'carrot' (*zanahoria* in Spanish) does not correspond to the image. As seen in the two examples, the performance of saccades movements is different.

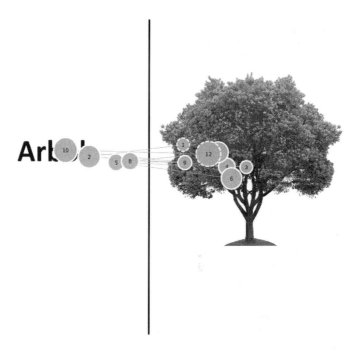

FIGURE 1.19   Example of response when there is coherence in the stimulus.

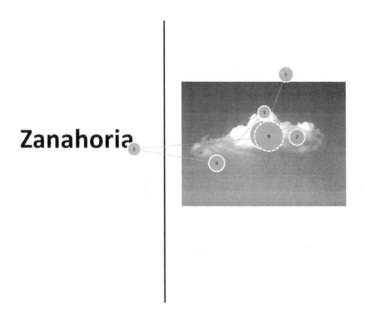

FIGURE 1.20 Example of response when there is no coherence in the stimulus.

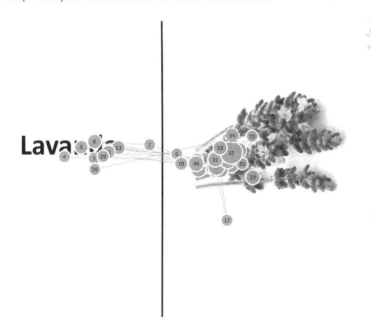

FIGURE 1.21 Example of response when there is coherence in the stimulus.

The calibration stage, including the coherence analysis, is thought provoking because it opens the door to the design of new methodologies to study different textures of cosmetic products, including emulsions, anhydrous products, rinse-off products and fragrances. Importantly, it can also be applied in the evaluation of concepts, brand names and product advertising. Figure 1.21 shows the response when the elements that make up the stimulus are coherent since the name lavender corresponds to the image and the fragrance that was presented to the panelist to smell. Figure 1.22 presents the type of response when there is no coherence in the stimuli since the name lavender corresponds to the image, but in this case, the panelist was presented with an olfactory note of coffee. As seen in Figures 1.21 and 1.22, the responses are different.

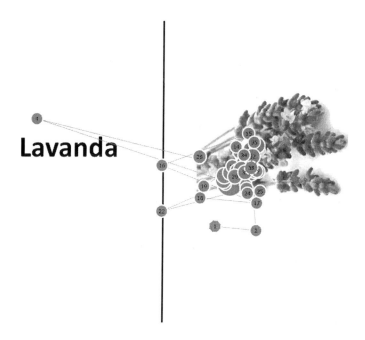

FIGURE 1.22   Example of response when there is no coherence in the stimulus

The coherence methodology can also be applied to evaluate the performance of products. We recently applied this methodology for the evaluation of soft-focus effect in anti-aging formulations (Jiménez and Guzmán, 2017). The objective was to develop an evaluation method that is quick and thus allows us to validate and discard formulations in less time. In this case, the prior coherence analysis was carried out with high-resolution photographs of the before and after anti-wrinkle effect. Leather pieces were used for the evaluation of the soft-focus effect. Figure 1.23 presents an example of a reading in one of the samples considered as blank (without product application), where the leather portion was divided into two equal parts: the distribution of the gaze points is the same on both sides. Figure 1.24 presents an example of a reading of a formulation where the right side of the leather matrix is observed as an immediate decrease in wrinkles due to the soft-focus effect. This effect changes the distribution of the gaze points in comparison to the blank.

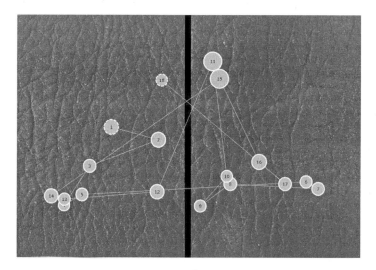

FIGURE 1.23   Example of eye-tracking readings for calibration panel. (Image courtesy of *Cosmetics & Toiletries*.)

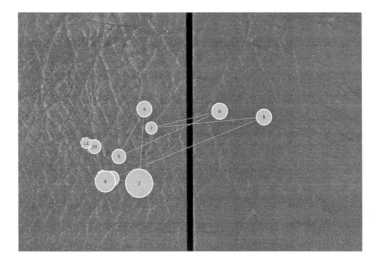

FIGURE 1.24   Example of eye-tracking readings in a sample evaluation. (Image courtesy of *Cosmetics & Toiletries*.)

Examples of Eye-Tracking and EEG Application in Hair Care Products

Hair care is a category where the application of neuroscience techniques is also fascinating. We applied the concordance/discordance methodology for the evaluation of the performance of hair conditioning agents (Guzmán and Jiménez 2016). In this case, the coherence was related to images where the hair was in very good condition and others where the hair was in poor condition (it had been previously washed with a 15% solution of SLS), as shown in Figure 1.25. Figure 1.26 presents an example of the reading for the two different conditioning agents. The performance in this case is related to a smaller range of gaze coverage by the gaze points.

In this example, the electroencephalography readings allow us to know the motivations that the panelists feel when faced with stimuli. Figure 1.27 shows an example of the response curve at the time when the panelist is combing the hair strands after the product is applied. It is interesting that one can see which of the study formulations generates more interest and commitment and which one reduces frustration levels in the consumer when evaluating the performance of the formulation. In this methodology, it is important to analyze the inflection points (points 1 and 2 in the figure) since these are a clear example of an immediate change in perception. This type of evaluation allows us to learn in greater depth which formulation has the better performance and therefore provides some assurance of which formula should have better evaluation in a later clinical efficacy or dermatological study or in a test with consumers.

Synesthesia in Design

The origin of the term *synesthesia* comes from the fusion of two Greek words that mean 'together' and 'sensation', therefore, it is understood as the perception of the same stimulus by two different senses. Synesthesia seeks to mix the sensations of different senses and in some cases feelings, and is the reason it has been widely used in music, painting, literature and poetic metaphors:

- Phrases in everyday language such as 'bitter cold', 'loud wallpaper', 'warm smile', 'sweet caress'.
- 'Without colour, he heard nothing. He filled notebooks with the sound of yellow and red' from Holly Payne's novel *The Sound of Blue.*
- A woman's heart is made to 'drink the pale drug of silence' in George Meredith's text *Modern Love.*
- 'Your name tastes like grass' from the song 'Tu nombre me sabe a yerba' by Joan Manuel Serrat.

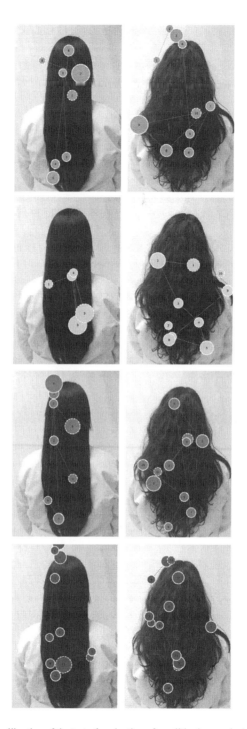

FIGURE 1.25   Images for the calibration of the test of evaluation of conditioning products.

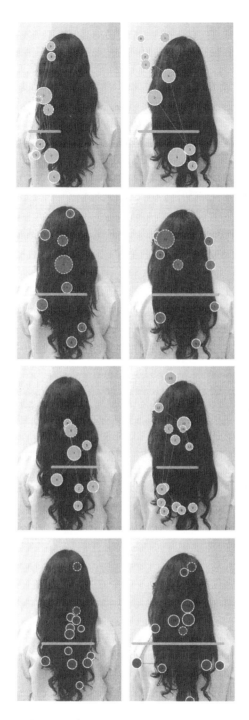

FIGURE 1.26 Example of eye-tracking results for controls.

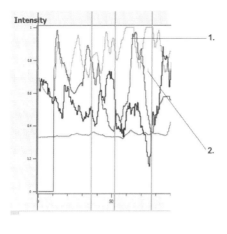

FIGURE 1.27   Example of EEG curve at the time of combing.

Synesthesia is a condition where experiences related to perception, such as a colour or taste, are caused by stimuli that would not normally be associated with experience. Synesthetic experiences have three characteristics: they are provoked by a stimulus, they are conscious perceptions and they are automatic (Stein, 2012).

Interesting studies on synesthesia have been conducted in the food science area. For example, Spence has created a large body of literature on multisensory perception, particularly as applied to food and beverages (Zampini and Spence, 2012; Velasco et al., 2018). He has assessed the role of visual and auditory cues in multisensory perception of flavour, for example showing that high-frequency sounds and curved shapes can improve the perceived sweetness of food and that a wine may taste better if it is preceded by the sound of uncorking a bottle. Involving several senses at once is beneficial, because brain activity is increased, and the experience becomes more rewarding.

Reinoso Carvalho et al. (2017) studied the effect that certain sounds can have in improving some attributes of basic flavors. The participants (65 females and 51 males) evaluated the creaminess of two samples of chocolate (without knowing the samples were identical) against different soundtracks. They perceived the chocolate to have greater creaminess and sweetness when they listened to the 'creamy' soundtrack (consonant-long flute notes) determined as 'creamy' compared to the 'rough' soundtrack (pizzicato short violin lines).

There are some quirky examples, related to cosmetic products, of the application of synesthetic concepts such as the famous Hotel Café Royal that had in its bar for a period of time, a menu of cocktails that were inspired by and tasted like some of the most aromatic notes of fine perfumery (LuxuryLaunches, 2016) and in Bart's bar in London the cocktails were mixed with specific aromatic notes, which were also edible and improve the mood (Eysenck, 2016).

Synesthesia drives us to think differently. Table 1.13 presents some product design ideas that could be leveraged in the synesthetic concept. The implicit evaluation methodologies are the most suitable ones for the design of synesthetic tests. In a recent investigation (Guzmán and Jimenez, 2018) we showed how music, in this case Vivaldi's *The Four Seasons*, can affect a different perception of certain olfactory notes given that each season has a different tempo and is associated with different sensations and emotions. We showed how these musical notes can change the sensory perception of hydration and softness for identical cream formulations. When four different formulations were evaluated with each of the seasons a silicone dispersion was rated best with the 'Summer' season (fast tempo evoking tension and force), a petrolatum dispersion and an emulsion rated equally best with the 'Autumn' season (moderate tempo evoking calmness), whilst the liquid hydrogel formulation rated best with the 'Winter' season (slow pace evoking contemplation). The authors also demonstrated smelling–hearing synesthesia with six fragrances and the music, and tasting–touching synesthesia between chocolate and the skin formulations.

TABLE 1.13

Product Ideas under a Synesthetic Approach

Category	Examples of Synesthetic Concepts
Emulsions	Can the texture of opera be developed?
Fragrances	Can a fragrance that smells like rock be developed?
Shampoo	Can a shampoo formula based on a song be developed?
Fragrances	Can an edible fragrance that tastes like cinnamon be developed?
Anti-age	Can an anti-age treatment that reminds of French gourmet food be developed?
Fragrances	Can the rhythms of a typical Latin American dance be translated into an aromatic note?
Body care	Can textures that allow one to touch oriental songs be developed?
Sun care	Can personality-based formulations be developed? A happy, optimistic or melancholic formulation?

The use of synesthetic design in cosmetic formulation offers a whole new world of developments. The application of neuroscience methods will continue to play a vital role in the development of new evaluation methodologies for cosmetic products that allow the cosmetic scientist to design products for and quantify the multisensory experience. Ultimately this will optimize the consumer experience and continue to drive innovation and success in cosmetic product development.

REFERENCES

Barden, P. (2013). *Decoded. The science behind why we buy.* United Kingdom, John Wiley & Sons Ltd.

Butler, H. (2000). "Cosmetics through the ages." In *Poucher's perfumes, cosmetics and soaps,* 10th ed. Dordrecht, Kluwer Academic Publishers: 13–64.

Chan, K. W. and R. Mauborgne (2015). *Blue ocean strategy. How to create uncontested market space and make the competition irrelevant.* Boston, Harvard Business Review Press.

Dalton, J. T. (2016). "What is insight? The 5 principles of insight definition." Retrieved July 30, 2018, from https://thrivethinking.com/2016/03/28/what-is-insight-definition/.

Eysenck, J. (2016). "I tried London's new 'happiness-boosting' cocktails – But did they work?" Retrieved February 25, 2016, from http://www.telegraph.co.uk/food-and-drink/cocktails/i-tried-londons-new-happiness-boosting-cocktails-but-did-they/.

Foro Marketing. (2018). "Verbatim." Retrieved July 30, 2018, from https://www.foromarketing.com/diccionario/verbatim/.

Gobé, M. (2005). *Branding emocional.* España, Divine Egg Publicaciones.

Guzmán, M. (2016). "Applying neuroscience to generating new applications in hair care products." *EURO Cosmetics Magazine* **24**: 10–14.

Guzmán, M. and J. Jimenez (2018). "Intersecting the senses. Synesthesia to connect cosmetics with emotion." *Cosmetics & Toiletries Magazine* **133**(3): 32–51.

Heller, E. (2011). *Psicología del color. Cómo actúan los colores sobre los sentimientos y la razón.* Barcelona, Editorial Gili.

Jiménez, J. (2015). "Application of eye-tracking methodology for fragrance evaluation." *FSCC Magazine* **8**(2): 23–26.

Jiménez, J. (2017). "Use of neuroscience tools during the R&D process of cosmetic products." Retrieved January 25, 2018, from https://knowledge.ulprospector.com/7532/pcc-use-neuroscience-tools-rd-process-cosmetic-products/.

Jiménez, J. and M. Guzmán (2017). "Soft-focus for the selfie-obsessed." *Cosmetics & Toiletries Magazine* **132**(3): 24–36.

Jiménez, J. and D. Sánchez (2010). "De la acción antiarrugas a la experiencia de marca." *Cosmeticos & Tecnologia* **1**(6): 34–42.

Jiménez, J. and D. Sánchez (2014). "O poder da aditividade sensorial." *Cosmetics & Toiletries Brasil* **26**(4): 62–68.

Jiménez, J., J. C. Pérez and M. Guzman (2016). "Neuro characterization of emulsions sensory profile: Use of electroencephalography (EEG) for evaluating pick-up, rub-out and after-feel." *SOFW Journal* **142**(4): 44–52.

Keane, L. (2018). "Ad campaigns inspired by powerful consumer insights." Retrieved July 30, 2018, from https://blog.globalwebindex.com/marketing/powerful-consumer-insights/.

Knutson, B., S. Rick, G. E. Wimmer, D. Prelec and G. Loewenstein (2007). "Neural predictors of purchases." *Neuron* **53**(1): 147–156.

Lindstrom, M. (2010). *Buyology. Truth and lies about why we buy.* New York, Crown Business.

Lindstrom, M. (2011). *Brandwashed.* New York, Crown Business.

LuxuryLaunches. (2016). Retrieved February 25, 2016, from http://www.luxuryretail.es/cocteles-inspirados-perfumes-givenchy/.

Malfitano, O. (2007). *Neuromarketing. Cerebrando negocios y servicios.* Argentina, Ediciones Granica.

Millenial Marketing. (2018). "That's an insight?!" Retrieved July 30, 2018, from http://www.millennialmarketing.com/2009/10/thats-an-insight/.

Poucher, W. A. (1941). *Perfumes, cosmetics and soaps.* London, Chapman & Hall.

Pradeep, A. K. (2010). *The buying brain.* Hoboken, NJ, Wiley.

Proyectacolor, P. (2018). "Fisiología del color." Retrieved accessed July 30, 2018, from http://proyectacolor.cl/teoria-de-los-colores/fisiologia-del-color/.

Quiñones, C. (2015). *Desnudando la mente del consumidor. Consumer insights en el marketing.* Lima, Editorial Planeta Perú.

Reinoso Carvalho, F., Q. J. Wang, R. van Ee, D. Persoone and C. Spence (2017). "'Smooth operator': Music modulates the perceived creaminess, sweetness, and bitterness of chocolate." *Appetite* **108**: 383–390.

Stein, B. (2012). *The new handbook of multisensory processing.* Cambridge, MA, Massachusetts Institute of Technology.

Velasco, C., M. Obrist, O. Petit and C. Spence (2018). "Multisensory technology for flavor augmentation: A mini review." *Frontiers in Psychology* **9**: 26.

Zampini, M. and C. Spence (2012). "Assessing the role of visual and auditory cues in multisensory perception of flavor." In *The neural bases of multisensory processes*, M. M. Murray and M. T. Wallace (eds.). Boca Raton, FL, CRC Press/Taylor & Francis: 739–758.

2

Skin Morphology, Development and Physiology

Kenneth A. Walters and Michael S. Roberts

CONTENTS

Introduction

The skin is the largest organ of the body, and it accounts for more than 10% of body mass, and the one that enables the body to interact most intimately with its environment. Figure 2.1 shows a diagrammatic illustration of the skin. In essence, the skin consists of four layers: the stratum corneum (often referred to as the nonviable epidermis), the remaining layers of the epidermis (viable epidermis), dermis and subcutaneous tissues. There are also a number of associated appendages: hair follicles, sweat ducts, apocrine glands and nails. Many of the functions of the skin can be classified as essential to survival of the body bulk of mammals and humans in a relatively hostile environment. In a general context, these functions may be classified as either protective, maintaining homeostasis or sensing. The importance of the protective and homeostatic role of the skin is illustrated in one context by its barrier property. This allows for the survival of humans in an environment of variable temperature, water content (humidity and bathing), presence of environmental dangers such as chemicals, bacteria, allergens, fungi and radiation. In a second context, the skin is a major organ for maintenance of the homeostasis of the body especially in terms of its composition, heat regulation, blood pressure control and excretory roles. Third, the skin is a major sensory organ in terms of sensing environmental influences such as heat, pressure, pain, allergen and microorganism entry. Finally, the skin is an organ that is in a continual state of regeneration and repair. In order to fulfil each of these functions, the skin must be tough, robust and flexible with effective communication between each of its intrinsic components.

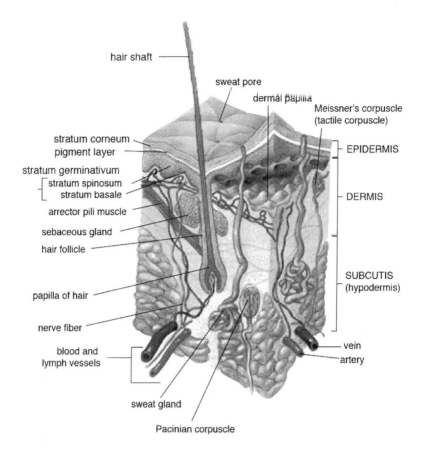

hair shaft

sweat pore

dermal papilla

Meissner's corpuscle
(tactile corpuscle)

stratum corneum
pigment layer

EPIDERMIS

stratum germinativum
  stratum spinosum
    stratum basale

DERMIS

arrector pili muscle

sebaceous gland

hair follicle

SUBCUTIS
(hypodermis)

papilla of hair

nerve fiber

blood and
lymph vessels

vein

artery

sweat gland

Pacinian corpuscle

FIGURE 2.1   (See Colour Insert.) Diagram of full thickness human skin showing the major strata, subcutis, and various other features. The skin has two principle layers, the epidermis and the dermis, which, over most of the body, is thicker than 1 mm. The epidermis (approximately 100 μm thick) is further split into two layers, the stratum corneum and the stratum germinativum (or viable epidermis). A number of appendages are also shown and these include the hair follicle and shaft, the associated sebaceous glands and erector muscles, and sweat ducts. The dermis contains a dense network of blood vessels, which provides constant support for the germinative tissue and important body temperature regulation. There are also nociceptors (pain receptors) and tactile nerve endings (Meissner's corpuscles). (From upload.wikimedia. org/Wikipedia/commons/2/27/skin.png.)

Structure of the Skin

Whereas Figure 2.1 provides an overview of the gross structure of the skin, Figure 2.2 represents the skin components in terms of the various functions they perform. It needs to be emphasised that the protection, homeostatic and sensing functions of the skin are both overlapping and integrated. For instance, the skin's barrier characteristics to a chemical entity involves resistance to its entry (where the barrier is provided by the stratum corneum), metabolism for that proportion of the entity bypassing the stratum corneum (in the viable epidermis), sensing of and attention to damage caused by entry (inflammatory mediator release in epidermis with involvement of dermis), and removal of entity from site by the dermal blood supply and distribution into those body organs specifically responsible for elimination of the entity by metabolism (liver) and excretion (kidney). Heat regulation occurs through the use of the subcutaneous fat pad, physiological regulation of blood flow to effect for instance, heat loss by vasodilation and cooling by perspiration.

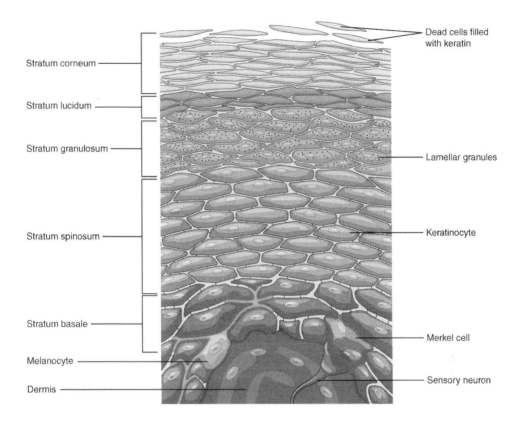

FIGURE 2.2 **(See Colour Insert.)** Cell strata of the epidermis are arranged in layers that are all derived from germinative cells in the basal layer (the stratum basale that forms the barrier between the dermis and the epidermis). This layer contains the skin's stem cells that generate the daughter cells that are forced outwards from the basal layer. As these cells mature they build protein bridges between them and this generates cell–cell communication and gives the cells spiny contours (the stratum spinosum). They also start to synthesise keratins. Further outwards the cells create packets of lipids, which give the cells a granular appearance (the stratum granulosum). The lipid granules are eventually exocytosed and the lipid contents reform to create the intercellular lipid lamellae. The cells subsequently lose their nuclei and become clear (the stratum lucidum) and the cells are no longer considered viable as they form the stratum corneum. Other cells contained in the epidermis are the pigment-producing melanocytes and the sensory Merkel cells. (From opentextbc.ca.)

The Epidermis

The epidermis performs a number of functions as shown in Figure 2.2, one of the most important being the generation of the stratum corneum as described later. The stratum corneum is the heterogeneous outermost layer of the epidermis and is approximately 10–20 μm thick. It is a non-viable epidermis and consists of 15 to 25 flattened, stacked, hexagonal and cornified cells embedded in a mortar of intercellular lipid. Each cell is approximately 40 μm in diameter and 0.5 μm thick. The thickness varies, however, and may be an order of magnitude larger in areas such as the palms of the hand and soles of the feet, areas of the body associated with frequent direct and substantial physical interaction with the physical environment. The stratum corneum barrier properties may be related in part to its very high density (1.4 g/cm^3 in the dry state), its low hydration of 15–20% compared to the usual 70% for the total body mass and its low surface area for solute transport. It is currently recognised that most applied compounds enter the body via the less than 0.1 μm wide intercellular regions of the stratum corneum. Each stratum corneum cell is composed mainly of insoluble bundled keratins (~70%) and lipid (~20%) encased in a cell envelope (Matsui and Amagai, 2015). The intercellular region of the stratum corneum consists mainly of lipids and desmosomes for corneocyte cohesion as briefly discussed later. The lipids of the stratum corneum are discussed in detail in Chapter 3.

The continuous terminal differentiation of the epidermis leading to the formation of the stratum corneum results, eventually, in desquamation of this horny layer with a total turnover of the stratum corneum occurring once every two to three weeks. Very lipophilic agents such as sunscreens and substances binding to the horny layer (e.g. hexachlorophane) may be less well absorbed into the body than would be indicated by the initial partitioning of the agents into the horny layer from an applied vehicle. The stratum corneum also functions as a barrier to prevent the loss of internal body components, particularly water, to the external environment. It is estimated that the efficiency of this barrier is such that water loss due to 'insensible perspiration' is restricted to 0.5 μl/cm²/hr, or 250 ml of water per day for a normal adult. Disorders of epithelialisation such as psoriasis lead to a faster skin turnover, sometimes being reduced to 2–4 days, with reduced stratum corneum barrier function formation, the latter offering the possibility of improved therapeutic outcomes (Namjoshi and Benson, 2010).

The cells of the stratum corneum originate in the viable epidermis and undergo many morphological changes before desquamation. Thus the epidermis consists of several cell strata at varying levels of differentiation (Figure 2.3). The origins of the cells of the epidermis lie in the basal lamina between the dermis and viable epidermis. In this layer there are melanocytes, Langerhans cells, Merkel cells and two major keratinic cell types – the first functioning as stem cells having the capacity to divide and produce new cells, the second serving to anchor the epidermis to the basement membrane (Lavker and Sun, 1982). The basement membrane is 50 to 70 nm thick and consists of two layers, the lamina densa and lamina lucida, which are comprised mainly of proteins such as type IV collagen, laminin-5, nidogen (entactin) and fibronectin. Type IV collagen is responsible for the mechanical stability of the basement membrane, whereas laminin-5 and fibronectin are involved with the attachment between the basement membrane and the basal keratinocytes (Schneider et al., 2007; El-Domyati et al., 2015).

The cells of the basal lamina are attached to the basement membrane by hemidesmosomes, which are found on the ventral surface of basal keratinocytes (Borradori and Sonnenberg, 1999; Walko et al., 2015). Hemidesmosomes appear to be comprised of distinct protein groups, two of which are bullous pemphigoid antigens (BPAG1e and BPAG2) and another group consists of the epithelial cell-specific integrins α6β4 and P1α. BPAG1e is associated with the organization of the cytoskeletal structure and forms a link between the hemidesmosome structure and the keratin intermediate filaments. BPAG2 is a transmembrane protein. Integrin α6β4 binds to the extracellular matrix protein laminin-5, and integrin P1α forms a bridge to the cytoplasmic keratin intermediate filaments. Integrin α6β4 and BPAG2 are thus the major hemidesmosomal protein contributors to binding the keratinocyte to the basal lamina, spanning from the keratin intermediate filament, through the lamina lucida, to the lamina densa. In the lamina densa these membrane-spanning proteins interact with the protein laminin-5, which, in turn, is linked to collagen VII, the major constituent of the anchoring fibrils within the dermal matrix.

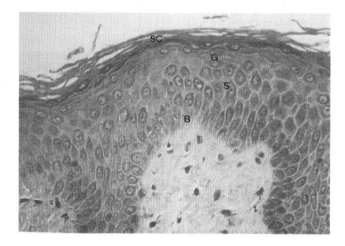

FIGURE 2.3 Cross-section through human skin (haematoxylin and eosin stained) showing the appearance of the various epidermal strata. SC, stratum corneum; G, stratum granulosum; S, stratum spinosum; B, stratum basale.

The importance of maintaining a secure link between the basal lamina cells and the basement membrane is obvious and the absence of this connection results in chronic blistering diseases such as pemphigus and epidermolysis bullosa.

In addition to hemidesmosome cell-matrix binding, another site for adhesion of the cells of the epidermal basal layer and the basal membrane is the adherens junction (Kaiser et al., 1993; Niessen, 2007), in which there are two basic adhesive units. Nectins are IgG-like adhesion receptors that possess a transmembrane region and form an adhesive unit with the actin-binding protein, afadin. In addition to this unit, cadherins are transmembrane proteins that link to α-, β- and p120 catenins. Adherens junctions appear more frequently in the upper epidermis (Ishiko et al., 2003).

The cohesiveness of and communication between the viable epidermal cells, the cell–cell interaction, is maintained in a similar fashion to the cell–matrix connection except that desmosomes replace hemidesmosomes. At the desmosomal junction, there are two transmembrane glycoprotein cadherins: desmogleins and desmocollins, which are associated with the cytoplasmic plaque proteins, desmoplakin, plectin, periplakin, envoplakin and plakoglobin (Figure 2.4), and provide a linkage to keratin intermediate filaments (Tariq et al., 2015). Thus, in the epidermis, the desmosomes are responsible for interconnecting individual cell keratin cytoskeletal structures and thereby create a tissue very resistant to shearing forces.

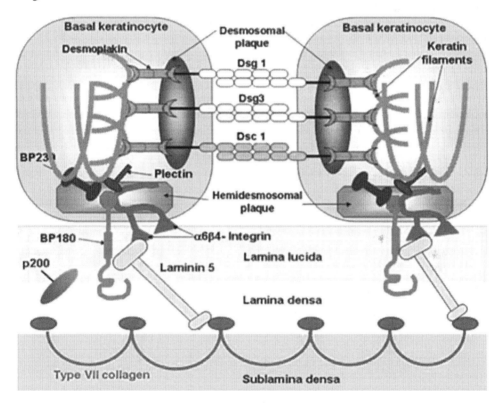

FIGURE 2.4 Diagram of the desmosome and the dermal–epidermal junction. Only structural proteins that function as autoantigens in autoimmune bullous skin diseases are shown. Neighbouring keratinocytes are associated via the extracellular portions of desmosomal cadherins. The homophilic interactions between desmoglein 1, desmoglein 3 and desmocollin 1 are depicted. Their intracellular portions bind to desmosomal plaque proteins that mediate the interaction of desmosomes with keratin filaments. Keratin filaments also bind to bullous pemphigoid antigen 230 (BP230) and plectin, the main intracellular constituents of the hemidesmosomes. BP230 and plectin function as ligands for transmembrane hemidesomosomal proteins, type XVII collagen (BP180) and $\alpha6\beta4$ integrin. These may connect the hemidesmosomes to laminin 5, which in addition to type IV collagen, is a major component of the lamina densa. Laminin 5 is a known ligand for type VII collagen, the major constituent of the anchoring fibrils, which connect lamina densa to the collagen bundles of the upper dermis. (Reproduced from Mihai S, Sitaru C, 2007, *J Cell Mol Med* 11: 462–481, with permission from John Wiley & Sons.)

Another cell type found in the epidermal basal layer, and throughout the epidermis and dermis, are the dendritic cells. There are several distinct subsets of the dendritic cells, with the Langerhans cells being located within the epidermis. These cells are the prominent antigen-presenting cells of the skin immune system (Klechevsky and Banchereau, 2013; Banchereau et al., 2012). These cells are continuously migrating to draining lymph nodes and their main function is to pick up contact allergens in the skin and present these agents to T lymphocytes, thus they play an important role in contact sensitisation. The ability of Langerhans cells to migrate from bone marrow and localize in a specific region of the epidermis and further migrate when activated suggests that there is some mechanism for accumulation in the epidermis, adhesion to keratinocytes and the basement membrane, and for disruption of the adhesive bond. Migration into the epidermis may be mediated by granulocyte-macrophage colony-stimulating factor (GM-CSF), tumour necrosis factor α (TNF-α), interleukin 6, transforming growth factor β, chemotactic cytokines such as monocyte chemotactic protein, and cutaneous lymphocyte-associated antigen (Nakamura et al., 1999). The adhesive bonds within the epidermis appear to be formed by interaction of Langerhans cells with extracellular matrix proteins, such as fibronectin and laminin, via β1 integrins. Detachment of Langerhans cells from keratinocytes and the basement membrane following skin sensitisation may be mediated by epidermal cytokines including GM-CSF and TNFα, while cell maturation, which occurs during transit to the local lymph nodes, may be mediated by GM-CSF. It has been demonstrated that IL-1β and TNF-α act directly on Langerhans cells to reduce adhesion to keratinocytes and the basement membrane by down-regulating the binding protein E-cadherin (Jakob and Udey, 1998). A comprehensive review of human dendritic cells has been provided by Boltjes and van Wijk (2014).

Melanocytes are a further functional cell type of the epidermal basal layer and are also present in hair and eyes. The main function of these cells is to produce melanins, high molecular weight polymers of indole quinone, which affect pigmentation of the skin, hair and eyes (Jimbow et al., 1986; Mort et al., 2015). Melanin is produced in the melanosomes, membrane-bound organelles which are transferred to keratinocytes, probably by phagocytosis, to provide a uniform distribution of pigmentation. Intracellular movement of melanosomes is possibly mediated via actin and microtubules. Visible pigmentation is dependent not only on the number, shape and size of the melanosomes but also on the chemical nature of the melanin. Hair colour is governed by melanocytes that reside in the hair bulbs within the dermis (Slominski et al., 2005; Tobin, 2008). Melanosomes are transferred to the growing hair shaft. The major function of skin pigmentation is to provide protection against harmful environmental effects such as ultraviolet radiation, especially in the proliferating basal layers where the mutagenic effects of this type of insult have particularly serious implications. Melanocytes remain attached to the basal layer and are thought to exist in a non-proliferative state when in contact with undifferentiated keratinocytes.

Regulation of melanogenesis is a complex process involving some 80 genetic loci and there may also be circadian rhythm involvement (Hardman et al., 2015). It is mutations of some of the involved genes that lead to pathological states such as albinism, vitiligo and Waardenburg syndrome. The initial substrate for melanin is tyrosine, which is hydroxylated to dihydroxyphenylalanine (DOPA) and from there may be processed via several routes to produce either the eumelanins (5,6-dihydroxyindole melanin and 5,6-dihydroxyindole-2-carboxylic acid melanin) or pheomelanins (Figure 2.5). Eumelanins are brown/black in colour, whereas pheomelanins are yellow/red. It is thought that interactions between the tridecapeptide α-melanocyte-stimulating hormone (α-MSH) and agouti signal protein are responsible for governing the type of melanogenesis pathway followed and, that in conditions in which α-MSH dominates, eumelanins are produced. UV radiation appears to increase production of the precursor hormone proopiomelanocortin, which increases α-MSH production, resulting in increased levels of eumelanins (Chakraborty et al., 1995; Suzuki et al., 1997; Nasti and Timares, 2015).

The final type of cell found in the basal layer of the stratum corneum is the Merkel cell. These cells, which appear to originate in the epidermis (Morrison et al., 2009), can be distinguished from keratinocytes by their clear cytoplasm and lack of tonofilaments. Merkel cells possess the features of sensory receptor cells and make 'synapse-like' contacts with low-threshold mechanoreceptors, present on the other side of the basement membrane, both of which suggest they function as tactile mediators (Maksimovic et al., 2014; Zimmerman et al., 2014; Walsh et al., 2015). Further evidence that a major function of Merkel cells is to participate in touch reception is provided by the observation that they express sensory ion channels and neurotransmitters (Lumpkin and Caterina, 2007; Kwan et al., 2009).

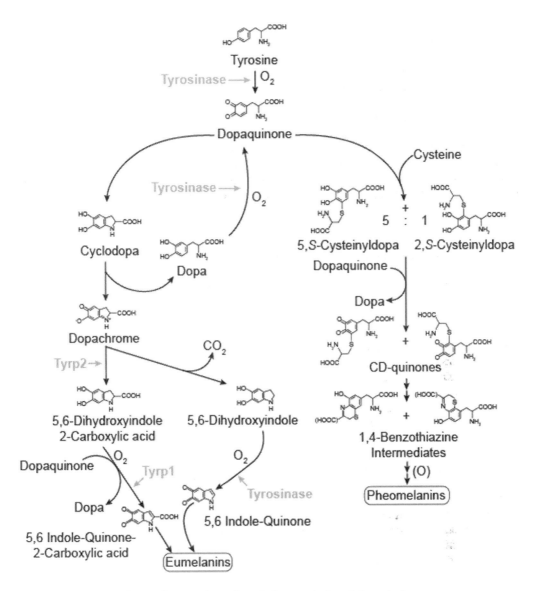

FIGURE 2.5 The synthetic route from tyrosine to the melanins, eumelanin and pheomelanin.

The Dermis

The dermis, a critical component of the body, not only provides the nutritive, immune and other support systems for the epidermis through a thin papillary layer adjacent to the epidermis but also plays a role in temperature, pressure and pain regulation. The dermis is about 0.1 to 0.5 cm thick and consists of collagenous fibres (70%), providing a scaffold of support and cushioning, and elastic connective tissue, providing elasticity, in a semi-gel matrix of mucopolysaccharides. In general, the dermis has a sparse cell population. The main cells present are the fibroblasts, which produce the connective tissue components of collagen, laminin, fibronectin and vitronectin; mast cells, which are involved in the immune and inflammatory responses; telocytes, which may be involved with skin regeneration and repair (Ceafalen et al., 2012; Kang et al., 2015); and melanocytes involved in the production of the pigment melanin.

Sensory nerve endings within the dermis include nociceptors, Pacinian corpuscles and Meissner corpuscles. Nociceptors are sensory neurones that respond to stimuli by sending signals to the spinal cord and brain that cause the perception of pain. They can respond to thermal, mechanical or chemical stimuli. Pacinian corpuscles, sometimes known as lamellar corpuscles, are located deep in the dermis and are responsible for sensitivity to vibration and pressure. Meissner corpuscles are another type of mechanoreceptor, located near the dermal–epidermal junction, which are responsible for sensitivity to light touch.

Contained within the dermis is an extensive vascular network providing skin nutrition, repair and immune responses and, for the rest of the body, heat exchange, immune response and thermal regulation. The blood flow rate to the skin is about 0.05 ml/min per cc of skin, providing a vascular exchange area equivalent to that of the skin surface area. Skin blood vessels derive from those in the subcutaneous tissues with an arterial network supplying the papillary layer, the hair follicles, the sweat and apocrine glands, and the subcutaneous area as well as the dermis itself. These arteries feed into arterioles, capillaries, venules and, thence, into veins. Of particular importance in this vascular network is the presence of arteriovenous anastomoses at all levels in the skin. These arteriovenous anastomoses, which allow a direct shunting of up to 60% of the skin blood flow between the arteries and veins and thus avoiding the fine capillary network, are critical to the skin's functions of heat regulation and blood vessel control. Blood flow changes are most evident in the skin in relation to various physiological responses and include psychological effects such as shock ('draining of colour from the skin') and embarrassment ('blushing'), temperature effects and physiological responses to exercise, haemorrhage, and alcohol consumption.

The lymphatic system is an important component of the skin in regulating its interstitial pressure, mobilisation of defense mechanisms and in waste removal. It exists as a dense flat meshwork in the papillary layers of the dermis and extends into the deeper regions of the dermis. Cross and Roberts (1993) have shown that whereas blood flow determines the clearance of small solutes, such as water and lidocaine, that have permeated across the skin, lymphatic flow is an important determinant in the dermal removal of larger solutes such as interferon.

The Subcutis

The deepest layer of the skin is the subcutaneous tissue or hypodermis. The hypodermis acts as a heat insulator, shock absorber and energy storage region. This layer is a network of fat cells (adipocytes) that are linked to the dermis by interconnecting collagen and elastin fibres. The adipocytes store globules of fat, which consist mainly of triglycerides and cholesteryl esters. As well as the adipocytes, the other main cells in the hypodermis are fibroblasts and macrophages. One of the major roles of the hypodermis is to carry the vascular and neural systems for the skin. It also anchors the skin to underlying muscle. Fibroblasts and adipocytes can be stimulated by the accumulation of interstitial and lymphatic fluid within the skin and subcutaneous tissue (Szuba and Rockson, 1997; see also Breslin, 2014).

Skin Appendages

There are four skin appendages: the hair follicles with their associated sebaceous glands, eccrine sweat glands, apocrine sweat glands and the nails. Each appendage has a different function. The hair follicles are distributed across the entire skin surface with the exception of the soles of the feet, the palms of the hand and the lips. A smooth muscle, the erector pilorum, attaches the follicle to the dermal tissue and enables hair to stand up in response to fear and cold. Each follicle has an associated sebaceous gland, which varies in size from 200 to 2000 μm in diameter. The sebum secreted by this gland consists of triglycerides, free fatty acids and waxes, and protects and lubricates the skin as well as maintaining a surface pH of about 5. The fractional area for these is slightly more than 1/1000 of the total skin surface. The eccrine or sweat glands and apocrine glands account for about two-thirds and one-third of all glands, respectively. The eccrine glands are epidermal structures that are simple, coiled tubes arising from a coiled ball of approximately 100 μm in diameter located in the lower dermis. It secretes a dilute salt solution with a pH of about 5, this secretion being stimulated by temperature-controlling determinants, such as exercise and high environmental temperature, as well as emotional stress through the autonomic (sympathetic) nervous system. These glands have a total surface area of about 1/10,000 of the total body surface. The apocrine glands are limited to specific body regions and are also coiled tubes. These glands are about 10 times the size of the eccrine ducts, extend as low as the subcutaneous tissues and are paired with hair follicles. In-depth analysis of sweat glands and hair follicles, together with a discussion of disorders of the skin appendages can be found in the reviews of Lu and Fuchs (2014), Rompolas and Greco (2014), and Vary (2015).

The structure, properties and function of the nail plate are comprehensively described in Chapter 5.

Development of the Stratum Corneum

The ability to control both the loss of water and the influx of potentially harmful chemicals and microorganisms is the result of the evolution of a unique mixture of protein and lipid materials which collectively form the stratum corneum, a coherent membrane composed of morphologically distinct domains. These domains are principally proteinaceous (the keratinocytes) or lipophilic (the intercellular spaces).

Epidermal Differentiation

The development of the stratum corneum from the keratinocytes of the basal layer involves several steps of cell differentiation, which has resulted in a structure-based classification of the layers above the basal layer (the stratum basale). Thus, the cells progress from the stratum basale through the stratum spinosum, the stratum granulosum, the stratum lucidium to the stratum corneum (Figure 2.3). Cell turnover, from the stratum basale to the stratum corneum, is estimated to be on the order of 21 days.

The exact mechanism whereby keratinocytes in the basal layer are stimulated to initiate the differentiation process is not fully understood. Certainly, epidermal stem cells are located in the basal epidermis and also in the hair follicles and sebaceous glands. The stem cells are responsible for generating the different cell types of the epidermis (Eckert et al., 2013). It is known that protein kinase C and several keratinocyte-derived cytokines may play a regulatory role in the differentiation process (Jerome-Morais et al., 2009; Chae et al., 2016). Thus, for example, interleukin-1 (IL-1) stimulates the production of other cytokines, in both autocrine and paracrine fashion, which act to induce proliferation and chemotaxis (Figure 2.6). These inducing cytokines include GM-GSF, transforming growth factor α (TGF-α), nerve growth factor, amphiregulin, IL-6 and IL-8. On the other hand, transforming growth factor β (TGF-β), production of which is also initiated by IL-1, suppresses keratinocyte growth but stimulates keratinocyte migration, the latter possibly as a result of modulation of hemidesmosomal and desmosomal integrins. It has also been suggested that urokinase-type plasminogen activator (uPA) may activate growth factors and stimulate epidermal proliferation (Jensen and Lavker, 1999). Generation and activation of the serine protease plasmin from plasminogen are induced by uPA, and the activated plasmin instigates localized extracellular proteolysis of cell surface adherent proteins and eventual disruption of the hemidesmosomes. The tumour suppressor protein TIG3 (tazarotene-induced gene 3) may regulate terminal differentiation by activating type 1 transglutaminase (Scharadin and Eckert, 2014).

The stratum spinosum (prickle cell layer), which lies immediately above the basal layer, consists of several layers of cells that are connected to each other and to the stratum basale cells by desmosomes and contain prominent keratin tonofilaments. The cells of the stratum spinosum have a larger cytoplasm than those of the stratum basale. Within the cytoplasm are numerous organelles and filaments. It is clear that the α-keratins of the stratum spinosum are somewhat different to those found in the stratum basale (Skerrow and Skerrow, 1983), indicating that, although mitosis has ceased and a phase of terminal differentiation has been initiated, the cell still maintains a capacity to alter the transcriptional expression of its genes. In the outer cell layers of the stratum spinosum, intracellular membrane-coating granules (100–300 nm in diameter) appear within the cytosol marking the transition between the stratum spinosum and stratum granulosum.

Although further keratin differentiation occurs in the stratum granulosum (Baden, 1979; Eckhart et al., 2013), new keratin synthesis stops. The most characteristic features of this layer are the presence of many intracellular keratohyalin granules and membrane-coating granules, the assembly of the latter appearing to take place in the endoplasmic reticulum and Golgi regions (Landmann, 1988; Manabe and O'Guin, 1992). Within these granules lamellar subunits arranged in parallel stacks are observed. These are believed to be the precursors of the intercellular lipid lamellae of the stratum corneum. Also present

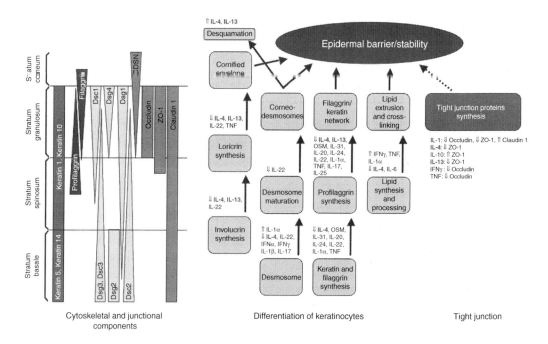

FIGURE 2.6   Schematic diagram of four layers of the epidermis, expression profiles of keratins, filaggrin, desmosomal and tight junction proteins, five terminal differentiation processes, and regulation by cytokines. The left panel shows four distinct epidermal layers and expression patterns of four keratins, filaggrin, four Dsgs, three Dscs, corneodesmosin, and tight junction proteins. The right panel shows detailed information for five distinct epidermal differentiation processes, as well as various cytokines, which upregulate or downregulate the epidermal differentiation. IL-4 and IL-13 are shown in bolded letters. (Reproduced from Tsuchisaka A, Furumura M, Hashimoto T, 2014, *J Invest Dermatol* 134: 1194–1196, with permission from Elsevier.)

in the lamellar granules are hydrolytic enzymes, the most important of which is stratum corneum chymotryptic enzyme (SCCE). SCCE is a serine proteinase, which, because of its ability to locate at desmosomal regions in the intercellular space, has been implicated in the desquamation process (Egelrud et al., 1993). In the outermost layers of the stratum granulosum, the lamellar granules migrate to the apical plasma membrane where they fuse and eventually extrude their contents into the intercellular space (Manabe and O'Guin, 1992). At this stage in the differentiation process, as a result of the release of selective lysing enzymes, the keratinocytes lose their nuclei and other cytoplasmic organelles, become flattened and compacted to form the stratum lucidum, which eventually forms the stratum corneum. The extrusion of the contents of lamellar granules is a fundamental requirement for the formation of the epidermal permeability barrier, and disturbances in this process have been implicated in various dermatological disorders.

The entire process of epidermal terminal differentiation is geared towards the generation of the specific chemical morphology of the stratum corneum. Thus, the end products of this process are the intracellular protein matrix and the intercellular lipid lamellae.

Cornified Cell Envelope

The cornified cell envelope (about 15 nm thick and 10% mass of the stratum corneum) is the outermost layer of a corneocyte, which mainly consists of tightly bundled keratin filaments aligned parallel to the main face of the corneocyte. The envelope, which replaces the plasma membrane of differentiating keratinocytes, consists of both protein and lipid components (Candi et al., 2005). The protein envelope (~10 nm thick) is a covalent cross-linking of several proteins as a result of actions by sulphydryl oxidases and transglutaminases, whereas the lipid envelope (~5 nm thick) comprises lipid attached covalently to the protein envelope. Sulphydryl oxidases and transglutaminases lead to the formation of disulphide and isopeptide bonds, respectively (Nemes and Steinert, 1999). The envelope lies adjacent to the interior surface of the plasma membrane. In addition to the predominant protein loricrin, several other envelope precursor proteins have been identified, including periplakin and envoplakin (for reviews, see Brown and McLean, 2012; Kypriotou et al., 2012; Abhishek and Palamadal Krishnan, 2016; Boczonadi and Määttä, 2016). The predominance of the structural proteins in the cornified envelope are involucrin (2–5%), loricin (80%), small proline-rich proteins (a family of 11–14 closely related proteins including cornifins and pancornulins, 3–5%), and cystalin A or keratolinin (2–5%). There is also a range of proteins with an expression of less than 1%, including elafin, profilaggrin, keratin intermediate filaments, desmoplakin I and II, S100 proteins and annexin I (also called lipocortin I).

Formation of the envelope is believed to occur in two stages (for a brief review of the formation of the cornified envelope, see Elias et al., 2014). In the first stage, soluble proteins, such as involucrin and cystatin-α, form a scaffold to which other insoluble precursors, including loricrin, are added in the latter stage. Thus the cornified envelope is formed by the sequential deposition of consecutively expressed proteins starting with the fixation of involucrin as a scaffold on the intracellular surface of the plasma membrane in a calcium and phospholipid-dependent manner. It is cross-linked to desmoplakin and envoplakin and also covalently bound to ω-hydroxyceramides. Other proteins then reinforce the envelope by attaching (Nemes and Steinert, 1999). The cross-linked protein complex of the corneocyte envelope is very insoluble and chemically resistant. Cornified cell envelopes are also present in the hair follicle and nail matrix but, although morphologically similar, the pattern and types of precursor are slightly different to those of the epidermis (Baden and Kvedor, 1993).

Stratum Corneum Intercellular Lipids

It is currently proposed that the corneocyte protein envelope plays an important role in the structural assembly of the intercellular lipid lamellae of the stratum corneum. The intercellular lipids of the stratum corneum are fully discussed in the chapter by Mojumdar and Bouwstra (Chapter 3) and will not be discussed here.

Stratum Corneum Proteins

The poor stratum corneum protein solubility is a consequence in part from the extensive cross-linking of both the cell envelope and intracellular proteins. The majority of the intracellular protein in the stratum corneum is composed of keratin filaments, which are cross-linked by intermolecular disulphide bridges (Steinert et al., 1994; Bragulla and Homberger, 2009), and the components of the cornified cell envelope. In the terminal stages of differentiation, the keratinocytes contain keratin intermediate filaments (keratins 1 and 10, which are derived from keratins 5 and 14 present in basal keratinocytes) together with several other proteins, including involucrin, loricrin and profilaggrin (Robinson et al., 1996; Candi et al., 2005). As mentioned earlier, loricrin and involucrin are major components of the cornified cell envelope, whereas profilaggrin is implicated both in the alignment of the keratin filaments. Profilaggrin is a large highly phosphorylated protein, consisting of multiple (10–12) filaggrin units that first appear in keratohyalin granules in the stratum granulosum. The profilaggrin molecule is processed in a calcium-dependent manner by dephosphorylation and proteolysis into individual filaggrin molecules that serve to aggregate keratin filaments. There is evidence of a role for profilaggrin in the pathogenesis of atopic dermatitis (Thyssen and Kezic, 2014).

Desquamation

The mechanisms underlying the desquamation of stratum corneum cells are complex. Suzuki et al. (1984) originally suggested that the action of two types of serine protease in the degradation of desmosomes led to desquamation. Certainly, serine proteases (kallikreines), aspartic proteases, glycosidases and cysteine proteases (cathepsins) play a key role in the process (Brocklehurst and Philpott, 2013). It has been suggested that desquamation may be regulated by the extent of activation of protease precursors and changes in the pH of the stratum corneum intercellular space (Ekholm et al., 2000).

In the stratum granulosum, desmosomes are transformed into corneodesmosomes or corneosomes, which form the main intercellular adhesive structures in the stratum corneum (Ishida-Yamamoto and Igawa, 2015). Much emphasis has been placed on the protein corneodesmosin, which is located in the extracellular part of the corneodesmosome and adjacent parts of the cornified cell envelope. It has been suggested that this protein is continuously degraded, providing an explanation for the gradient of increased corneocyte cohesiveness from the skin surface towards deeper layers (Lundstrom et al., 1994).

Scaly skin diseases and impaired barrier function can be a consequence of a disrupted desquamation process. It has been known for some time that desquamation is associated with a conversion of cholesterol sulphate to cholesterol (Long et al., 1985) and X-linked ichthyosis, a scaly disease characterized by a disrupted desquamation process, is identified with a lack of the enzyme cholesterol sulphatase (Koppe et al., 1978). Later work by Sato et al. (1998) demonstrated that hyperkeratosis attributable to desmosomes is associated with an increased content of cholesterol sulphate in patients with X-linked ichthyosis. It is apparent that cholesterol sulphate retards desquamation by acting as a serine protease inhibitor. Mutations in the corneodesmosin gene can lead to peeling skin syndrome (Israeli et al., 2011), and mutations in the gene encoding for a serine protease inhibitor can lead to Netherton syndrome, in which atopy and premature desquamation occur (Deraison et al., 2007; Igawa et al., 2013).

Epidermal Repair Mechanisms

The Effects of Hydration

It has been known for some time that the hydration level of the stratum corneum can have a profound effect on its barrier properties (reviewed in Rawlings and Harding, 2004). The mechanisms involved in skin hydration are not fully defined, although it is likely that it is the result of a combination of the presence of hygroscopic agents, collectively known as the natural moisturising factor, and the orderly arrangement of the intercellular lipids (Verdier-Sévrain and Bonté, 2007). The natural moisturising factor is the result of the breakdown of the filaggrin monomer and comprises a pool of amino acids and derivatives. The hygroscopic compounds glycerol and urea have also been located within the stratum corneum and serve as natural humectants (Björklund et al., 2013a,b).

In the normal state, the stratum corneum holds between 15% and 20% (dry weight) water, most of which appears to be associated with intracellular keratin. Stratum corneum water can be increased up to ~400% (dry weight) following excessive soaking. Swelling of corneocytes is possibly due to increased uptake of water, which then interacts with keratin to expand the spatial orientation of the protein. The observation that the corneocytes of the nail plate and hair do not swell to the same extent as those of the stratum corneum following excessive hydration indicates that the degree of interaction between water and keratin is a function of the positioning and stability of disulphide bonds in the peptide. Thus where the α-helix keratin filaments are loosely packed and more flexible, as in the stratum corneum keratinocytes, there is a greater ability to alter conformation to accommodate water.

In fully hydrated stratum corneum the corneocytes are swollen, with pools of water displacing and separating keratin filaments (van Hal et al., 1996).

As mentioned earlier, it is well recognised that natural moisturising factors (NMFs) can make up to 10% of the corneocyte dry weight and that these materials can absorb water extensively. Rawlings et al. (1994) pointed out that the amino acids to which filaggrin is proteolysed to are themselves precursors for the natural moisturising factor. Glutamine is converted to the potent humectant pyrrolidone carboxylic acid, a major component of NMFs, whereas histidine is converted to urocanic acid. Interestingly, filaggrin is only converted to NMF when the water activity is between 0.70 and 0.95, filaggrin being stable at higher water activities and proteolysis being impeded by low water activity. Hence, under occlusive conditions, the stratum corneum NMF level decreases to close to zero, and all corneocytes contain filaggrin. The result of this homeostatic mechanism is that the skin has prevented itself from being 'over-hydrated'.

There appears to be an absence of NMF in severe dry flaking skin and in both psoriasis and ichthyosis vulgaris. Mutations in the gene encoding for filaggrin have certainly been shown to be the cause of common skin conditions such as ichthyosis vulgaris (Smith et al., 2006; Sandilands et al., 2009). The link between these loss-of-function mutations, which are carried by up to 10% of the population, and atopic eczema has been established (McLean, 2016).

In conclusion, the current observations suggest that stratum corneum hydration does not lead to an overall decrease in intercellular lipid order and only small amounts of water are present in the intercellular polar head group regions. It is therefore tempting to revisit a possible mechanism by which hydration promotes percutaneous absorption and which was raised in an earlier review (Roberts and Walker, 1993). In that model, swelling of the keratin is akin to the 'bricks' becoming swollen in the 'bricks-and-mortar' model of the stratum corneum with a loosening of the intercellular lipid 'mortar'. The overall effect should be an increase in the mobility of the chains and in permeability, without an effect on the lipid ordering.

Chemical Damage

When the stratum corneum is perturbed, several localised biochemical events occur which result in rapid reconstitution of barrier function (Taljebini et al., 1996). Thus, in extreme cases of stratum corneum damage such as acetone-induced delipidation or tape stripping, there appears to be a biphasic pattern of recovery: a rapid phase of repair followed by a slower phase of normalisation. The initial rapid phase of barrier recovery involves the expeditious secretion of preformed lamellar bodies from the granular cells into the intercellular space, an increase in epidermal cholesterol and fatty acid synthesis, and accelerated production and secretion, into the intercellular space, of new lamellar bodies. The subsequent and slower phase of barrier repair involves an increase in ceramide synthesis and an increase in deoxyribonucleic acid synthesis leading to epidermal hyperplasia. It is remarkable that the initial perturbation, which occurs in the outermost layers of the stratum corneum can rapidly stimulate biochemical events in the stratum granulosum and lower levels of the epidermis.

Although the exact mechanisms stimulating these events are unknown, it has been known for some time that a change in the rate of transepidermal water loss (TEWL), induced by barrier alterations, plays a role (Grubauer et al., 1989). This increase in TEWL may lead to focal changes in the concentration of certain ions, mainly Ca^{++} in the outer epidermis (Behne et al., 2011; Celli et al., 2016). In the normal state, there is more Ca^{++} stored in the endoplasmic reticulum of the cells of outer layers of the epidermis than the inner. Following barrier disruption the Ca^{++} is released creating a major shift in the gradient. The shift of the Ca^{++} gradient may increase lamellar body secretion. Thus, although there are still many uncertainties regarding the biochemistry of barrier repair, there is much evidence that suggests that ion concentration and the induction of lipid-producing enzymes play a major role.

Perturbation of barrier function sometimes, but not always, also induces an inflammatory response that results in irritation. It is important to appreciate that 'irritation' is used to describe skin reactions that can range from a mild and transient erythema and/or itch to serious vesiculation. The application of the surfactant sodium dodecyl sulphate (SDS) almost always results in an irritant response (Patil et al., 1994; Angelova-Fischer et al., 2016). SDS intercalates with the lamellae and increases fluidity in this region resulting in an increase in TEWL. Furthermore, although other surface-active agents, such as sodium laurate and polysorbates, can increase TEWL to similar levels as SDS, the resultant irritation is much less and in some cases, not significantly different to untreated skin. It follows that irritation subsequent to exposure to SDS must be a result of factors other than an increase in water transport and the stimulation of lipogenesis.

That surface-active agents can cause skin irritation is well established and has been so for many years (reviewed in Prottey, 1978; Basketter et al., 2008). It is also well known that, whereas ionic surfactants can cause severe irritation, nonionic surfactants are considered virtually non-irritant in normal use. Thus much of the research on surfactant-induced skin irritation has involved studies on SDS. The collective data suggests that SDS can interact with both lipid and protein structures in the stratum corneum. As mentioned earlier, interaction with lipids will increase lipid fluidity and thereby enhance skin permeability. This alone, however, apart from increasing its own permeation, will not account for the irritation caused by SDS. It is known that SDS can penetrate into the corneocyte and interact with the protein structure such that α-keratin is uncoiled (Imokawa et al., 1975), but it is difficult to relate this aspect to an irritant response. A more likely explanation for the irritation induced by SDS is its capacity to stimulate keratinocyte production of inflammatory mediators such as interleukin 1 and prostaglandin E_2. Whether this induction is secondary to some interaction between SDS and the corneocyte cell membrane is unknown.

Biochemical Abnormalities

There are a large number of diseases that can affect epidermal barrier function, but it is beyond the realm of this chapter to consider any of these in detail. Some of the diseases affect the formation of the corneo-cyte ('broken brick syndrome'), whereas others affect the intercellular lipid ('weak mortar syndrome'). For instance, Nemes and Steinert (1999) refer to more than 10 different diseases involving genes that encode keratin intermediate filaments, including Unna-Thost disease and tylosis. Other genetic diseases are related to defects in the genes associated with the structural proteins of the cornified envelope or transglutaminases. For instance, a genetic defect in TGM1, the gene that encodes the transglutaminase I enzyme, leads to the life-threatening disease lamellar ichthyosis.

A reduction of the effective intercellular lipid barrier properties can lead to deficiencies ranging from dry skin (depletion of lipids due to excessive use of detergents) to hyperproliferation and abnormal scaling. Causes include essential fatty acid deficiency, abnormal intercellular deposition of various lipids, accumulation of cholesterol sulphate in X-linked ichthyosis, and genetic defects of lipid metabolism (e.g. Refsum disease and Sjogren-Larsson syndrome due to phytanoyl-CoA hydroxylase and fatty aldehyde deficiencies, respectively).

Concluding Remarks

The aim of this chapter has been to introduce the reader to the basic morphology of skin, to outline the stages in the development of the barrier layer, to define the major proteins of the stratum corneum and to illustrate repair mechanisms following barrier disruption. What is evident is that the skin is more than another simple biological barrier membrane onto which various cosmetic products may be applied for long or short periods, and into and through which therapeutic agents can be delivered. Rather, the skin should be viewed as an extremely selective semipermeable membrane overlying a powerful immune system ready to react to any given insult.

REFERENCES

Abhishek S, Palamadai Krishnan S (2016) Epidermal differentiation complex: A review on its epigenetic regulation and potential drug targets. *Cell J* 18: 1–6.

Angelova-Fischer I, Stilla TR, Kezic S, Fischer TW, Zillikens D (2016) Barrier function and natural moisturizing factor levels after cumulative exposure to short-chain aliphatic alcohols and detergents: Results of occlusion-modified tandem repeated irritation test. *Acta Derm Venereol* 96(7): 880–884. doi: 10.2340/00015555-2363.

Baden HP (1979) Keratinization in the epidermis. *Pharm Ther* 7: 393–411.

Baden HP, Kvedar JC (1993) Epithelial cornified envelope precursors are in the hair follicle and nail. *J Invest Dermatol* 101: 72S–74S.

Banchereau J, Thompson-Snipes L, Zurawski S, Blanck JP, Cao Y, Clayton S, Gorvel JP, Zurawski G, Klechevsky E (2012) The differential production of cytokines by human Langerhans cells and dermal CD14(+) DCs controls CTL priming. *Blood* 119: 5742–5749.

Basketter DA, English JS, Wakelin SH, White IR (2008) Enzymes, detergents and skin: Facts and fantasies. *Br J Dermatol* 158: 1177–1181.

Behne MJ, Sanchez S, Barry NP, Kirschner N, Meyer W, Mauro TM, Moll I, Gratton E (2011) Major translocation of calcium upon epidermal barrier insult: Imaging and quantification via FLIM/Fourier vector analysis. *Arch Dermatol Res* 303: 103–115.

Björklund S, Engblom J, Thuresson K, Sparr E (2013a) Glycerol and urea can be used to increase skin permeability in reduced hydration conditions. *Eur J Pharm Sci* 50: 638–645.

Björklund S, Ruzgas T, Nowacka A, Dahi I, Topgaard D, Sparr E, Engblom J (2013b) Skin membrane electrical impedance properties under the influence of varying water gradient. *Biophys J* 104: 2639–2650.

Boczonadi V, Määttä A (2016) Functional analysis of periplakin and envoplakin, cytoskeletal linkers, and cornified envelope precursor proteins. *Methods Enzymol* 569: 309–329.

Boltjes A, van Wijk F (2014) Human dendritic cell functional specialization in steady-state and inflammation. *Front Immunol* 5: 131. doi: 10.3389/fimmu.2014.00131.

Borradori L, Sonnenberg A (1999) Structure and function of hemidesmosomes: More than simple adhesion complexes. *J Invest Dermatol* 112: 411–418.

Bragulla HH, Homberger DG (2009) Structure and functions of keratin proteins in simple, stratified, keratinized and cornified epithelia. *J Anat* 214: 516–559.

Breslin JW (2014) Mechanical forces and lymphatic transport. *Microvasc Res* 96: 46–54.

Brocklehurst K, Philpott MP (2013) Cysteine proteases: Mode of action and role in epidermal differentiation. *Cell Tissue Res* 351: 237–244.

Brown SJ, McLean WH (2012) One remarkable molecule: Filaggrin. *J Invest Dermatol* 132: 751–762.

Candi E, Schmidt R, Melino G (2005) The cornified envelope: A model of cell death in the skin. *Nat Rev Mol Cell Biol* 6: 328–340.

Ceafalan L, Gherghiceanu M, Popescu LM, Simionescu O (2012) Telocytes in human skin – Are they involved in skin regeneration? *J Cell Mol Med* 16: 1405–1420.

Celli A, Crumrine D, Meyer JM, Mauro TM (2016) Endoplasmic reticulum calcium regulates epidermal barrier response and desmosomal structure. *J Invest Dermatol* 136: 1840–1847.

Chae M, Jung JY, Bae IH, Kim HJ, Lee TR, Shin DW (2016) Lipin-1 expression is critical for keratinocyte differentiation. *J Lipid Res* 57: 563–573.

Chakraborty A, Slominski A, Ermak G, Hwang J, Pawelek J (1995) Ultraviolet B and melanocyte-stimulating hormone (MSH) stimulate mRNA production for α-MSH receptors and proopiomelanocortin-derived peptides in mouse melanoma cells and transformed keratinocytes. *J Invest Dermatol* 105: 655–659.

Cross SE, Roberts MS (1993) Subcutaneous absorption kinetics of interferon and other solutes. *J Pharm Pharmacol* 45: 606–609.

Deraison C, Bonnart C, Lopez F, Besson C, Robinson R, Jayakumar A, Wagberg F, Brattsand M, Hachem JP, Leonardsson G, Hovnanian A (2007) LEKTI fragments specifically inhibit KLK5, KLK7, and KLK14 and control desquamation through a pH-dependent interaction. *Mol Biol Cell* 18: 3607–3619.

Eckert RL, Adhikary G, Balasubramanian S, Rorke EA, Vemuri MC, Boucher SE, Bickenbach JR, Kerr C (2013) Biochemistry of epidermal stem cells. *Biochim Biophys Acta* 1830: 2427–2434.

Eckhart L, Lippens S, Tschachler E, Declercq W (2013) Cell death by cornification. *Biochim Biophys Acta* 1833: 3471–3480.

Egelrud T, Régnier M, Sondell B, Shroot B, Schmidt R (1993) Expression of stratum corneum chymotryptic enzyme in reconstructed human epidermis and its suppression by retinoic acid. *Acta Derm Venereol* 73: 181–184.

Ekholm IE, Brattsand M, Egelrud T (2000) Stratum corneum tryptic enzyme in normal epidermis: A missing link in the desquamation process? *J Invest Dermatol* 114: 56–63.

El-Domyati M, Abdel-Wahab H, Ahmad H (2015) Immunohistochemical localization of basement membrane laminin 5 and collagen IV in adult linear IgA disease. *Int J Dermatol* 54: 922–928.

Elias PM, Gruber R, Crumrine D, Menon G, Williams ML, Wakefield JS, Holleran WM, Uchida Y (2014) Formation and functions of the corneocyte lipid envelope (CLE). *Biochim Biophys Acta* 1841: 314–318.

Grubauer G, Elias PM, Feingold KR (1989) Transepidermal water loss: The signal for recovery of barrier structure and function. *J Lipid Res* 30: 323–334.

Hardman JA, Tobin DJ, Haslam IS, Farjo N, Farjo B, Al-Nuaimi Y, Grimaldi B, Paus R (2015) The peripheral clock regulates human pigmentation. *J Invest Dermatol* 135: 1053–1064.

Igawa S, Kishibe M, Honma M, Murakami M, Mizuno Y, Suga Y, Seishima M, Ohguchi Y, Akiyama M, Hirose K, Ishida-Yamamoto A, Iizuka H (2013) Aberrant distribution patterns of corneodesmosomal components of tape-stripped corneocytes in atopic dermatitis and related skin conditions (ichthyosis vulgaris, netherton syndrome and peeling skin syndrome type B). *J Dermatol Sci* 72: 54–60.

Imokawa G, Sumura K, Katsumi M (1975) Study on skin roughness caused by surfactants II Correlation between protein denaturation and skin roughness. *J Am Oil Chem Soc* 52: 484–489.

Ishida-Yamamoto A, Igawa S (2015) The biology and regulation of corneodesmosomes. *Cell Tissue Res* 360: 477–482.

Ishiko A, Matsunaga Y, Matsunaga T, Aiso S, Nishikawa T, Shimizu H (2003) Immunomolecular mapping of adherens junction and desmosomal components in normal human epidermis. *Exp Dermatol* 12: 747–754.

Israeli S, Zamir H, Sarig O, Bergman R, Sprecher E (2011) Inflammatory peeling skin syndrome caused by a mutation in CDSN encoding corneodesmosin. *J Invest Dermatol* 131: 779–781.

Jakob T, Udey MC (1998) Regulation of E-cadherin-mediated adhesion in Langerhan cell-like dendritic cells by inflammatory mediators that mobilize Langerhans cells in vivo. *J Immunol* 160: 4067–4073.

Jensen PJ, Lavker RM (1999) Urokinase is a positive regulator of epidermal proliferation in vivo. *J Invest Dermatol* 112: 240–244.

Jerome-Morais A, Rahn HR, Tibudan SS, Denning MF (2009) Role for protein kinase C-alpha in keratinocyte growth arrest. *J Invest Dermatol* 129: 2365–2375.

Jimbow K, Fitzpatrick TB, Quevedo WC (1986) Formation, chemical composition and functions of melanin pigments in mammals. In: *Biology of the Integument*, Vol 2 (Matolsky A, ed), Springer-Verlag, New York, pp. 278–296.

Kaiser HW, Ness W, Jungblut I, Briggaman RA, Kreysel HW, O'Keefe EJ (1993) Adherens junctions: Demonstration in human epidermis. *J Invest Dermatol* 100: 180–185.

Kang Y, Zhu Z, Zheng Y, Wan W, Manole CG, Zhang Q (2015) Skin telocytes versus fibroblasts: Two distinct dermal cell populations. *J Cell Mol Med* 19: 2530–2539.

Klechevsky E, Banchereau J (2013) Human dendritic cells subsets as targets and vectors for therapy. *Ann NY Acad Sci* 1284: 24–30.

Koppe JG, Marinkovic-Ilsen A, Rijken Y, DeGroot WP, Jobsis AC (1978) X-linked ichthyosis. A sulfatase deficiency. *Arch Dis Child* 53: 803–806.

Kwan KY, Glazer JM, Corey DP, Rice FL, Stucky CL (2009) TRPA1 modulates mechanotransduction in cutaneous sensory neurones. *J Neurosci* 29: 4808–4819.

Kypriotou M, Huber M, Hohl D (2012) The human epidermal differentiation complex: Cornified envelope precursors, S100 proteins and the 'fused gene' family. *Exp Dermatol* 21: 643–649.

Landmann L (1988) The epidermal permeability barrier. *Anat Embryol (Berl)* 178: 1–13.

Lavker RM, Sun T (1982) Heterogeneity in epidermal basal keratinocytes: Morphological and functional correlations. *Science* 215: 1239–1241.

Long SA, Wertz PW, Strauss JS, Downing DT (1985) Human stratum corneum polar lipids and desquamation. *Arch Dermatol Res* 277: 284–287.

Lu C, Fuchs E (2014) Sweat gland progenitors in development, homeostasis, and wound repair. *Cold Spring Harb Perspect Med* 4: a015222.

Lumpkin EA, Caterina MJ (2007) Mechanisms of sensory transduction in the skin. *Nature* 445: 858–865.

Lundstrom A, Serre G, Haftek M, Egelrud T (1994) Evidence for a role of corneodesmosin, a protein which may serve to modify desmosomes during cornification, in stratum corneum cell cohesion and desquamation. *Arch Dermatol Res* 286: 369–375.

Maksimovic S, Nakatani M, Baba Y, Nelson AM, Marshall KL, Wellnitz SA, Firozi P, Woo SH, Ranade S, Patapoutian A, Lumpkin EA (2014) Epidermal Merkel cells are mechanosensory cells that tune mammalian touch receptors. *Nature* 509: 617–621.

Manabe M, O'Guin WM (1992) Keratohyalin, trichohyalin and keratohyalin-trichohyalin hybrid granules: An overview. *J Dermatol* 19: 749–755.

Matsui T, Amagai M (2015) Dissecting the formation, structure and barrier function of the stratum corneum. *Int Immunol* 27: 269–280.

McLean WH (2016) Filaggrin failure – From ichthyosis vulgaris to atopic eczema and beyond. *Br J Dermatol* 175(suppl 2): 4–7.

Mihai S, Sitaru C (2007) Immunopathology and molecular diagnosis of autoimmune bullous diseases. *J Cell Mol Med* 11: 462–481.

Morrison KM, Miesegaes GR, Lumpkin EA, Maricich SM (2009) Mammalian Merkel cells are descended from the epidermal lineage. *Dev Biol* 336: 76–83.

Mort RL, Jackson IJ, Patton EE (2015) The melanocyte lineage in development and disease. *Development* 142: 620–632.

Nakamura K, Saitoh A, Yasaka N, Furue M, Tamaki K (1999) Molecular mechanisms involved in the migration of epidermal dendritic cells in the skin. *J Invest Dermatol Symp Proc* 4: 169–172.

Namjoshi S, Benson HA (2010) Cyclic peptides as potential therapeutic agents for skin disorders. *Biopolymers* 94: 673–680.

Nasti TH, Timares L (2015) Invited review MC1R, eumelanin and pheomelanin: Their role in determining the susceptibility to skin cancer. *Photochem Photobiol* 91: 188–200.

Nemes Z, Steinert PM (1999) Bricks and mortar of the epidermal barrier. *Exp Mol Med* 31: 5–19.

Niessen CM (2007) Tight junctions/adherens junctions: Basic structure and function. *J Invest Dermatol* 127: 2525–2532.

Patil SM, Singh P, Maibach HI (1994) Cumulative irritancy in man to sodium lauryl sulfate: The overlap phenomenon. *Int J Pharmaceut* 110: 147–154.

Prottey C (1978) The molecular basis of skin irritation. In: *Cosmetic Science*, Vol 1 (Breuer MM, ed), Academic Press, London, pp. 275–349.

Rawlings AV, Harding CR (2004) Moisturization and skin barrier function. *Dermatol Ther* 17(suppl 1): 43–48.

Rawlings AV, Scott IR, Harding CR, Bowser PA (1994) Stratum corneum moisturization at the molecular level. *J Invest Dermatol* 103: 731–741.

Roberts MS, Walker M (1993) Water – The most natural penetration enhancer. In: *Pharmaceutical Skin Penetration Enhancement* (Walters KA and Hadgraft J, eds) Marcel Dekker, New York, pp. 1–30.

Robinson NA, LaCelle PT, Eckert RL (1996) Involucrin is a covalently crosslinked constituent of highly purified epidermal corneocytes: Evidence for a common pattern of involucrin crosslinking in vivo and in vitro. *J Invest Dermatol* 107: 101–107.

Rompolas P, Greco V (2014) Stem cell dynamics in the hair follicle niche. *Semin Cell Dev Biol* 25–26: 34–42.

Sandilands A, Sutherland C, Irvine AD, McLean WHI (2009) Filaggrin in the frontline: Role in skin barrier function and disease. *J Cell Sci* 122: 1285–1294.

Sato J, Denda M, Nakanishi J, Nomura J, Koyama J (1998) Cholesterol sulfate inhibits proteases that are involved in desquamation of stratum corneum. *J Invest Dermatol* 111: 189–193.

Scharadin TM, Eckert RL (2014) TIG3: An important regulator of keratinocyte proliferation and survival. *J Invest Dermatol* 134, 1811–1816.

Schneider H, Mühle C, Pacho F (2007) Biological function of laminin-5 and pathogenic impact of its deficiency. *Eur J Cell Biol* 86: 701–717.

Skerrow D, Skerrow CJ (1983) Tonofilament differentiation in human epidermis: Isolation and polypeptide chain composition of keratinocyte subpopulations. *Exp Cell Res* 143: 27–35.

Slominski A, Wortsman J, Plonka PM, Schallreuter KU, Paus R, Tobin DJ (2005) Hair follicle pigmentation. *J Invest Dermatol* 124: 13–21.

Smith FJ, Irvine AD, Terron-Kwiatkowski A, Sandilands A, Campbell LE, Zhao Y, Liao H, Evans AT, Goudie DR, Lewis-Jones S, Arseculetatne G, Munro CS, Sergeant A, O'Regan G, Bale SG, Compton JG, DiGiovanna JJ, Presland RB, Fleckman P, McLean WH (2006) Loss-of-function mutations in the gene encoding filaggrin cause ichthyosis vulgaris. *Nat Genet* 38: 337–342.

Steinert PM, North AC, Parry DA (1994) Structural features of keratin intermediate filaments. *J Invest Dermatol* 103: 19S–24S.

Suzuki Y, Nomura J, Koyama J, Horii I (1984) The role of proteases in stratum corneum: Involvement in stratum corneum desquamation. *Arch Dermatol Res* 286: 249–253.

Suzuki I, Tada A, Ollmann MM, Barsh GS, Im S, Lamoreux ML, Hearing VJ, Nordlund JJ, Abdel-Malek ZA (1997) Agouti signaling protein inhibits melanogenesis and the response of human melanocytes to α-melanotropin. *J Invest Dermatol* 108: 838–842.

Szuba A, Rockson SG (1997) Lymphedema: Anatomy, physiology and pathogenesis. *Vasc Med* 2: 321–326.

Taljebini M, Warren R, Mao-Qiang M, Lane E, Elias PM, Feingold KR (1996) Cutaneous permeability barrier repair following various types of insults: Kinetics and effects of occlusion. *Skin Pharmacol* 9: 111–119.

Tariq H, Bella J, Jowitt TA, Holmes DF, Rouhi M, Nie Z, Baldock C, Garrod D, Tabernero L (2015) Cadherin flexibility provides a key difference between desmosomes and adherens junctions. *Proc Natl Acad Sci USA* 112: 5395–5400.

Thyssen JP, Kezic S (2014) Causes of epidermal filaggrin reduction and their role in the pathogenesis of atopic dermatitis. *J Allergy Clin Immunol* 134: 792–799.

Tobin DJ (2008) Human hair pigmentation – Biological aspects. *Int J Cosmet Sci* 30: 233–257.

Tsuchisaka A, Furumura M, Hashimoto T (2014) Cytokine regulation during epidermal differentiation and barrier formation. *J Invest Dermatol* 134: 1194–1196.

van Hal DA, Jeremiasse E, Junginger HE, Spies F, Bouwstra JA (1996) Structure of fully hydrated human stratum corneum: A freeze-fracture electron microscopy study. *J Invest Dermatol* 106: 89–95.

Vary JC (2015) Selected disorders of skin appendages – Acne, alopecia, hyperhidrosis. *Med Clin North Am* 99: 1195–1211.

Verdier-Sévrain S, Bonté F (2007) Skin hydration: A review of its molecular mechanisms. *J Cosmet Dermatol* 6: 75–82.

Walko G, Castanon MJ, Wiche G (2015) Molecular architecture and function of the hemidesmosome. *Cell Tissue Res* 360: 529–544.

Walsh CM, Bautista DM, Lumpkin EA (2015) Mammalian touch catches up. *Curr Opin Neurobiol* 34: 133–139.

Zimmerman A, Bai L, Ginty DD (2014) The gentle touch receptors of mammalian skin. *Science* 346: 950–954.

3

Stratum Corneum Lipid Composition and Organization

Enamul Haque Mojumdar and Joke A. Bouwstra

CONTENTS

The Structure and Function of the Stratum Corneum

The stratum corneum (SC) is a non-viable layer consisting of corneocytes (dead cells filled with keratin and water) embedded in a highly ordered lipid matrix. It is about 10–15 μm thick and consists of 10–20 layers of corneocytes (Holbrook and Odland, 1974; Russell et al., 2008). The corneocytes have a diameter of about 20–30 μm and a thickness of about 0.5 μm, and absorb water when hydrating the skin (Bouwstra et al., 2003). The corneocytes are oriented parallel to the skin surface. By nature, the corneocytes have a limited permeability of most compounds due to the cornified envelope (densely cross-linked layer of proteins) wrapped around these cells (Simonetti et al., 1995). The intercellular space between the corneocytes is filled with a crystalline lipid matrix. Due to the low permeability of the cornified envelope, topically applied molecules preferably diffuse along the intercellular lipid matrix (Boddé et al., 1991; Talreja et al., 2001). The architecture of the SC therefore is often described as a simplified 'brick and mortar' structure with the corneocytes being the bricks and the lipids being the mortar (Michaels et al., 1975). Although the corneocytes in the SC are dead, flattened cells filled with keratin and water, the SC is a dynamic tissue because of its high enzymatic activity which can regulate (1) the desquamation process of the outermost corneocytes, (2) the formation of a matured cornified envelope and (3) the hydration level in the SC. The hydration level is affected by the presence of the natural moisturizing factor (NMF) composed of, for example, (chemically modified) amino acids, glycerol, urea and salts. A part of the NMF compounds is metabolic products of the degradation of filaggrin, an important barrier protein (O'Regan et al., 2010). The SC serves as the main barrier against the intruding pathogens and the penetration of substances across the skin (Blank and Scheuplein 1969; Scheuplein and Blank, 1971; Elias and Menon, 1991; Wertz and van den Bergh, 1998).

A drug compound may take two different routes to penetrate the skin: the transappendage route (transport via the sweat glands, hair follicles and sebaceous glands) and the transepidermal route. The transappendage route accounts for 0.1% of the total skin surface area (Barry, 1983). Although there is much debate in literature about the contribution of hair follicles to the permeation of substances, it is considered to be less important when it comes to skin permeation than transport across the transepidermal route. The transepidermal route may follow either the intercellular or the intracellular (transcellular) pathway depending on the nature of the molecules. As mentioned earlier, the cornified envelope minimizes the uptake of most compounds into corneocytes, thus the intracellular pathway may be less prominent. For this reason it has been thought that the intercellular tortuous pathway along the intercellular lipid matrix is an important route of compound penetration through the skin (Williams and Elias 1987; Boddé et al 1991; Johnson et al., 1997; Meuwissen et al., 1998). Studies have shown that after lipid extraction, the permeability of the SC has increased several folds (Rastogi and Singh, 2001), which may suggest that indeed the intercellular lipid matrix plays a pivotal role in the skin barrier.

Lipid Composition in Human Stratum Corneum

The intercellular lipid matrix present in the SC consists of three main lipid classes: ceramides (CERs), cholesterol (CHOL) and free fatty acids (FFAs). The ratio between these lipids in the human SC is approximately equimolar (Wertz et al., 1985; Robson et al., 1994; Stewart and Downing, 1999; Weerheim and Ponec, 2001; Ponec et al., 2003; Masukawa et al., 2008). The CERs in the SC are of exceptional molecular architecture (Figure 3.1). They are composed of a long hydrocarbon acyl chain linked to a sphingoid base through an amide linkage. The head-group moiety of the sphingoid base can vary in structure; sphingosine (S), phytosphingosine (P), 6-hydroxysphingosine (H) and dihydrosphingosine (dS) are the four different sphingoid bases. The acyl chain moiety also varies. The acyl chains are non-hydroxylated (N), α-hydroxylated (A) or ω-hydroxylated (O). The latter can be linked to another fatty acid chain (e.g. linoleate [C18:2]) through an ester linkage referred to as esterified ω-hydroxy (EO) fatty acid. The EO subclasses, also known as acyl-CER, have an exceptionally long acyl chain that is extraordinary in the SC lipids. The different combinations of both the variation in acyl chain architecture and the variation in sphingoid head-group lead to 12 different CER subclasses. Low levels of three O-CERs, that is the EO CER without the linoleate chain, has also been detected in SC (t'Kindt et al., 2012) (not depicted in Figure 3.1) Two new subclasses have also been recently identified and are referred to as 1-O-acylceramides (Rabionet et al., 2014). These CERs consist of saturated long acyl chains in both the N- and O- position, hence they are named as 1-O-ENS and 1-O-EAS CERs, respectively (the letter E in the acronym describes the additional FFA esterified to the CER). Therefore, to date, 17 different subclasses of CERs have been identified in the human SC (Wertz et al., 1985; Masukawa et al., 2008; Ponec et al., 2003; van Smeden et al., 2011; Rabionet et al., 2014; t'Kindt et al., 2012) (Figure 3.1). Furthermore, apart from variation in the subclasses, the CERs exhibit a wide distribution in their total chain length, that is, the total chain length of the acyl chain plus the sphingoid has been reported to be between 34 and 72 carbon atoms in healthy individuals (Farwanah et al., 2005a; Masukawa et al., 2008).

When focusing on the FFAs, a wide variation in the chain length has been observed. The FFAs reported in the human SC are mainly saturated and ranges in chain length between C14 to C34 with an average chain length of C20 to C22 (Ansari et al., 1970; Norlén et al., 1998; Jennemann et al., 2012; Park et al., 2012; van Smeden et al., 2014). In addition to the saturated FFA, low levels of unsaturated FFA (e.g. oleic acid, linoleic acid) and hydroxy FFA are also present in SC (Ansari et al., 1970; Bouwstra et al., 1996; Norlén et al., 1998).

Apart from CERs and FFAs, another lipid class that is abundantly present is CHOL. Some other derivatives of CHOL, CHOL sulphate and CHOL esters, are also present in the SC, some of them in trace amounts (Gray and Yardley, 1975; Gray and White, 1978; Yardley and Summerly, 1981).

The Lipid Organization in Human Stratum Corneum

The extracellular lipid matrix present in the SC is very unique and shows an exceptional three-dimensional ordering of the lipids. The lipids are organized in stacks of lamellae referred to as the lamellar organization. The lamellae are oriented approximately parallel to the skin surface. Two lamellar phases coexist in the lipid matrix of human SC. The length of the repeating unit of these lamellar phases as examined by X-ray diffraction is about 13 nm and 6 nm, being referred to as the long periodicity phase

Non-Hydroxylated CERs

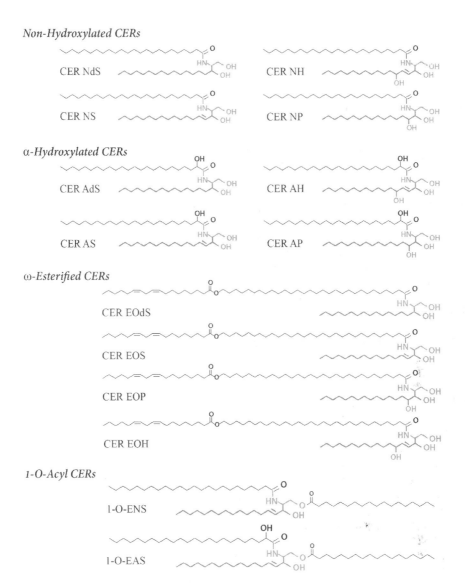

CER NdS

CER NH

CER NS

CER NP

α-Hydroxylated CERs

CER AdS

CER AH

CER AS

CER AP

ω-Esterified CERs

CER EOdS

CER EOS

CER EOP

CER EOH

1-O-Acyl CERs

1-O-ENS

1-O-EAS

FIGURE 3.1 Molecular structure of the CERs present in human SC. The CERs can be sub-grouped based on their hydroxylation and esterification in the fatty acid and sphingoid moiety as presented in the figure.

(LPP) and the short periodicity phase (SPP), respectively (Madison et al., 1987; White et al., 1988; Bouwstra et al., 1991; Hatta et al., 2006) (Figure 3.2, top right). Unlike the SPP that is arranged in a bilayer structure (Mojumdar et al., 2013), the repeat distance of the LPP is very exceptional compared to other lipid systems, and this phase is only detected in SC. This may suggest that the LPP plays a key role in the skin barrier function (Bouwstra, 2000).

The presence of lamellar phases in SC was first monitored by freeze-fracture electron microscopy, but with this method, it is difficult to examine the orientation and stacking of the lipids (Breathnach et al., 1973). Subsequently, the broad–narrow–broad electron translucent bands were reported using electron microscopy with RuO$_4$ post-fixation (Madison et al., 1987; Hou et al., 1991). These studies revealed for the first time regular stacks of lipid lamellae between the corneocytes (Figure 3.3). Using small-angle X-ray diffraction (SAXD), the presence of the LPP with a periodicity of about 13 nm has been detected in human, pig and mouse SC (White et al., 1988; Bouwstra et al., 1991, 1994, 1995). Using cryo-electron microscopy the lamellar organization was visualized. The repeat distance observed was different

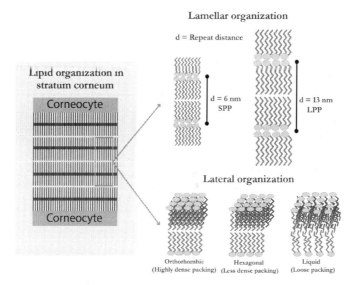

FIGURE 3.2   SC lipid organization. The lipids can be assembled in two lamellar phases (shown top right part of the figure): LPP with a repeating unit (D) of ~13 nm and SPP with a repeating unit of ~6 nm. The lateral organization (bottom right) describes the way the lipids in the lipid matrix are packed in the plane perpendicular to the direction of the lamellae. The lateral packing can be orthorhombic (highly dense), hexagonal (less dense) or liquid (loose packing) as depicted in the figure.

from that obtained by X-ray diffraction, but could also not be explained by a normal bilayer structure (Iwai et al., 2012).

However, the arrangement of the lamellar phases is only the organization in one direction in the three-dimensional space. In the other two dimensions, that is in the plane perpendicular to the direction of the arrangement of the lamellar phases, the lipids assemble in the so-called lateral organization. Not only the lamellar phases but also this lateral organization in the lipid matrix is of crucial importance for the skin barrier function. Depending on the lipid composition, the lipids adopt an orthorhombic (dense) packing, hexagonal (less dense) packing or liquid (loose) packing. In Figure 3.2, a schematic drawing of the three possible arrangements is shown. In the early 1990s, the lateral packing of human skin was detected with

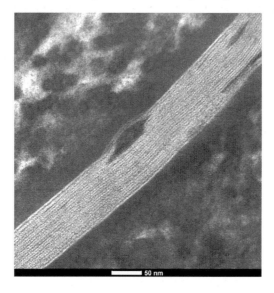

FIGURE 3.3   The lamellar organization as visualized by freeze fracture electron microscopy combined with ruthenium tetroxide staining. A

wide-angle X-ray diffraction as well as with Fourier transform infrared spectroscopy (FTIR). The lipids prevail mainly in orthorhombic lateral packing with a small subpopulation of lipids forming a hexagonal packing (Bommannan et al., 1990; Ongpipattanakul et al., 1994). Whether the liquid lateral packing coexists with the crystalline packing is difficult to detect with X-ray diffraction, as the broad reflection of the liquid packing is obscured by the reflections from the keratin present in the corneocytes (Bouwstra et al., 2001). Interestingly, an orthorhombic to hexagonal phase transition has been observed just above the skin temperature in a temperature region of 30–45°C, also depending on the water content in SC. As it has been shown that a hexagonal packing results in higher transepidermal water loss (TEWL) values than an orthorhombic packing, this is a very interesting observation (Damien and Boncheva, 2009). An increase in the environmental temperature may induce this phase transition, resulting in a higher water loss, thereby consuming a substantial amount of free energy (more precisely enthalpy of evaporation) that may help to maintain the body temperature even in a high-temperature environment.

Lipid Composition and Organization in the Diseased Skin and Dry Skin

In several skin diseases, such as atopic dermatitis, lamellar ichthyosis, Netherton syndrome, autosomal recessive congenital ichthyosis and lesional skin from psoriasis, the skin barrier function is significantly reduced (Imokawa et al., 1991; Motta et al., 1994; Paige et al., 1994; Bonnart et al., 2010; Ishikawa, 2010; Jungersted et al., 2010; Park et al., 2012; Radner et al., 2013; Sassa, 2013; Tawada et al., 2014). Besides other factors that play a role in this reduced barrier function, the lipid composition in the SC of these patients shows significant deviations from that of healthy skin. Atopic dermatitis is a prevailing inflammatory skin disease with an impaired skin barrier function. The lipid composition of SC of patients suffering from atopic dermatitis is marked by a significant reduction in the level of EO CER subclasses in the lipid composition (Imokawa et al., 1991; Pilgram et al., 2001). The reduction in the level of EO CERs together with the increased level of the short-chain CERs resulted in a reduction in the mean CER chain length in SC of atopic dermatitis patients compared to controls. This, in turn, resulted in an increased level of lipids adopting a hexagonal lateral packing. The reduced chain length and increased presence of the hexagonal lateral packing both correlated excellently with the impaired skin barrier function in atopic dermatitis (Janssens et al., 2012). Besides the contribution of EO CER to the increased formation of the hexagonal lateral packing, the level of EO CER subclasses also contributes to the formation of the LPP. As the LPP has been shown to be important for the skin lipid barrier, this may also contribute to the impaired skin barrier in atopic dermatitis patients (Janssens et al., 2012; Jennemann et al., 2012). Apart from the changes in CER composition, an increased level of short-chain FFAs, a change in the CER/CHOL/FFA ratio and an increased level of monounsaturated fatty acids (MUFAs) are also reported in the SC of these patients (Macheleidt et al., 2002; Takigawa et al., 2005; Janssens et al., 2012, 2014; van Smeden et al., 2014).

An elevated level of MUFAs has also been observed in another skin disorder known as Netherton syndrome. In some patients, the level of MUFA can approach up to 80–90% of the total FFAs (van Smeden et al., 2014). In addition, a decrease in FFA chain lengths, an increase in the level of short-chain CERs, an increase in the level of unsaturated CERs and a reduction in the level of EO CERs are also observed in these patients (van Smeden et al., 2014). Consequently, an altered lipid organization in the SC of most of these patients has been reported (van Smeden et al., 2014). This includes a change in the lamellar phases or even an absence and a reduced ordering of the lipid chains in the lateral packing.

As far as psoriasis is concerned, several changes have been observed in a number of studies. Psoriasis is a chronically inflammatory skin disease. Due to an abnormal epidermal proliferation and differentiation, lesional skin shows abnormalities in the lipid composition and organization. Motta et al. 1993, 1995) reported an altered CER subclass composition with a decrease in EO CERs. Furthermore, an increase in the TEWL levels was monitored on psoriatic lesions. Another more recent paper showed that in non-lesional skin there was no change in CER composition in SC of psoriasis patients (Farwanah et al., 2005b). As far as the lipid organization is concerned, little data are available. Recently, it was shown that lesional skin has marked differences in the lamellar organization, whereas non-lesional skin lipid organization was not different from that of control skin (van Smeden et al., 2014). Furthermore, transmission electron microscopy studies reveal an aberrant SC lipid structure in the psoriatic skin (Ghadially et al., 1996). These changes in lamellar organization may be related to the reduced level of EO CER in psoriasis skin observed by Motta et al. (1993, 1995).

In lamellar ichthyosis, the overall level of FFAs is significantly reduced compared to the healthy SC (Lavrijsen et al., 1995). This may enhance the increased presence of lipids adopting the less dense hexagonal lateral packing in this disease (Pilgram et al., 2001). However, as no information on chain length of the FFA has been reported, it is not known whether altered chain length distribution of FFA or CERs may also play a role. X-ray diffraction studies revealed changes in the lamellar organization compared to the healthy SC, in which an altered CER composition as well as a reduction in chain length may play a role (Lavrijsen et al., 1995).

Another ichthyosis skin disease, autosomal recessive congenital ichthyosis, also showed a change in lipid composition. In this disease ceramide synthase 3, responsible for the linking of the very long fatty acid chain to the sphingoid base, is absent due to a genetic mutation. This results in a reduction of the level of all non-hydroxylated CERs with an acyl chain length of more than 22 carbon atoms, while the EO CERs and the bound CERs reduce in level beyond a chain length of 26 carbon atoms (Eckl et al., 2013). These drastic changes in CER composition are expected to result in dramatic changes in the lipid organization.

Finally, recessive X-linked ichthyosis suffers from the elevated levels of CHOL sulphate in the SC due to the deficiency of the enzyme CHOL sulphatase (Rehfeld et al., 1988; Zettersten et al., 1998). An increment in CHOL sulphate resulted in a fluid phase formation in the lipid model mixtures, and therefore elevated cholesterol sulphate levels may contribute to a reduced skin barrier in this disease (Bouwstra et al., 1999).

Whether similar compositional changes are also detected in cosmetically dry skin is not known yet. Only a few studies reported the lipid composition in cosmetically dry skin, but there is no information on the chain length distribution. In two publications it was shown that in the winter season in the EO CER, primarily CER EOS, the oleate moiety is increased dramatically at the expense of linoleate moiety (Conti et al., 1996; Rogers et al., 1996). Although in that publication it is reported that this change in lipid composition may be normalized by topical application of unsaturated fatty acids and triglycerides, from the knowledge we have gained during the years on SC lipid phase behaviour, this is only possible when the unsaturated linoleate will be linked to the very long acyl chain of EO CER at the expense of the chemically linked oleate. The authors of this chapter doubt whether this indeed occurs. In another publication (Schreiner et al., 2000) it was shown that the CER EOS level could be related to the presence of the LPP. However, in this study, the number of persons in each group including young and old dry skin was too small to observe any statistical differences between the various groups. However, it was recently reported that in atopic eczema patients a reduced level of EO CERs correlated with a reduced presence of lipids forming the LPP (van Smeden et al., 2014).

Lipid Model Systems as a Tool to Examine the Relation between Lipid Composition and Organization in More Detail

Lipid Model Systems to Elucidate the Lipid Phase Behaviour

Examining how changes in lipid composition and organization are related to the skin barrier function is often very difficult to study when performing clinical studies. In the best-case scenario, correlations will be observed. However, this does not provide information on whether the change in lipid composition or organization is also the cause of the impaired skin barrier function. In order to study this, SC lipid models have been developed. This allows studying of the lipid properties without interference of the corneocytes. Furthermore, the lipid composition can be changed on demand. Therefore, such models facilitate the interpretation when focusing on the lipid organization and hence can provide novel information on the role of the lipid classes and subclasses in the lipid phase behaviour. The most simplistic models of this category are mono, di, ternary or quaternary lipid mixtures of CERs, CHOL and FFA (Rerek et al., 2005; Chen et al., 2007; Kessner et al., 2008a,b; Ruettinger et al., 2008; Brief et al., 2009; Schröter et al., 2009).

These models are particularly useful to study the basic interactions between different lipids. Although these models provide important information regarding lipid phase behaviour, nonetheless, they often form phases different from those in SC. Furthermore, neutron and X-ray diffraction studies reveal that these model lipid mixtures often lack the ability to form the LPP and SPP. This makes the simplified model systems less suitable for identifying the role of various lipid subclasses in the lipid phase behaviour in the SC when interpreting dry skin and diseased skin data focusing on the relation between reduced skin barrier and changes in lipid composition (Kessner et al., 2008; Schröter et al., 2009).

As opposed to simple lipid mixtures, the CERs isolated from human or pig SC have also been used together with CHOL and FFA in order to examine the lipid phase behaviour (Bouwstra et al., 1996, 2001). The mixtures of isolated human CERs and CHOL revealed the presence of two lamellar phases with repeat distances of 5.4 and 12.8 nm lamellae, closely mimicking the lamellar phases present in human and pig SC (Bouwstra et al., 2001). However, there was one clear difference. These lipid mixtures exhibited hexagonal lateral packing instead of orthorhombic, which is dominantly present in the SC (Bommannan et al., 1990; Ongpipattanakul et al., 1994). The addition of FFA mixture with long and variable chain lengths resulted not only in the formation of the two prominent lamellar phases, but the lipids also assembled in an orthorhombic lateral packing (Bouwstra et al., 1998a). Further studies revealed that the absence of CER EOS in the isolated CER mixture dramatically reduced the formation of the LPP, signifying the impact of CER EOS for a proper lipid organization in the SC (McIntosh et al., 1996; Bouwstra et al., 1998b, 2001). The effect of the degree of unsaturation of the fatty acid that is ester bound to the 30 C acyl chain of CER EOS was also investigated using equimolar FFA, CHOL and isolated human CERs. In these studies, the CER EOS was replaced by synthetic CER EOS linoleate (di-unsaturated in the linoleate moiety), CER EOS oleate (mono-unsaturated) or CER EOS stearate (saturated) (Bouwstra et al., 2002). These studies revealed that the degree of unsaturation in the fatty acid chain highly affects the lipid lamellar organization. The lipid mixtures prepared with CER EOS linoleate forms dominantly the LPP lamellae compared to the CER EOS oleate. In the case of CER EOS stearate, the LPP was even absent (de Sousa et al., 2011). When extrapolating this to the higher level of CER EOS oleate found in cosmetically dry skin, one can expect a reduced level of lipids adopting the LPP (Conti et al., 1996).

As human skin is not readily available and isolation of CERs from SC is a time-consuming process, as an alternative synthetic CERs can be used (de Jager et al., 2005). The advantage of synthetic CERs is that the composition can be changed on demand, facilitating study of the role the various lipid subclasses play in the lipid organization. The lipid mixtures of equimolar synthetic CER, CHOL and FFA results in the formation of the LPP and SPP similarly to that present in SC, but with slightly shorter periodicities of 12.2 and 5.4 nm along with the orthorhombic lateral packing (de Jager et al., 2005). This phase behaviour is very similar to that obtained with the isolated CERs. In the lipid mixtures, the phase-separated crystalline CHOL domains are also observed similar to that in the SC. Point to be noted here, in the case of synthetic CERs, the LPP formation is only possible in the presence of FFA which contrast with the observations made from natural CER mixtures. The difference in phase behaviour between natural and synthetic CER mixtures might be due to the limited acyl chain length variation in the latter case.

The Stratum Corneum Substitute (SCS) as a Tool to Investigate the Lipid Barrier Function

As described earlier, the lipid mixtures prepared from synthetic CERs, CHOL and FFA can be used to study the lipid phase behaviour and offer an excellent tool to investigate the role of individual CER subclasses or lipid classes in the SC lipid organization. Using these lipid mixtures so far, the relation between the lipid composition and organization has been examined. However, in order to understand the skin barrier function, the effect of lipid composition and organization on the lipid barrier function needs to be established. For this reason, a lipid membrane device was developed, referred to as stratum corneum substitute (SCS) (de Jager et al., 2006; Groen et al., 2008). The SCS was prepared by casting lipids (mixtures of synthetic CERs, CHOL and FFAs) on a porous support (for example, spraying lipids on top of polycarbonate filter membrane). It was observed that when selecting the proper composition, the SCS very closely mimics the SC lipid composition and organization (de Jager et al., 2004, 2006). To examine the barrier functionality and to relate the lipid barrier to the lipid composition and organization, the SCS can be clamped in a diffusion cell, which allows examining the transport of (model) compounds by performing *in vitro* permeation studies. Various model drug compounds have been selected to perform permeation studies. These compounds varied in their physical properties: benzoic acid, para aminobenzoic acid (PABA), ethyl PABA, butyl PABA, hydrocortisone (de Jager et al., 2006; Groen et al., 2008, 2011a; Basse et al., 2013). Among these benzoic acids is the least lipophilic drug compound. The permeation profile of all these compounds across the SCS revealed a similar flux as observed in human SC, demonstrating the suitability of the SCS model to investigate the lipid barrier when comparing with SC.

As mentioned earlier, the reduction in the level of CER EO and the increased level of short chain CERs and FFAs have been reported in SC of diseased skin, *in vitro* permeation studies were performed using SCS model membrane where the lipid composition was changed to mimic various aspects of the lipid composition in SC of atopic dermatitis (Groen et al., 2011; Basse et al., 2013). Studies performed with lipid mixtures in the absence of CER EOS have reported a reduced lipid barrier, indicating the importance of CER EOS not only in the lipid phase behaviour but also for proper skin lipid barrier function (de Jager et al., 2006). In other studies, the effect of reduced chain length of FFAs has also been examined using benzoic acid as the model drug. A reduction in the FFA chain length impairs the lipid barrier compared to the SC, demonstrating the importance of FFA chain length in the skin barrier (Groen et al., 2011b).

In subsequent studies (Mojumdar et al., 2014a), the effect of chain length distribution of the CERs was also examined. For this reason, SCS prepared from synthetic CERs (total chain length [fatty acid + sphingoid base] 34, 48 or 68 carbon atoms) was compared with CERs isolated from pig SC. In pig CERs, the total chain length varied between 31 carbon atoms and 74 carbon atoms. When comparing the steady-state fluxes of hydrocortisone, the flux across SCS prepared with the pig CERs was approximately 12 times higher compared to SCS prepared with synthetic CERs in approximately the same CER subclass composition (Figure 3.4). As this is a drastic difference, these studies show that chain length distribution drastically affects the skin lipid barrier. Very recently, this was confirmed in a study in which CERs with an acyl chain of 24 C atoms were replaced by that of CERs with an acyl chain of 16 C atoms. This resulted in an impaired lipid barrier and an altered lipid organization as examined by X-ray diffraction and infrared spectroscopy (Pullmannová et al., 2017).

When extrapolating these findings to diseased skin, in which often a reduced chain length distribution is observed simultaneously with a wider chain length distribution, it is expected that these changes in chain length distribution contribute to the impaired skin barrier function. In another study, from the same authors, the effect of monounsaturated FFA (MUFAs) on the permeability has been addressed (Mojumdar et al., 2014b). In these studies, saturated FFAs were partially replaced by the MUFA counterparts. The studies demonstrated that the presence of MUFAs results in a higher flux of hydrocortisone,

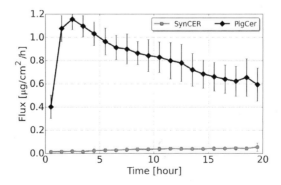

FIGURE 3.4 Hydrocortisone flux across an SCS prepared from pig CERs, CHOL and FFAs in an equimolar ratio; or synthetic CERs, CHOL and FFAs in an equimolar ratio. The CER subclass composition is very similar to that observed in pig CERs.

but also to a higher level of TEWL. This demonstrates that besides chain length distribution of CERs, an increase in unsaturation of FFAs may play an important role in the impaired skin barrier in skin diseases. Important to note is that in these SCS, a clear increase in the level of lipids that adopt a hexagonal packing has been detected at the expense of lipids assembled in an orthorhombic packing. This makes it very obvious that besides a reduction in the formation of the LPP, an increased presence of lipids adopting a hexagonal lateral packing plays an important role in the impaired skin barrier in diseased skin.

REFERENCES

Ansari, M. N. A., N. Nicolaides, and H. C. Fu. 1970. Fatty acid composition of the living layer and stratum corneum lipids of human sole skin epidermis. *Lipids* 5 (10):838–845.

Barry, B. W. 1983. Structure, function, diseases and topical treatment of human skin. In *Dermatological Formulations: Percutaneous Absorption*, New York: Marcel Dekker, pp. 1–48.

Basse, Line Hollesen, Daniël Groen, and Joke A. Bouwstra. 2013. Permeability and lipid organization of a novel psoriasis stratum corneum substitute. *Int J Pharm* 457 (1):275–282.

Blank, I. H., and R. J. Scheuplein. 1969. Transport into and within skin. *Br J Dermatol* 81 (s4):4–10.

Boddé, H. E., I. van den Brink, H. K. Koerten, and F. H. N. de Haan. 1991. Visualization of in vitro percutaneous penetration of mercuric chloride; transport through intercellular space versus cellular uptake through desmosomes. *J Control Rel* 15 (3):227–236.

Bommannan, D., Russell O. Potts, and Richard H. Guy. 1990. Examination of stratum corneum barrier function in vivo by infrared spectroscopy. *J Invest Dermatol* 95 (4):403–408.

Bonnart, Chrystelle, C. Deraison, Matthieu Lacroix, Yoshikazu Uchida, C. Besson, Aur Robin, Ana Briot et al. 2010. Elastase 2 is expressed in human and mouse epidermis and impairs skin barrier function in Netherton syndrome through filaggrin and lipid misprocessing. *J Clin Invest* 120 (3):871–882.

Bouwstra, J. A., G. S. Gooris, J. A. van der Spek, and W. Bras. 1991. Structural investigations of human stratum corneum by small-angle X-ray scattering. *J Invest Dermatol* 97 (6):1005–1012.

Bouwstra, J. A., G. S. Gooris, J. A. van der Spek, S. Lavrijsen, and W. Bras. 1994. The lipid and protein structure of mouse stratum corneum: A wide and small angle diffraction study. *Biochim Biophys Acta BBA Lipids Lipid Metab* 1212 (2):183–192.

Bouwstra, J. A., G. S. Gooris, W. Bras, and D. T. Downing. 1995. Lipid organization in pig stratum corneum. *J Lipid Res* 36 (4):685–695.

Bouwstra, J. A., G. S. Gooris, K. Cheng, A. Weerheim, W. Bras, and M. Ponec. 1996. Phase behavior of isolated skin lipids. *J Lipid Res* 37 (5):999–1011.

Bouwstra, J. A., G. S. Gooris, F. E. R. Dubbelaar, A. M. Weerheim, A. P. Ijzerman, and M. Ponec. 1998a. Role of ceramide 1 in the molecular organization of the stratum corneum lipids. *J Lipid Res* 39 (1):186–196.

Bouwstra, J. A., G. S. Gooris, F. E. R. Dubbelaar, A. M. Weerheim, and M. Ponec. 1998b. pH, cholesterol sulfate, and fatty acids affect the stratum corneum lipid organization. *J Invest Derm Symp P* 3 (2):69–74.

Bouwstra, J. A., G. S. Gooris, F. E. R. Dubbelaar, and M. Ponec. 1999. Cholesterol sulfate and calcium affect stratum corneum lipid organization over a wide temperature range. *J Lipid Res* 40 (12):2303–2312.

Bouwstra, J. A., F. E. R. Dubbelaar, G. S. Gooris, and M. Ponec. 2000. The lipid organisation in the skin barrier. *Acta Dermvenereol* 80:23–30.

Bouwstra, J. A., G. S. Gooris, F. E. R. Dubbelaar, and M. Ponec. 2001. Phase behavior of lipid mixtures based on human ceramides: Coexistence of crystalline and liquid phases. *J Lipid Res* 42 (11):1759–1770.

Bouwstra, J. A., G. S. Gooris, F. E. R. Dubbelaar, and M. Ponec. 2002. Phase behavior of stratum corneum lipid mixtures based on human ceramides: The role of natural and synthetic ceramide 1. *J Invest Dermatol* 118 (4):606–617.

Bouwstra, J. A., A. de Graaff, G. S. Gooris, J. Nijsse, J. W. Wiechers, and A. C. van Aelst. 2003. Water distribution and related morphology in human stratum corneum at different hydration levels. *J Invest Dermatol* 120 (5):750–758.

Breathnach, A. S., T. Goodman, C. Stolinski, and M. Gross. 1973. Freeze-fracture replication of cells of stratum corneum of human epidermis. *J Anat* 114 (1):65–81.

Brief, Elana, Sungjong Kwak, John T. J. Cheng, Neil Kitson, Jenifer Thewalt, and Michel Lafleur. 2009. Phase behavior of an equimolar mixture of N-Palmitoyl-d-erythro-sphingosine, cholesterol, and palmitic acid, a mixture with optimized hydrophobic matching. *Langmuir* 25 (13):7523–7532.

Mojumdar, Enamul H., Richard W. J. Helder, Gert S. Gooris, and Joke A. Bouwstra. 2014b. Monounsaturated fatty acids reduce the barrier of stratum corneum lipid membranes by enhancing the formation of a hexagonal lateral packing. *Langmuir* 30 (22):6534–6543.

Motta, Stefania, Marcello Monti, Silvia Sesana, Ruggero Caputo, Stephana Carelli, and Riccardo Ghidoni. 1993. Ceramide composition of the psoriatic scale. *Biochim Biophys Acta BBA Mol Basis Dis* 1182 (2):147–151.

Motta, S., M. Monti, S. Sesana, L. Mellesi, R. Ghidoni, and R. Caputo. 1994. Abnormality of water barrier function in psoriasis: Role of ceramide fractions. *Arch Dermatol* 130 (4):452–456.

Motta, S., S. Sesana, R. Ghidoni, and M. Monti. 1995. Content of the different lipid classes in psoriatic scale. *Arch Dermatol Res* 287 (7):691–694.

Norlén, L., Ingrid Nicander, Anders Lundsjö, Tomas Cronholm, and Bo Forslind. 1998. A new HPLC-based method for the quantitative analysis of inner stratum corneum lipids with special reference to the free fatty acid fraction. *Arch Dermatol Res* 290 (9):508–516.

O'Regan, Gráinne M., Patrick M. J. H. Kemperman, Aileen Sandilands, Huijia Chen, Linda E. Campbell, Karin Kroboth, Rosemarie Watson et al. 2010. Raman profiles of the stratum corneum define 3 filaggrin genotype–determined atopic dermatitis endophenotypes. *J Allergy Clin Immunol* 126 (3):574–580.e1.

Ongpipattanakul, Boonsri, Michael L. Francoeur, and Russell O. Potts. 1994. Polymorphism in stratum corneum lipids. *Biochim Biophys Acta BBA Biomembr* 1190 (1):115–122.

Paige, D. G., N. Morse-Fisher, and J. I. Harper. 1994. Quantification of stratum corneum ceramides and lipid envelope ceramides in the hereditary ichthyoses. *Br J Dermatol* 131 (1):23–27.

Park, Yang-Hui, Won-Hee Jang, Jung A. Seo, Miyoung Park, Tae Ryong Lee, Young-Ho Park, Dae Kyong Kim et al. 2012. Decrease of ceramides with very long-chain fatty acids and downregulation of elongases in a murine atopic dermatitis model. *J Invest Dermatol* 132 (2):476–479.

Pilgram, Gonneke S. K., Debby C. J. Vissers, Hans van der Meulen, Stan Pavel, Sjan P. M. Lavrijsen, Joke A. Bouwstra, and Henk K. Koerten. 2001. Aberrant lipid organization in stratum corneum of patients with atopic dermatitis and lamellar ichthyosis. *J Invest Dermatol* 117 (3):710–717.

Ponec, Maria, Arij Weerheim, Peter Lankhorst, and Phil Wertz. 2003. New Acylceramide in native and reconstructed epidermis. *J Invest Dermatol* 120 (4):581–588.

Pullmannová, Petra, Ludmila Pavlíková, Andrej Kováčik, Michaela Sochorová, Barbora Školová, Petr Slepička, Jaroslav Maixner et al. 2017. Permeability and microstructure of model stratum corneum lipid membranes containing ceramides with long (C16) and very long (C24) acyl chains. *Biophys Chem* 224: 20–31.

Rabionet, Mariona, Karin Gorgas, and Roger Sandhoff. 2014. Ceramide synthesis in the epidermis. *Biochim Biophys Acta BBA Mol Cell Biol Lipids* 1841 (3):422–434.

Radner, Franz P. W., Slaheddine Marrakchi, Peter Kirchmeier, Gwang-Jin Kim, Florence Ribierre, Bourane Kamoun, Leila Abid et al. 2013. Mutations in CERS3 cause autosomal recessive congenital ichthyosis in humans. *Plosgentics* 9 (6):1–11.

Rastogi, Sumeet K., and Jagdish Singh. 2001. Lipid extraction and transport of hydrophilic solutes through porcine epidermis. *Int J Pharm* 225 (1–2):75–82.

Rehfeld, Selwyn J., William Z. Plachy, Mary L. Williams, and Peter M. Elias. 1988. Calorimetric and electron spin resonance examination of lipid phase transitions in human stratum corneum: Molecular basis for normal cohesion and abnormal desquamation in recessive X-linked ichthyosis. *J Invest Dermatol* 91 (5):499–505.

Rerek, Mark E., Dina Van Wyck, Richard Mendelsohn, and David J. Moore. 2005. FTIR spectroscopic studies of lipid dynamics in phytosphingosine ceramide models of the stratum corneum lipid matrix. *Chem Phys Lipids* 134 (1):51–58.

Robson, K. J., M. E. Stewart, S. Michelsen, N. D. Lazo, and D. T. Downing. 1994. 6-Hydroxy-4-sphingenine in human epidermal ceramides. *J Lipid Res* 35 (11):2060–2068.

Rogers, J., C. Harding, A. Mayo, J. Banks, and A. Rawlings. 1996. Stratum corneum lipids: The effect of ageing and the seasons. *Arch Dermatol Res* 288 (12):765–770.

Ruettinger, A., M. A. Kiselev, T. Hauss, S. Dante, A. M. Balagurov, and R. H. Neubert. 2008. Fatty acid interdigitation in stratum corneum model membranes: A neutron diffraction study. *Eur Biophys J* 37 (6):759–771.

Russell, Lisa M., Sandra Wiedersberg, and M. B. Delgado-Charro. 2008. The determination of stratum corneum thickness an alternative approach. *Eur J Pharm Biopharm* 69 (3):861–870.

Sassa, Takayuki, Yusuke Ohno, Shotaro Suzuki, Toshifumi Nomura, Chieko Nishioka, Toshiki Kashiwagi, Taisuke Hirayama et al. 2013. Impaired epidermal permeability barrier in mice lacking Elovl1, the gene responsible for very-long-chain fatty acid production. *Mol Cell Biol* 33 (14):2787–2796.

Scheuplein, R. J., and I. H. Blank. 1971. Permeability of the skin. *Physiol Rev* 51 (4):702–747.

Schreiner, Volker, Gert S. Gooris, Stephan Pfeiffer, Ghita Lanzendorfer, Horst Wenck, Walter Diembeck, Ehrhardt Proksch et al. 2000. Barrier characteristics of different human skin types investigated with X-ray diffraction, lipid analysis, and electron microscopy imaging. *J Invest Dermatol* 114 (4):654–660.

Schröter, Annett, Doreen Kessner, Mikhail A. Kiselev, Thomas Hauß, Silva Dante, and Reinhard H. H. Neubert. 2009. Basic nanostructure of stratum corneum lipid matrices based on ceramides [EOS] and [AP]: A neutron diffraction study. *Biophys J* 97 (4):1104–1114.

Simonetti, O., J. A. Kempenaar, M. Ponec, A. J. Hoogstraate, W. Bialik, A. H. G. J. Schrijvers, and H. E. Boddé. 1995. Visualization of diffusion pathways across the stratum corneum of native and in-vitro-reconstructed epidermis by confocal laser scanning microscopy. *Arch Dermatol Res* 287 (5):465–473.

Stewart, Mary Ellen, and Donald Talbot Downing. 1999. A new 6-hydroxy-4-sphingenine-containing ceramide in human skin. *J Lipid Res* 40 (8):1434–1439.

Takigawa, Hirofumi, Hidemi Nakagawa, Michiya Kuzukawa, Hajime Mori, and Genji Imokawa. 2005. Deficient production of hexadecenoic acid in the skin is associated in part with the vulnerability of atopic dermatitis patients to colonization by *Staphylococcus aureus*. *Dermatology* 211 (3):240–248.

Talreja, Priva, Gerald Kasting, Nancy Kleene, William Pickens, and Tsuo-Feng Wang. 2001. Visualization of the lipid barrier and measurement of lipid pathlength in human stratum corneum. *AAPS J* 3 (2):48–56.

Tawada, Chisato, Hiroyuki Kanoh, Mitsuhiro Nakamura, Yoko Mizutani, Tomomi Fujisawa, Yoshiko Banno, and Mariko Seishima. 2014. Interferon-c decreases ceramides with long-chain fatty acids: Possible involvement in atopic dermatitis and psoriasis. *J Invest Dermatol* 134 (3): 712–718.

t'Kindt, R., L. Jorge, E. Dumont, P. Couturon, F. David, P. Sandra, and K. Sandra. 2012. Profiling and characterizing skin ceramides using reversed-phase liquid chromatography-quadrupole time-of-flight mass spectrometry. *Anal Chem* 84 (1):403–411.

van Smeden, Jeroen, Louise Hoppel, Rob van der Heijden, Thomas Hankemeier, Rob J. Vreeken, and Joke A. Bouwstra. 2011. LC/MS analysis of stratum corneum lipids: Ceramide profiling and discovery. *J Lipid Res* 52 (6):1211–1221.

van Smeden, J., M. Janssens, G. S. Gooris, and J. A. Bouwstra. 2014a. The important role of stratum corneum lipids for the cutaneous barrier function. *Biochim Biophys Acta Mol Cell Biol Lipids* 1841 (3):295–313.

van Smeden, Jeroen, Walter A. Boiten, Thomas Hankemeier, Robert Rissmann, Joke A. Bouwstra, and Rob J. Vreeken. 2014b. Combined LC/MS-platform for analysis of all major stratum corneum lipids, and the profiling of skin substitutes. *Biochim Biophys Acta Mol Cell Biol Lipids* 1841 (1):70–79.

van Smeden, Jeroen, Michelle Janssens, Walter A. Boiten, Vincent van Drongelen, Laetitia Furio, Rob J. Vreeken, Alain Hovnanian et al. 2014c. Intercellular skin barrier lipid composition and organization in netherton syndrome patients. *J Invest Dermatol* 134 (5):1238–1245.

van Smeden, Jeroen, Michelle Janssens, Edward C. J. Kaye, Peter J. Caspers, Adriana P. Lavrijsen, Rob J. Vreeken, and Joke A. Bouwstra. 2014d. The importance of free fatty acid chain length for the skin barrier function in atopic eczema patients. *Exp Dermatol* 23 (1):45–52.

Weerheim, A., and M. Ponec. 2001. Determination of stratum corneum lipid profile by tape stripping in combination with high-performance thin-layer chromatography. *Arch Dermatol Res* 293 (4):191–199.

Wertz, Philip W., and Benedicte van den Bergh. 1998. The physical, chemical and functional properties of lipids in the skin and other biological barriers. *Chem Phys Lipids* 91 (2):85–96.

Wertz, Philip W., Marion C. Miethke, Sherri A. Long, John S. Strauss, and Donald T. Downing. 1985. The composition of the ceramides from human stratum corneum and from comedones. *J Invest Dermatol* 84 (5):410–412.

White, Stephen H., Dorla Mirejovsky, and Glen I. King. 1988. Structure of lamellar lipid domains and corneocyte envelopes of murine stratum corneum. An x-ray diffraction study. *Biochemistry* 27 (10):3725–3732.

Williams, M. L., and P. M. Elias. 1987. The extracellular matrix of stratum corneum: Role of lipids in normal and pathological function. *Crit Rev Ther Drug Carrier Syst* 3 (2):95–122.

Yardley, H. J., and R. Summerly. 1981. Lipid composition and metabolism in normal and diseased epidermis. *Pharmac Ther* 13 (2):357–383.

Zettersten, Elizabeth, Mao-Qiang Man, Junko Sato, Mitsuhiro Denda, Angela Farrell, Ruby Ghadially, Mary L. Williams et al. 1998. Recessive X-linked ichthyosis: Role of cholesterol-sulfate accumulation in the barrier abnormality. *J Invest Dermatol* 111 (5):784–790.

4

Immunology of Skin and Reactivity

Krishna Telaprolu, Heather A.E. Benson, Jeffrey E. Grice,
Michael S. Roberts and Philip L. Tong

CONTENTS

Skin Immune Network

The skin is the first line of defense against the entry of exogenous substances and microorganisms from the environment, and this defense is both structural as well as immunological in nature. As outlined in detail in Chapter 2, the skin structure consists of an outer epidermis and dermis attached to the basement membrane. The skin has a barrier composed of four different layers of protection. The first layer of the defense comprises a stratified stratum corneum along with tight junctions in the stratum granulosum. The second line of defense is a chemical barrier provided by the antimicrobial proteins present in the epidermis. The microbial flora present in the skin provides the third layer of protection by inhibiting entry of skin pathogens such as *Staphylococcus aureus* and other microbes. The fourth layer of protection is the immunological barrier, which is comprised of both the innate and adaptive immune network that provide immediate as well as long-lasting immunity against invading pathogens (Kuo et al., 2013a; Heath and Carbone, 2013). The skin immune network contains cells from both hematopoietic and non-hematopoietic origin (Figure 4.1). The epidermis contains mainly keratinocytes, and other cells such as melanocytes and immune cells, Langerhans cells. The collagen matrix of the dermis provides a micro-environment in which dermal dendritic cells, αβ-T cells, γδ-T cells, mast cells and macrophages reside

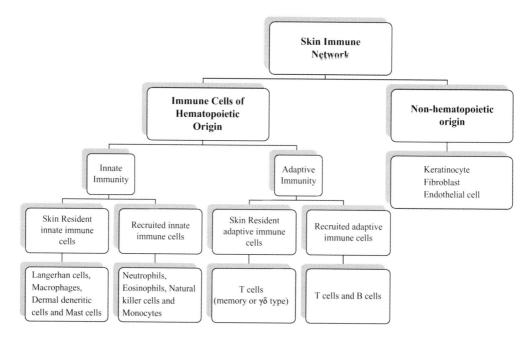

FIGURE 4.1　Skin immune network.

(Di Meglio et al., 2011). Cells that are also recruited during certain inflammatory states include natural killer cells, B cells, eosinophils and circulating T cells.

Immunological Composition of the Epidermis

Keratinocytes

Keratinocytes, the primary cell type distributed in the epidermis, play a vital role in maintaining the immune homeostasis in the skin by contributing to both innate and adaptive immunity. Apart from their immune functions, they are also crucial mediators of skin growth, differentiation, regulation and wound healing (Steinhoff et al., 2001).

Keratinocytes in Innate Immunity

Keratinocytes constitute 90% of the epidermal cells and play a key role in maintaining the physical barrier of the skin. They actively participate in innate immunity by expressing antimicrobial peptides (AMPs) such as human β-defensins (HBD-2 and -3) and human cathelicidin (hCAP18/LL-37). These AMPs and others including psoriasin, RNase 7 and dermicidin demonstrate broad antimicrobial activity, including inhibiting pathogen entry and colonization (Brown and McLean, 2012; Mann, 2014). These AMPs bind to the microbial membrane by their positively charged hydrophobic surface and kill microbes by forming pores in their membrane (Bangert et al., 2011). In addition to cytotoxic properties, AMPs act as chemotactic agents for inflammatory cells by releasing proinflammatory agents and attract adaptive immune cells. The AMPs also cause the degranulation of mast cells and production of cytokines and chemokines from keratinocytes during acute inflammation (Steinstraesser et al., 2008). Keratinocytes recognize and defend against the entry of the pathogens by producing pattern recognizing receptors (PRRs), which include Toll-like receptors (TLRs) for responding to pathogen components and Nod-like receptors (NLRs) Nod1 and Nod2 for responding to bacterial peptidoglycan, and viral and fungal elements. Keratinocytes express the retinoic acid-inducible gene (RIG)–like receptors (RLRs) for responding to virus infection and C-type lectin receptors (CLRs) for responding to fungal infection. PRRs can recognize specific components of the pathogens such as lipoproteins, flagella and cell wall components. During homeostasis, the keratinocytes express several of these antimicrobial peptides, cytokines and chemokines. Upon microbial entry into the skin, the PRRs are activated and a surge in AMP production is seen, along with recruitment and crosstalk with other immune cells. Through these actions, the keratinocytes generate immediate protection from the entry of pathogens and further stimulate humoral and adaptive immune responses (Wang et al., 2017; Kuo et al., 2013b; Gallo and Nizet, 2008).

Keratinocytes in Adaptive Immunity

Keratinocytes also participate in the adaptive immune response that results in a brisk and specific memory response on subsequent re-exposure. The quality of the adaptive immune response is highly dependent on the initial innate immunity response. Keratinocytes produce cytokines, interleukins, colony-stimulating factors, growth factors and chemokines. These chemical messengers are stored in the cell during normal conditions and are released in response to external stimuli such as ultraviolet (UV) radiation, bacterial infection and chemical substances. Keratinocytes participate in adaptive immunity by acting as antigen-presenting cells, and the antigen is finally killed by T cells and B cell-derived antibodies. Keratinocytes produce a plethora of signalling proteins including proinflammatory mediators IL-1, -6, -8 and TNF; T-cell trophic interleukins IL-7 and IL-15; immunomodulatory interleukins IL-10, IL-12, IL-18; and other interleukins which act as ligands to some of the receptors. Any physical or chemical stress disrupts the membranes of keratinocytes causing the immediate release of preformed proinflammatory mediator IL-1α, which upon release stimulates the production of another proinflammatory mediator IL-1β. These two IL-1 interleukins further trigger the production of other proinflammatory mediators including tumour necrosis factor-α (TNF-α), chemotactic cytokine IL-8, growth-promoting factors IL-6, IL-7, IL-15, cytokines IL-10 and IL-12, which regulate humoral and cellular immunity (Coquette et al., 2003; Lee et al., 2013). The IL-1α activates fibroblasts to release other mediators such as CXCL9, CXCL10, CXCL11 and CCL20. The hair follicle keratinocytes demonstrate distinct compartmentalization, with different expression of chemokines depending on the micro-anatomical location, and regulate the entry of Langerhans cells, a key resident leukocyte in the epidermis, during epidermal repopulation (Nagao et al., 2012; Wang et al., 2017; Grone, 2002; Steinhoff et al., 2001).

Melanocytes

Melanocytes are melanin-producing cells that are embryonically derived from the neural crest, and situated in hair follicles and epidermis. In the epidermis, they reside in the basal layer of the skin and are surrounded by keratinocytes which together form the epidermal melanin unit first proposed by Fitzpatrick and Breathnach in 1963 (Nordlund, 2007; Fitzpatrick and Breathnach, 1963). The strategic location and the large surface area covered by these cells results in protection of the skin by the production of melanin pigment, which colours the skin and hair follicles and plays a fundamental role in skin physiology by protecting the cells from UV radiation-induced structural changes in DNA. During melanin production, several intermediates produced have antimicrobial properties, which along with melanin can trap and kill invading pathogens. Melanin also has an immunoregulatory role by inhibiting the production of proinflammatory cytokines by T lymphocytes, monocytes, fibroblasts and endothelial cells (Wood et al., 1999). The melanosomes from the melanocytes have acidic properties transferred to the surrounding keratinocytes, which upon differentiation produce acidified stratum corneum and enhance the barrier protection (Wood et al., 1999). Melanocytes participate in the innate response by producing inducible nitric oxide synthase (iNOS), expressing Toll-like receptors TLR 2–5, 7, 9 and 10, which can recognize the molecular patterns of pathogens and induce production of proinflammatory cytokines and chemokines, which further modulate the recruitment of the immune cells into the skin. Melanocytes also produce an alpha-melanocyte stimulating hormone (α-MSH), which has anti-inflammatory and immunomodulatory activity. These melanocyte activities suggest that they are active participants in cutaneous immunity and not just a physical barrier to microbial entry (Hong et al., 2015).

Langerhans Cells (LCs)

Langerhans cells (LCs) were first recognized and classified as skin nerves by Paul Langerhans in 1868 (Shamoto, 1983). The ontogeny of LCs was always a debatable topic, but recent *in vivo* lineage tracing experiments in mice demonstrate that adult LCs have a dual origin, predominantly from fetal liver monocyte-derived cells, with a minor contribution of yolk sac-derived primitive myeloid progenitors (Hoeffel et al., 2012). Earlier studies have always described their origin from hematopoietic stem cells in the bone marrow (Katz et al., 1979). LCs are myeloid hematopoietic cells and constitute only 3–8% of the mammalian epidermal cells. Steinman et al. (Steinman and Cohn, 1973) discovered LCs in the spleen of mice and classified them as a member of the dendritic cells (DCs). LCs were initially identified by the presence of cytoplasmic granules called Birbeck granules induced by the lectin langerin (CD207); later langerin was also seen to be expressed in dermal dendritic cells (dDCs). LCs are categorized as classic dendritic immune cells by the detection of Fc and C3 receptors, CD45, E-cadherin, epithelial cell adhesion molecules and major histocompatibility complex (MHC) II antigens on their surface (Birbeck et al., 1961; Bangert et al., 2011; Tamaki et al., 1980; Stingl et al., 1977). The additional molecules CD1a, the MHC II antigens human leukocyte antigen (HLA)-DR, HLA-DQ and HLA-DP, and CD39 allow the identification of LCs within the normal state of the epidermis. LCs adhere to keratinocytes by E-cadherin–mediated adhesion (Tang et al., 1993). LCs act as antigen-presenting cells that control the induction of adaptive immune responses by draining into the lymph nodes and presenting the antigens to T cells. By using various types of receptors on their cell surface, LCs recognize the invaded pathogens and microbial products. During this period of antigen recognition, LCs synthesize, load and rearrange the MHCII molecules to the cell surface, and upregulate other co-stimulatory factors and maturation markers. After draining into lymph nodes, they secrete T-cell–attracting chemokines to attract the naïve T cells and thus contribute to the adaptive immunity against invading pathogens (Elbe-Burger and Schuster, 2010). Under homeostatic conditions, LCs also migrate to present their self-peptides to maintain tolerance by inducing regulatory T cells (Loser et al., 2006). The cytokines released from both LCs and surrounding keratinocytes control their maturation and migration to lymph nodes.

Immunological Composition of the Dermis

Dermal Dendritic Cells (dDCs)

In contrast to epidermal dendritic cells (i.e. Langerhans cells) which are sessile, the dermal dendritic cells (dDCs) are highly mobile and continuously migrate through the dermis (Ng et al., 2008). DCs originate from bone marrow-derived hematopoietic stem cells (HSCs). Immunostaining first identified DCs in the skin by staining for Factor XIIIa. Later, it was recognized that dDCs could be differentiated from epidermal DCs (Langerhans cells) by the expression of CD14 and CD1a markers on the dDCs and Langerin expression and Birbeck granules in Langerhans cells. The different markers expressed on the surface of dDCs include CD1b, CD1c (blood DC antigen [BCDA]-1), CD11c, CD36, CD205 and MHC II (Meunier et al., 1993). The CD14+ marker is expressed in both dDCs and macrophages, and dDCs can be differentiated based on their migration behaviour, morphology, and so forth (Haniffa et al., 2015). The primary types of myeloid dDCs in humans are CD14+, CD1a+ and 6-Sulpho LacNAc+ (slan) DCs (Nestle et al., 1993). Among them, CD14+ DCs are weak stimulators of T cells and activate the humoral immune mechanism. The CD14+ DCs may differentiate into LCs upon stimulation by transforming growth factor (TGF β) and can enter the epidermis (Larregina et al., 2001; Chu et al., 2012). A subpopulation of CD14+ DCs express the thrombomodulin markers and differ from other DCs lacking CD14 expression and lower expression of CD11c (Haniffa et al., 2012). These CD141+ DCs actively regulate the immune functions by producing IL-10 (Chu et al., 2012). The CD1a+ DCs are potent activators of CD8+ T and CD4+ cells in the immune system (Pasparakis et al., 2014). 6-Sulpho LacNAc+ (slan) DCs are efficient producers of IL-23 and IL-1β, which are important for the pathogenesis of diseases such as psoriasis (Poltorak and Zielinski, 2017). The different subsets of myeloid DCs play roles in various types of inflammatory disease. In human psoriatic skin, DCs have been found to produce TNF, inducible nitric oxide synthase, IL-20 and IL-23. In other inflammatory diseases such as atopic dermatitis, inflammatory DCs have been shown to produce CCL3, IL-1, IL-12p70, IL-16 and IL-18 (Lowes et al., 2005; Wollenberg et al., 1996). Plasmacytoid dendritic cells (pDCs) are another type of DC present in the skin. These have a limited antigen-presenting function and act as mediators of antiviral immunity by recognizing the entered pathogen RNA and DNA by their Toll-like receptors (TLR) 7 and TLR 9 and secrete type 1 interferon (Gilliet et al., 2008; Pasparakis et al., 2014). The pDCs are phenotypically identified by the expression of CD123, BDCA-2 (CD303), BDCA-4 (neuropilin-1), and immunoglobulin-like transcript 7. Once recruited into the skin, pDCs become localized in perivascular clusters with T cells, and depending on their activation and maturation status, can upregulate MHC-II and generate distinct T-cell responses (Bangert et al., 2011). Plasmacytoid DCs are not normally found in steady-state conditions but are recruited in disease states such as systemic lupus erythematosus and psoriasis, and are reported to participate in cytolysis of tumour cells through granzyme B (Drobits et al., 2012).

Mast Cells (MCs)

Mast cells (MCs) are found throughout the dermis but also prominently in the proximity of blood vessels and are best known for their Th2 immune responses in the skin. The different mast cells in human tissues are classified based on their proteinase content: MC_{TC} (contains tryptase, chymase, carboxypeptidase and a cathepsin G–like proteinase), MC_T (contains tryptase) and MC_C (contains chymase and carboxypeptidase). The mast cells in human skin mostly belong to the MC_{TC} subtype. Tryptase augments the inflammatory response through its proangiogenic activity and action on fibronectin to assist in the recruitment of immune-competent cells such as neutrophils, mononuclear cells, and T lymphocytes to enter the epidermis during an antimicrobial response (Harvima and Nilsson, 2011). The chymase proteinase is responsible for the proinflammatory action of mast cells by its effect on IL-1β and IL-18 (Harvima and Nilsson, 2011). In addition to the flexibility of mast cells, the two enzymes give the ability to regulate the immune response by disintegrating the proinflammatory mediators such as chemokines and cytokines (Zhao et al., 2005). Along with indirectly modulating the immune response by the secreted enzymes, mast cells directly modulate the immune response by cell-to-cell contact and cytokine production. Mast cells interact with T cells including regulatory T cells via OX-40L and TNF-α production. Furthermore, mast cells appear to play a role in psoriasis and atopic dermatitis through IFN-ɣ and IL-4, respectively (Horsmanheimo et al., 1994; Ackermann et al., 1999). Mast cells are generally activated via cross-linking of high-affinity IgE-receptors (Fc&RI) on the mast cell surface by IgE-bound allergens. After activation, mast cells release their preformed histamine (a potent vasodilator) and other preformed mediators, and in turn promote tissue inflammation such as urticaria and angioedema. Along with participation in Th2 responses, tissue resident mast cells are involved in contact hypersensitivity response and wound healing by performing both pro- and anti-inflammatory functions (Harvima and Nilsson, 2011; Galli and Tsai, 2012). Other than binding to cell surface immunoglobulin, various microbial substances bind to Toll-like receptors (TLR2-4, TLR6-7 and TLR9) present on the mast cells, and initiate the release of mediators such as cytokines and degranulation of mast cells. When any double-stranded RNA or DNA virus analogue binds to TLRs on mast cells, proinflammatory chemokines and cytokines are released to potentially assist in recruitment and activation of innate immune responses. Mast cells also have some antigen-presenting cell (APC) functions, similar to LCs and DCs. Mast cells express MHC-I and –II, and costimulatory molecules CD80 and CD86, and were shown to migrate to lymph nodes to influence T cell accumulation (Hershko and Rivera, 2010). Through longitudinal intravital multiphoton microscopy studies, mast cells have recently been shown to interact with DCs during the later stages of skin inflammation to result in DC-to-MC molecule transfers, including MHC-II proteins. Importantly, the extent of MHC-II transfer to MCs correlates with T cell priming efficiency: this expands the current understanding of mast cell function as an integral cell type in the innate immune system of the skin (Dudeck et al., 2017).

T Cells

T cells have long been known to form the adaptive arm of the skin immune system, although more recently new classes of T cells have demonstrated more innate functions. Circulating T cells and resident T cells protect the host from *de novo* infection (Boyman et al., 2007). The resident T cells hold a strategic position near the LCs for rapid protection of skin from pathogen infection (Schenkel and Masopust, 2014). The dermis predominantly contains CD4 memory T cells and some CD8+ cells located near the dermal–epidermal junction, while the epidermis contains CD8+ αβ memory T cells that reside for extended periods in the epidermis and are commonly referred to as tissue-resident memory T cells (T_{rem}) (Gebhardt et al., 2009; Mueller et al., 2014). The T_{rem} cells express a unique set of markers, including CD69 (expressed in skin T cells independent of antigen recognition), VLA-1 and CD103 (Gebhardt et al., 2009). These T cells enter the dermis from blood and are recruited into the epidermis during inflammation by the expression of the CXCR3 ligand of chemokine receptors (Mackay et al., 2013). The skin also consists of 5% of regulatory T cells (T_{reg}), which actively circulate between lymph nodes and skin, and regulate the T cell responses, some functions of APCs such as DCs and macrophages, and neutrophil accumulation during early stages of inflammation. T_{reg} induce the anti-inflammatory functional profile in macrophages by inhibiting the production of TNF-α and have immunosuppressive activity related to cancer. The skin also contains γδ T cells, which are responsible for cutaneous immunosurveillance against pathogens, wound healing and contact hypersensitivity (Macleod and Havran, 2011; Sumaria et al., 2011). Recently γδ T cells were shown to be major producers of IL17, which is now a novel target in the management of psoriasis (Langley et al., 2014; Cai et al., 2011).

Macrophages

The various macrophages present in the skin are M1 proinflammatory macrophages that are activated by IFN-γ or by microbial products, and contribute to both acute and chronic inflammation in the skin. Proinflammatory macrophages expressing CD163 are found in psoriasis, atopic dermatitis and cutaneous T cell lymphoma (Sugaya et al., 2012). M2 wound-healing macrophages are known to regulate the skin inflammatory response through IL-10 and transforming growth factor-β (TGFβ). M2 macrophages are induced by IL-4 and -13 and promote type 2 immune responses by capturing antigens and reducing their availability to lymphocytes. The skin resident macrophages along with the circulating monocytes support the DCs in immunosurveillance and in the transport of antigens to the draining lymph nodes (Fuentes-Duculan et al., 2010).

Innate Lymphoid Cells

Innate lymphoid cells (ILCs) are a newly described and diverse group of immune cells which coordinate inflammatory responses in the skin and other tissues by cytokine production. ILCs in a range of tissues are divided into three subsets (ILC1, ILC2 and ILC3) based on their expression of cytokines and transcription factors. ILC1 consists of natural killer (NK) cells and other ILC1 cells that participate in viral immunity and are triggered by IL-12 and IL-18. The NK cells are the only ILC1 subset identified in healthy human skin and in increased number in psoriasis skin (Ebert et al., 2006). ILC2 cells were recently identified in human skin, and are reported to accumulate in the skin of atopic dermatitis and play a role in the induction of T_H2 type dermatitis (Roediger et al., 2013). ILC3 is the dominant type of ILC present in the skin, and natural cytotoxicity triggering receptor positive ILC3 was found in circulation in healthy skin and in increased number in non-lesional psoriatic skin (Villanova et al., 2014).

Implications of the Skin Immune System in Cosmetic Science

Cosmetics and personal care products by nature and purpose are required to be in contact with human skin, often for prolonged and repeated periods. As a result, they can cause unwanted side effects, including both irritant and allergic contact dermatitis (CD). The industry has over time developed methods to overcome these issues through laboratory testing with the use of gas chromatography–mass spectrometry (Debonneville and Chaintreau, 2004; Rastogi, 1995) for screening for potential cosmetic allergens as well as *in vitro* skin models, amongst many models, as a means of non-animal alternatives for assessing sensitization potential in cosmetic products (Sharma et al., 2012). These testing processes are described in Chapters 25 and 26.

The immune system can also be a potential target with the advent of cosmeceuticals, a term that refers to cosmetic products that have medicinal or drug-like benefits. These products are applied topically and contain active compounds that have the potential to affect the cellular function in skin. If these active compounds reach their cellular target in sufficient concentration, they can have a clinically meaningful effect in areas such as skin anti-aging and whitening. Although clinical studies in cosmeceuticals are not subjected to the same review processes as medicines, this is an active area of ongoing research.

Contact Dermatitis

In the formulation design and development of a new or improved cosmetic product, the primary considerations are efficacy, safety and toxicity as a requirement for the regulatory authorities prior to approval for consumer use. The penetration of allergens or toxic molecules into the skin can cause an immune reaction. The different classes of skin reactions are allergic contact dermatitis (ACD), irritant contact dermatitis (ICD) and photocontact dermatitis. Allergic contact dermatitis is an adaptive immune response to allergen exposure, often to chemicals as an initial step of sensitization, followed by a response elicited by re-exposure to the same allergen. The allergen binds to skin proteins and acts as a hapten that is internalized by the innate cells of the immune system, and the initiation of the immune response begins (Alenius et al., 2008; Adler et al., 2011). Exposing the skin to harsh chemical substances such as surfactants, alkaline and acid solutions has the potential to cause irritant contact dermatitis. The interaction of photoreactive chemicals with ultraviolet radiation-A (UVA) can result in phototoxic compounds against skin cells that cause photocontact dermatitis (Johansen et al., 2015). Allergy induced by cosmetic ingredients has been found to account for 19.4% of all dermatitis incidence in patients tested between 1990 and 2014. The most common cosmetic allergens with an incidence of dermatitis higher than 20% are toluene-2,5-diamine, *Evernia prunastri* (oakmoss), and methylisothiazolinone 2000 ppm (preservative) (Aerts et al., 2014). Ammonium persulfate, linalool hydroperoxides, propolis cera and a fragrance mixture (amyl cinnamal, cinnamal, cinnamyl alcohol, hydroxycitronellal, eugenol, isoeugenol, geraniol and oakmoss extract) have also been reported to be allergens of significance to the cosmetic industry (Goossens, 2011). Euxyl K400® is a preservative composed of a mixture of methyldibromo glutaronitrile and phenoxyethanol that is no longer approved in the European Union (EU) because of increased incidence of skin sensitization (Jackson and Fowler, 1998). Polyhexamethylene biguanide is another preservative used in shampoos, wet wipes, facial tissues and cleansers that has been reported to cause skin sensitization (Leysen et al., 2014). Other widely used preservatives include thiomersal, found to have a high incidence of positive findings in patch testing of cases of suspected allergic contact dermatitis, and parabens which act as weak allergens on damaged skin. Methyl acrylates are common adhesive ingredients in nail gel formulations that cause allergic contact dermatitis on the hands and sometimes at remote sites such as the face, neck or eyelids (Ramos et al., 2014). Dyes are commonly incorporated into cosmetic ingredients. Among them, paraphenylenediamine (PPD) is an aromatic amine compound used in colouring shampoos, textiles and hair dyes that can cause secondary allergy to other dyes. Many dyes are reported to cause phototoxicity and photoallergic dermatitis when exposed to ultraviolet radiation. For example, eosin was removed from American and European markets due to its phototoxicity (Zukiewicz-Sobczak et al., 2013). The UVA sunscreen filter avobenzone that is widely used in cosmetic products has been shown to undergo significant degradation and can cause phototoxicity that is dependent on dose and exposure time to UV radiation. Sunscreen products are often comprised of multiple UV filters to provide a broad spectrum of UVA and UVB protection. Consequently, testing of individual ingredients of these and other products is insufficient to assess the risk of phototoxicity to consumers. Gaspar et al. (2013) demonstrated that a combination of the sunscreen agents octyl methoxycinnamate, avobenzone and 4-methylbenzilidene camphor had a synergistic effect in causing acute phototoxicity in a number of *in vitro* phototoxicity tests, although they concluded that *in vitro* testing alone is insufficient to predict the true incidence of phototoxic reactions in humans. Table 4.1 shows the list of different ingredients causing various type of dermatitis.

TABLE 4.1

List of Chemicals Causing Various Types of Dermatitis

Allergic Contact Dermatitis	Irritant Contact Dermatitis	Photocontact Dermatitis
Plant-based essential oils and extracts • Fragrance mixture amyl cinnamal, cinnamal, cinnamyl alcohol, hydroxycitronellal, eugenol, isoeugenol, geraniol and oakmoss extract • Propolis (1.4 to 3.5%) (Oliver et al., 2015)	**Agricultural products** • Chlorothalonil • (fungicide), and aliphatic amine epoxy catalysts (Lushniak, 2004) **Surfactants** • Anionic and cationic surfactants (Effendy and Maibach, 1995)	**Phototoxic agents** Dyes • Eosin (Zukiewicz-Sobczak et al., 2013) **Pharmaceutical drugs** • Sulfonamides • Doxycycline • Voriconazole • Amiodarone hydrochloride • Norfloxacin (OECD, 2004)
Cosmetic chemical ingredients • Glyceryl thioglycolate • Methylisothiazolinone • Ammonium persulfate • Linalool • Polyhexamethylene biguanide • Methyldibromo glutaronitrile and phenoxyethanol • Hydroperoxides • Paraphenylenediamine (Verhulst and Goossens, 2016)	**Plants** Dieffenbachias, philodendrons, agaves, and daffodils, which are rich in oxalic acids (Lushniak, 2004)	**Photoallergic agents** **Sunscreen agents** • Avobenzone • Methoxycinnamate, avobenzone and 4-methylbenzilidene camphor combination (Gaspar et al., 2013) **Fragrances** • 6-Methylcoumarin • Musk ambrette • Sandalwood oil (Maibach and Honari, 2014), (Neumann et al., 2005)

Pathogenesis of Dermatitis

Irritant contact dermatitis is the result of activation of the innate immune system and its response to the penetrated chemical. It is initiated by disruption of the skin barrier by irritants or harsh chemicals such as detergents, surfactants, or solvents such as alcohols that are present in daily use products, along with environmental factors. The disruption of the skin barrier may occur by disorganization of stratum corneum lipids by the penetrated chemical and can be characterized by an increase in water loss from the skin. Damage to keratinocytes may follow barrier disruption, resulting in the release of proinflammatory cytokines, newly synthesized cytokines and MHC class II adhesion molecules, further promoting the immune response. Mast cells are activated and release proinflammatory molecules and cause vasodilation, the first sign of acute irritant contact dermatitis (Fonacier and Boguniewicz, 2010; Rustemeyer et al., 2011). An important distinction between acute irritant contact dermatitis and allergic contact dermatitis is that anyone can develop irritant contact dermatitis given sufficient exposure to the irritant, whilst allergic contact dermatitis results from a host immune response after sensitization.

Thus, allergic contact dermatitis is a type 4 hypersensitivity reaction towards an allergen that penetrates the skin and sensitizes the immune system (Figure 4.2). More specifically, irritant and allergic contact dermatitis differ in the involvement of T cells and the development of immunological memory to the allergen. The allergen binds to the skin carrier proteins and acts as a hapten, causing sensitization of the immune system, and release of proinflammatory mediators IL-1β and TNF-α that are essential for the maturation of DCs. Mature DCs phagocytize the hapten and break it into small fragments, then load them onto MHC molecules which will be recognized by various T cells. MHC class I molecules will be recognized by CD8+ cytotoxic T cells and MHC class II by CD4+ T helper cells. MHC class III molecules (i.e. CD1d molecules) form complexes with synthetic lipid antigen which activate natural killer T cells (Alenius et al., 2008). The activated T cells migrate to the circulation and lymphatic system, and remain until subsequent exposure. This sensitization phase occurs in 10 to 14 days. When the skin is exposed successively to the same chemical, the elicitation of skin immunity occurs by the recognition of the hapten on Langerhans cells by the CD8+ T cells, followed by the release of IL-1, IL-2, interferon-γ

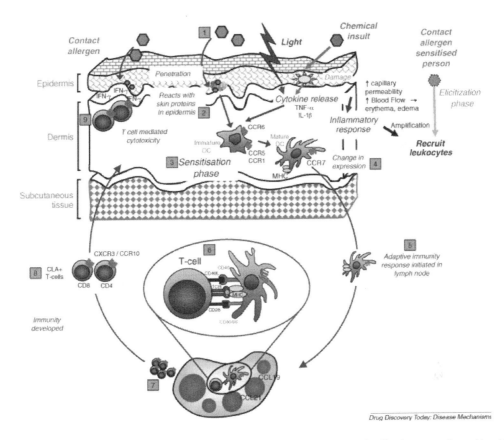

FIGURE 4.2 Schematic representation of type 4 hypersensitivity reaction as seen in allergic contact dermatitis with initial sensitization and subsequent recruitment of various immune cells upon repeated allergen exposure. (From Alenius, Harri, et al., 2008, *Drug Discov Today Dis Mech* 5 (2):e211–e220.)

and TNF-α. The role of CD4+ T cells is less understood. The released cytokines initiate production of cell adhesion molecules and slow the movement of leukocytes in blood vessels and aid their transfer to the skin, bringing about skin edema (Flores and Maibach, 2010; Krasteva et al., 1998). The clinical signs of acute contact dermatitis are heat, edema, itching and vesicles caused by the leukocyte infiltration in the skin. The duration and extent of acute contact dermatitis depend on regulatory T cells (T_{reg}), which show a suppressive effect by inhibiting the activity of T cells through direct cell contact and by the release of cytokines such as IL-10 and TGF-β. While the inflammation in acute contact dermatitis is terminating, activated T cells undergo apoptosis by the interaction of Fas molecules on T cells with the Fas ligand on DCs (Fyhrquist-Vanni et al., 2007; O'Garra and Vieira, 2004).

In photoallergic dermatitis, photosensitive agents are exposed to UV radiation resulting in the initiation of a phototoxic reaction by the generation of free radical species, followed by cellular toxicity and photoallergic reactions which elicit immunological consequences by the involvement of different immune cells. After the phototoxic reaction, the chemicals bind to skin protein and form hapten, loaded on to MHC class II molecules of Langerhans cells and elicit the antigen-specific T cell-mediated photoallergy (Tokura, 2000).

Cosmeceuticals Targeting Components of the Skin Immune System

Hyperpigmentation

Dyschromia, characterized by skin discolouration or patches of uneven colour, can be a challenging presentation, particularly among persons of colour. In Australia, dyschromia is often the result of chronic sun exposure resulting in Poikiloderma of Civatte (a scarf-like patch of dyschromia), atrophy and telan giectasia. Uneven skin tone has traditionally been managed with hydroquinone or corticosteroids, and although highly efficient, the treatment raises a number of safety concerns for local and systemic effects with long-term use.

A number of cosmeceutical agents, some naturally occurring, inhibit melanin synthesis, reduce melanin transfer into keratinocytes and remove epidermal melanin (Gonzalez and Perez, 2016). Of the many natural ingredients addressing skin pigmentation, the largest body of *in vivo* evidence is available for the use of soy and niacinamide (Leyden et al., 2011). Soy is an inhibitor of the protease-activated receptor 2 (PAR-2), acting to prevent keratinocyte–melanocyte contact and reduction in melanosome phagocytosis by keratinocytes. It has the ability to both de-pigment and prevent UV-induced pigmentation (Paine et al., 2001). Niacinamide is the active amide form of niacin and has also been demonstrated to affect pigmentation through inhibition of melanosome transfer to keratinocytes, as well as melanogenesis (Hakozaki et al., 2002; Greatens et al., 2005). Its utility has been confirmed in clinical trials (Kimball et al., 2010; Navarrete-Solis et al., 2011).

Photoprotection

Chronic UV radiation can cause prominent histological changes in the extracellular matrix with the disintegration of elastic fibres and deposition of abnormal collagen (Yaar and Gilchrest, 2007). Such changes are thought to be a result of reactive oxygen species (ROS), which in excess deplete the skin's own antioxidant enzymes such as superoxide dismutase, catalase and glutathione peroxidase (Pillai et al., 2005). As a consequence, the remaining free oxygen radicals can generate matrix metalloproteinases and suppress collagen gene expression, leading to abnormal matrix degradation and resulting in the clinical phenotype of dermatoheliosis (Kim et al., 2009; Pillai et al., 2005). Currently, there are a number of obstacles to the delivery of sufficiently high concentrations of antioxidants to the skin, such that an effective cosmeceutical product that can deliver meaningful clinical efficacy has yet to be developed.

Although a diet rich in antioxidants is advocated to enhance skin protection and health by prevention of oxidative stress (Saluja and Fabi, 2017), it is likely that only a small proportion of the consumed antioxidants will reach the skin. Nevertheless, there are *in vitro* and *in vivo* studies on green tea polyphenols (GTPs) that demonstrate an attenuation of UVB induced premature skin ageing (Vayalil et al., 2004). Furthermore, studies in laboratory animals demonstrate that GTPs can prevent UVB-induced local, as well as systemic, immune suppression. GTPs have also been shown to inhibit UVB-induced erythema, oxidative stress, and infiltration of inflammatory leukocytes in human skin (Katiyar, 2003). Topical application of GTPs has proved similarly efficacious in human clinical studies, with the mechanism unlikely to be due to a sunscreen effect of the topical preparation, given its low sun protection factor of 1 (Camouse et al., 2009; Elmets et al., 2001). Other antioxidants such as vitamins C and E have been shown to provide additive protection from UVB exposure when applied in combination to pig skin, with the greatest contribution due to vitamin E (Darr et al., 1996). On the other hand, vitamin C provided significantly better protection than vitamin E against UVA-mediated phototoxicity. Moreover, the addition of vitamin C or vitamin C+E to a formulation containing the UVA sunscreen oxybenzone produced a synergistic response against UVA phototoxicity, compared to oxybenzone alone (Darr et al., 1996).

The discovery and development of new compounds that confer antioxidant properties continues to develop with compounds such as patchouli oil, traditionally used in Chinese medicines, found to inhibit the effects of UV-induced skin photoaging, such as wrinkle formation and reduced skin elasticity, in an *in vivo* model (Lin et al., 2014). There is no doubt that many other compounds will be developed in the future.

Conclusion

The cosmetic field is driven by multiple forces including new biological discoveries, new compounds or applications of existing compounds, innovations in formulation, and technological advancements to aid the consumer in achieving their skin care requirements and expectations. The consumer plays a vital role in informing the cosmetic industry of their requirements and expectations through market data on sales volume, online feedback and various Internet forums that allow the industry to perform post-market surveillance on their products. Prior to a new product launch, rigorous testing is essential to ensure the product is formulated for high tolerability and a low risk of irritation and sensitization.

A good understanding of the immune system in the skin is an essential consideration in cosmetic formulation. It is a complex mix of multiple cell types and chemical mediators that work together to protect the skin and the body through the immune response. In this chapter, we discussed how the innate and adaptive immune systems provide a comprehensive structural, chemical, and immunological barrier to external agents and microbiological pathogens. There is also a microanatomical specialization of the immunological barrier with distinct roles for leukocytes residing in the epidermis and dermis (Tong et al., 2015).

Added to this is an emerging understanding of the skin microbiome and the additional role it plays in maintaining skin and immune defence homeostasis, an area likely to lead to new opportunities for innovative therapeutic intervention (Sanford and Gallo, 2013). Monitoring of the skin microbiome will see it being increasingly used as a 'microbiological barometer' to deliver information on skin conditions such as eczema, psoriasis, rosacea and acne, as well as in the skin aging process (Shibagaki et al., 2017).

We have covered the issue of irritant and allergic contact dermatitis that can occur with all forms of cosmetics applied to the skin. This highlights the need for formulators to be aware of the current and emerging allergens in the form of dyes, preservatives and fragrances. Indeed, the American Contact Dermatitis Society brings this awareness to the medical community through the 'Allergen of the Year Award' to highlight common allergens that are on the rise and could be easily missed. In 2017, it was awarded to alkyl glucosides, a common ingredient made from palm or coconut oil, and increasingly used in cosmetic formulations because of its perceived eco-friendly and biodegradable properties (Alfalah et al., 2017).

The immune system in the skin can also be the source for innovation and therapeutic targets for common cosmetic considerations such as hyperpigmentation and photoprotection. It is anticipated that the field of cosmetic science will continue to learn from new advances in skin immunology to target or protect against erythema, dyschromia and photoaging.

REFERENCES

Ackermann, L., I. T. Harvima, J. Pelkonen, V. Salo Ritamäki, A. Naukkarinen, R. J. Harvima, and M. Horsmanheimo. 1999. "Mast cells in psoriatic skin are strongly positive for interferon-gamma." *Br J Dermatol* 140 (4):624–633. doi: 10.1046/j.1365-2133.1999.02760.x.

Adler, S., D. Basketter, S. Creton, O. Pelkonen, J. van Benthem, V. Zuang, K. E. Andersen et al. 2011. "Alternative (non-animal) methods for cosmetics testing: Current status and future prospects-2010." *Arch Toxicol* 85 (5):367–485. doi: 10.1007/s00204-011-0693-2.

Aerts, O., M. Baeck, L. Constandt, B. Dezfoulian, M. C. Jacobs, S. Kerre, H. Lapeere et al. 2014. "The dramatic increase in the rate of methylisothiazolinone contact allergy in Belgium: A multicentre study." *Contact Dermatitis* 71 (1):41–48. doi: 10.1111/cod.12249.

Alenius, H., D. W. Roberts, Y. Tokura, A. Lauerma, G. Patlewicz, and M. S. Roberts. 2008. "Skin, drug and chemical reactions." *Drug Discov Today Dis Mech* 5 (2):e211–e220. doi: 10.1016/j.ddmec.2008.06.001.

Alfalah, M., C. Loranger, and D. Sasseville. 2017. "Alkyl glucosides." *Dermatitis* 28 (1):3–4. doi: 10.1097/DER.0000000000000234.

Bangert, C., P. M. Brunner, and G. Stingl. 2011. "Immune functions of the skin." *Clin Dermatol* 29 (4):360–376. doi: 10.1016/j.clindermatol.2011.01.006.

Birbeck, M. S., A. S. Breathnach, and J. D. Everall. 1961. "An electron microscope study of basal melanocytes and high-level clear cells (Langerhans cells) in vitiligo." *J Invest Dermatol* 37 (1):51–64. doi: 10.1038/jid.1961.80.

Boyman, O., C. Conrad, G. Tonel, M. Gilliet, and F. O. Nestle. 2007. "The pathogenic role of tissue-resident immune cells in psoriasis." *Trends Immunol* 28 (2):51–57. doi: 10.1016/j.it.2006.12.005.

Brown, S. J., and W. H. McLean. 2012. "One remarkable molecule: Filaggrin." *J Invest Dermatol* 132 (3 Pt 2):751–762. doi: 10.1038/jid.2011.393.

Cai, Y., X. Shen, C. Ding, C. Qi, K. Li, X. Li, V. R. Jala et al. 2011. "Pivotal role of dermal IL-17-producing gammadelta T cells in skin inflammation." *Immunity* 35 (4):596–610. doi: 10.1016/j.immuni.2011.08.001.

Camouse, M. M., D. S. Domingo, F. R. Swain, E. P. Conrad, M. S. Matsui, D. Maes, L. Declercq et al. 2009. "Topical application of green and white tea extracts provides protection from solar-simulated ultraviolet light in human skin." *Exp Dermatol* 18 (6):522–526. doi: 10.1111/j.1600-0625.2008.00018.x.

Chu, C. C., N. Ali, P. Karagiannis, P. Di Meglio, A. Skowera, L. Napolitano, G. Barinaga et al. 2012. "Resident CD141 (BDCA3)+ dendritic cells in human skin produce IL-10 and induce regulatory T cells that suppress skin inflammation." *J Exp Med* 209 (5):935–945. doi: 10.1084/jem.20112583.

Coquette, A., N. Berna, A. Vandenbosch, M. Rosdy, B. De Wever, and Y. Poumay. 2003. "Analysis of interleukin-1alpha (IL-1alpha) and interleukin-8 (IL-8) expression and release in in vitro reconstructed human epidermis for the prediction of in vivo skin irritation and/or sensitization." *Toxicol In Vitro* 17 (3):311–321.

Darr, D., S. Dunston, H. Faust, and S. Pinnell. 1996. "Effectiveness of antioxidants (vitamin C and E) with and without sunscreens as topical photoprotectants." *Acta Derm Venereol* 76 (4):264–268.

Debonneville, C., and A. Chaintreau. 2004. "Quantitation of suspected allergens in fragrances: Part II. Evaluation of comprehensive gas chromatography–conventional mass spectrometry." *Journal of Chromatography A* 1027 (1):109–115. doi: 10.1016/j.chroma.2003.08.080.

Di Meglio, P., G. K. Perera, and F. O. Nestle. 2011. "The multitasking organ: Recent insights into skin immune function." *Immunity* 35 (6):857–869. doi: 10.1016/j.immuni.2011.12.003.

Drobits, B., M. Holcmann, N. Amberg, M. Swiecki, R. Grundtner, M. Hammer, M. Colonna, and M. Sibilia. 2012. "Imiquimod clears tumors in mice independent of adaptive immunity by converting pDCs into tumor-killing effector cells." *J Clin Invest* 122 (2):575–585. doi: 10.1172/JCI61034.

Dudeck, J., A. Medyukhina, J. Fröbel, C-M. Svensson, J. Kotrba, M. Gerlach, A-C. Gradtke et al. 2017. "Mast cells acquire MHCII from dendritic cells during skin inflammation." *J Exp Med* 214 (12):3791–3811.

Ebert, L. M., S. Meuter, and B. Moser. 2006. "Homing and function of human skin γδ T cells and NK cells: Relevance for tumor surveillance." *J Immunol* 176 (7):4331–4336. doi: 10.4049/jimmunol.176.7.4331.

Effendy, I., and H. I. Maibach. 1995. "Surfactants and experimental irritant contact dermatitis." *Contact Dermatitis* 33 (4):217–225.

Elbe-Burger, A., and C. Schuster. 2010. "Development of the prenatal cutaneous antigen-presenting cell network." *Immunol Cell Biol* 88 (4):393–399. doi: 10.1038/icb.2010.13.

Elmets, C. A., D. Singh, K. Tubesing, M. Matsui, S. Katiyar, and H. Mukhtar. 2001. "Cutaneous photoprotection from ultraviolet injury by green tea polyphenols." *J Am Acad Dermatol* 44 (3):425–432. doi: 10.1067/mjd.2001.112919.

Fitzpatrick, T. B., and A. S. Breathnach. 1963. "The epidermal melanin unit system." *Dermatol Wochenschr* 147:481–489.

Flores, S., and H. Maibach. 2010. "Chapter 25 – Allergic contact dermatitis." In *Hayes' Handbook of Pesticide Toxicology* (3rd edition), edited by Robert Krieger, 669–675. New York: Academic Press.

Fonacier, L., and M. Boguniewicz. 2016. "Chapter 53 - Contact Dermatitis." In *Pediatric Allergy: Principles and Practice* (3rd edition), edited by D.Y.M. Leung, F.A. Bonillo, H.A. Sampson, S.J. Szefler and C.A. Akdis, 467–481. Amsterdam: Elsevier.

Fuentes-Duculan, J., M. Suárez-Fariñas, L. C. Zaba, K. E. Nograles, K. C. Pierson, H. Mitsui, C. A. Pensabene et al. 2010. "A subpopulation of CD163-positive macrophages is classically activated in psoriasis." *J Invest Dermatol* 130 (10):2412–2422. doi: 10.1038/jid.2010.165.

Fyhrquist-Vanni, N., H. Alenius, and A. Lauerma. 2007. "Contact dermatitis." *Dermatol Clin* 25 (4):613–623. doi: 10.1016/j.det.2007.06.002.

Galli, S. J., and M. Tsai. 2012. "IgE and mast cells in allergic disease." *Nat Med* 18 (5):693–704. doi: 10.1038/nm.2755.

Gallo, R. L., and V. Nizet. 2008. "Innate barriers against infection and associated disorders." *Drug Discov Today Dis Mech* 5 (2):145–152. doi: 10.1016/j.ddmec.2008.04.009.

Gaspar, L. R., J. Tharmann, P. M. B. G. M. Campos, and M. Liebsch. 2013. "Skin phototoxicity of cosmetic formulations containing photounstable and photostable UV-filters and vitamin A palmitate." *Toxicol In Vitro* 27 (1):418–425. doi: 10.1016/j.tiv.2012.08.006.

Gebhardt, T., L. M. Wakim, L. Eidsmo, P. C. Reading, W. R. Heath, and F. R. Carbone. 2009. "Memory T cells in nonlymphoid tissue that provide enhanced local immunity during infection with herpes simplex virus." *Nat Immunol* 10 (5):524–530. doi: 10.1038/ni.1718.

Gilliet, M., W. Cao, and Y. J. Liu. 2008. "Plasmacytoid dendritic cells: Sensing nucleic acids in viral infection and autoimmune diseases." *Nat Rev Immunol* 8 (8):594–606. doi: 10.1038/nri2358.

Gonzalez, N., and M. Perez. 2016. "Natural cosmeceutical ingredients for hyperpigmentation." *J Drugs Dermatol* 15 (1):26–34.

Goossens, A. 2011. "Contact-allergic reactions to cosmetics." *J Allergy (Cairo)* 2011:467071. doi: 10.1155/2011/467071.

Greatens, A., T. Hakozaki, A. Koshoffer, H. Epstein, S. Schwemberger, G. Babcock, D. Bissett et al. 2005. "Effective inhibition of melanosome transfer to keratinocytes by lectins and niacinamide is reversible." *Exp Dermatol* 14 (7):498–508. doi: 10.1111/j.0906-6705.2005.00309.x.

Grone, A. 2002. "Keratinocytes and cytokines." *Vet Immunol Immunopathol* 88 (1–2):1–12.

Hakozaki, T., L. Minwalla, J. Zhuang, M. Chhoa, A. Matsubara, K. Miyamoto, A. Greatens et al. 2002. "The effect of niacinamide on reducing cutaneous pigmentation and suppression of melanosome transfer." *Br J Dermatol* 147 (1):20–31.

Haniffa, M., A. Shin, V. Bigley, N. McGovern, P. Teo, P. See, P. S. Wasan et al. 2012. "Human tissues contain CD141hi cross-presenting dendritic cells with functional homology to mouse CD103+ nonlymphoid dendritic cells." *Immunity* 37 (1):60–73. doi: 10.1016/j.immuni.2012.04.012.

Haniffa, M., M. Gunawan, and L. Jardine. 2015. "Human skin dendritic cells in health and disease." *J Dermatol Sci* 77 (2):85–92. doi: 10.1016/j.jdermsci.2014.08.012.

Harvima, I. T., and G. Nilsson. 2011. "Mast cells as regulators of skin inflammation and immunity." *Acta Derm Venereol* 91 (6):644–650. doi: 10.2340/00015555-1197.

Heath, W. R., and F. R. Carbone. 2013. "The skin-resident and migratory immune system in steady state and memory: Innate lymphocytes, dendritic cells and T cells." *Nat Immunol* 14 (10):978–985. doi: 10.1038/ni.2680.

Hershko, A. Y., and J. Rivera. 2010. "Mast cell and T cell communication; amplification and control of adaptive immunity." *Immunol Lett* 128 (2):98–104. doi: 10.1016/j.imlet.2009.10.013.

Hoeffel, G., Y. Wang, M. Greter, P. See, P. Teo, B. Malleret, M. Leboeuf et al. 2012. "Adult Langerhans cells derive predominantly from embryonic fetal liver monocytes with a minor contribution of yolk sac-derived macrophages." *J Exp Med* 209 (6):1167–1181. doi: 10.1084/jem.20120340.

Hong, Y., B. Song, H. D. Chen, and X. H. Gao. 2015. "Melanocytes and skin immunity." *J Investig Dermatol Symp Proc* 17 (1):37–39. doi: 10.1038/jidsymp.2015.14.

Horsmanheimo, L., I. T. Harvima, A. Jarvikallio, R. J. Harvima, A. Naukkarinen, and M. Horsmanheimo. 1994. "Mast cells are one major source of interleukin-4 in atopic dermatitis." *Br J Dermatol* 131 (3):348–353.

Jackson, J. Mark, and Joseph F. Fowler. 1998. "Methyldibromoglutaronitrile (Euxyl K400): A new and important sensitizer in the United States?" *J Am Acad Dermatol* 38 (6):934–937. doi: 10.1016/S0190-9622(98)70589-4.

Johansen, J. D., K. Aalto-Korte, T. Agner, K. E. Andersen, A. Bircher, M. Bruze, A. Cannavo et al. 2015. "European Society of Contact Dermatitis guideline for diagnostic patch testing – Recommendations on best practice." *Contact Dermatitis* 73 (4):195–221. doi: 10.1111/cod.12432.

Katiyar, S. K. 2003. "Skin photoprotection by green tea: Antioxidant and immunomodulatory effects." *Curr Drug Targets Immune Endocr Metabol Disord* 3 (3):234–242.

Katz, S. I., K. Tamaki, and D. H. Sachs. 1979. "Epidermal Langerhans cells are derived from cells originating in bone marrow." *Nature* 282 (5736):324–326.

Kim, Y. G., M. Sumiyoshi, M. Sakanaka, and Y. Kimura. 2009. "Effects of ginseng saponins isolated from red ginseng on ultraviolet B-induced skin aging in hairless mice." *Eur J Pharmacol* 602 (1):148–156. doi: 10.1016/j.ejphar.2008.11.021.

Kimball, A. B., J. R. Kaczvinsky, J. Li, L. R. Robinson, P. J. Matts, C. A. Berge, K. Miyamoto, and D. L. Bissett. 2010. "Reduction in the appearance of facial hyperpigmentation after use of moisturizers with a combination of topical niacinamide and N-acetyl glucosamine: Results of a randomized, double-blind, vehicle-controlled trial." *Br J Dermatol* 162 (2):435–441. doi: 10.1111/j.1365-2133.2009.09477.x.

Krasteva, M., J. Kehren, F. Horand, H. Akiba, G. Choquet, M. T. Ducluzeau, R. Tedone et al. 1998. "Dual role of dendritic cells in the induction and down-regulation of antigen-specific cutaneous inflammation." *J Immunol* 160 (3):1181–1190.

Kuo, I. H., T. Yoshida, A. De Benedetto, and L. A. Beck. 2013a. "The cutaneous innate immune response in patients with atopic dermatitis." *J Allergy Clin Immunol* 131 (2):266–278. doi: 10.1016/j.jaci.2012.12.1563.

Kuo, I. Hsin, Takeshi Yoshida, Anna De Benedetto, and Lisa A. Beck. 2013b. "The cutaneous innate immune response in patients with atopic dermatitis." *J Allergy Clin Immunol* 131 (2):266–278. doi: 10.1016/j.jaci.2012.12.1563.

Langley, R. G., B. E. Elewski, M. Lebwohl, K. Reich, C. E. Griffiths, K. Papp, L. Puig et al. 2014. "Secukinumab in plaque psoriasis – Results of two phase 3 trials." *N Engl J Med* 371 (4):326–338. doi: 10.1056/NEJMoa1314258.

Larregina, A. T., A. E. Morelli, L. A. Spencer, A. I. Logan, S. C. Watkins, A. W. Thomson, and L. D. Falo Jr. 2001. "Dermal resident CD14+ cells differentiate into Langerhans cells." *Nat Immunol* 2:1151. doi: 10.1038/ni731.

Lee, H. Y., M. Stieger, N. Yawalkar, and M. Kakeda. 2013. "Cytokines and chemokines in irritant contact dermatitis." *Mediators Inflamm* 2013:916497. doi: 10.1155/2013/916497.

Leyden, J. J., B. Shergill, G. Micali, J. Downie, and W. Wallo. 2011. "Natural options for the management of hyperpigmentation." *J Eur Acad Dermatol Venereol* 25 (10):1140–1145. doi: 10.1111/j.1468-3083.2011.04130.x.

Leysen, J., A. Goossens, J. Lambert, and O. Aerts. 2014. "Polyhexamethylene biguanide is a relevant sensitizer in wet wipes." *Contact Dermatitis* 70 (5):323–325. doi: 10.1111/cod.12208.

Lin, R-F., X-X. Feng, C-W. Li, X-J. Zhang, X-T. Yu, J-Y. Zhou, X. Zhang et al. 2014. "Prevention of UV radiation-induced cutaneous photoaging in mice by topical administration of patchouli oil." *J Ethnopharmacol* 154 (2):408–418. doi: 10.1016/j.jep.2014.04.020.

Loser, K., A. Mehling, S. Loeser, J. Apelt, A. Kuhn, S. Grabbe, T. Schwarz et al. 2006. "Epidermal RANKL controls regulatory T-cell numbers via activation of dendritic cells." *Nat Med* 12 (12):1372–1379. doi: 10.1038/nm1518.

Lowes, M. A., F. Chamian, M. V. Abello, J. Fuentes-Duculan, S. L. Lin, R. Nussbaum, I. Novitskaya et al. 2005. "Increase in TNF-alpha and inducible nitric oxide synthase-expressing dendritic cells in psoriasis and reduction with efalizumab (anti-CD11a)." *Proc Natl Acad Sci USA* 102 (52):19057–19062. doi: 10.1073/pnas.0509736102.

Lushniak, B. D. 2004. "Occupational contact dermatitis." *Dermatol Ther* 17 (3):272–277. doi: 10.1111/j.1396-0296.2004.04032.x.

Mackay, L. K., A. Rahimpour, J. Z. Ma, N. Collins, A. T. Stock, M. L. Hafon, J. Vega-Ramos et al. 2013. "The developmental pathway for CD103(+)CD8+ tissue-resident memory T cells of skin." *Nat Immunol* 14 (12):1294–1301. doi: 10.1038/ni.2744.

Macleod, A. S., and W. L. Havran. 2011. "Functions of skin-resident gammadelta T cells." *Cell Mol Life Sci* 68 (14):2399–2408. doi: 10.1007/s00018-011-0702-x.

Maibach, H., and G. Honari. 2014. "Chapter 3 – Photoirritation (phototoxicity): Clinical aspects." In *Applied Dermatotoxicology*, edited by Howard Maibach and Golara Honari, 41–56. Boston: Academic Press.

Mann, E. R. 2014. "Review: Skin and the immune system." *J Clin Exp Dermatol Res* 4 (3). doi: 10.4172/2155-9554.S2-003.

Meunier, L., A. Gonzalez-Ramos, K. D. Cooper. 1993. "Heterogeneous populations of class II MHC+ cells in human dermal cell suspensions. Identification of a small subset responsible for potent dermal antigen-presenting cell activity with features analogous to Langerhans cells." *J Immunol* 151 (8):4067–4080.

Mueller, S. N., A. Zaid, and F. R. Carbone. 2014. "Tissue-resident T cells: Dynamic players in skin immunity." *Front Immunol* 5:332. doi: 10.3389/fimmu.2014.00332.

Nagao, K., T. Kobayashi, K. Moro, M. Ohyama, T. Adachi, D. Y. Kitashima, S. Ueha et al. 2012. "Stress-induced production of chemokines by hair follicles regulates the trafficking of dendritic cells in skin." *Nat Immunol* 13 (8):744–752. doi: 10.1038/ni.2353.

Navarrete-Solis, J., J. P. Castanedo-Cazares, B. Torres-Alvarez, C. Oros-Ovalle, C. Fuentes-Ahumada, F. J. Gonzalez, J. D. Martinez-Ramirez, et al. 2011. "A double-blind, randomized clinical trial of niacinamide 4% versus hydroquinone 4% in the treatment of melasma." *Dermatol Res Pract* 2011:379173. doi: 10.1155/2011/379173.

Nestle, F. O., X. G. Zheng, C. B. Thompson, L. A. Turka, and B. J. Nickoloff. 1993. "Characterization of dermal dendritic cells obtained from normal human skin reveals phenotypic and functionally distinctive subsets." *J Immunol* 151 (11):6535–6545.

Neumann, N. J., A. Blotz, G. Wasinska-Kempka, M. Rosenbruch, P. Lehmann, H. J. Ahr, and H.-W. Vohr. 2005. "Evaluation of phototoxic and photoallergic potentials of 13 compounds by different in vitro and in vivo methods." *J Photochem Photobiol B* 79 (1):25–34. doi: 10.1016/j.jphotobiol.2004.11.014.

Ng, L. G., A. Hsu, M. A. Mandell, B. Roediger, C. Hoeller, P. Mrass, A. Iparraguirre, et al. 2008. "Migratory dermal dendritic cells act as rapid sensors of protozoan parasites." *PLOS Pathog* 4 (11):e1000222. doi: 10.1371/journal.ppat.1000222.

Nordlund, J. J. 2007. "The melanocyte and the epidermal melanin unit: An expanded concept." *Dermatol Clin* 25 (3):271–281. doi: 10.1016/j.det.2007.04.001.

O'Garra, A., and P. Vieira. 2004. "Regulatory T cells and mechanisms of immune system control." *Nat Med* 10 (8):801. doi: 10.1038/nm0804-801.

OECD. 2004. "Test No. 432: In Vitro 3T3 NRU phototoxicity test." In *OECD Guidelines for the Testing of Chemical, Section 4*. Paris: OECD Publishing.

Oliver, B., S. Krishnan, M. R. Pardo, and A. Ehrlich. 2015. "Cosmeceutical contact dermatitis—Cautions to herbals." *Curr Treat Options Allergy* 2 (4):307–321. doi: 10.1007/s40521-015-0066-9.

Paine, C., E. Sharlow, F. Liebel, M. Eisinger, S. Shapiro, and M. Seiberg. 2001. "An alternative approach to depigmentation by soybean extracts via inhibition of the PAR-2 pathway." *J Invest Dermatol* 116 (4):587–595. doi: 10.1046/j.1523-1747.2001.01291.x.

Pasparakis, M., I. Haase, and F. O. Nestle. 2014. "Mechanisms regulating skin immunity and inflammation." *Nat Rev Immunol* 14 (5):289–301. doi: 10.1038/nri3646.

Pillai, S., C. Oresajo, and J. Hayward. 2005. "Ultraviolet radiation and skin aging: Roles of reactive oxygen species, inflammation and protease activation, and strategies for prevention of inflammation-induced matrix degradation – A review." *Int J Cosmet Sci* 27 (1):17–34. doi: 10.1111/j.1467-2494.2004.00241.x.

Poltorak, M. P., and C. E. Zielinski. 2017. "Hierarchical governance of cytokine production by 6-sulfo LacNAc (slan) dendritic cells for the control of psoriasis pathogenesis." *Exp Dermatol* 26 (4):317–318. doi: 10.1111/exd.13170.

Ramos, L., R. Cabral, and M. Goncalo. 2014. "Allergic contact dermatitis caused by acrylates and methacrylates – A 7-year study." *Contact Dermatitis* 71 (2):102–107. doi: 10.1111/cod.12266.

Rastogi, S. C., 1995. "Analysis of fragrances in cosmetics by gas chromatography–mass spectrometry." *J High Resolut Chromatogr* 18 (10):653–658. doi: 10.1002/jhrc.1240181008.

Roediger, B., R. Kyle, K. H. Yip, N. Sumaria, T. V. Guy, B. S. Kim, A. J. Mitchell et al. 2013. "Cutaneous immunosurveillance and regulation of inflammation by group 2 innate lymphoid cells." *Nat Immunol* 14 (6):564–573. doi: 10.1038/ni.2584.

Rustemeyer, T., I. M. W. van Hoogstraten, B. Mary E. von Blomberg, S. Gibbs, and R. J. Scheper. 2011. "Mechanisms of irritant and allergic contact dermatitis." In *Contact Dermatitis*, edited by Jeanne Duus Johansen, Peter J. Frosch and Jean-Pierre Lepoittevin, 43–90. Berlin, Heidelberg: Springer Berlin Heidelberg.

Saluja, S. S., and S. G. Fabi. 2017. "A holistic approach to antiaging as an adjunct to antiaging procedures: A review of the literature." *Dermatol Surg* 43 (4):475–484. doi: 10.1097/DSS.0000000000001027.

Sanford, J. A., and R. L. Gallo. 2013. "Functions of the skin microbiota in health and disease." *Semin Immunol* 25 (5):370–377. doi: 10.1016/j.smim.2013.09.005.

Schenkel, J. M., and D. Masopust. 2014. "Tissue-resident memory T cells." *Immunity* 41 (6):886–897. doi: 10.1016/j.immuni.2014.12.007.

Shamoto, M. 1983. "Langerhans cells increase in the dermal lesions of adult T cell leukaemia in Japan." *J Clin Pathol* 36 (3):307.

Sharma, N. S., R. Jindal, B. Mitra, S. Lee, L. Li, T. J. Maguire, R. Schloss et al. 2012. "Perspectives on non-animal alternatives for assessing sensitization potential in allergic contact dermatitis." *Cell Mol Bioeng* 5 (1):52–72. doi: 10.1007/s12195-011-0189-4.

Shibagaki, N., W. Suda, C. Clavaud, P. Bastien, L. Takayasu, E. Iioka, R. Kurokawa et al. 2017. "Aging-related changes in the diversity of women's skin microbiomes associated with oral bacteria." *Sci Rep* 7 (1):10567. doi: 10.1038/s41598-017-10834-9.

Steinhoff, M., T. Brzoska, and T. A. Luger. 2001. "Keratinocytes in epidermal immune responses." *Curr Opin Allergy Clin Immunol* 1 (5):469–476.

Steinman, R. M., and Z. A. Cohn. 1973. "Identification of a novel cell type in peripheral lymphoid organs of mice. I. Morphology, quantitation, tissue distribution." *J Exp Med* 137 (5):1142–1162.

Steinstraesser, L., T. Koehler, F. Jacobsen, A. Daigeler, O. Goertz, S. Langer, M. Kesting et al. 2008. "Host defense peptides in wound healing." *Mol Med* 14 (7–8):528–537. doi: 10.2119/2008-00002.Steinstraesser.

Stingl, G., E. C. Wolff-Schreiner, W. J. Pichler, F. Gschnait, W. Knapp, and K. Wolff. 1977. "Epidermal Langerhans cells bear Fc and C3 receptors." *Nature* 268 (5617):245–246.

Sugaya, M., T. Miyagaki, H. Ohmatsu, H. Suga, H. Kai, M. Kamata, H. Fujita et al. 2012. "Association of the numbers of CD163(+) cells in lesional skin and serum levels of soluble CD163 with disease progression of cutaneous T cell lymphoma." *J Dermatol Sci* 68 (1):45–51. doi: 10.1016/j.jdermsci.2012.07.007.

Sumaria, N., B. Roediger, L. G. Ng, J. Qin, R. Pinto, L. L. Cavanagh, E. Shklovskaya et al. 2011. "Cutaneous immunosurveillance by self-renewing dermal γδ T cells." *J Exp Med* 208 (3):505.

Tamaki, K., G. Stingl, and S. I. Katz. 1980. "The origin of Langerhans cells." *J Invest Dermatol* 74 (5):309–311. doi: 10.1111/1523-1747.ep12543533.

Tang, A., M. Amagai, L. G. Granger, J. R. Stanley, and M. C. Udey. 1993. "Adhesion of epidermal Langerhans cells to keratinocytes mediated by E-cadherin." *Nature* 361 (6407):82–85. doi: 10.1038/361082a0.

Tokura, Y. 2000. "Immune responses to photohaptens: Implications for the mechanisms of photosensitivity to exogenous agents." *J Dermatol Sci* 23 (Suppl 1):S6–S9. doi: 10.1016/S0923-1811(99)00071-7.

Tong, P. L., B. Roediger, N. Kolesnikoff, M. Biro, S. S. Tay, R. Jain, L. E. Shaw et al. 2015. "The skin immune atlas: Three-dimensional analysis of cutaneous leukocyte subsets by multiphoton microscopy." *J Invest Dermatol* 135 (1):84–93. doi: 10.1038/jid.2014.289.

Vayalil, P. K., A. Mittal, Y. Hara, C. A. Elmets, and S. K. Katiyar. 2004. "Green tea polyphenols prevent ultraviolet light-induced oxidative damage and matrix metalloproteinases expression in mouse skin." *J Invest Dermatol* 122 (6):1480–1487. doi: 10.1111/j.0022-202X.2004.22622.x.

Verhulst, L., and A. Goossens. 2016. "Cosmetic components causing contact urticaria: A review and update." *Contact Dermatitis* 75 (6):333–344. doi: 10.1111/cod.12679.

Villanova, F., B. Flutter, I. Tosi, K. Grys, H. Sreeneebus, G. K. Perera, A. Chapman et al. 2014. "Characterization of innate lymphoid cells in human skin and blood demonstrates increase of NKp44+ ILC3 in psoriasis." *J Invest Dermatol* 134 (4):984–991. doi: 10.1038/jid.2013.477.

Wang, X., X-H. Gao, X. Zhang, L. Zhou, Q-S. Mi, Y. Hong, B. Song et al. 2017. "Cells in the skin." In *Practical Immunodermatology*, edited by Xing-Hua Gao and Hong-Duo Chen, 63–113. Dordrecht: Springer Netherlands.

Wollenberg, A., S. Kraft, D. Hanau, and T. Bieber. 1996. "Immunomorphological and ultrastructural characterization of Langerhans cells and a novel, inflammatory dendritic epidermal cell (IDEC) population in lesional skin of atopic eczema." *J Invest Dermatol* 106 (3):446–453.

Wood, J. M., K. Jimbow, R. E. Boissy, A. Slominski, P. M. Plonka, J. Slawinski, J. Wortsman, et al. 1999. "What's the use of generating melanin?" *Exp Dermatol* 8 (2):153–164.

Yaar, M., and B. A. Gilchrest. 2007. "Photoageing: Mechanism, prevention and therapy." *Br J Dermatol* 157 (5):874–887. doi: 10.1111/j.1365-2133.2007.08108.x.

Zhao, W., C. A. Oskeritzian, A. L. Pozez, and L. B. Schwartz. 2005. "Cytokine production by skin-derived mast cells: Endogenous proteases are responsible for degradation of cytokines." *J Immunol* 175 (4):2635–2642.

Zukiewicz-Sobczak, W. A., P. Adamczuk, P. Wroblewska, J. Zwolinski, J. Chmielewska-Badora, E. Krasowska, E. M. Galinska et al. 2013. "Allergy to selected cosmetic ingredients." *Postepy Dermatol Alergol* 30 (5):307–310. doi: 10.5114/pdia.2013.38360.

5

The Human Nail: Structure, Properties, Therapy and Grooming

Kenneth A. Walters and Majella E. Lane

CONTENTS

Introduction

This chapter describes the morphology and function of the human nail structure. The nail plate is the most prominent feature of the nail and has been the focus of attention for cosmetic scientists, mainly for decorative purposes. An accompanying chapter will describe the art and science of nail cosmetics while this chapter will concentrate on the properties and pathologies of the nail. The nail plate is a highly keratinised tissue, and the physiological and morphological properties of the nail that are responsible for its mechanical strength and barrier properties will be discussed. Major emphasis will be given to the biophysical and bioanalytical approaches that have contributed to our knowledge of nail composition and function. The methodologies that have been explored as diagnostic tools of the nail will also be considered. Finally, with the understanding that nail diseases are frequently difficult to cure and require long periods of treatment (Rich and Scher, 2005), common pathologies and topical therapeutic modalities will be described.

Nail Structure and Physiology

The human nail plate is made up of layers of flattened keratinised cells that are fused into a dense but somewhat elastic mass. The plate is surrounded on three sides by the periungual grooves, more commonly referred to as nail folds, which are an extension of the epidermis and composed of cornified tissue. A further extension of the epidermis is the nail matrix, a germinative tissue located at the proximal end of each nail plate beneath and behind the plate. The nail matrix is formed of dorsal (apical) and ventral sections, both of which exhibit a mixed pattern of epithelial and hair keratin. The ventral matrix has been designated a pre-keratogenous zone (Perrin et al., 2004) that is the main source of the nail plate, whereas the apical matrix is thought to produce the thin dorsal parts of the nail plate. The hardened cells of the nail plate grow distally from the nail matrix at a rate of about 2–3 mm per month. The ventral matrix is partially visible as a slightly opaque convex margin beneath the proximal plate, an area known as the lunula. Macroscopic features of the human nail are shown in Figure 5.1. During keratinisation, cells undergo shape and other changes similar to those experienced by the epidermal cells forming the stratum corneum. As a result, the nail plate cells contain no organelles or nuclei and comprise keratin fibrils embedded in a protein matrix.

There are three very tightly knit keratinised layers within the nail plate: a hard, thin outermost dorsal lamina, a softer yet thicker intermediate lamina and an innermost ventral layer (Runne and Orfanos, 1981). Collectively it appears that the expression of different keratin types in different regions of the nail matrix and nail bed may be responsible for the variable physical characteristics of the nail plate layers (Bragulla and Homberger, 2009; Nogueiras-Nieto et al., 2011). For example, the eponychium, hyponychium and nail bed only express epidermal keratins (Perrin, 2004).

The nail bed epithelium is a specialized, epithelial structure with a true basal layer, two or three epithelial layers but no granular layer. The nail bed exhibits significant proliferative activity. The interlocking between the inferior border of the nail plate and the surface of the nail bed explains the strong attachment between the two tissues (Perrin, 2007). The transitional zone between the nail bed and the hyponychium exhibits a different pattern of keratinisation to the hyponychium itself. Here the cornified layer of the nail bed adheres strongly to the undulating inferior surface of the nail plate generating an effective seal and preventing onycholysis.

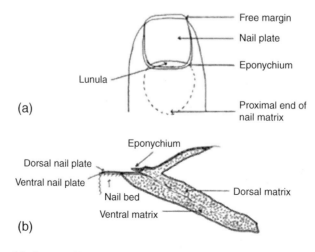

FIGURE 5.1 Features of the human nail plate.

Perrin (2008) further examined the distal region of the nail unit, defining the transition of the nail bed to the hyponychium, known clinically as the onychodermal band. Perrin designated the transitional zone between the nail bed and the hyponychium as the nail isthmus. The nail isthmus was described morphologically by two features: 'a stair-like appearance of the epithelium of the distal nail bed with a marked depression of the epithelium below the inferior surface of the nail plate' and 'a specialized mode of attachment of its horny layer to nail plate via a horizontal mode of differentiation' (Perrin, 2008). He found that the ventral surface of the nail plate was slightly serrated in longitudinal planes and that the nail isthmus belonged functionally to the nail bed. The strong adherence between the nail plate and the nail bed in this region was replaced by a seal between the parakeratotic extension of the nail isthmus and the nail plate. The nail bed and the nail plate grow forward together. The nail plate is generated in epithelial bulbs in the ventral matrix, which are oriented to ensure the oblique growth of the new plate and the nail bed moves slowly, and parallel to, the direction of the nail growth. Unlike the viable epidermis that generates the stratum corneum, the nail matrix does not contain a granular layer (de Berker et al., 2007) and this has a profound influence on the amount of intercellular lipid in the nail plate as discussed later.

The eponychium and hyponychium (nail folds) seal the nail plate at the proximal and lateral edges. These structures do not contain pilosebaceous units but otherwise are morphologically similar to other skin regions.

As a cornified epithelial structure, the chemical composition of the nail plate has many similarities to hair (Baden et al., 1973). Although the majority of the protein in hair and nail is classified as 'hard' trichocyte keratins, both hard and soft keratins have been detected in the nail matrix, and both forms of the protein have been found in the nail plate. Keratins 1 and 10 (soft) are found in the matrix together with the hard keratin Ha-1 (de Berker et al., 2000) and several further hard keratins have also been shown to be present in the nail matrix, nail bed and nail plate (Perrin et al., 2004; Perrin, 2007). Intermediate filament-associated proteins and trichohyalin are also found within the nail (Cashman and Sloan, 2010). The sulphur content of the nail amounts to about 10% and is mainly present within the cystine disulphide bonds that contribute to nail tensile strength by linking the keratin fibres.

The nail contains significant amounts of phospholipid, mainly in the dorsal and intermediate layers, that contribute to its flexibility. Glycolic and stearic acids are also found in the nail. The total lipid content of the nail plate is between 0.1% and 1%, which is considerably different to that of the stratum corneum (~10%). The principal plasticizer of the nail is water, the content of which varies widely dependent on prevailing relative humidity (Farran et al., 2008), but it is normally present at around 18%. When the water content is less than 16% the nail plate becomes brittle, and when the nail is hydrated to water levels of ~25% it becomes soft. Minerals are also important constituents of nails. Lack of selenium and magnesium can have a profound effect on the health of the nail plate (Bauer and Stevens, 1983; Kien and Ganther, 1983).

Biophysical Characterisation of the Nail Plate

Many biophysical and bioengineering techniques that were originally developed to study the skin have subsequently been investigated for their utility to probe the nail and its constituents. The types of features probed at a molecular level have generally been the keratin composition, structure and packing, water content, and hydration effects. Some techniques have also been used to examine nail elemental composition with proposed applications in the use of this tissue as diagnostic of other underlying conditions.

Near-Infrared Raman Spectroscopy

Near-infrared Fourier transform (NIR-FT) Raman spectroscopy exploits an effect in which a small portion of monochromatic light scattered by a substance has a frequency that is different from that of the incoming beam by an amount equal to the vibration frequency of the chemical bond. Frequency shifts of the scattered light can be analyzed and presented as spectra. The bands represent vibrations characteristic for chemical bonds within the molecules of the examined sample. NIR-FT Raman spectroscopy is especially suited to analysis of biological material as it may be used for complex, impure samples and only minute amounts of tissue are required.

The significant Raman vibrations sensitive for protein conformation are those of the peptide bonds and two of several vibrations, designated as amide I and amide III, are prominent in nail as well as skin, hair and stratum corneum (Figure 5.2) (Gniadecka et al., 1997).

Gniadecka et al. (1998) used near-infrared Fourier transform Raman spectroscopy to investigate skin, hair and nail samples collected from 10 human subjects. The spectral range was from 0 to 3500 cm^{-1}. Combining the frequencies of amide I and amide III intensity maxima indicated that the majority of the proteins in the nail were in the α-helix conformation, similar to whole skin, hair and stratum corneum. A strong C-C stretch band at ~935 cm^{-1}, which is typical of the α-helix, was also present in these spectra. The S-S band for nail tissue was prominent and located at around 510 cm^{-1} indicating that these bonds are primarily in the gauche–gauche–gauche conformation; however, a shoulder at 520 cm^{-1} indicated that some S-S bonds were in the gauche–gauche–trans conformation. The mean wave numbers of the CH$_3$ asymmetric band for nail compared with whole skin suggested that, in nail, the protein is highly folded and interacted with its surroundings to only a minimal degree. A band at 1447 cm^{-1} representing scissoring vibrations was also present in nail spectra.

Prominent lipid peaks observed in nail were at 1130 cm^{-1} and 1030 cm^{-1}, implying that lipid structure is highly ordered in nail. The S_{lat} ratio for the CH group vibrations in the 2850–2890 cm^{-1} region was 1, indicating that the majority of lipids are in the crystal state, suggested to be a lamellar liquid crystalline form by the authors. This also suggests a low degree of intermolecular interactions between lipids and proteins in nail. The best estimate of total hydrogen bonded water was obtained by calculating I_{2940}/I_{3250} and was reported to be slightly under 20%. The hydrogen bonded water content for nail was comparable to that for stratum corneum. The 180 cm^{-1} water band did not appear, indicating that, in nail, the majority of water molecules are bound to other components and do not form tetrahedron structures.

In a later study by the same group (Wessel et al., 1999), NIR-FT Raman spectroscopy was used to investigate hydration effects on human nails. Raman spectra were obtained both *in vitro* from nail samples and *in vivo*, before and after soaking in water. No major differences were observed between the spectra for the *in vivo* and *in vitro* distal nail measurements. Water uptake was smaller in fingernails *in vitro* (27%) compared with toenails *in vitro* (45%). For *in vivo* studies, the adherent nail plate showed a slightly different Raman spectrum in comparison with the unadhered distal part. Unsurprisingly, the relative water content of the adherent nail plate (including the underlying tissue) was much higher than that for the free part. Overall the results indicate that the nail has a limited water-holding capacity and that the adsorbed water causes changes in protein geometry.

Near Infrared Diffuse Reflectance (NIR-DR) Spectroscopy

Preliminary studies with near-infrared diffuse reflectance (NIR-DR) on nail were first reported by Sowa and co-workers (1995). Egawa and co-workers (2003) investigated its potential to determine the water content of human nail plates. The water-holding capacity of cut nail plates was determined by immersing

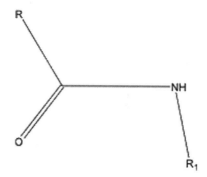

FIGURE 5.2 Peptide bond in proteins.

samples in distilled water for up to 60 min and measuring their NIR spectra. NIR-DR spectra were measured *in vivo* for the free edges of fingernails, before and immediately after washing, for three women aged 26–32 years (mean 29.7). The water content of cut nail plates changed from 1.4 wt% to 33.9 wt% in the relative humidity range 1–99%. Water content increased markedly when ambient relative humidity increased above 80%. Nail thickness appeared to have little effect on the amount of water in nails. The authors observed a broad feature in the 1500–1530 nm regions, which they postulated might be due to hydrogen bonded water and keratin. With higher water content, the intensity of a band at 1420 nm, associated with free water, was observed to increase. At very high relative humidity a new peak at 1488 nm was detected and may have reflected changes in the states of hydrogen bonds in cut nail plates under these conditions. Cut nail plates with a water content of about 7–9 wt% before immersion had a water content as high as 24–33 wt% immediately after the end of the immersion period. *In vivo* water content ranged from ~15% to 21%. Good correlations for *in vitro* water content were obtained between the NIR approach, Karl Fischer analysis and nuclear magnetic resonance measurements.

The same authors (Egawa et al., 2005) investigated seasonal variation of nail water content. Spectra were collected *in vivo* both in summer and in winter under controlled conditions of temperature and humidity (23.8–25.0°C, 40–41% RH). The water content of the nail plate varied between subjects but not between the nails of individual subjects. Average water content was significantly ($p < 0.05$) lower in winter than in summer. Significant differences in spectra were not observed between finger and toenails.

Dynamic Vapour Sorption and Near-Infrared (DVS-NIR) Spectroscopy

Dynamic vapour sorption is a gravimetric method for studying the sorption properties of materials under controlled conditions of temperature and humidity. Recently we evaluated the sorption properties of human nails in a specialised DVS apparatus, which was also fitted with an NIR probe (Walters et al., 2012). Nail clippings from six volunteers were subjected to a fixed temperature of 32°C and a range of relative humidity values, ranging from 40% to 90%, with 10% increments. The increase in the relative humidity (40–90%) was associated with a linear change in mass profile. The NIR spectra also indicated increases in band absorbance (second derivative spectra) at 1422 nm and 1908 nm. Good correlations (Figure 5.3) were obtained for the gravimetric changes and band absorbances at 1422 nm and 1908 nm.

Attenuated Total Reflectance Fourier Transform Infrared (ATR FTIR) Spectroscopy

Sowa et al. (1995) examined nails *in vivo* using ATR FTIR. The presence of lipid on the nail surface was confirmed from the appearance of the lipid ester carbonyl peak at 1745 cm⁻¹ and the lipid methylene stretching vibrations at 2924 and 2853 cm⁻¹, which are typical of unordered lipid acyl chains. Characteristic protein absorptions were also identified at 1650, 1540 and 1250 cm⁻¹. In a more recent application, ATR FTIR was used to study secondary protein structures in 41 healthy donors and 65 patients with chronic fatigue syndrome (CFS) (Sakudo et al., 2009). Spectra in the region 4000–6000

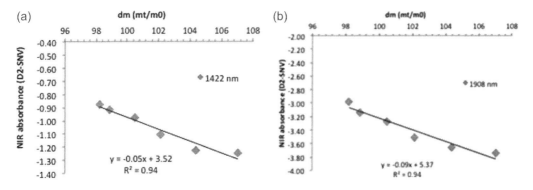

FIGURE 5.3 Correlation between increase in mass change from 40% to 90% RH with band intensity at (a) 1422 nm, free water; and (b) 1908 nm, bound water.

cm⁻¹ at 4 cm⁻¹ resolution were obtained. The α-helix, β-sheet, β turn and random coil contents of nails were estimated from the amide I region of the ATR FTIR spectra after correction for the refractive index of the material and Fourier deconvolution and curve fitting. The secondary protein structure of healthy fingernails was observed to consist of 11.74% α-helices, 37.58% β-sheets, 25.01% β turns and 25.67% random coils. The CF3 nail samples contained fewer α-helices and greater levels of β-sheet, which was attributed to reduced levels of normal elements of the nail plate; however, evidence for this was not provided.

Confocal Raman Spectroscopy (CRS)

In vivo CRS combines Raman spectroscopy with confocal laser scanning microscopy (CLSM) to accurately determine the skin depth from which the Raman signal is collected (Caspers et al., 1998, 2001). However, its application to the nail has not been explored. Recently, we reported the first investigation of the application of this technique in the measurement of nail hydration (Walters et al., 2012). Healthy human subjects with no history of nail disease or application of cosmetics to the nail were studied. Fingernails were soaked in water for 0, 2.5, 5, 10 and 15 minutes before the measurements were taken. Fingernails were also hydrated for 2 hours with water, and measurements were collected at 0, 10, 30 and 60 minutes. *In vivo* measurements were conducted using a CRS microscope operating at two wavelengths, 671 nm (high wave number region) and 785 nm (fingerprint region). Nail water content measurements were obtained from the 671 nm laser. The baseline water content at 0 μm within the nail was ~8% w/w (Figure 5.4). This value increased with depth to ~21% w/w at a depth of 30 μm into the nail, at which it showed a plateau phase. Hydrating the nail for 2.5 minutes elevated the water content within the nail to ~15% w/w at 0 μm and ~24% w/w at 30 μm. The greater the hydration time, the higher the water content within the nail up to 15 minutes after which there was no further increase even after 2 hours of hydration. The water content within the nail returned back

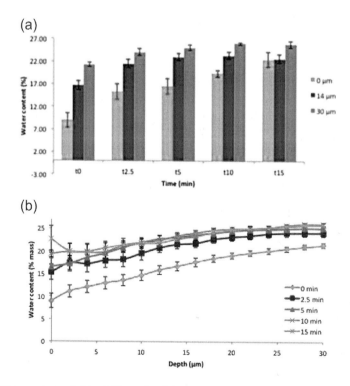

FIGURE 5.4   (a) Water content at 0, 14 and 30 μm after 2.5, 5, 10 and 15 min soaking in water. (b) Nail water content *in vivo* measured 2 hours after hydration for 0, 2.5, 5, 10 and 15 min soaking in water ($n = 6$; Mean ± SD).

to the baseline 30 minutes after 2 hours hydration. The values are in reasonable agreement with the vNIR-FT Raman spectroscopy measurements (Gniadecka et al., 1998) and NIR DR values (Egawa et al., 2003) discussed earlier.

Atomic Absorption Spectrometry (AAS)

Because of a proposed correlation between systemic levels of trace elements with nail levels, there are a number of AAS reports on nail in the literature. Harrison and Tyree (1971) were the first to report the application of flame atomic absorption to determine five elements (Ca, Zn, Mg, Cu and Fe) in fingernail clippings from seven females and ten males. Barnett and Kahn (1972) also reported a method to determine copper levels in fingernail clippings from newborn infants. However, despite early optimism and many publications on this application, nail sampling has not been developed as a routine non-invasive method to assess mineral status of individuals possibly because of the wide inter-subject variation in trace element levels.

Inductively Coupled Plasma Mass Spectrometry (ICP-MS)

ICP-MS utilises a gas that contains a sufficient concentration of ions and electrons to make the vapour electrically conductive. The plasma used in an ICP-MS is made by partially ionizing argon gas using an electrical current. After the sample is introduced into the plasma, the extreme temperature causes the sample to separate into individual atoms. For coupling to mass spectrometry, the ions from the plasma are extracted through a series of cones into a mass spectrometer. Goullé et al. (2009) have reported its use for multi-elemental analysis of fingernail and toenail clippings collected from healthy volunteers. Li, Be, B, Al, V, Cr, Mn, Co, Ni, Cu, Zn, Ga, Ge, As, Se, Rb, Sr, Mo, Pd, Ag, Cd, Sn, Sb, Te, Ba, La, Gd, W, Pt, Hg, Tl, Pb, Bi and U were detected with validated measurements. The authors suggested the application of ICP-MS as a non-invasive technique for detection of industrial, domestic or environmental metal exposure. Button et al. (2009) used this technique to assess arsenic uptake in toenail clippings and proposed this approach as a biomarker of exposure to environments with elevated arsenic.

Opto-Thermal Transient Emission Radiometry (OTTER)

OTTER uses pulsed laser excitation to induce temperature jumps of the order of a few degrees Celsius in the top few microns of the material under study. These temperature jumps decay within microseconds and do not materially increase the average substrate temperature or the rate of diffusion under study. They are observed with a high-speed infrared detector sensitive to the heat radiation emitted by the surface (Bindra et al., 1992). For bio-tissue, this radiation is strongest in the mid-infrared 6–13 µm band of wavelengths. The measurement captures the decay dynamics of this transient component of the heat radiation and relates it to the physical properties of the near-surface layers through appropriate mathematical models. In such models, depth resolution is linked to the time parameter of the transients and chemical specificity to the absorption spectra of the molecules of interest (Imhof et al., 1994). OTTER has been used to measure nail water content, nail water concentration depth profiles and topically applied solvent penetration (Figure 5.5) through nail (Xiao et al., 2010, 2011). Combining the water content results with transonychial water (TOWL) flux through nail also allows the water diffusion coefficient of nail to be calculated (see section 'Transonychial Water Loss (TOWL)').

Optical Coherence Tomography (OCT)

OCT is a non-invasive optical imaging technology that provides cross-sectional, tomographic images of tissue *in situ* and in real time (Fujimoto, 2003). In this technique, the reflection of infrared light from the tissue of interest is measured and the intensity is imaged as a function of position. It is possible to obtain images of tissue *in situ* with a high axial resolution. A representative image of the nail and proximal tissue is shown in Figure 5.6. Mogensen et al. (2007) investigated nail morphology and thickness in

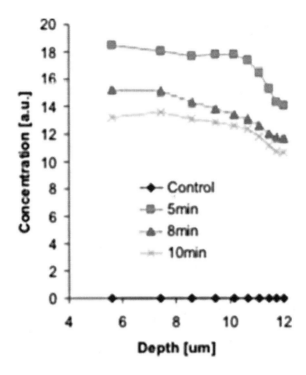

FIGURE 5.5   Glycerol depth profiles in human nail *in vivo* expressed in arbitrary units.

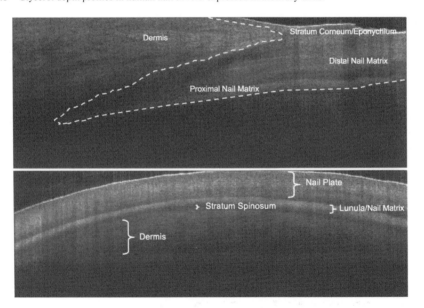

FIGURE 5.6   (a) Representative OCT scan of nail and proximal tissue. (b) OCT scan of fingernail with scanning axis perpendicular to direction of nail growth. (VivoSight OCT scanner, Michelson Diagnostics, United Kingdom.)

healthy volunteers using OCT images and compared the results with high-frequency ultrasound (HFUS) images and measurements with callipers and polarization-sensitive (PS) OCT. In standard OCT the nail plate appeared as a layered structure containing a varying number of horizontal homogeneous bands of varying intensity and thickness. PS-OCT images of the nail plate also showed a layered structure. It was possible to measure nail thickness using OCT and subtle changes not detected by HFUS could be discriminated by OCT. More recently, Sattler and co-workers (2011) compared OCT, CLSM and TOWL as tools for study of the nail. Healthy nails were investigated using OCT and CLSM. In a separate study, healthy volunteers washed the middle finger for 30 minutes at 38°C for 2 weeks. OCT, TOWL and CLSM measurements were conducted before and after the washing period. The penetration depth was about 2 mm with a resolution of 10 µm. OCT confirmed that nail thickness was significantly higher for the middle finger (mean value 465.8 µm) of the dominating hand of subjects. As for the earlier OCT study, the band-like structure of the healthy nail plate was confirmed. The washing experiment did not result in any significant increase in nail thickness as measured with OCT.

Confocal Laser Scanning Microscopy (CLSM)

CLSM utilises a laser and a pair of pinhole apertures for imaging and optical sectioning of tissue at high resolution. One of the first applications of the technique to study the nail was described by Hongcharu et al. (2000). Both *in vivo* and *in vitro* analyses were conducted on the nails of a patient suffering from onychomycosis. Full-thickness nail clippings were collected and imaged. Virtual sections were obtained *in vivo* by focusing the confocal microscope inside the nail plate. *In vivo* confocal images from just below the surface of the nail plate revealed a network of branched hyphae. The authors hypothesized that CLSM may be a faster and more accurate method of diagnosing onychomycosis compared with fungal culture, conventional microscopy or chemical hydrolysis.

Sattler and co-workers (2011) utilised CLSM along with other methods to evaluate healthy nails. The major advantage of CLSM compared with OCT was the ability to examine the microscopic structure of the nail as corneocyte borders could be visualised and their integrity examined. The authors achieved a depth of 400–500 um. The possibility of diagnosing leukonychia and mycotic infection was also demonstrated with individual images of diseased nails.

Transonychial Water Loss (TOWL)

TOWL is a measure of water flux through the nail and is frequently used with other biophysical techniques to characterise nail barrier function or permeability. Jemec and co-workers (1989) measured TOWL in healthy volunteers with an evaporimeter to establish the usefulness of this technique and to study various parameters such as sex, age and nail-plate thickness. The median TOWL was higher than for transepidermal water loss and it decreased with increasing age of the subject. No significant correlation between nail-plate thickness, as measured by ultrasound, was found. Krönauer et al. (2001) investigated TOWL values in healthy subjects and patients with atopic eczema, psoriasis and onychomycosis. TOWL values were significantly higher ($p < 0.05$) in healthy patients compared with those with atopic eczema and onychomycosis; only patients with nail psoriasis showed significantly lower TOWL compared with healthy subjects.

Exploiting the Permeability Characteristics of the Nail Plate

Our increasing knowledge concerning the permeability characteristics of the nail plate has served to increase our confidence that diseases of the nail can be treated topically. Novel and efficacious systems for ungual delivery are discussed in this section. For a review of the actual models used to evaluate such experimental techniques, we refer the reader to the comprehensive discussion of the subject by Nair et al. (2012) and recent publications in the area (Palliyil et al., 2014; Baraldi et al., 2015).

Current Approaches and Clinical Evidence

For many years dermatologists believed that topical treatment for anything other than the most superficial infections of the nail plate was futile (Zaias and Serrano, 1989). The nail plate was often viewed as an impenetrable barrier and the only means of treating diseases of the nail unit was by delivering drugs via the blood supply to the nail matrix or by removal of the diseased nail prior to topical application of the therapy. The former involved prolonged oral dosing with powerful antifungal agents and other drugs (see, for example, Roberton and Hosking, 1983), and the latter was only considered if only a few nails were involved or oral therapy was contraindicated (Hettinger and Valinsky, 1991). The dermatologist's armamentarium has been supplemented with nail lacquers and other formulations including those containing efinaconazole, tioconazole, tavaborole, amorolfine, ciclopirox, clobetasol 17-propionate, calcipotriol and tazarotene. However, it is well known that the true test of any potential therapeutic regimen is the ability to cure the disease and to prevent remission.

Nakano et al. (2006) performed a pilot study to assess the safety and efficacy of pulse therapy with oral terbinafine in 66 patients with onychomycosis. Each pulse consisted of oral terbinafine (500 mg/day) for 1 week followed by a 3-week interval. Topical 1% terbinafine cream was applied daily. Efficacy was assessed 1 year after treatment initiation. There was a complete cure in 51 patients (approximately 77%). It was concluded that terbinafine pulse therapy in combination with topical application of terbinafine cream was safe and effective, but it was not possible to determine whether the topical applications improved the outcome. An earlier study (Alpsov et al., 1996) using a similar oral treatment regimen but with no supplemental topical therapy reported a 74% cure rate, suggesting that topical application was of little additive benefit. However, it is important to appreciate that the application vehicle is very important for achieving successful delivery of the drug into the nail plate and it is possible that the cream used in the Nakano study had not been optimized in this respect. Furthermore, the highly lipophilic nature of terbinafine suggests that it would not penetrate into the nail to any great extent. Perhaps the use of a more innovative delivery vehicle would generate a clearer picture of the effectiveness of dual therapy. For example, Ghannoum et al. (2011) have incorporated terbinafine within transfersomes (described by Ghannoum et al. as 'carrier vehicles [consisting of] responsive, composite lipid aggregates that are highly deformable and have high surface hydrophilicity') and demonstrated that such vehicles, at an active concentration of 1.5% w/v, were considerably more potent (with MIC_{50} values against *Trichophyton rubrum* and *Trichophyton mentagrophytes* being 30-fold and 60-fold lower, respectively) than a 1% terbinafine spray *in vitro*.

The usefulness of supplemental topical treatment with oral therapy was also investigated by Rigopoulos et al. (2003) who evaluated a combination of systemic and topical antifungals to improve the cure rates and reduce the duration of systemic treatment for onychomycoses. Itraconazole pulse therapy was combined with amorolfine lacquer and compared with itraconazole alone in the treatment of nail Candida. Although statistical analysis showed no statistically significant difference in therapeutic effectiveness ($P > 0.1$) between the two treatment groups, the combination of topical amorolfine and oral itraconazole was shown to exhibit considerable synergy.

In a multicentre, randomised, open-label, parallel group study, Baran et al. (2007) evaluated topical amorolfine HCl 5% nail lacquer once weekly for 12 months plus terbinafine 250 mg once daily for 3 months (AT group) against terbinafine alone once daily for 3 months (T group). A significantly higher success rate was observed for patients in the AT group relative to those in the T group at 18 months. These results demonstrated that amorolfine nail lacquer in combination with oral terbinafine was more effective in the treatment of toenail onychomycosis that included matrix involvement. On the other hand, a study using a similar protocol to that of Baran et al. (2007) suggested that terbinafine pulse therapy was effective and safe in the treatment of onychomycosis, and that additional therapy with topical ciclopirox olamine (8%) or amorolfine HCl (5%) did not show any significant difference in efficacy in comparison to monotherapy with oral terbinafine (Jaiswal et al., 2007).

Monotherapy with topical treatment has also shown some success. Baran and Coquard (2005) treated onychomycotic patients with a solution of 1% fluconazole and 20% urea in an ethanol–water mixture, applied once daily. There was complete resolution of the disease in four cases and four patients demonstrated a 90% improvement. Gupta et al. (2000) reviewed the efficacy and safety of 8% ciclopirox nail lacquer in the treatment of onychomycoses and found that the data demonstrated that ciclopirox nail lacquer was significantly more effective than placebo in the treatment of onychomycosis. The lacquer demonstrated a broad spectrum of activity showing efficacy against Candida species and some nondermatophytes. In subsequent studies, topical ciclopirox efficacy has been confirmed (Sardana et al., 2006).

As pointed out previously, not all nail diseases are fungal or bacterial infections. Psoriasis can affect the entire nail plate and is a common feature in psoriasis patients. A lacquer formulation containing 8% clobetasol 17-propionate was evaluated for efficacy and safety (Sanchez Regana et al., 2005). Patients, with both nail bed and nail matrix psoriasis, were treated with the nail lacquer applied once daily for 21 days and subsequently twice weekly for 9 months. There was a reduction of all the nail alterations, including nail pain, within 4 weeks of initiating therapy and responses were directly related to the length of treatment. Tazarotene gel has also been evaluated as a therapy for nail psoriasis (Scher et al., 2001). Tazarotene treatment resulted in a significantly greater reduction in onycholysis (loosening of the nail plate–nail bed connection) in both occluded and nonoccluded nails together with a significantly greater reduction in pitting in occluded nails. Nail psoriasis was also improved following topical application of calcipotriol and betamethasone dipropionate ointment applied once daily for 12 weeks (Rigopoulos et al., 2009).

Brittle nail syndrome is a common problem and refers to nails that exhibit surface roughness, raggedness and peeling. Sherber et al. (2011) evaluated the usefulness of tazarotene cream (0.1%) for the treatment of brittle nails. Patients applied tazarotene cream to the nails twice daily for 24 weeks. The study showed that tazarotene was effective at reducing the symptoms of brittle nail syndrome with minimal to no irritation. However, it is difficult to rationalize how tazarotene, a retinoid, would be effective in brittle nail syndrome, and it is possible that some other ingredient in the cream formulation was responsible for the beneficial effects. There was, however, no reported placebo data.

Experimental Approaches

The major experimental enhancement approaches either focus on compounds that disrupt the nail barrier or vehicles such as lacquers with improved residence time on the nail (Table 5.1).

Chemical Enhancement: Barrier Disruption

Palliyil et al. (2013) have outlined a preformulation strategy for the selection of nail permeation enhancers. Their screening process included the ability of putative penetration enhancers to improve solubility and *in vitro* nail plate penetration of ciclopirox olamine. *In vitro* skin penetration and permeation was also assessed. Of the enhancers evaluated, thiourea was the preferred nail penetration enhancer.

TABLE 5.1

Chemical Agents which Disrupt/Damage the Nail Barrier

Chemical/Biochemical Classification	Compound	Mechanism of Action
Beta hydroxy acid	Salicylic acid, sodium salicylate	Keratolytic
Carbamide	Urea	Protein denaturant
Imine	Guanidine hydrochloride	Protein denaturant
Mercaptans, thiols	N-acetyl cysteine	Reduce disulphide links in nail keratin matrix
	Dithiothreitol	
	Mercaptoethanol	
	8-Mercaptomenthone	
	N-[2-Mercaptopropionyl] glycine	
	Meso-2,3-Dimercaptosuccinic acid	
	Sodium pyrithone, zinc pyrithone	
	Thioglycolic acid	
Enzyme	Papain	Protease
Sulphite, bisulphite	Sodium metabisulphite, sodium sulphite	Reduce disulphide links in nail keratin matrix

In earlier studies, Quintanar-Guerrero et al. (1998) investigated the effects of three keratolytics (papain, urea, salicylic acid) on the permeability of three imidazole antimycotics (miconazole nitrate, ketoconazole, itraconazole) through healthy human nails samples mounted in side-by-side diffusion cells. Only the combined effects of papain (15% for 1 day) and salicylic acid (20% for 10 days) were capable of enhancing the permeability of the antimycotics.

Kobayashi et al. (1998) assessed the enhancing effect of urea and sodium salicylate on the transungual delivery of a hydrophilic model drug (5-fluorouracil) using human nail samples in modified side-by-side diffusion cells. Aqueous suspensions containing either urea or sodium salicylate decreased drug flux compared with suspensions containing no keratolytic. The disulphide-reducing agents acetyl cysteine and mercaptoethanol were also evaluated for their effects on 5-FU and tolnaftate permeation. Suspensions of the drugs were formulated with acetyl cysteine or mercaptoethanol in ethanol:water or ethanol:isopropyl myristate vehicles. The ethanol:water vehicles promoted up to 16-fold greater 5-FU flux compared with aqueous suspensions, and the ethanol:isopropyl myristate vehicles enhanced 5-FU permeation up to 8-fold. The permeation of tolnaftate was also enhanced in the same vehicles but to a much lesser extent.

Using human nail samples in a Franz cell setup, Malhotra and Zatz (2002b) screened a range of molecules with known ability to interact with keratin. The compounds studied included mercaptans (N-[2-mercaptopropionyl] glycine, zinc and sodium pyrithone, 8-mercaptomenthone, meso-2,3-dimercapto succinic acid), sodium metabisulphite, keratolytic agents (salicylic acid, urea, guanidine hydrochloride). Gels were prepared containing enhancers either alone, or in combination with each other, in a vehicle that was aqueous, hydroalcoholic or one containing dimethyl sulphoxide (DMSO). The results indicated that the chemical structure of the modifier is most important in determining its ability to enhance penetration. The best enhancement effect was obtained using N-(2-mercaptopropionyl) glycine in combination with urea. Mercaptan compounds contain sulphydryl groups (SH), and the primary mechanism for their enhancement of nail penetration is therefore reduction of disulphide linkages in the nail keratin matrix. Generally, the barrier integrity of nails was compromised irreversibly after treatment with effective chemical modifiers.

Hui et al. (2004) prepared a novel ethanolic-gel formulation of ciclopirox-containing urea and propylene glycol, and evaluated drug permeation through nail samples and amount of drug localised in nail samples at the end of the experiment. A commercial gel and a lacquer formulation (without keratolytics) were also examined. Ciclopirox delivery into and through the nail was significantly greater from the commercial gel than from either the experimental gel or the nail lacquer.

Chemical Enhancement: Vehicle Effects

Although ethanol is a widely used penetration enhancer in topical and transdermal delivery, the delivery of 5-fluorouracil in menthol:ethanol:water or lactic acid:ethanol:isopropyl myristate vehicles did not enhance drug permeation compared with the aqueous control (Kobiyashi et al., 1998).

The ability of a skin penetration enhancer, 2-*n*-nonyl-1,3-dioxolane, to promote transungual delivery of econazole was evaluated by Hui et al. (2003). The test topical lacquer formulation contained 5% (w/w) econazole, Eudragit® RL/PO, ethanol, and 2-*n*-nonyl-1,3-dioxolane (18% w/w). A commercial control had the same formulation with the exception of 2-*n*-nonyl-1,3-dioxolane. Ungual permeation was evaluated in human nail samples in a modified diffusion cell setup. The addition of 2-*n*-nonyl-1,3-dioxolane to econazole nail lacquer delivered six times more antifungal drug through human nail than an identical lacquer–drug formulation without enhancer. Concentrations of econazole in the deep nail layer and nail bed were significantly higher in the test group than in the control group. The authors suggested that the dioxolane had plasticizing effects on the lacquer formulation and this may have contributed to its efficacy.

Monti et al. (2005) investigated the transungual permeation of ciclopirox from a water-soluble hydroxy-propyl chitosan-based formulation compared with a commercial water-insoluble nail lacquer (Penlac®). Formulations were evaluated using bovine hoof samples mounted in vertical diffusion cells. The flux values at steady state were of a similar order of magnitude for both vehicles. The percentage of drug permeated was significantly higher for the water-soluble polymer-based formulation. The percentage of drug retained in the membranes at the end of the experiment was also comparable for both vehicles. The authors suggested that the chitosan-based formulation might have better adhesion properties to the membrane than the commercial lacquer. More recently, Monti et al. (2010), compared permeation of ciclopirox in bovine hoof membranes with that of amorolfine in the same hydroxypropyl chitosan-based nail lacquer described earlier and with a non water-soluble reference (Loceryl®). Amorolfine flux from the water-soluble formulation was significantly higher when compared with Loceryl. Ciclopirox was able to permeate hoof membranes more easily compared with amorolfine, which was not explained by the different amounts of actives in the respective formulations.

Hui et al. (2007a) investigated a range of different nail lacquers and a co-solvent vehicle for delivery of a novel oxaborole antifungal to the nail. Lacquers were formulated with a film-forming agent, solvent, and, optionally, a penetration enhancer. Ethanol was the solvent, poly (vinyl methyl ether alt maleic acid monobutyl ester), poly (2-hydroxyethyl methacrylate), poly (vinyl acetate) were the film formers. The co-solvent vehicle consisted of ethanol and propylene glycol. Approximately the same amount of the test compound penetrated the nail plates regardless of the vehicle. The co-solvent vehicle showed a superior permeation capability into and through the normal human nail plate *in vitro* compared with a commercial formulation of ciclopirox (Penlac®, ciclopirox 8%).

The same group (Hui et al., 2007b) compared the ungual penetration of panthenol in a non-lacquer film forming nail formulation (ethanol, acrylates co-polymer, phytantriol) with an aqueous solution formulation. The nail treatment formulation showed better effectiveness on enhancement of transungual panthenol delivery into/through the deeper layers of the human nail plate than the aqueous solution of panthenol.

Cosmetic Nail Formulations

A number of cosmetic products are available for maintenance of nail health, and application and removal of nail colour. These include various nail polish preparations (enamel, lacquer, varnish), acrylate resin–based mixtures, light-cured gels, sprays and simple solvents.

Nail Polish

Nail polish formulations broadly encompass classic nail polishes for painting colour on the nail, base coats, topcoats, hardeners, varnishes for brittle nails, varnishes for ridged nails and hypoallergenic nail polishes. For the nail formulations containing pigments (mineral and organic), the main constituent that remains on the nail after drying is nitrocellulose, which serves as the film-forming polymer (Wimmer and Schlossmann, 1992). Film modifiers (e.g. toluene sulphonamide/formaldehyde resin) are required to improve adherence and shine of the formulation. Other components are also present including plasticizers (e.g. castor oil, glyceryl tribenzoate), solvents (e.g. ethyl acetate, butyl acetate), thinners (e.g. ethanol, isopropanol, butanol, toluene) and thixotropic agents (e.g. stearalkonium hectorite).

As base coats function to increase adherence of the varnish to the nail they contain more resin than formulations containing pigments. The base coat also acts as a protective barrier against staining of the nail plate hence the absence of pigment. Topcoats contain more nitrocellulose and plasticizers than the underlying nail varnish-containing pigment to ensure resistance of the latter. In addition, the topcoat may contain UV-absorbing materials, such as benzophenone 1 and 3, to help protect the underlying coloured coats. Base coats and topcoats are formulated with different proportions of thinners and solvents than the colour-containing preparation to facilitate application and drying. Finally, hardeners, varnishes for brittle nails, varnishes for ridged nails are essentially base coats which additionally may contain nylon fibres, acrylic resins or formaldehyde (Schoon, 2005).

Acrylate Resin Preparations

These formulations are used to effectively 'sculpture' artificial nails and consist of an acrylic resin obtained by blending a methyl, ethyl, or isobutyl methacrylate monomer and a polymethyl or ethyl methacrylate polymer. The monomer contains a stabilizer such as hydroquinone and N,N-dimethyl-p-toluidine as an accelerator, and the polymer also contains benzoylperoxide as a polymerization initiator. Other components such as plasticizers, solvents, accelerators and pigments may also be included (Kanerva et al., 1996).

Light-Cured Gels

These preparations are supplied as a pre-prepared mixture of acrylic monomers and polymers. There are two main types of light-cured gels for application of colour to the nail: acrylic and cyanoacrylate. For acrylic light-cured gels the hardening is obtained by exposure to light, which causes polymerization. These gels may contain (meth)acrylated urethanes, triethylene glycol dimethacrylate, methacrylated epoxy resin, or hydroxyl-functional methacrylates and a photoinitiator (Hemmer et al., 1996). Polymerization is obtained for cyanoacrylate gel by spraying or brushing on an activator.

Nail Polish Remover

The application of nail polish remover to the nail will dissolve residual nitrocellulose and also remove lipids. These products generally consist of large amounts of organic solvents, with small amounts of oils. The solvents used are typically acetone, butyl acetate, ethyl acetate or ethoxyethanol; representative oils include castor oil or lanolin oil. Formulations may also include dyes, fragrances, preservatives, vitamins and UV absorbers (de Groot et al., 1994).

Outlook and Summary

A diverse number of biophysical and bioengineering techniques have been used by scientists and clinicians to date to study the nail and to gain insight into how its properties may be influenced. The potential of many of these approaches in the treatment of diseases of the nail has still not been fully exploited. The modest success of current clinical therapies should be a spur to innovation in the field. Conventional formulations are largely lacquer based, suggesting that, as for skin, residence time of both active and excipients in or on the nail is critical for efficacy. Experimental approaches which clearly damage the nail must be viewed with caution and are likely to be faced with significant regulatory hurdles.

REFERENCES

Alpsoy E, Yilmaz E, Basaran E. 1996. Intermittent therapy with terbinafine for dermatophyte toe-onychomycosis: A new approach. *J Dermatol* 23:259–262.

Baden HP, Goldsmith LA, Fleming B. 1973. A comparative study of the physicochemical properties of human keratinized tissues. *Biochim Biophys Acta* 322:269–278.

Baraldi A, Jones SA, Guesné S, Traynor MJ, McAuley WJ, Brown MB, Murdan S. 2015. Human nail plate modifications induced by onychomycoses: Implications for topical therapy. *Pharm Res* 32:1626–1633.

Baran R, Coquard F. 2005. Combination of fluconazole and urea in a nail lacquer for treating onychomycosis. *J Dermatol Treat* 16:52–55.

Baran R, Sigurgeirsson B, de Berker D, Kaufmann R, Lecha M, Faergemann J, Kerrouche N et al. 2007. A multicentre, randomized, controlled study of the efficacy, safety and cost-effectiveness of a combination therapy with amorolfine nail lacquer and oral terbinafine compared with oral terbinafine alone for the treatment of onychomycosis with matrix involvement. *Br J Dermatol* 157:149–157.

Barnett WB, Kahn HL. 1972. Determination of copper in fingernails by atomic absorption with the graphite furnace. *Clin Chem* 18(9):923–927.

Bauer F, Stevens B. 1983. Investigations of trace metal content of normal and diseased nails. *Australas J Dermatol* 24:127–129.

Bindra RMS, Imhof RE, Eccleston GM, Birch DJS. 1992. Monitoring changes in skin condition due to the application of DMSO, using an opto-thermal technique. *SPIE Proc* 1643:299–309.

Bragulla HH, Homberger DG. 2009. Structure and functions of keratin proteins in simple, stratified, keratinized and cornified epithelia. *J Anat* 214:516–559.

Button M, Jenkin GR, Harrington CF, Watts MJ. 2009. Human toenails as a biomarker of exposure to elevated environmental arsenic. *J Environ Monit* 11(3):610–617.

Cashman MW, Sloan SB. 2010. Nutrition and nail disease. *Clin Dermatol* 28:420–425.

Caspers PJ, Lucassen GW, Wolthuis R, Bruining HA, Puppels GJ. 1998. In vitro and in vivo Raman spectroscopy of human skin. *Biospectroscopy* 4(5 Suppl):S31–S39.

Caspers PJ, Lucassen GW, Carter EA, Bruining HA, Puppels GJ. 2001. In vivo confocal Raman microspectroscopy of the skin: Noninvasive determination of molecular concentration profiles. *J Invest Dermatol* 116(3):434–442.

de Berker D, Wojnarowska F, Sviland L, Westgate GE, Dawber RP, Leigh IM. 2000. Keratin expression in the normal nail unit: Markers of regional differentiation. *Br J Dermatol* 142:89–96.

de Berker DAR, André J, Baran R. 2007. Nail biology and nail science. *Int J Cosmet Sci* 29:241–275.

de Groot AC, Weyland JW, Nater JP. 1994. Nail cosmetics. In: *Unwanted Effects of Cosmetics and Drugs Used in Dermatology*, 3rd ed. Elsevier Science, Amsterdam, 524–529.

Egawa M, Fukuhara T, Takahashi M, Ozaki Y. 2003. Determining water content in human nails with a portable near-infrared spectrophotometer. *Appl Spectrosc* 57:473–478.

Egawa M, Ozaki Y, Takahashi M. 2005. In vivo measurement of water content of the fingernail and its seasonal change. *Skin Res Technol* 12:126–132.

Farran L, Ennos AR, Eichhorn J. 2008. The effect of humidity on the fracture properties of human fingernails. *J Exp Biol* 211:3677–3681.

Fujimoto JG. 2003. Optical coherence tomography for ultrahigh resolution *in vivo* imaging. *Nat Biotechnol* 21:1361–1367.

Ghannoum M, Isham N, Herbert J, Henry W, Yurdakul S. 2011. Activity of TDT 067 (terbinafine in transfersome) against agents of onychomycosis, as determined by minimum inhibitory and fungicidal concentrations. *J Clin Microbiol* 49:1716–1720.

Gniadecka M, Wulf HC, Nielsen OF, Christensen DH, Hercogova D. 1997. Distinctive molecular abnormalities in benign and malignant skin lesions: Studies by Raman spectroscopy. *Photochem Photobiol* 66:418–423.

Gniadecka M, Ole Faurskov Nielsen, Daniel Hojgaard Christensen, Hans Christian Wulf. 1998. Structure of water, proteins and lipids in intact human skin, hair and nail. *J Invest Dermatol* 110:393–398.

Goullé JP, Saussereau E, Mahieu L, Bouige D, Groenwont S, Guerbet M, Lacroix C. 2009. Application of inductively coupled plasma mass spectrometry multielement analysis in fingernail and toenail as a biomarker of metal exposure. *J Anal Toxicol* 33(2):92–98.

Gupta AK, Fleckman P, Baran R. 2000. Ciclopirox nail lacquer topical solution 8% in the treatment of toenail onychomycosis. *J Am Acad Dermatol* 43(4 Suppl): S70–S80.

Harrison WW, Tyree AB. 1971. The determination of trace elements in human fingernails by atomic absorption spectroscopy. *Clin Chim Acta* 31:63–73.

Hemmer W, Focke M, Wantke F, Götz M, Jarisch R. 1996. Allergic contact dermatitis to artificial fingernails prepared from UV light-cured acrylates. *J Am Acad Dermatol* 35:377–380.

Hettinger DF, Valinsky MS. 1991. Treatment of onychomycosis with nail avulsion and topical ketoconazole. *J Am Podiatr Med Assoc* 81:28–32.

Hongcharu W, Dwyer P, Gonzalez S, Anderson RR. 2000. Confirmation of onychomycosis by *in vivo* confocal microscopy. *J Am Acad Dermatol* 42(2 Pt 1):214–216.

Hui X, Chan TCK, Barbadillo S, Lee C, Maibach HI, Wester RC. 2003. Enhanced econazole penetration into human nail by 2-n-nonyl-1,3-dioxolane. *J Pharm Sci* 92:142–148.

Hui X, Wester RC, Barbadillo S, Lee C, Patel B, Wortzmman M, Gans EH et al. 2004. Ciclopirox delivery to the human nail plate. *J Pharm Sci* 93:2545–2548.

Hui X, Baker SJ, Webster RC, Barbadillo S, Cashmore AK, Sanders V, Hold KM et al. 2007a. *In vitro* penetration of novel oxaborole antifungal (AN2690) into the human nail plate. *J Pharm Sci* 96:2622–2631.

Hui X, Hornby SB, Wester RC, Barbadillo S, Appa Y, Maibach H. 2007b. *In vitro* human nail penetration and kinetics of panthenol. *Int J Cosm Sci* 29:277–282.

Jaiswal A, Sharma RP, Garg AP. 2007. An open randomized comparative study to test the efficacy and safety of oral terbinafine pulse as a monotherapy and in combination with topical ciclopirox olamine 8% or topical amorolfine hydrochloride 5% in the treatment of onychomycosis. *Int J Dermatol Venereol Leprol* 73:393–396.

Jemec GB, Agner T, Serup J. 1989. Transonychial water loss: Relation to sex, age and nail-plate thickness. *Br J Dermatol* 121(4):443–446.

Kanerva L, Lauerma A, Estlander T, Alanko K, Henrick-Eckerman M-L, Jolanki R. 1996. Occupational allergic contact dermatitis caused by photobonded sculptured nails and a review of (meth)acrylates in nail cosmetics. *Am J Contact Dermatitis* 7:109–115.

Kien CL, Ganther HE. 1983. Manifestations of chronic selenium deficiency in a child receiving total parenteral nutrition. *Am J Clin Nutr* 37:319–328.

Kobayashi Y, Miyamoto M, Sugibayashi K, Morimoto Y. 1998. Enhancing effect of N-acetyl-L-cysteine or 2-mercaptoethanol on the *in vitro* permeation of 5-fluorouracil or tolnaftate through the human nail plate. *Chem Pharm Bull* 46:1797–1802.

Krönauer C, Gfesser M, Ring J, Abeck D. 2001. Transonychial water loss in healthy and diseased nails. *Acta Derm Venereol* 81(3):175–177.

Malhotra GG, Zatz JL. 2002a. Characterization of the physical factors affecting nail permeation using water as a probe. *J Cosmet Sci* 51:367–377.

Malhotra GG, Zatz JL. 2002b. Investigation of nail permeation enhancement by chemical modification using water as a probe. *J Pharm Sci* 91:312–323.

Mogensen M, Thomsen JB, Skovgaard LT, Jemec GB. 2007. Nail thickness measurements using optical coherence tomography and 20-MHz ultrasonography. *Br J Dermatol* 157(5):894–900.

Monti D, Saccomani L, Chetoni P, Burgalassi S, Saettone MF, Mailland F. 2005. *In vitro* transungual permeation of ciclopirox from a hydroxypropyl chitosan-based, water-soluble nail lacquer. *Drug Dev Ind Pharm* 31:11–17.

Monti D, Saccomani L, Chetoni P, Burgalassi S, Senesi S, Ghelardi E, Mailland F. 2010. Hydrosoluble medicated nail lacquers: In vitro drug permeation and corresponding antimycotic activity. *Br J Dermatol* 162:311–317.

Nakano N, Hiruma M, Shiraki Y, Chen X, Porgpermdee S, Ikeda S. 2006. Combination of pulse therapy with terbinafine tablets and topical terbinafine cream for the treatment of dermatophyte onychomycosis: A pilot study. *J Dermatol* 33:753–758.

Nair A, Lane ME, Murthy SN. 2012. *In vitro* and *in vivo* models for transungual drug delivery studies. In: *Nail Structure and Formulations to Target the Nail*, SN Murthy, ed. CRC Press, Boca Raton, FL, 123–148.

Nogueiras-Nieto L, Gómez-Amoza JL, Delgado-Charro MB, Otero-Espinar FJ. 2011. Hydration and N-acetyl-l-cysteine alter the microstructure of the human nail and bovine hoof: Implications for drug delivery. *J Contr Rel* 156:337–344.

Palliyil BB, Lebo DB, Patel PR. 2013. A preformulation strategy for the selection of penetration enhancers for a transungual formulation. *AAPS PharmSciTech* 14:682–691.

Palliyil BB, Li C, Owaisat S, Lebo DB. 2014. Lateral drug diffusion in human nails. *AAPS PharmSciTech* 15:1429–1438.

Perrin C. 2007. Expression of follicular sheath keratins in the normal nail with special reference to the morphological analysis of the distal nail unit. *Am J Dermatopathol* 29:543–550.

Perrin C. 2008. The 2 clinical subbands of the distal nail unit and the nail isthmus. Anatomical explanation and new physiological observations in relation to the nail growth. *Am J Dermatopathol* 30:216–221.

Perrin C, Langbein L, Schweizer J. 2004. Expression of hair keratins in the adult nail unit: An immunohistochemical analysis of the onychogenesis in the proximal nail fold, matrix and nail bed. *Br J Dermatol* 151:362–371.

Quintanar-Guerrero D, Ganem-Quintanar A, Tapia-Olguin P, Kalia YN, Buri P. 1998. The effect of keratolytic agents on the permeability of three imidazole antimycotic drugs through the human nail. *Drug Dev Ind Pharm* 24:685–690.

Rich P, Scher RK. 2005. *An Atlas of Diseases of the Nail* (Encyclopedia of Visual Medicine Series). Parthenon Publishing Group, New York.

Rigopoulos D, Katoulis AC, Ionnides D, Georgaia S, Kalogeromitros D, Bolbasis I, Karistinou A et al. 2003. A randomised trial of amorolfine 5% solution nail lacquer in association with itraconazole pulse therapy compared with itraconazole alone in the treatment of Candida fingernail onychomycosis. *Br J Dermatol* 149:151–156.

Rigopoulos D, Gregoriou S, Danielli CR, Belyayeva H, Larios G, Verra P, Stamou C et al. 2009. Treatment of nail psoriasis with a two-compound formulation of calcipotriol plus betamethasone dipropionate ointment. *Dermatology* 218:338–341.

Roberton DM, Hosking CS. 1983. Ketoconazole treatment of nail infection in chronic mucocutaneous candidiasis. *Aust Paediatr J* 19:178–181.

Runne U, Orfanos CE. 1981. The human nail – Structure, growth and pathological changes. *Curr Probl Dermatol* 9:102–149.

Sakudo A, Kuratsune H, Kato YH, Ikuta K. 2009. Secondary structural changes of proteins in fingernails of chronic fatigue syndrome patients from Fourier-transform infrared spectra. *Clin Chim Acta* 402:75–78.

Sanchez Regana M, Martin Ezquerra G, Umbert Millet P, Llambi Mateos F. 2005. Treatment of nail psoriasis with 8% clobetasol nail lacquer: Positive experience in 10 patients. *J Eur Acad Dermatol Venereol* 19:573–577.

Sattler E, Kaestle R, Rothmund G, Welzel J. 2011. Confocal laser microscopy, optical coherence tomography and transonychial water loss for in vivo investigation of nails. *Br J Dermatol* 166(4):740–746.

Scher RK, Stiller M, Zhu YI. 2001. Tazarotene 0.1% gel in the treatment of fingernail psoriasis: A double-blind, randomised, vehicle-controlled study. *Cutis* 68(5):355–358.

Schoon D. 2005. Nail varnish formulation. In: *Textbook of Cosmetic Dermatology*, 3rd ed., Baran R, Maibach HI, eds. Martin Dunitz, London, 313–318.

Sherber NS, Hoch AM, Coppola CA, Carter EL, Chang HL, Barsanti FR, Mackay-Wiggan JM. 2011. Efficacy and safety study of tazarotene cream 0.1% for the treatment of brittle nail syndrome. *Cutis* 87:96–103.

Sowa MG, Wang J, Schultz CP, Ahmed MK, Mantsch HH. 1995. Infrared spectroscopic investigation of in vivo and ex vivo human nails. *Vib Spectrosc* 10:49–56.

Walters KA, Abdalghafor HM, Lane ME. 2012. The human nail: Barrier characterisation and permeation enhancement. *Int J Pharm* 435:10–21.

Wessel S, Gniadecka M, Jemec GBE, Wulf HC. 1999. Hydration of human nails investigated by NIR-FT-Raman spectroscopy. *Biochim Biophys Acta* 1433:210–216.

Wimmer EP, Schlossman ML. 1992. The history of nail polish. *Cosmet Toilet* 107:115–120.

Xiao P, Ciortea LI, Singh H, Berg EP, Imhof RE. 2010. Opto-thermal radiometry for in-vivo nail measurements. *J Phys Conf Ser* 214:012008.

Xiao P, Zheng X, Imhof RE, Hirata K, McAuley WJ, Mateus R, Hadgraft J et al. 2011. Opto-Thermal Transient Emission Radiometry (OTTER) to image diffusion in nails in vivo. *Int J Pharm* 406(1–2):111–113.

Zaias N, Serrano L. 1989. The successful treatment of finger Trichophyton rubrum onychomycosis with oral terbinafine. *Clin Exp Dermatol* 14:120–123.

6

Hair Morphology, Biogenesis, Heterogeneity, Pathophysiology and Hair Follicle Penetration

Alexa Patzelt and Jürgen Lademann

CONTENTS

Introduction

Although not visible at first sight, the hair and the hair follicle provide multiple functions in humans. First, the hair is a sensory organ. The autonomous nervous system can induce the erection of the hair, which is controlled by the arrector pili muscle leading to so-called goosebumps which are mainly triggered by coldness, anxiety or emotions (Benedek and Kaernbach, 2011). Animals use this effect to appear larger and more threatening to their enemies; for men, this function seems to play a subordinate and sometimes even annoying role (Benedek and Kaernbach, 2011). Moreover, hairs provide several protective functions. Eyelashes and eyebrows, for example, can prevent sweat and foreign particles from reaching the eyes. Scalp hair offers efficient sun protection (Parisi et al., 2009) as can be seen in the high incidence of skin cancer on the scalp of bald men. Nowadays, scalp hair is a very important instrument of psychosocial communication. A loss of scalp hair frequently induces severe psychological strain (Katoulis et al., 2015).

As part of the pilosebaceous unit, the hair follicle is an important organ involved in sebum secretion (Matsui et al., 2016). The sebum is released from the sebaceous gland, enters the hair follicle and then reaches the skin surface, where it is homogeneously distributed and significantly contributes to the efficiency of the skin barrier.

As the hair follicle represents a disruption of the skin barrier, the hair follicles also play an important role as a penetration pathway for percutaneous absorption and as a long-term reservoir for topically applied particulate substances (Lademann et al., 2006).

Moreover, several skin disorders are associated with the hair follicles and thus the hair follicles can be considered as a target for several therapeutic strategies.

Summarizing all these functions, it becomes obvious that the hairs and hair follicles are important structures in the skin. Different aspects of the hairs and hair follicles will be discussed in the present chapter.

The Morphology of Hair and Hair Follicles

The hair follicles are appendages of the skin and can be found at high density in most mammals. Humans possess approximately 5 million hair follicles (Krause and Foitzik, 2006) of different densities and sizes depending on the different skin regions (Otberg et al., 2004). Next to the scalp, the forehead is characterized by a very high follicular density. Here, the hair follicles are rather small. A lower follicular density is found in the region of the calf, whereas the size of the hair follicles is significantly higher in this region (Otberg et al., 2004).

The pilosebaceous unit, which is the superordinate structure of hair follicle, hair shaft, arrector pili muscle and sebaceous gland, is a very complex, three-dimensional structure where more than 20 different cell populations are interacting (Meidan et al., 2005; Rogers, 2004). The hair follicle is schematically depicted in Figure 6.1.

The hair shaft is composed of different layers. The inner layer of the hair is the medulla, which is surrounded by the cortex being responsible for the elasticity of the hair. The outer layer is represented by the cuticula, which consists of flat cornified cells that overlay one another and give the hair the typical roof tile-like structure (Buffoli et al., 2014). In humans, there are mainly two different hair types: the larger terminal hairs and the smaller vellus hairs. Terminal hairs are long and thick (>50 μm) and grow from a hair follicle that extends deeply into the subcutaneous fat tissue (>3000 μm). The vellus hairs are unpigmented, shorter than 2 cm, and thinner than 30 μm. The corresponding hair follicles extend only into the dermis (<1000 μm) (Meidan et al., 2005; Vogt et al., 2007).

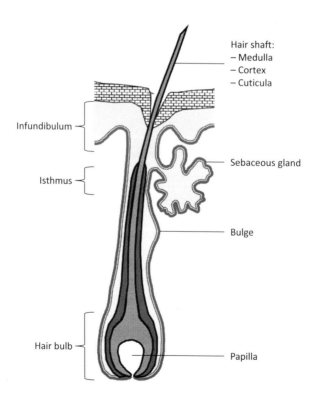

FIGURE 6.1   Schematic drawing of a hair follicle.

The hair follicle consists of a permanent and a transient portion. Whereas the transient portion is involved in the hair cycle, the permanent portion remains more or less unaffected.

The permanent part of the hair follicle combines the infundibulum, the isthmus and the bulge region. The infundibulum is described as the part of the hair follicle between the skin surface and the sebaceous gland, and provides at least in the upper part an intact keratinized epidermis. In the lower infundibulum, the differentiation pattern changes from epidermal to trichilemmal with reduced barrier function in the region of the outer root sheath (Blume-Peytavi and Vogt, 2011). In its lower part, the hair follicle's barrier function is provided predominantly by tight junctions. In the region of the hair bulb, no barrier can be detected (Mathes et al., 2016). The isthmus is characterized as the region between the sebaceous gland and the bulge region. The latter is the host of stem cells (Vogt and Blume-Peytavi, 2003) and skin mast cell precursors (Kumamoto et al., 2003). The transient part of the hair follicle is the region between the bulge and the hair bulb. The hair bulb is located around the papilla and contains hair matrix cells and germinative cells. The papilla is responsible for the nutrition of the hair follicle, as it is the only part of the hair follicle that is supplied with blood vessels (Krause and Foitzit, 2004). The anagen hair bulb shows a reduced expression of MHC-I molecules and a decreased number of T cells and Langerhans cells (Christoph et al., 2000), and thus has been identified as an area of relative and cycle-dependent immune privilege (Paus et al., 2005). In contrast, probably as the hair follicle represents a disruption of the skin barrier, a high number of antigen-presenting cells is located in the region around the infundibulum.

The Hair Cycle

Different signal transduction pathways, growth factors and cytokines are responsible for the regulation of the hair growth cycle, which consists of three different stages (Vogt and Blume-Peytavi, 2003). During the anagen phase, the matrix keratinocytes proliferate in the bulb region and migrate to form the hair shaft (Cotsarelis, 2006). During the following catagen phase, the inferior part of the hair follicle declines and the hair follicle enters the telogen phase where the hair is shed (Meidan et al., 2005). Subsequently, another anagen phase starts with dividing hair matrix cells and redevelopment of the inferior part of the hair follicle. The duration of the single stages is phase-specific and shows high interindividual differences. The anagen phase lasts 2–6 years, whereas the catagen phase is quite short at only 2 weeks. The telogen phase persists for 2–4 months. Depending on the individual duration of the anagen phase, the maximum hair length can be very different.

Ethnic Heterogeneity in Hair and Hair Follicle Morphology

Obviously, hair morphology is different in various ethnic groups. Whereas the hair fibres in Asian people are relatively wide and rounded, Caucasian hair is more elliptical. The hair fibres of African Americans are even more elliptical, with an almost flat cross-section in their curly structure and with a high susceptibility to breakage and knotting (McMichael, 2003). The scalp hair density likewise differs from highest values in Caucasians and lowest in Asians. Ethnic differences in hair density and hair follicle morphology were also detected for other body regions. In comparison to Caucasians, the follicular density on the forehead is significantly lower in Asians and African Americans. Smaller values for follicular volume, surface and follicular orifices and hair shaft diameters on the thigh and calf regions were also detected in these ethnic groups. The follicular reservoir was demonstrated to be generally higher in Caucasians (Mangelsdorf et al., 2006).

Hair Follicle Pathophysiology

Among the variety of skin diseases, several diseases are hair follicle-associated and go along with hair loss. Hair loss can either be reversible or can be accompanied by an irreversible destruction of cell populations with regenerative potential, which has a significant impact on the quality of life of affected patients.

Reversible Hair Loss

Reversible hair loss can be found in several hair diseases such as in telogen effluvium, premature anagen arrest, androgenic and androgenetic alopecia, and alopecia areata.

Telogen Effluvium and Premature Anagen Arrest

The symptom of reversible hair loss is relatively frequent among the population and can be induced by various factors that are able to interfere with the hair cycle. This interference with the hair cycle is responsible for an increased number of hair follicles prematurely entering the telogen phase, which is associated with increased hair shedding (Sperling, 2001). This situation is called telogen effluvium and can be induced also by systemic diseases such as endocrinopathies (Freinkel and Freinkel, 1972) or certain drugs (Gollnick et al., 1990). In contrast, when the metabolic and mitotic activity is completely suppressed due to certain chemotherapeutics or inflammatory skin diseases, this circumstance is called premature anagen arrest (Sperling, 2001). If the outer conditions normalize, hair loss is reversible.

Androgen-Associated Disorders

Changes in androgen or androgen receptor levels can have implications on hair growth and other hair follicle-associated diseases such as acne. Androgens normally induce the enlargement of hair follicles in androgen-dependent areas such as beards, axillae and pubic regions (Sperling, 2001). In the region of the scalp, however, androgens can promote miniaturization of hair follicles and a reduction of the anagen phase resulting in androgenetic alopecia (Piraccini and Alessandrini, 2014).

Autoimmunological-Induced Hair Loss

Alopecia areata is an autoimmune hair loss in well-demarcated areas of the scalp or in rarer cases on the rest of the body (Wasserman et al., 2007). The complete absence of scalp and body hair is also possible (alopecia areata universalis). The pathogenesis has not yet been clarified in detail, but the increased prevalence of autoimmune comorbidities, increased levels of autoantibodies, the association of lymphocyctic infiltrates around the hair follicles, increases in CD4/CD8, and several cytokine abnormalities indicate that alopecia areata is an organ-specific autoimmune disorder (Wasserman et al., 2007).

Inflammatory Hair Loss

Inflammatory diseases also can affect the hair follicles, for example, by bacterial, fungal or viral infections (Edlich et al., 2005; Gupta and Summerbell, 2000; Weinberg et al., 1997). It is known that the hair follicle represents a very efficient reservoir for a diversity of pathogens. A previous study demonstrated that about 25% of the skin-resident pathogens are located within the hair follicles and thus are not affected by conventional skin antiseptics, as they are not able to sufficiently penetrate deep into the hair follicles (Lange-Asschenfeldt et al., 2011). This results in a rapid recolonization of the skin surface after skin antisepsis, which might be the reason for many surgical site infections. Main representatives of the skin-resident flora are staphylococcus, proprioni bacteria and *Malassezia furfur* (Leeming et al., 1984).

Irreversible Hair Loss

Different physical, chemical, inflammatory and traumatic events are able to destroy the hair follicles, which can lead to scarring alopecia. In these hair follicles, hair growth cannot occur again. It is assumed that a persistent disturbance of bulge-papilla communication results in irreversible alopecia (Hermes and Paus, 1998). This can be due to several inflammatory diseases, such chronic or discoid lupus erythematodes (Tebbe, 2004), lichen planopilaris (d'Ovidio et al., 2007), dermatomyositis, graft-versus-host diseases, lichen sclerosus et atrophicus, sarcoidosis and scleroderma (Oremovic et al., 2006). Infectious diseases, such as lupus vulgaris, syphilis and Hansen's disease can also lead to scarring alopecia (Jaworsky and Gilliam, 1999).

Hair Follicles and Skin Tumours

Several benign, semi-malignant or malignant skin tumours have their origin in hair follicles. Basal carcinoma, for example, has mutations of the patched gene. This gene is a tumour suppressor associated with the Sonic Hedgehog–patched signalling pathway, which is active in hair follicle morphogenesis (Misago et al., 2003). Basal carcinoma is thus considered to be a malignant neoplasm of abnormal follicular germinative cells (Misago et al., 2003). Other skin tumours with follicular origin are seborrheic keratosis, trichoepithelioma, trichoblastoma, pilomatricoma and tricholemmal carcinoma (Headington, 1976).

Hair Follicle Penetration

Only in the last two decades have hair follicles been accepted as a relevant penetration pathway for topically applied substances. Previously, the intercellular penetration pathway along the lipid layers around the corneocytes of the stratum corneum was considered the only relevant possibility to overcome this potent skin barrier. All penetration pathways are presented schematically in Figure 6.2. The hair follicles were regarded to be irrelevant, as they were supposed to cover only a small area of the skin surface. Furthermore, it was difficult to investigate follicular penetration due to a lack of method and models, which certainly need spatial resolution.

Investigations performed by Otberg et al. (2004) demonstrated that the contribution of the hair follicles especially on some body sites cannot be neglected. They calculated that the stratum corneum

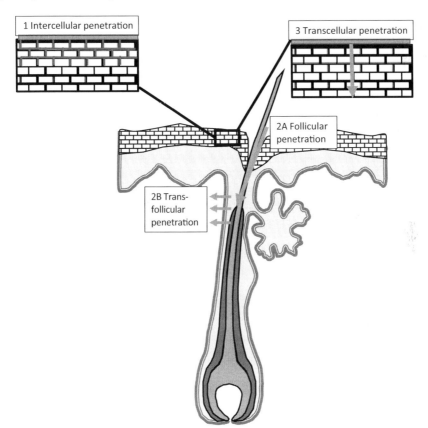

FIGURE 6.2 Schematic drawing of the different penetration pathways. (1) The intercellular penetration pathway along the lipid layers around the corneocytes. (2) The follicular penetration pathways, which as to be divided into (A) the follicular pathway and (B) the transfollicular pathway. (3) The transcellular penetration pathway, which seems to be of minor importance for most topically applied substances.

reservoir and the follicular reservoir even correlate in some body regions, such as the forehead and calf, where either many but tiny, or large but fewer hair follicles are situated, respectively.

In principle, the follicular penetration process has to be distinguished into two stages. In stage one, the substance penetrates only into the hair follicle, where it is either stored or, in a second stage, can penetrate transfollicularly into the hair follicle surrounding epidermis or dermis. Whether follicular and transfollicular penetration occurs clearly depends on the physiochemical properties of the applied substance (Patzelt and Lademann, 2013).

Methods to Investigate Follicular Penetration

Whereas several methods have been established for the investigation of the intercellular penetration pathway, such as tape stripping, diffusion cells, dermal microdialysis and Raman microscopy (Holmgaard et al., 2014), the investigation of the follicular penetration pathway is more complicated as it requires spatial resolution and a quantitative model system that is follicle free but offers the same structural, biochemical and barrier properties of normal skin (Meidan et al., 2005). Additionally, it should be able to distinguish between both stages of follicular penetration. In recent years, several methods to investigate follicular penetration have been introduced which roughly fulfil these criteria.

Differential stripping represents a valuable method to investigate the follicular penetration, however, without considering the transfollicular penetration process (Teichmann et al., 2005). Differential stripping is a combined method of tape stripping and cyanoacrylate skin surface biopsy. After topical application and penetration of a substance, the stratum corneum is removed by tape stripping. Subsequently, the substance is only located in the hair follicles, and the follicular content can be removed by cyanoacrylate skin surface biopsy. This method allows the quantitative analysis of the amount of substance penetrated into the stratum corneum and the hair follicles. This method can be applied to *in vivo* and *ex vivo* skin models. However, it has to be taken into consideration that the follicular reservoir is significantly reduced by 90% in excised human skin due to skin contraction occurring after skin excision (Patzelt et al., 2008). In porcine ear skin, hair follicles are fixed more tightly within the skin and cannot be removed completely by cyanoacrylate skin surface biopsies, which likewise has to be considered during quantification.

Differential stripping revealed that the hair follicles represent long-term reservoirs for topically applied substances. In a previous study, it was demonstrated that nanocarriers were stored within the hair follicles for a period of 10 days, whereas the portion of nanocarriers, which was located on the skin surface, had already been removed completely from the stratum corneum after 1 day (Lademann et al., 2006).

The selective follicular closing technique also has become a valuable tool to investigate the transfollicular penetration process (Teichmann et al., 2006). The selective closing of each hair follicle opening within a test area allows the determination of the portion of the hair follicle on the total penetration process of topically applied substances. The methods can be performed *in vivo* (Otberg et al., 2008) and *ex vivo* in combination with diffusion cell experiments (Trauer et al., 2010) and dermal microdialysis. *In vivo* investigations revealed that topically applied caffeine becomes bioavailable after 5 min if the hair follicles are open, but only after 20 min in the case of occluded hair follicles (Otberg et al., 2008).

Additionally, confocal laser scanning microscopy can be used to determine the follicular penetration depth of topically applied substances in the micrometer range, provided these substances are labelled with fluorescent markers (Patzelt et al., 2011).

The Mechanism of Follicular Penetration

The follicular and transfollicular penetration efficiency strongly depends on the physicochemical properties of the penetrating substance.

Whereas small molecules such as caffeine show a strong tendency to penetrate transfolliculary into the living tissue (Otberg et al., 2008), particulate substances predominantly accumulate within the hair follicles (Lademann et al., 2007). Only recently, the mechanism of the follicular penetration process has been partly elucidated. Both, *ex vivo* and *in silico* data demonstrate that nanocarriers are transported by the movement of the hair inside the hair follicles comparable to a ratchet mechanism (Radtke et al., 2017). The follicular penetration depth was shown to be more efficient if the nanocarriers are sized around 600 nm (Patzelt et al., 2011).

The penetration of smaller and larger nanocarriers was significantly inefficient. Radial movement of the hair seems to be more effective than axial movement, although during massage application a mixture of both movement directions is realistic (Radtke et al., 2017). Without any hair movement, the follicular penetration depth is negligible (Lademann et al., 2007). Moreover, the latest data demonstrate that the frequency of hair movement is also of relevance, whereby slower frequencies seem to be superior.

Knowledge about the follicular penetration mechanism is of importance in order to optimize nanocarriers or other substances for this specific transportation pathway.

Nanocarriers as Transporters for Skin Delivery

Nanocarriers represent a heterogeneous group of particulate substances which are expected to improve drug delivery significantly in many fields of medicine, pharmacology and cosmetics. Although successful in some of these fields, such as tumour targeting, drug delivery via the skin with nanocarriers still represents a challenge, as nanocarriers are not able to overcome the very potent skin barrier by themselves due to their size. Instead, the nanocarriers are entrapped within the hair follicles and at first sight cannot contribute to a better bioavailability of topically applied substances (Patzelt and Lademann, 2013).

However, the advantages of nanocarriers are convincing. They can transport significantly higher drug concentrations, and are able to improve the solubility of hydrophobic substances and the physical and chemical stability of drugs. Moreover, they are able to release their drugs in a controlled manner (Goyal et al., 2016). In order to use the positive properties of nanocarriers for improved drug delivery, new strategies for the effective use of nanocarriers have to be developed.

A very promising strategy is to use nanocarriers exclusively as transporters for drugs in the hair follicles. Once located within the hair follicle, the nanocarriers are designed to release their drug upon an exogenous or endogenous trigger signal. The released drug can then be active in the hair follicle, for example, in acne therapy, or translocate transfollicularly on its own. In this case, the bioavailability of topically applied drugs could be increased significantly (Patzelt et al., 2017). A scheme of this nanocarrier-assisted triggered release drug delivery approach is presented in Figure 6.3. Several approaches to realize the triggered release of drugs from nanocarriers within the hair follicle are currently under research.

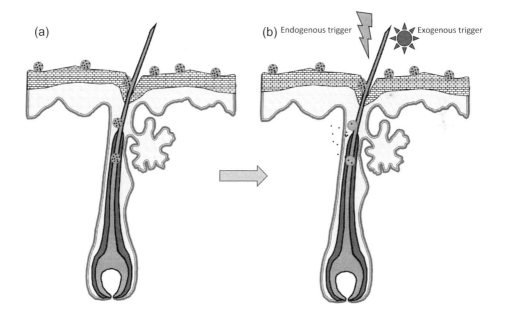

FIGURE 6.3 Schematic representation of the triggered release approach for topical drug delivery. (a) Nanocarriers are used to transport drugs into the hair follicle. (b) Upon an endogenous or exogenous trigger signal, the drug is released from the nanocarrier within the hair follicle and can translocate independently into the hair follicle surrounding tissue and becomes bioavailable.

In principle, two different types of trigger signals have to be distinguished. On the one hand, exogenous trigger signals are feasible such as enzymes (Mak et al., 2011), external light sources (Lademann et al., 2016) or ultrasound (Mak et al., 2011). The idea is that the nanocarriers deliver the drug into the hair follicle and that the release only occurs after irradiation or additional enzyme application. The advantage is that the applicant can precisely determine the time point of action. The disadvantage is that a second component is always necessary to realize the release, which might be a hurdle in application for some clinical indications.

On the other hand, endogenous trigger signals can be used, which have the clear advantage that a second component is not necessary. Endogenous trigger signals can be pH changes. This approach is very promising, as physiologically a pH gradient exists within the hair follicle (Dimde et al., 2017), which can be used to keep the nanocarriers stable at the skin surface and trigger release at higher pH in a distal part of the hair follicle. Temperature changes are also conceivable as a trigger signal. Previous studies with thermoresponsive nanogels have demonstrated that higher temperatures resulted in increased follicular penetration of a released model dye (Jung et al., 2018). In the clinical situation, this might be difficult, as temperature fluctuations during storage and transport of the drugs could lead to a premature release of the drug.

A new approach is the utilization of redox-sensitive nanocarriers which will be able to release their drugs in skin areas with higher concentrations of free radicals such as in inflammatory skin diseases. This would allow selective treatment of only inflammatory skin areas, whereas healthy skin areas will remain unaffected by drugs and side effects.

In conclusion, these new strategies are very promising, but further research work will have to be done before these smart nanocarrier systems will become commercially available for cosmetic, personal care or pharmaceutical products.

REFERENCES

Benedek M, Kaernbach C: Physiological correlates and emotional specificity of human piloerection. *Biol Psychol* 2011;86:320–329.

Blume-Peytavi U, Vogt A: Human hair follicle: Reservoir function and selective targeting. *Br J Dermatol* 2011;165 Suppl 2:13–17.

Buffoli B, Rinaldi F, Labanca M, Sorbellini E, Trink A, Guanziroli E, Rezzani R et al.: The human hair: From anatomy to physiology. *Int J Dermatol* 2014;53:331–341.

Christoph T, Muller-Rover S, Audring H, Tobin DJ, Hermes B, Cotsarelis G, Ruckert R et al.: The human hair follicle immune system: Cellular composition and immune privilege. *Br J Dermatol* 2000;142:862–873.

Cotsarelis G: Epithelial stem cells: A folliculocentric view. *J Invest Dermatol* 2006;126:1459–1468.

Dimde M, Sahle FF, Wycisk V, Steinhilber D, Camacho LC, Licha K, Lademann J, Haag R: Synthesis and validation of functional nanogels as pH-sensors in the hair follicle. *Macromol Biosci* 2017;17:1600505.

Edlich RF, Winters KL, Britt LD, Long WB, 3rd: Bacterial diseases of the skin. *J Long Term Eff Med Implants* 2005;15:499–510.

Freinkel RK, Freinkel N: Hair growth and alopecia in hypothyroidism. *Arch Dermatol* 1972;106:349–352.

Gollnick H, Blume U, Orfanos CE: Adverse drug reactions on hair. *Z Hautkr* 1990;65:1128–1134.

Goyal R, Macri LK, Kaplan HM, Kohn J: Nanoparticles and nanofibers for topical drug delivery. *J Control Release* 2016;240:77–92.

Gupta AK, Summerbell RC: Tinea capitis. *Med Mycol* 2000;38:255–287.

Headington JT: Tumors of the hair follicle. A review. *Am J Pathol* 1976;85:479–514.

Hermes B, Paus R: Scar forming alopecia. Comments on classification, differential diagnosis and pathobiology. *Hautarzt* 1998;49:462–472.

Holmgaard R, Benfeldt E, Nielsen JB: Percutaneous penetration – Methodological considerations. *Basic Clin Pharmacol Toxicol* 2014;115:101–109.

Jaworsky C, Gilliam AC: Immunopathology of the human hair follicle. *Dermatol Clin* 1999;17:561–568.

Jung S, Nagel G, Giulbudagian M, Calderon M, Patzelt A, Knorr F, Lademann J: Temperature-enhanced follicular penetration of thermoresponsive nanogels. *Z Phys Chem* 2018;232:805–817.

Katoulis AC, Christodoulou C, Liakou AI, Kouris A, Korkoliakou P, Kaloudi E, Kanellea A et al.: Rigopoulos D: Quality of life and psychosocial impact of scarring and non-scarring alopecia in women. *J Dtsch Dermatol Ges* 2015;13:137–142.

Krause K, Foitzik K: Biology of the hair follicle: The basics. *Semin Cutan Med Surg* 2006;25:2–10.

Kumamoto T, Shalhevet D, Matsue H, Mummert ME, Ward BR, Jester JV, Takashima A: Hair follicles serve as local reservoirs of skin mast cell precursors. *Blood* 2003;102:1654–1660.

Lademann J, Richter H, Knorr F, Patzelt A, Darvin ME, Ruhl E, Cheung KY et al.: Triggered release of model drug from AuNP-doped BSA nanocarriers in hair follicles using IRA radiation. *Acta Biomater* 2016;30:388–396.

Lademann J, Richter H, Schaefer UF, Blume-Peytavi U, Teichmann A, Otberg N, Sterry W: Hair follicles – A long-term reservoir for drug delivery. *Skin Pharmacol Physiol* 2006;19:232–236.

Lademann J, Richter H, Teichmann A, Otberg N, Blume-Peytavi U, Luengo J, Weiss B et al.: Nanoparticles – An efficient carrier for drug delivery into the hair follicles. *Eur J Pharm Biopharm* 2007;66:159–164.

Lange-Asschenfeldt B, Marenbach D, Lang C, Patzelt A, Ulrich M, Maltusch A, Terhorst D et al.: Distribution of bacteria in the epidermal layers and hair follicles of the human skin. *Skin Pharmacol Physiol* 2011;24:305–311.

Leeming JP, Holland KT, Cunliffe WJ: The microbial ecology of pilosebaceous units isolated from human skin. *J Gen Microbiol* 1984;130:803–807.

Mak WC, Richter H, Patzelt A, Sterry W, Lai KK, Renneberg R, Lademann J: Drug delivery into the skin by degradable particles. *Eur J Pharm Biopharm* 2011;79:23–27.

Mangelsdorf S, Otberg N, Maibach HI, Sinkgraven R, Sterry W, Lademann J: Ethnic variation in vellus hair follicle size and distribution. *Skin Pharmacol Physiol* 2006;19:159–167.

Mathes C, Brandner JM, Laue M, Raesch SS, Hansen S, Failla AV, Vidal S et al.: Tight junctions form a barrier in porcine hair follicles. *Eur J Cell Biol* 2016;95:89–99.

Matsui MS, Pelle E, Dong K, Pernodet N: Biological rhythms in the skin. *Int J Mol Sci* 2016;17:801.

McMichael AJ: Ethnic hair update: Past and present. *J Am Acad Dermatol* 2003;48:S127–133.

Meidan VM, Bonner MC, Michniak BB: Transfollicular drug delivery – Is it a reality? *Int J Pharm* 2005;306:1–14.

Misago N, Satoh T, Narisawa Y: Basal cell carcinoma with tricholemmal (at the lower portion) differentiation within seborrheic keratosis. *J Cutan Pathol* 2003;30:196–201.

Oremovic L, Lugovic L, Vucic M, Buljan M, Ozanic-Bulic S: Cicatricial alopecia as a manifestation of different dermatoses. *Acta Dermatovenerol Croat* 2006;14:246–252.

Otberg N, Patzelt A, Rasulev U, Hagemeister T, Linscheid M, Sinkgraven R, Sterry W et al.: The role of hair follicles in the percutaneous absorption of caffeine. *Br J Clin Pharmacol* 2008;65:488–492.

Otberg N, Richter H, Schaefer H, Blume-Peytavi U, Sterry W, Lademann J: Variations of hair follicle size and distribution in different body sites. *J Invest Dermatol* 2004;122:14–19.

d'Ovidio R, Sgarra C, Conserva A, Angelotti UF, Erriquez R, Foti C: Alterated integrin expression in lichen planopilaris. *Head Face Med* 2007;3:11.

Parisi AV, Smith D, Schouten P, Turnbull DJ: Solar ultraviolet protection provided by human head hair. *Photochem Photobiol* 2009;85:250–254.

Patzelt A, Lademann J: Drug delivery to hair follicles. *Expert Opin Drug Deliv* 2013;10:787–797.

Patzelt A, Mak WC, Jung S, Knorr F, Meinke MC, Richter H, Ruhl E et al.: Do nanoparticles have a future in dermal drug delivery? *J Control Release* 2017;246:174–182.

Patzelt A, Richter H, Buettemeyer R, Huber HJ, Blume-Peytavi U, Sterry W, Lademann J: Differential stripping demonstrates a significant reduction of the hair follicle reservoir in vitro compared to in vivo. *Eur J Pharm Biopharm* 2008;70:234–238.

Patzelt A, Richter H, Knorr F, Schafer U, Lehr CM, Dahne L, Sterry W et al.: Selective follicular targeting by modification of the particle sizes. *J Control Release* 2011;150:45–48.

Paus R, Nickoloff BJ, Ito T: A 'hairy' privilege. *Trends Immunol* 2005;26:32–40.

Piraccini BM, Alessandrini A: Androgenetic alopecia. *Giorn Ital Dermat V* 2014;149:15–24.

Radtke M, Patzelt A, Knorr F, Lademann J, Netz RR: Ratchet effect for nanoparticle transport in hair follicles. *Eur J Pharm Biopharm* 2017;116:125–130.

Rogers GE: Hair follicle differentiation and regulation. *Int J Dev Biol* 2004;48:163–170.

Sperling LC: Hair and systemic disease. *Dermatol Clin* 2001;19:711–726, ix.

Tebbe B: Clinical course and prognosis of cutaneous lupus erythematosus. *Clin Dermatol* 2004;22:121–124.

Teichmann A, Jacobi U, Ossadnik M, Richter H, Koch S, Sterry W, Lademann J: Differential stripping: Determination of the amount of topically applied substances penetrated into the hair follicles. *J Invest Dermatol* 2005;125:264–269.

Teichmann A, Ossadnik M, Richter H, Sterry W, Lademann J: Semiquantitative determination of the penetration of a fluorescent hydrogel formulation into the hair follicle with and without follicular closure by microparticles by means of differential stripping. *Skin Pharmacol Physiol* 2006;19:101–105.

Trauer S, Lademann J, Knorr F, Richter H, Liebsch M, Rozycki C, Balizs G et al.: Development of an in vitro modified skin absorption test for the investigation of the follicular penetration pathway of caffeine. *Skin Pharmacol Physiol* 2010;23:320–327.

Vogt A, Blume-Peytavi U: Biology of the human hair follicle. New knowledge and the clinical significance. *Hautarzt* 2003;54:692–698.

Vogt A, Hadam S, Heiderhoff M, Audring H, Lademann J, Sterry W, Blume-Peytavi U: Morphometry of human terminal and vellus hair follicles. *Exp Dermatol* 2007;16:946–950.

Wasserman D, Guzman-Sanchez DA, Scott K, McMichael A: Alopecia areata. *Int J Dermatol* 2007;46:121–131.

Weinberg JM, Mysliwiec A, Turiansky GW, Redfield R, James WD: Viral folliculitis. Atypical presentations of herpes simplex, herpes zoster, and molluscum contagiosum. *Arch Dermatol* 1997;133:983–986.

7

Common Cosmetic Ingredients: Chemistry, Actions, Safety and Products

Brett MacFarlane

CONTENTS

Skin care is based on a regimen of cleansing, moisturising and barrier protection. Skin care products and cosmetics are composed of ingredients that are necessary to formulate the product and provide its efficacy and safety (e.g. water, oils, preservatives and surfactants), as well as ingredients to improve cosmetic acceptability (e.g. fragrances). Ingredients are listed on the product or packaging and appear in descending order of concentration. Consumers are generally not in a position to evaluate the safety and efficacy of products based on what can potentially be a complicated list of ingredients. Therefore they rely on the advice of professionals who are versed in evaluating ingredients lists.

Suitability of products for sensitive skin depends on the safety profile of all the individual ingredients. In general, skin care products and cosmetics are well tolerated in the wider community, however, they may cause adverse skin reactions in a small number of individuals. Adverse reactions include dermatitis, acneform reactions and allergy. Symptoms of adverse skin reactions include erythema, rash, itch, urticaria, stinging and burning. The most commonly reported adverse reactions are minor and will generally cease once the product is discontinued. In a small number of individuals there is potential for cross-reactivity between many different products (status cosmeticus). Except in the most severe of cases, adverse reactions do not occur with initial use of the product, but slowly develop over time with repeated, cumulative application. Therefore, even skin care products and cosmetics that consumers have been safely using for long periods of time can be the cause of sensitisation.

People who react to skin care products and cosmetics might use words such as 'hypersensitive' to describe themselves. There are a number of misconceptions in the community over the need to use special products on sensitive skin and the term 'hypoallergenic' is commonly used in marketing. Considering that the likelihood of suffering an adverse skin reaction from a skin care product or cosmetic is relatively low, almost all products could be considered to be hypoallergenic. Guidelines for the use of the term on skin care product labels require that the product must be tested on humans and found to cause less adverse skin reactions than reference substances. While there are a myriad of reports of adverse skin reactions to common ingredients, these generally follow application of concentrations far greater than those used in products for human use (Juliász and Marmur, 2014).

However, predicting how an individual might react to a skin care product or cosmetic is usually not possible. They may have reacted to products in the past or have a family history of skin sensitivity. In general, people with conditions involving compromised skin barrier function, for example, eczema, are the most likely to react. A few simple questions about their history may be beneficial in determining risk. Those who commonly suffer from adverse skin reactions may need to be assessed by an allergist to determine the exact ingredient that is the cause. The best treatment for skin sensitivity due to skin care products and cosmetics is complete avoidance of the sensitiser.

Cleansers, Surfactants and Soaps

Skin cleansing is a staple of the skin care regimen. Cleansers wash away dirt and oil, and help to rejuvenate the skin by removing dead skin cells. They contain surfactants (surface active agents), which are a group of chemicals of various structures. Surfactants (also known as emulsifying agents, detergents, stabilisers or soaps) prevent the aqueous and lipid miscible components of the preparation from separating, and provide the cleanser with its foaming action when mixed with water.

The chemical structure of surfactants can be

- Anionic (negatively charged)
- Cationic (positively charged)
- Non-ionic (neutral)
- Zwitterionic (carries both negative and positive charges within the one molecule)

Surfactants can cause water loss from normal and sensitive skin, leading to dryness and damage of the skin barrier. They are amongst the most common cause of contact allergy dermatitis due to skin care products and cosmetics (González-Muñoz et al., 2014). Surfactants may damage the skin by

- Causing disarray of the highly structured nature of skin lipids
- Interacting with cellular components, for example, keratin
- Causing skin inflammation

Early signs of skin incompatibility with surfactants include itching, dryness, erythema and tightness. Long-term effects of irritant surfactants include prolonged erythema, dryness, scaling and cracking (Tupker et al., 1999).

The surfactant can bind to proteins in the skin causing swelling, which leads to water loss and allows other irritants to further penetrate the skin. This causes immediate dryness, redness and itch. With prolonged use, chronic dryness may develop leading to compromised skin barrier function and longer-term adverse skin reactions.

There are a number of cleanser products on the market that claim to be 'soap free'. The term 'soap' generally refers to the surfactant sodium laurate, which is a major cause of skin irritation and is now rarely, if ever, used. Therefore most cleansers on the market today can be considered 'soap free'. The reality is if the cleanser has a foaming action then it contains a surfactant. So-called soap-free cleansers generally contain surfactants that are less irritating; however, sensitivity to these cleansers is still possible.

The best cleanser to choose is therefore one containing the least irritating surfactants. The ultimate cleanser will not only clean the skin, but also soften and protect the skin. Many of the newer moisturising cleansers contain ingredients that deposit on the skin in order to protect it from water loss by boosting the barrier function. Examples of beneficial cleanser ingredients include glycerol for hydration, paraffin for barrier protection and oils for softening. Some ingredients replace skin fatty acids such as cholesterol and stearic acid that are washed away by cleansing. Evidence suggests that moisturising washes are better than bars at depositing these protective ingredients on the skin (Ananthapadmanabhan et al., 2004).

The term 'soap' has become used generally for the anionic surfactants (in particular sodium laurate). These are salts of fatty acids and are alkaline in nature (up to pH 10). Synthetic surfactants (syndets) are less irritating, as they are generally of neutral or slightly acidic pH. The skin has an overall acidic pH, and alkaline skin pH is associated with loss of skin lipid, increased loss of water across the compromised barrier and low levels of hydration (Thune et al., 1988). Even neutral pH solutions have been shown to cause delayed healing of skin barrier function (Mauro et al., 1998). The pH of the cleanser is important when considering the potential of the product to cause skin irritation (Hachem et al., 2003).

Inflammatory skin conditions including eczema, dermatitis and seborrhoea have been associated with increases in skin pH. It has been suggested that increasing the pH of the skin may adversely affect certain enzymes responsible for the production of skin barrier lipids. Therefore, further increases in skin pH of inflamed and atopic skin, from neutral or alkaline pH skin care products and cosmetics, could further irritate already damaged skin (Hachem et al., 2003).

The size of the surfactant molecule is an important factor in skin irritation risk. Smaller molecular size surfactant molecules penetrate the skin to deeper layers that contain actively dividing cells. As a result they appear to cause more significant adverse skin reactions than larger molecular size surfactant molecules, which penetrate less (Bettley, 1963). The smaller molecular size potassium laurate penetrates farther into the skin and causes significantly more adverse skin reactions than the larger potassium palmitate or octanoate. While one compound may have a greater toxicity than another, it may not cause adverse skin reactions as readily if it does not penetrate the skin to a significant extent due to limitations of molecular size. It has also been suggested that skin sensitivity to surfactants may be due to impurities that develop as a result of chemical processing. However, this is very difficult to test for.

It is important to highlight that surfactant skin irritancy testing is commonly carried out with concentrations of the surfactant that are higher than those used in marketed products. Testing involves dissolving the surfactant in solvents that are not commonly used to formulate cleansers. Also cleanser products contain other ingredients that help to reduce the potential for skin irritancy caused by the surfactant components. Therefore surfactant irritancy data may overestimate the potential for development of clinical adverse skin reactions.

The newer glucoside surfactants have very low potential to cause skin irritation (Loffler and Happle, 2003). They are also significantly more expensive than older surfactants and therefore their use may be limited in skin care and cosmetic products due to cost restrictions. Decyl polyglucoside was found to be the least irritating of a group of surfactants tested in humans (Vozmediano et al., 2000). Similarly, lauryl glucoside did not cause significant moisture loss or redness (Baranay et al., 1999). Alkyl sulphonates can cause skin irritation at high concentration and high pH, and therefore these should be controlled when using these surfactants in skin formulations (Becker et al., 2010).

Amino acid derivative surfactants, such as lycinates and glycinates, are also less likely to irritate skin. Results from studies in skin cells grown in a laboratory indicate that they have less toxicity than other surfactants (Sanchez et al., 2006). Safety assessment of amino acid alkyl amides used as surfactants, such as lauroyl lysine and sodium lauroyl glutamate, supported their safety in concentrations used in skin care products and cosmetics when formulated to be non-irritating (Burnett et al., 2017).

Surfactants that have been linked to skin irritation include sodium cocoamphoacetate, sodium cocoyl isethionate, sodium laurate, sodium lauryl sulphate, sodium pareth sulphate (Baranay et al., 1999; Tupker et al., 1999), and cocamidopropyl betaine (Jacob and Amini, 2008).

Bath and Shower Oils

Bath and shower oils are used to aid in the management of dry, sensitive skin by forming a protective oily layer on the skin during or after cleansing. However, evidence indicates that while oils are effective, they may also increase skin deposition of lipid soluble irritants from the formulation (Loden et al., 2004). Also bath oils are often wiped off by towelling after bathing and therefore should be used in conjunction with a moisturiser to be of greatest benefit.

Moisturisers

Moisturisers are used extensively for damaged and dry skin conditions and in anti-aging products. Moisture content of the skin is important for maintenance of the barrier function. When a moisturiser is first applied, tiny cracks between skin cells on the surface become filled and smoothed over. The skin begins to look healthier, and feels softer and lubricated. Moisture content (aqua, water) from the cream is absorbed into the outer layers of the skin, rapidly improving dryness. However, only skin care products that contain humectants and barrier agents can promote the long-term improvement in skin hydration.

Skin barrier function is affected by the degree of hydration (water content). The ability of the skin to limit water loss is dependent on skin lipids and skin cell maturity. The skin also contains a group of substances, which when grouped together act as natural moisturisation factor (NMF). The components of NMF include amino acids and other small ions that are essential for skin hydration. Skin hydration is also dependant on the environment, yearly season and relative humidity.

Healthy skin limits water loss from the body and penetration of irritants from the environment. Damaged and inflamed skin have reduced barrier function and may lose larger than normal amounts of water, worsening skin dryness. In some skin conditions associated with an interrupted barrier, for example, eczema, dermatitis and psoriasis, the body attempts to prevent water loss by thickening the skin. However, the resultant skin cells are immature and the barrier function remains compromised.

The optimal use of moisturiser products involves the following:

1. *Light physical application* – The layer of a cream moisturiser left behind on the skin after rubbing compared to light application can be up to 50% less (Rhodes and Diffey, 1997). Vigorous rubbing and massaging results in suboptimal skin deposition of the moisturiser.

2. *Apply to dry skin* – The moisturiser will adhere better and be less likely to rub off onto clothing.

3. *Multiple daily application* – Moisture levels in the skin rise rapidly and substantially after moisturiser application, however they return to baseline within hours (Chiang and Eichenfield, 2009). Therefore, repeated application (at least twice daily) is required to promote maximal skin moisture levels.

4. *Focus on applying the moisturiser over the entire skin surface* – Thicker ointments are spread more consistently over the skin surface than thinner creams or lotions (Ivens et al., 2001). Creams and lotions are not spread to the edge of the application area as well as ointments, and therefore care must be taken to ensure creams and lotions are spread to the outer limits of the application area.

Barriers

Barrier agents are an important component of the management of dry skin, and for people who regularly handle irritant chemicals or frequently wash their hands in the home or workplace. Repeated exposure to water and irritant chemicals is an important factor in the development of chronic hand dermatitis. It often occurs in individuals constantly exposed to irritants during their occupation including hairdressers, childcare workers, and kitchen and domestic staff. Barrier agents form a protective layer over the surface of the skin to limit water loss and prevent contact with irritant chemicals.

Dimethicone is a mixture of siloxane polymers and can be used in creams at concentrations of up to 15% as a barrier agent. Dimethicone can prevent water loss across the skin and improve barrier thickness (Short et al., 2007). It also reduces melanin intensity, which may be an important part of photoaging, and improves dermatitis (Fowler, 2000). An extensive review of toxicity studies concluded that dimethicone and its variants were safe when used in cosmetic formulations (Nair, 2003).

Many protective skin care products contain paraffin (also called petrolatum or petroleum jelly), which is a purified semi-solid mixture of hydrocarbons from a mineral source. Paraffin exists as yellow and white variants (white soft paraffin being a bleached version of yellow soft paraffin). Thicker ointments contain soft paraffin, while thinner creams and lotions generally contain liquefied paraffin.

Paraffin is naturally devoid of water. It forms a barrier layer on the skin reducing the water that evaporates into the atmosphere. Also, paraffin can take the place of lost lipid in damaged skin (Ghadially et al., 1992). Therefore, it is effective at preventing skin dehydration and restoring barrier function.

Popular belief suggests that paraffin may not be appropriate to use in people with sensitive skin, as it is a by-product of petroleum distillation. There are actually a few different methods to manufacture paraffin. The natural form is indeed derived from petroleum distillation and may contain contaminants including polycyclic aromatic hydrocarbons. However, this form of paraffin is highly refined to remove these contaminants prior to incorporation in skin care formulations. On the other hand, synthetic paraffin can be produced free of petroleum contaminants and does not represent a risk to sensitive skin.

Popular belief also suggests that oils derived from minerals may cause adverse skin reactions, due to their chemical nature, as opposed to vegetable oils that are thought to be milder. There are certainly case reports of allergic contact dermatitis resulting from paraffin use (Tam and Elston, 2006; Kang et al., 2004; Kundu et al., 2004). However, the small number of adverse skin reactions reported over a substantial usage history indicates that paraffin is actually well tolerated. A study comparing a range of different paraffins found a number of possible causes of sensitisation due to differing origins, colour and purification procedures used (Dooms-Goossens and Degreef, 1983a). Patients with sensitive skin may benefit from synthetically produced paraffin, which is lower in chemical contaminants than 'natural' paraffin (Dooms-Goossens and Degreef, 1983b).

Dermal penetration studies of mineral oils and waxes indicate that they do not penetrate the skin past the level of the stratum corneum, and therefore cannot reach the circulation to cause adverse health outcomes (Petry et al., 2017).

Zinc oxide is a solid that is incorporated as a fine particle powder into ointment bases and used as a protective agent on damaged skin. Zinc oxide ointment provides the skin with a sound barrier against water loss. It is often used as a nappy rash treatment or in people with incontinence. A small study in humans found that zinc oxide could protect the skin against sodium lauryl sulphate irritation. However, it was not effective at improving skin hydration or preventing skin from becoming macerated due to constant exposure to irritants (Hoggarth et al., 2005).

Lanolin is produced from the wool of sheep and is an effective anhydrous barrier agent. It has long been thought to be a contact allergen with the primary allergen component thought to be wool alcohols (also called lanolin alcohol, wool wax and wool wax alcohol), however, this remains to be fully determined. The majority of reports of contact allergy to lanolin are in people with damaged skin barrier function, for example, eczema. However, a large study indicated that, even in people with eczema, the prevalence of lanolin allergy is as low as 1.7% (Wakelin et al., 2001).

Body Butters

Body butter is a thick, smooth, fragrant body product that provides the skin with softness and probably protection from water loss. After body butter use the skin feels smooth and supple. However, this is generally transitory. Body butter is composed of thick fatty acids found in natural sources such as shea butter or cocoa butter.

Cocoa butter has been added to soap to reduce its irritancy. However, this was unsuccessful as very little of the cocoa butter was deposited on the skin (Strube and Nicoll, 1987). It has been shown to cause an acneform reaction in an animal skin model (Nguyen et al., 2007).

Shea butter is produced from the fatty component of the seed of the shea tree (*Vitellaria paradoxa* Gaertner). Shea butter contains a number of different components depending on the origin of the tree it is extracted from, including

- Triglycerides
- Saturated and unsaturated fatty acids, for example, stearic acid and oleic acid

- Polycyclic triterpenes
- Alpha, beta, gamma and delta forms of tocopherol (vitamin E); shea butter produced from shea trees in hotter climates produce the greatest amount of vitamin E (Maranz and Wiesman, 2004)
- Polyphenols such as gallic acid, cathchin and quercetin (Maranz et al., 2003)

An experiment in humans found that shea butter did not improve skin irritation caused by sodium lauryl sulphate. Therefore, the components of shea butter probably do not improve the quality of damaged skin (Loden and Andersson, 1996).

Water (Aqua)

Water is an essential component of the majority of skin care products and cosmetics. In the case of moisturisers, it is generally the major component of the formulation. In healthy skin, plain water has been shown to be as irritating as some commonly used surfactants (Tupker et al., 1997). However, water is a necessary component of creams and lotions as it gives them their characteristic consistency and is vital for the ability of the formulation to improve skin moisture levels.

Denatured Alcohol

Alcohol (ethanol) is commonly used in skin care products and cosmetics as a solvent to improve solubility of some poorly soluble active ingredients and as a penetration enhancer. It has the added benefit of resisting microbial contamination. When alcohol is applied to the skin it can cause dryness and irritation, and increase the absorption of other contaminants. Alcohol is denatured with various compounds (e.g. methanol) to make the product unpalatable and discourage consumption. While ethanol itself is not likely to cause significant adverse reactions, the denaturants can be toxic, particularly when consumed orally (Cosmetic Ingredient Review Expert Panel, 2008).

Humectants

Humectants improve skin surface hydration by attracting water from the lower layers of skin and from skin care products, and holding it within the stratum corneum. Humectants bind water via the formation of hydrogen bonds between themselves and water. The water then remains bound within the humectant and is available to hydrate the stratum corneum over a longer period.

The most commonly used humectant in dry skin preparations is glycerol (10–20%). At low relative humidity, each gram of glycerol can bind approximately 0.125 grams of water (Cohen et al., 1993). This makes glycerol an ideal agent to attract water into the dry stratum corneum. This is particularly true during winter when the humidity is low and skin becomes dryer due to evaporation.

Glycerol has been shown to (Fluhr, 1999; Cohen et al., 1993; Andersen et al., 2006a; Andersen et al., 2006b)

- Produce long-lasting skin moisturisation
- Limit skin surface water evaporation
- Help maintain skin barrier function
- Prevent chemical skin irritation
- Aid in desquamation

As glycerol attracts and binds water from skin care products, it is only effective in hydrous products such as creams and lotions, and is not an effective humectant in anhydrous ointments and pastes.

Pyrrolidone carboxylic acid (PCA) is the prominent component of the skin's NMF and has been included in skin moisturisers as a humectant. Dry skin has been found to contain lower levels of PCA and other amino acids present in NMF (Feng et al., 2014).

Urea (5–10%) increases the level of skin hydration and prevents water loss. It can increase the suppleness of skin, making it feel less irritated and uncomfortable. It can also improve the appearance of cracked and scaly, dry skin. Urea (10%) can improve the level of hydration of the skin as well as limit the absorption of irritant substances from the environment (Loden, 1996). However, urea can cause significant inflammation and moisture loss, even from healthy skin, as the concentration increases to 20% (Agner, 1992).

Oils, Fatty Acids and Polymers

Oils (from mineral or vegetable sources) are major components of most skin care products and cosmetics. When mixed with water in the presence of an emulsifier they give the product its creamy consistency. The oil can absorb into or sit on the skin to improve elasticity, softness and suppleness; prevent water loss; and improve skin barrier function.

A study in humans with dry skin conditions found that virgin coconut oil improved skin dryness and scaliness (Agero and Verallo-Rowell, 2004). Assessment of the skin of participants found a significant increase in the level of hydration. The effectiveness was similar to that following topical mineral oil use. However, unlike mineral oil, some of the fatty components of coconut oil are actually found naturally in human skin. Importantly, no adverse skin reactions were witnessed in the study subjects. This indicates that virgin coconut oil could be used as an alternative to mineral oil for people seeking a plant-based option.

However, a toxicology study in mice found that acids present in coconut oil (coconut diethanolamide [CDEA]) could cause skin irritation when applied over long periods of time (National Toxicology Program, 2001). The acids caused thickening of skin cells and sebaceous glands as well as ulcers. CDEA is found in a number of skin care products and cosmetics. It is generally well tolerated in humans with only a small number of reported adverse skin reactions (Pinola et al., 1993). However, the irritation witnessed in mice indicates that long-term use of coconut oil may adversely affect the skin.

Evening primrose oil (EPO) is a source of gamma-linolenic and linoleic acids. The skin cannot synthesize gamma-linolenic acid and therefore this important fatty acid must be taken in the diet. An emulsion containing EPO added to the bath water has been shown to improve skin irritated by sodium lauryl sulphate (De Paepe et al., 2002). EPO was also found to improve dermatitis symptoms when applied as a water-in-oil emulsion (Gehring et al., 1999).

Peanut oil (arachis oil) is the refined fixed oil extracted from the kernels of *Arachis hypogaea*. Peanut oil is used in skin care products and cosmetics to soften and protect the skin. When ingested orally peanut proteins are a major cause of serious allergy in humans, including hives. Refined peanut oil is very low in the proteins that cause allergy. Case reports of allergy to topical peanut oil are rare. There is no reliable data to indicate that skin care products and cosmetics containing peanut oil cause allergy in sensitive individuals when used topically (Ring and Möhrenschlager, 2007).

Soy is mainly composed of phospholipids and essential fatty acids (Hausen et al., 1999). It also contains isoflavones (including phyto-oestrogens), saponins, essential amino acids, phytosterols and minerals. Topical soy oil has shown promise is lightening skin hyperpigmentation (Konda et al., 2012) and in the treatment of UV-induced skin aging and cancer (Bosch et al., 2015).

Sunflower oil has been shown to significantly reduce infant mortality due to poor skin barrier function (Darmstadt, 2008). It was suggested that the oil improved the barrier function of the skin thus reducing infection risk. However, a sunflower oil–based moisturiser did not significantly improve irritated skin in another study (Loden and Andersson, 1996).

Triglycerides can be divided into short-, medium- and long-chain examples. Medium-chain triglycerides contain between 5 and 20 carbon atoms. They are included in some skin care products and cosmetics to provide a waxy smooth character to the product. The medium chain triglycerides include palmitic acid (16 carbons), myristic acid (14 carbons), lauric acid (12 carbons), capric acid (10 carbons), caprylic acid (8 carbons) and caproic acid (6 carbons). In general, in concentrations regularly used in skin care products and cosmetic, the medium-chain triglycerides are well tolerated (Traul et al., 2000).

Polyethylene glycol (PEG) is a polymer that is used very regularly in skin care products and cosmetics. There is a large range of different-sized PEG molecules with different chemical activities. The molecules are listed as PEG followed by the number of subunits of the polymer they contain (e.g. PEG 32). PEG provides viscosity to skin care products and cosmetics. It therefore affects the way the product feels on the skin during regular use. In general, PEG is very well tolerated on the skin. It can be absorbed across healthy skin, and there are few reports of sensitivity (Fruijtier-Polloth, 2005). Alkyl PEG ethers are used as emulsion stabilisers and skin conditioning agents, and are well tolerated (Fiume, 2012).

Propylene glycol (PG) is widely used in skin care products and cosmetics in concentrations of 25–50% and in some case over 50%. In general PG is used as a solvent to dissolve other ingredients during manufacturing. While many reports of adverse skin reactions to products containing PG exist, subsequent testing has often failed to pinpoint PG as the cause. The number of adverse skin reactions to products with low PG concentration is actually very small. A study that compared a very large number of human studies reporting PG sensitivity reactions has found that less than 1% of people had moderate to serious cases of contact allergy, while up to 2% had weaker allergic type reactions. The vast majority of adverse reactions reported were simply skin irritancy. Many of the patients who reported skin irritancy were also sensitive to a range of other ingredients (Ananthapadmanabhan et al., 2004).

Preservatives

Chemical preservatives help to prevent the growth of microorganisms that can spoil the product and cause skin or eye infections. Microorganisms can be introduced to the product during the manufacturing process or via everyday use. Chemical preservatives should prevent the growth of microorganisms throughout the product shelf life without causing adverse skin reactions.

A small number of individuals are sensitive to preservatives. However, evaluation of a large skin irritancy dataset involving cumulative testing of preservatives commonly used in skin care products, at in-use concentrations, found preservatives do not appear to contribute to skin irritation in people with healthy skin (Walters et al., 2015).

Preservative-free products can be produced, however, they are expensive to manufacture. It is important that people who are sensitive to preservatives are knowledgeable about the potential for adverse reactions. Unfortunately, labelling of skin care products and cosmetics has been found to be incorrect or misleading with respect to the preservative content in up to 45% of cases (Rastogi, 2000).

Preservatives can be divided into formaldehyde (and formaldehyde releasing) and non-formaldehyde types.

Formaldehyde

Of the preservatives used in skin care products and cosmetics, formaldehyde has the greatest potential to cause skin allergy (Jong et al., 2007). Formaldehyde has been used to preserve skin care products, wart removers, nail products and in some cotton- and rayon-containing textiles and bedding. There is a concerning increase in the number of people sensitive to formaldehyde preservative, with one study finding a 400–500% increase in formaldehyde sensitivity in people in Israel (Zachariae et al., 2006).

Formaldehyde-Releasing Preservatives

Formaldehyde-releasing preservatives contain formaldehyde as a detachable part of their chemical structure. During normal use and shelf life, formaldehyde is slowly released. Formaldehyde-releasing preservatives are useful in products that prove difficult to preserve by other means. If individuals are sensitive to formaldehyde, it is suggested that they also avoid formaldehyde-releasing preservatives, as there is significant cross-sensitivity (Jong et al., 2007).

Bronopol (2-bromo-2-nitropropane-1,3-diol) appears to have a very low frequency of dermal irritation (Frosch et al., 1990). People who are sensitive to formaldehyde are also likely to be sensitive to bronopol. Research indicates that bronopol may cause toxicity to cells following UV irradiation indicating that it may be problematic if products containing bronopol are applied to the skin prior to sun exposure; however, this still remains to be determined (Placzek et al., 2005).

Cresols (p-chloro-m-cresol [PCMC or chlorocresol], sodium p-chloro-m-cresol, chlorothymol, mixed cresols, m-cresol, o-cresol, p-cresol, isopropyl cresol, carvacrol, thymol and o-Cymen-5-ol) can be absorbed across the skin when applied in high concentrations, as well as promote the absorption of other compounds. At high concentrations, the cresols can cause significant dermal irritation. At lower concentrations that are generally used in skin care products and cosmetics, significant dermal irritation has been recorded for a number of the cresols. However, the agents PCMC, thymol and o-Cymen-5-ol generally did not cause dermal irritation at low concentrations (Andersen, 2006).

Diazolidinyl urea frequency of sensitivity appears to be relatively low and stable (Zoller et al., 2006). However, reactions have included raised red rashes of the neck and arms, and widespread redness on the face (Zachariae et al., 2006). It has been suggested that a number of reactions to diazolidinyl urea are due to the release of formaldehyde as opposed to a direct effect of the preservative itself (Hectorne and Fransway, 1994).

DMDM hydantoin (dimethyloln dimethyl hydantoin) causes sensitivity in approximately one-third of people who are also sensitive to formaldehyde (de Groot et al., 1988).

Imidazolidinyl urea is reported to cause a significant number of adverse skin reactions, and the frequency appears to be increasing. This may be due to the increase in human exposure to cosmetics and skin care products around the world (Jong et al., 2007).

Quaternium-15 (Q15) causes a marked frequency of positive skin reactions (Jong et al., 2007). Individuals sensitive to formaldehyde and all other formaldehyde-releasing preservatives are also likely to be sensitive to Q15.

Non-Formaldehyde Preservatives

Benzalkonium chloride is a very widely used preservative in skin care products and cosmetics such as face washes, hand scrubs, aftershaves, cosmetics, detergents, deodorants and toothpastes. It has been implicated in the development of allergic contact dermatitis of the eyelid in a large study (Amin and Belsito, 2006).

Chloroxylenol is reported to have the lowest potential to cause allergy out of a group of 10 preservatives studied. The incidence of allergy with chloroxylenol was 10 times lower than for formaldehyde (Jong et al., 2007).

MCI/MI is a combination of 5-chloro-2-methyl-4-isothiazolin-3-one (methylchloroisothiazolinone) and 2-methyl-4-isothiazolin-3-one (methylisothiazolinone). It has been shown to be a significant cause of contact allergy in people with sensitive skin with a reported incidence as high as 3% (Menné et al., 1991).

Methyldibromo glutaronitrile (MDBGN) caused a significant increase in the frequency of adverse skin reactions in Europe in the 1990s (Wilkinson et al., 2002). The frequency of contact dermatitis has been reported to be as high as 4% (de Groot et al., 1996). The majority of these cases were considered to have significant severity. The main source of allergy to MDBGN was from cosmetics and moistened toilettes.

Parabens (methyl-, ethyl-, propyl-, butyl-, isobutyl- and isopropyl- esters of p-hydroxybenzoic acid) are the most widely used preservatives in skin care products and cosmetics. There is usually a combination of multiple different parabens in the one formulation. Historically, parabens have developed a poor reputation for causing severe skin allergies. Evidence suggests that the incidence of these reactions is actually very small; however, they may have a high incidence of causing skin irritation such as redness and itching (Jong et al., 2007). Irritation is less likely to occur in intact healthy skin and is generally only a problem is disorders of poor skin barrier function (Soni et al., 2005).

Fragrance

Of the commonly used ingredients in skin care products and cosmetics, fragrances cause the most frequent adverse skin reactions (González-Muñoz et al., 2014). Fragrances can cause contact allergy on the site of application and even on distant sites due to touch spread by the user (translocation) and evaporation. The prevalence of sensitivity to fragrances has been estimated to be a little less that 1 in 10 in people in their 20s to a little more than 1 in 10 in those in their 60s (Buckley et al., 2003). The incidence of the reporting of fragrance sensitivity was on the rise in some countries (Scheinman, 2002).

Fragrances can either be of man made or botanical origin. There are more than 3000 man-made fragrances and hundreds of fragrant botanicals used in skin care products and cosmetics. A study in the United Kingdom found 300 products available on the shelves stating they contained 'perfume'. Of the products tested, the top sensitising fragrances were found in the majority of these. Only 34 of the tested products did not contain the fragrances considered to be the most irritating (Buckley, 2007).

Adverse skin reactions to fragrance components are well documented. Workers in a fragrance factory have presented with reactions to fragrance components that range from simple skin irritation to more severe skin allergy. The main agents involved included extracts of lemon and orange, geraniol and cinnamic aldehyde (Schubert, 2006).

Fragrance ingredients include

- Alpha-isomethyl ionone
- Balsam of Peru
- Benzaldehyde
- Bisabolol
- Butyl phenyl methyl propional (p-tert-butyl alpha-methylhydrocinnamic aldehyde)
- Cinnamic acid derivatives
- Citral
- Citronellol
- Coumarin
- Evernia prunastri
- Farnesol
- Geraniol
- Hydroxycintronellol
- Hydroxy-isohexyl-3-cyclohexane carboxaldehyde (HICC)
- 4-hydroxy-4-methylpentyl cyclohexane-1-carboxaldehyde (HMPCC)
- Isoeugenol derivatives
- Jasmine absolute
- Limonene
- Linalool (and linalyl acetate)
- Narcissus absolute
- Sandalwood oil
- Ylang ylang oil

The fragrance component of a 'fragrance-free' product ingredient list may be hidden, as it may be botanical in nature. It is very important to understand that botanical agents such as rose oil, sandalwood oil, cedarwood oil, citrus oils, lavender oil, and other volatile oils and plant extracts are considered to be fragrances. Therefore, fragrance-sensitive individuals must also be careful with some 'fragrance-free' skin care products and cosmetics, as they may actually contain fragrances that are listed as botanical ingredients.

It can be difficult to determine an appropriate product to use by a fragrance-sensitive individual, as the ingredient list may simply read 'fragrance' or 'parfum' or 'other ingredients'. Some fragrant compounds present in the formulation may have other functions (e.g. preservatives). These compounds include benzyl alcohol, benzaldehyde, ethylene brassylate, phenoxyethanol and cyclopentadecanolide.

People who have known sensitivities to fragrance components may be able to use fragrance-containing products without problem. The adverse skin reactions are dependant on the formulation, concentration of the fragrance, the site and duration of application, and the health of the skin at the time of application. The most reliable method to determine if an individual is sensitive to fragrance ingredients is to undergo formal patch testing. Simple in-use testing involving application of the product to an inconspicuous site on the skin (e.g. inside of the arm) twice daily for one to two weeks can also give an indication of the potential for reaction.

As the ingredients list available to the consumer can be limited with respect to fragrance components, the most effective means to prevent adverse skin reactions to fragrances is to avoid fragrant products all together. That is, if the product has a fragrant odour, it should be avoided. There are some simple things consumers can do in order to minimise their risk of developing adverse skin reactions to fragrances. Avoid use of fragrance-containing skin products and cosmetics on

1. Damaged areas of skin including shaved areas, sunburn, windburn, dry skin, nappy rash and other rashes
2. Thinner areas of the skin that are more likely to absorb the fragrance, for example, eyelids and genitalia
3. Intertriginous areas (between skin folds), under the arms and between the buttocks
4. Children or the elderly
5. Skin with compromised barrier function, for example, eczema, dermatitis and psoriasis

Summary

Skin care products and cosmetics are composed of a number of different ingredients that are necessary to allow for the safe and effective formulation, usage and storage of the product. Some of these ingredients have been implicated in the development of adverse skin reactions. However in the main, if used as directed on healthy skin, they are very well tolerated. Their adverse effect profiles are well researched, generally minor and acceptable to most users. Some people with sensitive and damaged skin are more likely to react to skin care ingredients, particularly some fragrances, surfactants and preservatives. Sensitive individuals can be tested to determine which ingredients to avoid.

REFERENCES

Agero AL, Verallo-Rowell VM. 2004. A randomized double-blind controlled trial comparing extra virgin coconut oil with mineral oil as a moisturizer for mild to moderate xerosis. *Dermatitis* 15(3):109–16.

Agner T. 1992. An experimental study of irritant effects of urea in different vehicles. *Acta. Derm. Venereol. Suppl. (Stockh)* 177:44–46.

Amin KA, Belsito DV. 2006. The aetiology of eyelid dermatitis: A 10-year retrospective analysis. *Contact Dermatitis* 55(5):280–5.

Ananthapadmanabhan KP, Moore DJ, Subramanyan K, Misra M, Meyer F. 2004. Cleansing without compromise: The impact of cleansers on the skin barrier and the technology of mild cleansing. *Dermatol. Ther.* 17(Suppl 1):16–25.

Andersen A. 2006. Final report on the safety assessment of sodium p-chloro-m-cresol, p-chloro-m-cresol, chlorothymol, mixed cresols, m-cresol, o-cresol, p-cresol, isopropyl cresols, thymol, o-cymen-5-ol, and carvacrol. *Int. J. Toxicol.* 25(Suppl 1):29–127.

Andersen F, Hedegaard K, Petersen TK, Bindslev-Jensen C, Fullerton A, Andersen KE. 2006a. Anti-irritants I: Dose-response in acute irritation. *Contact Dermatitis* 55(3):148–54.

Andersen F, Hedegaard K, Fullerton A, Petersen TK, Bindslev-Jensen C, Andersen KE. 2006b. The hairless guinea-pig as a model for treatment of acute irritation in humans. *Skin Res. Technol.* 12(3):183–9.

Baranay E, Lindberg M, Loden M. 1999. Biophysical characterization of skin damage and recovery after exposure to different surfactants. *Contact Dermatitis* 40(2):98–103.

Becker LC, Bergfeld WF, Belsito DV, Hill RA, Klaassen CD, Liebler DC, et al. 2010. Amended safety assessment of dodecylbenzenesulfonate, decylbenzenesulfonate, and tridecylbenzenesulfonate salts as used in cosmetics. *Int. J. Toxicol.* 29(6 Suppl):288S–305S.

Bettley FR. 1963. The irritant effect of soap in relation to epidermal permeability. *Br. J. Dermatol.* 75:113–6.

Bosch R, Philips N, Suárez-Pérez JA, Juarranz A, Devmurari A, Chalensouk-Khaosaat J, et al. 2015. Mechanisms of photoaging and cutaneous photocarcinogenesis, and photoprotective strategies with phytochemicals. *Antioxidants (Basel)* 4(2):248–68.

Buckley DA, Rycroft RJ, White IR, McFadden JP. 2003. The frequency of fragrance allergy in patch-tested patients increases with their age. *Br. J. Dermatol.* 149(5):986–9.

Buckley DA. 2007. Fragrance ingredient labelling in products on sale in the U.K. *Br. J. Dermatol.* 57(2):295–300.

Burnett CL, Heldreth B, Bergfeld WF, Belsito DV, Hill RA, Klaassen CD, et al. 2017. Safety assessment of amino acid alkyl amides as used in cosmetics. *Int. J. Toxicol.* 36(1 Suppl):17S–56S.

Chiang C, Eichenfield LF. 2009. Quantitative assessment of combination bathing and moisturizing regimens on skin hydration in atopic dermatitis. *Pediatr. Dermatol.* 26(3):273–8.

Cohen S, Marcus Y, Migron Y, Dikstein S, Shafran A. 1993. Water sorption, binding and solubility of polyols. *J. Chem. Soc. Faraday Trans.* 89(17):3271–5.

Cosmetic Ingredient Review Expert Panel. 2008. Final report of the safety assessment of Alcohol Denat., including SD Alcohol 3-A, SD Alcohol 30, SD Alcohol 39, SD Alcohol 39-B, SD Alcohol 39-C, SD Alcohol 40, SD Alcohol 40-B, and SD Alcohol 40-C, and the denaturants, Quassin, Brucine Sulfate/Brucine, and Denatonium Benzoate. *Int. J. Toxicol.* 27(Suppl 1):1–43.

Darmstadt GL, Saha SK, Ahmed AS, Ahmed S, Chowdhury MA, Law PA, et al. 2008. Effect of skin barrier therapy on neonatal mortality rates in preterm infants in Bangladesh: A randomized, controlled, clinical trial. *Pediatrics* 121(3):522–9.

de Groot AC, Bruynzeel DP, Jagtman BA, Weyland JW. 1988. Contact allergy to diazolidinyl urea (Germall II). *Contact Dermatitis* 18(4):202–5.

de Groot AC, de Cock PA, Coenraads PJ, van Ginkel CJ, Jagtman BA, van Joost T, et al. 1996. Methyldibromoglutaronitrile is an important contact allergen in the Netherlands. *Contact Dermatitis* 34(2):118–20.

De Paepe K, Hachem JP, Vanpee E, Roseeuw D, Rogiers V. 2002. Effect of rice starch as a bath additive on the barrier function of healthy but SLS-damaged skin and skin of atopic patients. *Acta Derm. Venereol.* 82(3):184–6.

Dooms-Goossens A, Degreef H. 1983a. Contact allergy to petrolatums. (II). Attempts to identify the nature of the allergens. *Contact Dermatitis* 9(4):247–56.

Dooms-Goossens A, Degreef H. 1983b. Contact allergy to petrolatums. (I). Sensitizing capacity of different brands of yellow and white petrolatums. *Contact Dermatitis* 9(3):175–85.

Feng L, Chandar P, Lu N, Vincent C, Bajor J, McGuiness H. 2014. Characteristic differences in barrier and hygroscopic properties between normal and cosmetic dry skin. II. Depth profile of natural moisturizing factor and cohesivity. *Int. J. Cosmet. Sci.* 36(3):231–8.

Fiume MM, Heldreth B, Bergfeld WF, Belsito DV, Hill RA, Klaassen CD, et al. 2012. Safety assessment of alkyl PEG ethers as used in cosmetics. *Int. J. Toxicol.* 31(5 Suppl):169S–244S.

Fluhr JW, Gloor M, Lehmann L, Lazzerini S, Distante F, Berardesca E. 1999. Glycerol accelerates recovery of barrier function in vivo. *Acta Derm. Venereol.* 79(6):418–21.

Fowler JF Jr. 2000. Efficacy of a skin-protective foam in the treatment of chronic hand dermatitis. *Am. J. Contact Dermatitis* 11(3):165–9.

Frosch PJ, White IR, Rycroft RJ, Lahti A, Burrows D, Camarasa JG, et al. 1990. Contact allergy to Bronopol. *Contact Dermatitis* 22(1):24–26.

Fruijtier-Polloth C. 2005. Safety assessment on polyethylene glycols (PEGs) and their derivatives as used in cosmetic products. *Toxicology* 214(1–2):1–38.

Gehring W, Bopp R, Rippke F, Gloor M. 1999. Effect of topically applied evening primrose oil on epidermal barrier function in atopic dermatitis as a function of vehicle. *Arzneimittelforschung* 49(7):635–42.

Ghadially R, Halkier-Sorensen L, Elias PM. 1992. Effects of petrolatum on stratum corneum structure and function. *J. Am. Acad. Dermatol.* 26(3 Pt 2):387–96.

González-Muñoz P, Conde-Salazar L, Vañó-Galván S. 2014. Allergic contact dermatitis caused by cosmetic products. *Actas Dermosifiliogr.* 105(9):822–32.

Hachem JP, Crumrine D, Fluhr J, Brown BE, Feingold KR, Elias PM. 2003. pH directly regulates epidermal permeability barrier homeostasis, and stratum corneum integrity/cohesion. *J. Invest. Dermatol.* 121(2):345–53.

Hausen BM, Reichling J, Harkenthal M. 1999. Degradation products of monoterpenes are the sensitizing agents in tea tree oil. *Am. J. Contact Dermatitis* 10(2):68–77.

Hectorne KJ, Fransway AF. 1994. Diazolidinyl urea: Incidence of sensitivity, patterns of cross-reactivity and clinical relevance. *Contact Dermatitis* 30(1):16–19.

Hoggarth A, Waring M, Alexander J, Greenwood A, Callaghan T. 2005. A controlled, three-part trial to investigate the barrier function and skin hydration properties of six skin protectants. *Ostomy Wound Manage.* 51(12):30–42.

Ivens UI, Steinkjer B, Serup J, Tetens V. 2001. Ointment is evenly spread on the skin, in contrast to creams and solutions. *Br. J. Dermatol.* 145(2):264–7.

Jacob SE, Amini S. 2008. Cocamidopropyl betaine. *Dermatitis* 19(3):157–60.

Jong CT, Statham BN, Green CM, King CM, Gawkrodger DJ, Sansom JE, et al. 2007. Contact sensitivity to preservatives in the UK, 2004–2005: Results of multicentre study. *Contact Dermatitis* 57(3):165–8.

Juhász ML, Marmur ES. 2014. A review of selected chemical additives in cosmetic products. *Dermatol. Ther.* 27(6):317–22.

Kang H, Choi J, Lee AY. 2004. Allergic contact dermatitis to white petrolatum. *J. Dermatol.* 31(5):428–30.

Konda S, Geria AN, Halder RM. 2012. New horizons in treating disorders of hyperpigmentation in skin of color. *Semin. Cutan. Med. Surg.* 31(2):133–9.

Kundu RV, Scheman AJ, Gutmanovich A, Hernandez C. 2004. Contact dermatitis to white petrolatum. *Skinmed* 3(5):295–6.

Loden M. 1996. Urea-containing moisturizers influence barrier properties of normal skin. *Arch. Dermatol. Res.* 288(2):103–7.

Loden M, Andersson AC. 1996. Effect of topically applied lipids on surfactant-irritated skin. *Br. J. Dermatol.* 134(2):215–20.

Lodén M, Buraczewska I, Edlund F. 2004. Irritation potential of bath and shower oils before and after use: A double-blind randomized study. *Br. J. Dermatol.* 150(6):1142–7.

Loffler H, Happle R. 2003. Profile of irritant patch testing with detergents: Sodium lauryl sulfate, sodium laureth sulfate and alkyl polyglucoside. *Contact Dermatitis* 48(1):26–32.

Maranz S, Wiesman Z. 2004. Influence of climate on the tocopherol content of shea butter. *J. Agric. Food Chem.* 52(10):2934–7.

Maranz S, Wiesman Z, Garti N. 2003. Phenolic constituents of shea (Vitellaria paradoxa) kernels. *J. Agric. Food Chem.* 51(21):6268–73.

Mauro T, Holleran WM, Grayson S, Gao WN, Man MQ, Kriehuber E, et al. 1998. Barrier recovery is impeded at neutral pH, independent of ionic effects: Implications for extracellular lipid processing. *Arch. Dermatol. Res.* 290(4):215–22.

Menné T, Frosch PJ, Veien NK, Hannuksela M, Björkner B, Lachapelle JM, et al. 1991. Contact sensitization to 5-chloro-2-methyl-4-isothiazolin-3-one and 2-methyl-4-isothiazolin-3-one (MCI/MI). A European multicentre study. *Contact Dermatitis* 24(5):334–41.

Nair B, Cosmetic Ingredients Review Expert Panel. 2003. Final report on the safety assessment of stearoxy dimethicone, dimethicone, methicone, amino bispropyl dimethicone, aminopropyl dimethicone, amodimethicone, amodimethicone hydroxystearate, behenoxy dimethicone, C24-28 alkyl methicone, C30-45 alkyl methicone, C30-45 alkyl dimethicone, cetearyl methicone, cetyl dimethicone, dimethoxysilyl ethylenediaminopropyl dimethicone, hexyl methicone, hydroxypropyldimethicone, stearamidopropyl dimethicone, stearyl dimethicone, stearyl methicone, and vinyldimethicone. *Int. J. Toxicol.* 22(Suppl 2):11–35.

National Toxicology Program. 2001. Toxicology and carcinogenesis studies of coconut oil acid diethanolamine condensate (CAS No. 68603-42-9) in F344/N rats and B6C3F1 mice (dermal studies). *Natl. Toxicol. Program Tech. Rep. Ser.* 479:5–226.

Nguyen SH, Dang TP, Maibach HI. 2007. Comedogenicity in rabbit: Some cosmetic ingredients/vehicles. *Cutan. Ocul. Toxicol.* 26(4):287–92.

Petry T, Bury D, Fautz R, Hauser M, Huber B, Markowetz A, et al. 2017. Review of data on the dermal penetration of mineral oils and waxes used in cosmetic applications. *Toxicol. Lett.* 280:70–78.

Pinola A, Estlander T, Jolanki R, Tarvainen K, Kanerva L. 1993. Occupational allergic contact dermatitis due to coconut diethanolamide (cocamide DEA). *Contact Dermatitis* 29(5):262–5.

Placzek M, Krosta I, Gaube S, Eberlein-König B, Przybilla B. 2005. Evaluation of phototoxic properties of antimicrobials used in topical preparations by a photohaemolysis test. *Acta Derm. Venereol.* 85(1):13–16.

Rastogi SC. 2000. Analytical control of preservative labelling on skin creams. *Contact Dermatitis* 43(6):339–43.

Rhodes LE, Diffey BL. 1997. Fluorescence spectroscopy: A rapid, noninvasive method for measurement of skin surface thickness of topical agents. *Br. J. Dermatol.* 136(1):12–17.

Ring J, Möhrenschlager M. 2007. Allergy to peanut oil – Clinically relevant? *J. Eur. Acad. Dermatol. Venereol.* 21(4):452–5.

Sanchez L, Mitjans M, Infante MR, Vinardell MP. 2006. Potential irritation of lysine derivative surfactants by hemolysis and HaCaT cell viability. *Toxicol. Lett.* 161(1):53–60.

Scheinman PL. 2002. Prevalence of fragrance allergy. *Dermatology* 205(1):98–102.

Schubert HJ. 2006. Skin diseases in workers at a perfume factory. *Contact Dermatitis* 55(2):81–83.

Short RW, Chan JL, Choi JM, Egbert BM, Rehmus WE, Kimball AB. 2007. Effects of moisturization on epidermal homeostasis and differentiation. *Clin. Exp. Dermatol.* 32(1):88–90.

Soni MG, Carabin IG, Burdock GA. 2005. Safety assessment of esters of p-hydroxybenzoic acid (parabens). *Food Chem. Toxicol.* 43(7):985–1015.

Strube DD, Nicoll G. 1987. The irritancy of soaps and syndets. *Cutis* 39(6):544–5.

Tam CC, Elston DM. 2006. Allergic contact dermatitis caused by white petrolatum on damaged skin. *Dermatitis* 17(4):201–3.

Thune P, Nilsen T, Hanstad IK, Gustavsen T, Lövig Dahl H. 1988. The water barrier function of the skin in relation to the water content of stratum corneum, pH and skin lipids. The effect of alkaline soap and syndet on dry skin in elderly, non-atopic patients. *Acta Derm. Venereol.* 68(4):277–83.

Traul KA, Driedger A, Ingle DL, Nakhasi D. 2000. Review of the toxicologic properties of medium-chain triglycerides. *Food Chem. Toxicol.* 38(1):79–98.

Tupker RA, Bunte EE, Fidler V, Wiechers JW, Coenraads PJ. 1999. Irritancy ranking of anionic detergents using one-time occlusive, repeated occlusive and repeated open tests. *Contact Dermatitis* 40(6):316–22.

Tupker RA, Vermeulen K, Fidler V, Coenraads PJ. 1997. Irritancy testing of sodium laurate and other anionic detergents using an open exposure model. *Skin Res. Technol.* 3(2):133–6.

Vozmediano JM, Carbajo JM, Franco R, Milán VJ, Padilla M, Sarmiento C. 2000. Evaluation of the irritant capacity of decyl polyglucoside. *Int. J. Cosmet. Sci.* 22(1):73–81.

Wakelin SH, Smith H, White IR, Rycroft RJ, McFadden JP. 2001. A retrospective analysis of contact allergy to lanolin. *Br. J. Dermatol.* 145(1):28–31.

Walters RM, Khanna P, Hamilton M, Mays DA, Telofski L. 2015. Human cumulative irritation tests of common preservatives used in personal care products: A retrospective analysis of over 45 000 subjects. *Toxicol. Sci.* 148(1):101–7.

Wilkinson JD, Shaw S, Andersen KE, Brandao FM, Bruynzeel DP, Bruze M, et al. 2002. Monitoring levels of preservative sensitivity in Europe. A 10-year overview (1991–2000). *Contact Dermatitis* 46(4):207–10.

Zachariae C, Hall B, Cupferman S, Andersen KE, Menné T. 2006. ROAT: Morphology of ROAT on arm, neck and face in formaldehyde and diazolidinyl urea sensitive individuals. *Contact Dermatitis* 54(1):21–24.

Zoller L, Bergman R, Weltfriend S. 2006. Preservatives sensitivity in Israel: A 10-year overview (1995–2004). *Contact Dermatitis* 55(4):227–9.

8

Thickening Agents

Ricardo D'Agostino Garcia, Antony O'Lenick and Vânia Rodrigues Leite-Silva

CONTENTS

Introduction

Thickening agents, or viscosity-enhancing agents, are an important group of raw materials that deter mine the final form and function of a personal care formulation. Traditionally, their primary function was to increase the viscosity of the formulation, but nowadays thickeners can be considered multifunc tional ingredients that can contribute to several product attributes, such as they

- Modify the rheology and the appearance of a product
- Improve sensory and on-skin performance properties
- Suspend insoluble ingredients including pigments and pearling agents
- Stabilize emulsions and suspensions
- Modify foaming properties in surfactant based formulations

Selection of the most appropriate thickener must be done carefully, with the formulator considering multiple factors such as the type of rheology required, product appearance, ingredient compatibility, costs and benefits, marketing claims, specific formulation/final product characteristics, and packaging.

Thickeners are generally grouped into a number of categories, namely, polymers (synthetic or natural), minerals, waxes, electrolytes and nonionic materials. The properties and applications of these thickeners are discussed in the following sections.

Polymers

Cellulose-Based Agents

Hydroxyethylcellulose

Hydroxyethylcellulose (HEC) is a nonionic water-soluble polymer derived from cellulose, commercially available in a variety of molecular weights. HEC dissolves readily in water (cold or hot) to give clear, smooth, viscous solutions. HEC can thicken, suspend, bind, emulsify, form films, stabilize, disperse, retain water, and provide protective colloid action. It is readily soluble in hot or cold water, and can be used to prepare solutions with a wide range of viscosities. It has outstanding tolerance for dissolved electrolytes.

HEC differs from other cellulose-based polymers such as methylcellulose, hydroxypropylcellulose and ethylcellulose in that HEC is soluble in both cold and hot water; it does not display a 'cloud' or 'pre cipitation' point. It also differs from carboxymethylcellulose in that it is nonionic and less affected by pH change, presence of anions and organic co-solvents. HEC is used to produce solutions having a wide range of viscosity and such solutions are pseudoplastic, since they change in viscosity with rate of shear (i.e. they are thicker at rest and thinner when stirred). HEC is commonly used as a highly efficient thick ener and emulsion stabilizer, and in detergent systems it provides creamier and more lubricious foam.

Hydroxypropylcellulose

Hydroxypropylcellulose (HPC) is a nonionic water-soluble cellulose ether with a remarkable combina tion of properties that contribute to its utility in formulation. It combines organic solvent solubility, ther moplasticity, and surface activity with the aqueous thickening and stabilizing properties characteristic of other water-soluble cellulose polymers. The hydroxypropyl substitution confers increased lipophilicity on HPC than other water-soluble cellulose derivatives, a key difference when formulating with alcohol-based formulations. Accordingly, it is compatible with a wide range of anionic, nonionic, cationic and amphoteric surfactants. The viscosity of water-based solutions of HPC is not affected by changes in pH over the range of 2 to 11. Applications using HPC as a thickener include hairstyling aids, alcohol-based preparations, perfumes and colognes, shampoos, emulsion creams, and lotions.

Hydroxypropyl Methylcellulose

Hydroxypropyl methylcellulose (HPMC) is a modified natural carbohydrate polymer derived from cellulose. The hydroxypropyl and methyl group substituents impart water solubility and surface activity to the polymer. A variety of HPMC types are commercially available, and their properties determined by the degree of methyl and hydroxypropyl substitution, as well as their molecular weight. HPMC is a multifunctional additive that offers a broad range of impressive properties addressing multiple needs in rinse-off applications even at low use levels. The most important properties include a foam-boosting effect, a significantly improved skin feel, and viscosity control in cleansing and conditioning personal care formulations. The incorporation of HPMC reduces the required amount of certain ingredients, such as surfactants, other thickeners and emollients, without sacrificing performance.

Methylcellulose

Methylcellulose (MC) acts as a film former, suspension aid, lubricant, lather enhancer/stabilizer, emulsion stabilizer, gelling agent and dispersant. It delivers luxurious foam and stabilization to a variety of cleansing products. MC offers a unique feature called interfacial thermal gelation that reflects the combined effects of their surface activity and thermally gelling properties. It can be combined with xanthan gum or inorganic thickeners to achieve their suspension capabilities. MC delivers light, nongreasy films, an attractive benefit in creams/lotions, suncare products, styling gels and facial masks. It is used in body washes, facial cleansers, shampoos, liquid hand soaps, shaving creams and foams, hair styling mousses, and sulphate-free cleansing products. In solid products like makeup, it functions primarily as a binder, but also adds humectancy and film-forming properties to improve the product shelf life and texture.

Ethylcellulose

Ethylcellulose (EC) is an oil-thickening and film-forming additive. Due to its versatility and broad compatibility with a wide range of oils, it provides formulators with flexibility and new options for thickening and formulating with oils, ranging from a slightly thickened oil to a transparent oil gel. The film-forming properties of EC exhibit longer-lasting benefits compared to simple oil blends, thereby facilitating the inclusion and benefits of the oil properties in the final product. Applications include lipsticks and nail polishes (long-lasting brilliance), fragrance stabilizer, and thickener for perfumes and body creams.

Carboxymethylcellulose

Carboxymethylcellulose (CMC) is made by reacting sodium monochloroacetate with alkali cellulose under rigidly controlled conditions. It is synthesized by the alkali-catalyzed reaction of cellulose with chloroacetic acid. The polar (organic acid) carboxyl groups render the cellulose soluble and chemically reactive.

Following the initial reaction, the resultant mixture produces about 60% CMC plus 40% salts (sodium chloride and sodium glycolate). This product is the so-called technical CMC that is used in detergents. A further purification process is used to remove these salts to produce the pure CMC used for food, pharmaceutical and dentifrice (toothpaste) applications. An intermediate 'semi-purified' grade is also produced, typically used in paper applications such as restoration of archival documents.

The functional properties of CMC depend on the degree of substitution of the cellulose structure (i.e. how many of the hydroxyl groups have taken part in the substitution reaction), as well as the chain length of the cellulose backbone structure and the degree of clustering of the carboxymethyl substituents.

The resultant anionic polymer is soluble in hot or cold water, but is less tolerant to electrolytes and for this reason is not used in the formulation of detergent-based personal care products. CMC is used for water binding, synersis control, and to suspend pigments and active ingredients in solution. One of the main applications of CMC in the cosmetic industry is as a gelling agent in toothpaste formulation due to its water retention capacity and rheological profile.

Naturals

Xanthan Gum

Xanthan gum is a natural-derived, high molecular weight polysaccharide used to thicken and stabilize suspensions, emulsions and foams against separation. The high viscosity associated with xanthan gum solutions at low shear rates enables products to keep particles suspended or prevent oil droplets from coalescing. As the gum exhibits pseudoplastic rheology, the viscosity drops when shear is applied, therefore the end product can be easily scooped, poured or squeezed from its container. Once the force is removed, the solution regains its initial viscosity almost immediately, thereby stabilizing the formulation. Xanthan gum is soluble in hot or cold water; stable over a wide range of pH and temperatures; resistant to enzymatic degradation; and has excellent compatibility in the presence of anionic, amphoteric, nonionic surfactants and high concentrations of salt. Applications include body wash, shampoo, shower gel, hair gels, lotions, creams and conditioners.

Gellan Gum

Gellan gum is a water-soluble anionic polysaccharide manufactured by microbial fermentation using the bacterium *Sphinomonas elodea*. It is a gelling agent, effective at extremely low use levels, forming solid gels at concentrations as low as 0.1%. It can be used to form fluid gels, which represents a unique way of stabilizing suspensions and emulsions without adding viscosity. Higher levels will result in the formation of firm, transparent, mechanically robust and brittle gels. Gels are stable over a wide pH range; are compatible with anionic, amphoteric and nonionic surfactants; and are easily combined with most other rheology modifiers.

The polymer is available in two chemical forms: high and low acyl content. Low acyl products form hard, non-elastic gels. High acyl products form soft, elastic gels. Varying the ratio of the two forms produces a wide range of textures. Applications include shampoo and hair care products, skin lotions and creams, sunscreen and suncare products, emulsion and suspension stabilizer, toothpaste, and non-dissolving cosmetic films.

Carrageenans

Carrageenans are extracted from red seaweed and are available in three distinct forms based on the number of sulphate groups per disaccharide group in their chemical structure: kappa carrageenans form strong, brittle gels; iota carrageenans form soft, elastic gels; and lambda carrageenans are non-gelling thickeners. Carrageenan's ability to form stable gels under neutral and alkaline conditions can be used to impart texture and consistency in numerous cosmetic products.

Gelling grades of carrageenan can be formulated in very low concentrations to form fluid gels that stabilize emulsions and solid particles and give suspension properties to keep particles from settling, such as pigments in liquid preparations. Rheological profiles ranging from free-flowing liquids to thixotropic fluid gels, and self-supporting solid gels can be achieved dependent on the amount of carrageenan incorporated. Applications include skin care and lotions, hair care products, eye make-up, toothpaste, shaving foams, antiperspirant/deodorant sticks and air freshener gel.

Pectin

Pectins are a family of partially methylated esterified polysaccharides produced from citrus peel and sugar beet pulp. They are classified as high methoxyl (HM) pectin and low methoxyl (LM) pectin. HM pectin requires a minimum amount of soluble solids and a narrow pH range around 3.0 in order to form gels. LM pectin requires the presence of a controlled amount of calcium or other divalent cations to form a gel.

Apart from adding structure through gelation and viscosity buildup, pectin gels on the skin can provide moisture absorption while being skin friendly. When physically disrupted during processing, pectin gels can be used to formulate lotions and creams without the use of surfactants. Pectin is well known as the gel-forming component of fruits and as an ingredient of many food products. The inclusion of pectin on product labels is generally regarded favourably by consumers, as it is an easily recognizable natural ingredient. Applications include films and masks, lotions and creams, hair conditioners, and hairstyling products.

Starch Based

Hydroxypropyl Starch Phosphate

Hydroxypropyl starch phosphate is a specially processed pre-gelatinized starch that can create stable and elegant personal care emulsions. It is delivered as a powder, and the ease of use and immediate dispersability in cold water make it ideal for use in continuous manufacturing processes. No pre-mixes are needed, and it can be added to the oil phase or to adjust batch properties at the end of a production. Applied in any type of emulsion, this starch can aid in emulsion stabilization, enhance aesthetics and build viscosity. An emulsion with this polymer will have outstanding stability over a broad temperature range (−30°C up to 50°C). It also brings body to the formulation and a conditioning after feel. It is non-ionic in character, thereby offering broad compatibility that provides the formulator with the flexibility to formulate over a wide pH range with high amounts of mono- and polyvalent salts (up to 20%) and a large variety of raw materials.

Acrylic Acid Based: Carbomers

Carbomers are high molecular weight cross-linked polymers of acrylic acid, which when neutralized have the ability to absorb and retain water, resulting in a viscous gel or liquid. The dried carbomer comes in the form of a lightweight white powder. The resulting product of neutralization has a number of applications in cosmetics, with different carbomers being suited to different products. These products include gels, creams and lotions, shampoos, moisturizers, body washes and sunscreens. It provides viscosity, stabilization and suspension properties. They are an extremely efficient rheology modifier capable of providing high viscosity, and form sparkling clear, water-based or hydroalcoholic gels and creams.

Carbomers are tightly coiled acidic molecules. Once dispersed in water, the molecules begin to hydrate and partially uncoil, forming low viscosity solutions with an approximate pH range of 2.5–3.5 depending on the polymer concentration. These polymers must be neutralized in order to achieve maximum viscosity. The most common way to achieve maximum thickening from these polymers is by converting the acidic carbomers to a salt. Once a neutralizer is added to the dispersion, thickening gradually occurs and the optimum viscosity is typically achieved at a pH of 6.5–7.5. High viscosities can be achieved in pH ranges of 5.0–9.0. At pH 9.0 or higher the viscosity will begin to decrease, caused by the dampening of the electrostatic repulsion caused by the presence of excess electrolytes. It is possible to achieve high viscosity systems at pH values below 5 and above 9, but the amount of the carbomers must be increased to obtain these higher viscosity levels.

Neutralizing the polymer is achieved with a common base such as sodium hydroxide (NaOH), triethanolamine (TEA) or other neutralizers shown in Table 8.1.

A range of carbomers is commercially available and the selection of the most appropriate carbomer for a formulation depends on desired viscosity; rheology; suspension; sensorial and aesthetic properties; cost; and compatibility with surfactants, electrolytes or any other ingredient.

Amino Acid–Based Dibutyl Ethylhexanoyl Glutamide and Dibutyl Lauroyl Glutamide

Dibutyl ethylhexanoyl glutamide and dibutyl lauroyl glutamide are highly effective oil gelling agents based on l-glutamic acid. They provide hardness, strength and stability to the product, even at very high temperatures. These can form clear oil gels, as well as hard oil gel sticks. The sensory feel of the oil gels and oil sticks is the same as the oils. These thickening agents can be used in anhydrous products as well as emulsions. They form nano-sized fibre networks that can suspend particles.

Applications include skin care, hair care, sticks (antiperspirants/deodorants), sun care, makeup (lip gloss, lipstick, eyeshadow, mascara). Suggested use levels are 0.2–5.0 wt%.

TABLE 8.1

List of the Most Common Neutralizers and the Appropriate Ratio to Achieve Exact
Neutralization at a pH of 7.0

Neutralizer	Neutralization Ratio Base/Carbomer
Sodium hydroxide – 18%	2.3/1.0
Ammonium hydroxide – 28%	0.7/1.0
Potassium hydroxide – 18%	2.7/1.0
Aminomethyl propanol	0.9/1.0
Tetrahydroxypropyl ethylenediamine	2.3/1.0
Triethanolamine – 99%	1.5/1.0
Tromethamine – 40%	3.3/1.0
Diisopropanolamine	1.2/1.0
Triisopropanolamine	1.5/1.0

Source: Adapted from Lubrizol Advanced Materials Inc. TDS-237 (2009), Neutralizing
Carbopol™ and Pemulen™ Polymers in Aqueous and Hydroalcoholic Systems,
Cleveland, Ohio, The Lubrizol Corporation.

Polyamide-Based Agents

Polyamide-3 and Polyamide-8

Polyamide oil structuring polymers offer a broad range of functionality to personal care formulations as
film formers, water repellency agents, pigment and polymeric emulsion stabilizers, structuring agents
and rheology modifiers. These polymers are low colour and low odour, high-performance thermoplastic
solids proven to form crystal clear, thermo-reversible gels. The products work with a range of high to
low polarity oils, providing compatibility with an array of cosmetic ingredients. They also create novel
formats, from clear sticks and balms to sprayable gels and emulsions with real consumer benefits.

Mineral-Based Agents

Fumed Silica

Fumed silica is a synthetic, amorphous, colloidal silicon dioxide. It is an inorganic thickener with a
wide variety of cosmetics and personal care applications. With fumed silica, cosmetic oils can easily be
converted into highly viscous, largely transparent gels. There are some modified silica with hydrophobic properties which are perfect for creating or adjusting thixotropic properties, improving suspension
without significantly increasing viscosity, and imparting stability and water-resistance to emulsions,
especially for sunscreen and makeup formulations.

Fumed silica is a good thickener for non-polar systems, but it is not suitable for thickening aqueous
systems. It can be incorporated at room temperature, without activation and using high shear equipment that is customary in the cosmetic industry. Formulations thickened with fumed silica display low
sensitivity to temperature, electrolytes and pH. Depending on the desired effect, fumed silica can be
used in a concentration range from 0.5 to 10% w/w. Concentration levels above 3% w/w can specifically
counteract the oily or greasy skin feel of a formulation. Applications include toothpaste, dental powder,
antiperspirant (aerosol and solid), sunscreen products, fragrances, nail polish, powders, lipstick, face and
eye makeup, hair bleaching agents, and hair dyes. Hydrophilic types are very effective as drying agents,
while hydrophobic types are particularly helpful as powder flow regulators.

Hectorite

Hectorite is a hydrophilic clay with small-size, platelet-shape particles and large surface area to volume
ratio. It is a very effective thickener of water, forming strong, stable viscous gels. Hectorite has special
properties including lower iron content, lighter colour, higher swelling capacity, greater gel strength

and no crystalline silica. Hydrophilic clays are non-abrasive and have a pleasant silkiness that is quite different to that associated with the majority of cellulose, polymer and polysaccharide-based thickeners. Hectorite is not sensitive to temperature variations, but surfactants can influence the rheological behaviour of hydrophilic clays. Non-ionic surfactants can adsorb onto the clay surface if they are highly ethoxylated, but this rarely causes problems either in flow control or activity of the formulation. Anionic surfactants can act as dispersing agents and weaken the gel structure. Cationic surfactants interact strongly and should be avoided.

Hectorite can be reacted with special vegetable quaternary ammonium salts to produce organoclays that are able to thicken and gel organic liquids. Organoclays form thixotropic gels by developing hydrogen bonds between the edges of adjacent platelets.

Hydrophilic clays or organoclays need to be efficiently dispersed using high-shear equipment and then activated with the optimum level of a chemical activator, such as propylene carbonate. The benefits gained with these clays are thermostable viscosity control, thixotropic flow, suspension control of pigments and actives, emulsion stabilization and silky skin feel. Applications: emulsions, lipsticks, antiperspirant aerosols, mascara and UV sunscreen.

Magnesium Aluminium Silicate

Magnesium aluminium silicate is a naturally occurring smectite clay that has been water-washed to optimize purity and performance. It swells in the presence of water in a variety of aqueous compositions and is used as a thickener, and an emulsion and suspension stabilizer due to its colloidal structure. This hydrophilic clay has to be well dispersed with high-shear equipment in order to be effective. Applications include skin care lotions, creams, liquid soaps, shampoos, conditioners, make-up products and toothpaste.

Waxes

Waxes are a mixture of hydrophobic organic substances of medium-chain length. Waxes will melt at temperatures in the range of 40°C up to 140°C without decomposition and will re-solidify unchanged. Solubility and consistency of waxes is strongly temperature-dependent. Waxes are classified into animal, vegetable and mineral types depending on their source.

Animal

Beeswax

Beeswax is a secretion of the abdominal glands of the honeybee. When freshly secreted, beeswax is white and colourless, but will acquire a colour primarily by the bee picking up and storing pollen and honey. For cosmetic purposes, yellow-grade beeswax is defined as *cera flava* and white-grade beeswax as *cera alba*.

Beeswax has a relatively low melting point (61–65°C), is moderately hard, and somewhat sticky, plastic and kneadable at body temperature. It is one of the best oil-binding waxes known, therefore considerable amounts of beeswax are used in lipstick and lip balm preparations to produce creamy textures, favorable adhesion to the skin, and films that are well-received among consumers. The oil-gelling properties of beeswax are helpful for stabilizing water-in-oil emulsions for skin and hair applications and for texturizing effects. Beeswax is also used as a means of enhancing texture and adding volume in mascara and make-up.

Vegetable

Carnauba Wax

Carnauba wax is a secretion of the Brazilian palm *Copernicia cerifera*, which produces the wax in the cuticula of the fronds. The leaves are harvested from trees growing in the wild by cutting the leaves, drying them in the sun and threshing. Carnauba wax has an extremely narrow melting curve with a melting point of approximately 84°C. The wax, which exhibits high crystallinity and a high degree of contraction, is very hard and brittle. Carnauba wax also has highly favorable emulsification properties and an excellent capacity for binding ester oils and mineral oil. It raises the melting point of oil gels; therefore, carnauba wax is a preferred additive in lipsticks and lip balms, and is highly suitable for use in mascaras. It provides glossy and lustrous surfaces

Candelilla Wax

Candelilla wax is a secretion of wild *Euphorbia cerifera*, which is native to the northern Mexican desert. The wax is obtained from the above ground parts of the plant. The plant is dried, boiled in water and the wax is then skimmed off by decanting. Crude candelilla wax is dark brown and then refined to a pale yellow wax.

Candelilla wax is remarkably hard but exhibits low crystallinity and a degree of tackiness at higher temperatures. Its melting point of about 72°C lies between that of beeswax and carnauba wax. The resin and sitosterol it contains combine with esters to give candelilla wax its excellent capacity for binding ester oils. It provides high surface gloss when applied in lipsticks and exhibits satisfactory contraction properties – important for demolding lipsticks.

Sunflower Wax

The wax obtained from sunflower oil forms hard and very homogeneous thermally stable oil gels and is therefore excellent for light-coloured, practically odorless and tasteless lip balms, lipsticks and other oil gels. Chemically it has a long chain (about C60) wax ester mainly from monovalent alcohols and acids. Sunflower wax has a melting point of about 80°C and a narrow melting curve with no molten contents below 55°C. Consequently the oil gels are very heat resistant.

Rice Bran Wax

The main components are long-chain wax esters with chain lengths of around C60. Although it has similar chemistry to sunflower wax, its applications are quite different. Refined rice bran wax has a very hydrophobic character. It forms soft oil gels and is therefore highly preferred for emulsions like mascaras and skin care products.

Mineral

Microcrystalline Wax

Microcrystalline wax is a solid obtained by extracting the oil from petrolatum. It is a complex mixture composed mainly of C31–C70 isoparaffins. It has a microcrystalline structure, high adhesive power, good extensibility, is not susceptible to low temperatures and has a high melting point (60–85°C). When mixed with other waxes, it suppresses crystal formation making it useful in lipsticks and creams.

Ozokerite Wax

Ozokerite wax is generally a white, crystalline, odorless and tasteless solid. It is most often made from blends of paraffin and microcrystalline waxes, and when combined offers broader functionality than the constituent ingredients. It is used extensively in personal care and cosmetic formulations. It will increase viscosity, assist in emulsion stability and enhance gel strength in liquid and semi-solid systems. It is a hard wax with a relatively high melting point. These characteristics encourage its use in lipstick and lip care products to promote structure and stick strength.

Electrolytes

In commercial formulations, surfactants are often co-formulated with inorganic electrolytes, like sodium, potassium and magnesium salts that are included as inexpensive thickeners. Salts affect the surfactant's aggregative and functional behaviour. This thickening effect depends on the presence of an anionic surfactant and it works up to a concentration maximum. After this maximum is reached the viscosity collapses, as further increase in electrolytes leads to a destruction of the micelle network. These electrolytes thickeners should be added on the order of 1–2%, and generally in the form of an aqueous solution to prevent poor electrolyte distribution. The increase in viscosity is affected by several factors such as surfactant system, pH, electrolyte level and ionic strength.

Non-Ionic Thickeners

Typical non-ionic thickening agents for surfactant systems can be generally divided into two groups: hydrophobic and hydrophilic types. Hydrophobic monomeric or oligomeric types with a low molecular weight are mostly non-ionic surfactants, for example, glyceryl laurate, cocamide DEA, isostearamide MIPA and PPG-3 myristyl ether. Hydrophilic polymeric types with a high molecular weight are based on highly ethoxylated oleochemical derivatives, for example, PEG-120 methyl glucose dioleate, PEG-18 glyceryl oleate/cocoate and PEG-55 propylene glycol oleate. These two types of thickeners provide two important differences in performance: the flow behaviour and the temperature dependence of the viscosity. The hydrophobic thickeners provide a shear thinning flow. The viscosity decreases with increasing shear rate and at lower temperatures, but they have stable viscosity at higher temperatures. The hydrophilic thickeners provide a Newtonian flow behaviour, so their viscosity is independent of the shear rate. At lower temperatures the viscosity increases significantly, while at higher temperatures the viscosity decreases dramatically. Selection of the right thickening agent will provide a final product that has good consumer acceptance; stabilizing effects; better viscosity performance; reduction of the total amount of thickener; and additional benefits like conditioning, moisturizing, refatting and solubilizing.

BIBLIOGRAPHY

AkzoNobel Surface Chemistry Personal Care. (2014). STRUCTURE® XL Starch. AkzoNobel Surface Chemistry.
Becker, L. C., Bergfeld, W. F., Belsito, D. V., Hill, R. A., Klaassen, C. D., Liebler, D. C., Marks, J. G. Jr. et al. (2013). Safety assessment of ammonium hectorites as used in cosmetics. *International Journal of Toxicology*, *32*(6 suppl), 33S–40S.
Bradbeer, J. F., Hancocks, R., Spyropoulos, F., Norton, I. T. (2015). Low acyl gellan gum fluid gel formation and their subsequent response with acid to impact on satiety. *Food Hydrocolloids*, *43*, 501–509.
Carboxy methyl cellulose. https://en.wikipedia.org/wiki/Carboxymethyl_cellulose
Cosmetics, Toiletry and Fragrance Association (CTFA). (1986). Final report on the safety assessment of hydroxyethylcellulose, hydroxypropylcellulose, methylcellulose, hydroxypropyl methylcellulose, and cellulose gum. *Journal of the American College of Toxicology*, *5*, 1–59.
CP Kelco Inc. AB-43 (2004). Natural Biopolymers in Cosmetic and Personal Care Applications. CP Kelco Aps.
Endlein, E., Peleikis, K. H. (2011). Natural waxes – Properties, compositions and applications. *SÖFW-Journal*, *137*(4), 1–8.
Giancola, G., Schlossman, M. L. (2016). Decorative cosmetics. In Sivamani, R. K., Jagdeo, J. R., Elsner, P., Maibach, H. I. (eds.), *Cosmeceuticals and Active Cosmetics*, 3rd ed. CRC Press, Boca Raton FL, pp. 191–220.
Goswami, S., Naik, S. (2014). Natural gums and its pharmaceutical application. *Journal of Scientific and Innovative Research*, *3*, 112–121.
Imeson, A. P. (1997). *Thickening and Gelling Agents for Food*. Springer Science & Business Media, Dordrecht.

Imperatore, R., Vitiello, G., Ciccarelli, D., D'Errico, G. (2014). Effects of salts on the micellization of a short-tailed nonionic ethoxylated surfactant: An intradiffusion study. *Journal of Solution Chemistry*, *43*(1), 227–239.

Jani, G. K., Shah, D. P., Prajapati, V. D., Jain, V. C. (2009). Gums and mucilages: Versatile excipients for pharmaceutical formulations. *Asian Journal of Pharmaceutical Sciences*, *4*(5), 309–323.

Kadu, M., Vishwasrao, S., Singh, S. (2015). Review on natural lip balm. *International Journal of Research in Cosmetic Science*, *5*, 1–7.

Karsheva, M., Georgieva, S., Handjieva, S. (2007). The choice of the thickener – A way to improve the cosmetics sensory properties. *Journal of the University of Chemical Technology and Metallurgy*, *42*(2), 187–194.

Kirilov, P., Le Cong, A. K., Denis, A., Rabehi, H., Rum, S., Villa, C., Haftek, M., Pirot, F. (2015). Organogels for cosmetic and dermo-cosmetic applications – Classification, preparation and characterization of organogel formulations – Part 2. *Household and Personal Care Today*, *10*(4), 16–21.

Knowlton, J. L., Pearce, S. E. (2013). *Handbook of Cosmetic Science and Technology*. Elsevier, Oxford.

Kortemeier, U., Venzmer, J., Howe, A., Grüning, B., Herrwerth, S. (2010). Thickening agents for surfactant systems. *SÖFW Journal*, *136*(3), 30.

Lehmann, A., Volkert, B., Fischer, S., Schrader, A., Nerenz, H. (2008). Starch based thickening agents for personal care and surfactant systems. *Colloids and Surfaces A: Physicochemical and Engineering Aspects*, *331*(1), 150–154.

Lubrizol Advanced Materials Inc. TDS-103 (2002). Dispersion Techniques for Carbopol™. The Lubrizol Corporation, Cleveland, Ohio.

Lubrizol Advanced Materials Inc. TDS-237 (2009). Neutralizing Carbopol™ and Pemulen™ Polymers in Aqueous and Hydroalcoholic Systems. The Lubrizol Corporation, Cleveland, Ohio.

Lyapunov, A. N., Bezuglaya, E. P., Lyapunov, N. A., Kirilyuk, I. A. (2015). Studies of carbomer gels using rotational viscometry and spin probes. *Pharmaceutical Chemistry Journal*, *49*(9), 639–644.

Mehta, C., Bhatt, G., Kothiyal, P. (2016). A review on organogel for skin aging. *Indian Journal of Pharmacology*, *49*(3), 28–37.

Niederer, M., Stebler, T., Grob, K. (2015). Mineral oil and synthetic hydrocarbons in cosmetic lip products. *International Journal of Cosmetic Science*, *38*(2), 194–200.

Prajapati, V. D., Jani, G. K., Zala, B. S., Khutliwala, T. A. (2013). An insight into the emerging exopolysaccharide gellan gum as a novel polymer. *Carbohydrate Polymers*, *93*(2), 670–678.

Savary, G., Grisel, M., Picard, C. (2016). Cosmetics and personal care products. In Olatunji, O. (ed.), *Natural Polymers*. Springer International Publishing, Cham, Switzerland, pp. 219–261.

Shrestha, R. G., Shrestha, L. K., Aramaki, K. (2008). Wormlike micelles in mixed amino acid-based anionic/nonionic surfactant systems. *Journal of Colloid and Interface Science*, *322*(2), 596–604.

Somasundaran, P., Chakraborty, S., Qiang, Q., Deo, P., Wang, J., Zhang, R. (2004). Surfactants, polymers and their nanoparticles for personal care applications. *Journal of Cosmetic Science*, *55*, S1–S18.

9

Surfactants in Cosmetic Products

Ricardo Pedro and Kenneth A. Walters

CONTENTS

Introduction

The properties and characteristics of surfactants are mostly related to surface and interfacial phenomena, and the formation of aggregates (micelles, bilayers, liquid crystals, etc.), which are primarily a function of their concentration in use and their structural features.

Surfactants are used widely and can function as wetting agents, foamers or defoamers, detergents, dispersants, emulsifiers, solubilizers or carriers, hydrotropes, thickeners, stabilizers, softeners, emollients, antistatics, corrosion inhibitors, degreasers, lubricants, biocides and skin and eye irritation reducing agents. In general, surfactants perform more than one role in a process or formulation, being considered true multifunctional materials. Consequently, surfactants find application in several technological areas. In agrochemicals, for example, surfactants act as wetting agents, and enable the optimum spread of an active substance on the entire surface of a plant, specifically over the leaves. Medical, veterinary, dental and pharmaceutical products use surfactants to wet certain surfaces with water on contrast media, contact lenses and dentures. In cleaning products, surfactants can remove oil, fat and grease from dishes, floors, household appliances and fabrics. Environmental cleanup processes use surfactants for the dispersion of pollutants from oil and gas activities into water and enhance their biodegradation. In medicine, surfactants can be used in exogenous surfactant replacement therapy. A gentle detergency makes specific surfactants greatly valued for use in skin cleansers, shampoos and other cosmetic preparations.

The use of surfactants in cosmetics relies not only on their functional properties but also on how they can contribute to formulation stability, aesthetics and safety. On the one hand, the development of surfactants has led to great advances in the cosmetic industry, and on the other hand the challenges faced by cosmetic science have driven the development of surfactant science and the market.

A parallel between the definition of cosmetics and the applications, criteria and restrictions in the selection of surfactants can be made to illustrate how important these ingredients are for cosmetic products, from the simplest formulation to the most complex ones. The cosmetic market is one of the most globalized in the world and that is why attempts and initiatives for regulatory harmonization are a constant concern. The definition of cosmetics differs slightly throughout the world. The U.S. Federal Food, Drug, and Cosmetic Act (FDCA) defines cosmetics as '(1) Articles intended to be rubbed, poured, sprinkled, or sprayed on, introduced into, or otherwise applied to the human body, or any part thereof for cleansing, beautifying, promoting attractiveness, or altering the appearance, and (2) articles intended for use as a component of any such articles; except that such term shall not include soap' (U.S. Department of Health and Human Services, 2018). In Europe, 'a cosmetic product means any substance or mixture intended to be placed in contact with the external parts of the human body (epidermis, hair system, nails, lips and external genital organs) or with the teeth and the mucous membranes of the oral cavity with a view exclusively or mainly to cleaning them, perfuming them, changing their appearance, protecting them, keeping them in good condition or correcting body odours' (Official Journal of the European Union, 2018). In Japan, a cosmetic product is 'Any item having mild effects on the human body that is rubbed, spread, or otherwise applied in a similar manner for the purpose of cleansing, beautifying, or enhancing the attractiveness of the human body, to change physical appearance, or to maintain skin or hair in a healthy condition' (Hayashida, 2018). Additionally, some cosmetic preparations are intended for long-term use and therefore must be stable. Consumers also value products that are easy to use and that provide a pleasant sensory when applied to the different body sites in different formats.

Importance of Surfactants for Cosmetics

Considering the aforementioned cosmetics definitions and requirements, it is understandable why surfactants are so important to cosmetics. Not only are surfactants used as wetting, cleansing, foaming, solubilizing, conditioning and thickening agents, they are also useful in creating a broad range of different formats, such as solutions, gels, emulsions and suspensions.

Cosmetics that use surfactants include bar soaps, liquid skin cleansers, bath products, hand and body creams and lotions, shave products, deodorants, hair care products, oral hygiene products, and topical dermatological products, among many others.

Rheological properties can also be positively impacted by the use of surfactants, leading to different flow behaviours, designed to deliver sensory and practicality of use benefits. Thus, whether alone or in combination with other ingredients (especially with polymers), surfactants can perform or contribute to most of the cosmetics properties.

Surfactant characteristics are not expected to alter the organoleptic properties of the cosmetics and influence the consumer acceptance of finished products.

Surfactants optimize the efficacy of cosmetics, enabling the delivery of their active ingredients at desired rates and to specific target sites. The skin is covered by substances derived from cellular metabolism, protein degradation, environmental contaminants and cosmetics. Such substances are predominantly lipid in nature and form a barrier on the surface of the skin, contributing to its protection from external aggressions, but damaging it in some instances. Paradoxically, such a barrier hinders the skin penetration of active ingredients, which must reach the inner layers of skin to be effective. Thus, the main functions of surfactants include skin cleansing (by removing the lipid surface layer), modification of the skin surface and enhancement of the penetration of active ingredients.

Water is an innocuous, readily available and inexpensive solvent, which would make it an ideal vehicle for cosmetics. However, as a vehicle, water is not compatible with most of the active ingredients used in cosmetics, because they are mostly hydrophobic (they must be able to penetrate and diffuse into the inner parts of the skin). Most hydrophobic actives need to be delivered by the cosmetics across the stratum corneum and into the viable epidermal and dermal layer. On the one hand, water does not dissolve substances of lipid nature, and on the other does not spread the formulation on the skin to optimize its use. Surfactants allow the solubilization of the active substances in the cosmetic formulations and also modify the permeability properties of the skin.

Most surfactant properties are related to interfacial (and surface) phenomena, which are widely useful in the formulation and application of cosmetic products. Since water is the main ingredient of most cosmetics, surfactants are necessary to modify water properties and optimize the application, efficacy and stability of the formulation. Surfactants are needed in relatively high concentrations when compared to the other formulation ingredients. Consequently, the development of a stable and functional cosmetic depends on the selection of the surfactants and the ingredients that are compatible with them or even (ideally) optimize their properties.

Briefly, surfactants contribute to formulation structure and consumer acceptance. With regard to the latter aspect, safety in use, preferences and habits of consumers must be taken into account because consumers can decide to buy a certain product not only objectively based on their functionalities, but also on subjective aspects. Surfactants are often the 'heart' of a cosmetic formulation and, because they are a main component, they impart a great deal of their structural, functional or sensory characteristics on finished products. Colour, odour, appearance, foaming and texture are sensory characteristics greatly valued by consumers and must not be negatively impacted by surfactants.

This chapter aims to serve as a guide to the formulator, answering the usual questions and addressing the issues that arise during the development of cosmetic formulations in relation to the use of surfactants. Such questions include:

What are the roles of surfactants and why are they used in cosmetics?

What surfactants are used, and how do surfactants play a role in cosmetic formulations?

What interactions occur between surfactants and the skin and hair?

What problems are derived from the use of surfactants? What other aspects must be considered when using surfactants in cosmetics?

General Considerations of Surfactants

Let us start our discussion with some definitions. The word 'surfactant' is a contraction of the term 'surface active agent'. A surfactant is a substance that has the property of adsorbing onto the surfaces or interfaces of a given system, and remarkably alter the surface or interfacial free energies of those surfaces or interfaces (Rosen, 2004; Meyers, 2006). An interface can be considered as a boundary between any two immiscible phases, while the surface indicates an interface, represented by a gas (or usually air) in contact with a liquid or a solid. The interfacial free energy is the minimum amount of work required to create a unit area of the interface or to expand it by a unit area (Adamson and Gast, 1967). The interfacial free energy per unit area is what we measure when we determine the interfacial tension between two phases and is a measure of the difference in nature of two dissimilar phases. The greater the dissimilarity in their natures, the higher is the interfacial or surface tension between them (Rosen, 2004). When we rub a cosmetic on the skin or hair, energy has to be spent (work has to be done), especially because water, skin and hair surfaces have different natures.

Surfactants adsorb at most interfaces and surfaces in a given system, and consequently reduce the amount of work required to expand them and mitigate the application of a cosmetic onto skin or hair, the formation of foam, contact with 'water' and dirt, and the removal of the latter.

Surfactants are capable of these functions because they are amphiphilic substances, that is, they have groups with antagonistic characteristics in their molecular structure. All surfactant molecules possess a polar group that has affinity for water (and for other polar compounds), called the hydrophilic group. In the same molecule there is also a hydrophobic (or lipophilic) group, which, because of its non-polar nature, has no affinity for water but is compatible with lipid substances (or, generally, non-polar substances). Figure 9.1 shows a schematic representation of a generalised surfactant molecule.

All surfactants are thus composed of molecules exhibiting two distinct structural moieties, which exhibit opposite tendencies of solubility. The hydrophobic moiety is usually constituted of aliphatic or aromatic hydrocarbon chains, or both, whereas the hydrophilic moiety consists of polar groups, such as carboxylate, sulphate, sulphonate, quaternary ammonium, betaine, amine oxide, polyoxyethylene and polyglucoside chains.

The differentiation of each of the hydrophobic and hydrophilic moieties leads to a huge number of surfactants. Thus, for the same hydrophobic moiety, various surfactants can be obtained by varying the hydrophilic group. On the other hand, for the same hydrophilic moiety, various surfactants can be obtained by varying the hydrophobic part of the molecule.

Surfactants can be classified according to their use, their chemical structure or based on their physical properties. As mentioned before, surfactants can be grouped according to their functions, and, in cosmetics, they are used as solubilizers, emulsifiers, dispersants, detergents, foaming or antifoaming agents, conditioners and so forth. Although, the surfactant is a multifunctional molecule, which means it can exhibit more than one of the mentioned functions, there is always a prominent property which dictates its applications. Another very common classification of surfactants is based on the ionic character of its polar portion, its hydrophilic portion, allowing classification into anionic, cationic, amphoteric and non-ionic surfactants.

As mentioned before, surfactants are substances that modify surface and interfacial tensions, resulting in a number of other properties and applications. Their ability to aggregate and interact with other formulation ingredients results in complementary important properties.

A surfactant molecule can be 'tailored' to perform one or more functions in a formulation or process. In fact, there are hundreds of surfactants commercially available in the market. Some reaction schemes, related to the class of surfactants to be obtained, are illustrated later, and related to the functionality desired.

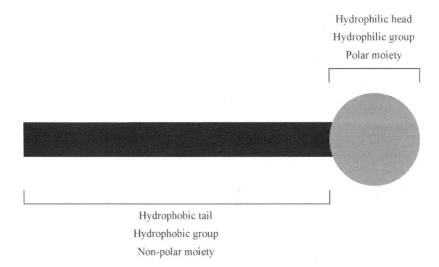

FIGURE 9.1   Schematic representation of a surfactant molecule. (From Falbe, J., 1987, *Surfactants in Consumer Products: Theory, Technology and Application*, Berlin: Springer-Verlag.)

Surfactants for Cosmetic Uses

Historical Considerations

The history of technological development goes from discovery by serendipity to highly structured research results, with very well-defined objectives motivated by the development or resolution of problems of humankind in the social, industrial, commercial, rural and other sectors. With regard to the chemical industry, specifically to the surfactant industry, the story is no different.

Soaps were the first surfactants reported by the literature and it is believed that they have been used for over 2800 years (Willcox, 2000). Nowadays, although natural-derived surfactants are quite popular, synthetic surfactants dominate the market. In the year 2016, synthetic surfactants have been around for 100 years. The first synthetic surfactants for general applications were developed in Germany to overcome the lack of natural raw materials during World War I. As the chemical industry progressed, new processes and raw materials led to the development of a wide variety of new surface-active compounds and manufacturing processes. In some regions of the world, the limiting factor to produce synthetic surfactants was almost always the availability of raw materials, but also the ease of processing, the production and distribution logistics, and the shelf life were important factors. Sulphated alcohols and alkyl benzene sulphonates were used until the early 1930s as cleaning agents but had little impact on the surfactants market. At the end of World War II, alkyl aryl sulphonates were mainly used for cleaning, while sulphated alcohols were preferably used in shampoo formulations and other personal care products. After World War II, an adduct of propylene tetramer (TP) and benzene became a predominant feedstock for the surfactant industry. However, by 1960, such ingredients were recognized as pollutants due to their persistent foam in rivers and lakes, attributed to their branched alkyl chain that hindered the action of microorganisms, resulting in their low biodegradability. In contrast, sulphated alcohols and fatty acids had high biodegradability, a property attributed to the fact that they are natural products of linear chains. The preference for linear alkyl chains began because of their higher biodegradability rate and greater ecological acceptability (Kogawa et al., 2017; Meyers, 2006).

Currently, although many industries that use surfactants are considered mature, ecological demands, population growth, fashion, raw materials sources and market demands continue to drive the technological developments and the growth of the range of surfactants, particularly in the cosmetics area.

Production and Types of Surfactants for Cosmetic Uses

Surfactant raw materials are in general high molecular weight and non-polar molecules or contain a non-polar part that will constitute the lipophilic group of the surfactant molecule. A series of reactions are required to 'functionalize' such molecules, making them surface active by linking polar moieties to them. Non-polar groups are mainly derived from petrochemical or oleochemical raw materials. Petrochemical raw materials are those derived exclusively from petroleum, while oleochemicals are derived from vegetable or animal sources. Examples of petrochemical-based surfactants are naphthalene sulphonates, ethoxylated nonylphenols, sulphated ethoxylated nonylphenols, among others. Examples of oleochemical surfactants are fatty carbonic chains (in general, alkyl chains containing more than six atoms of carbon) and include ethoxylated fatty alcohols, ethoxylated or non-ethoxylated fatty esters, fatty alkanolamides, ethoxylated or non-ethoxylated fatty acids, alkyl betaines, alkyl amine oxides, alkyl polyglucosides, fatty amino acid derivatives and the like.

Currently, oleochemical products have been experiencing greater market acceptance than those of petrochemical origin. It is important to note that products obtained from oleochemicals can raise controversial issues such as mad cow disease, bird flu and non-renewable resources, all of which can be addressed by certificates of origin (Kosher products, for example) and responsible care. Figure 9.2 shows the integration of oleochemistry into the personal care segment.

The types of surfactants that are of major interest in the cosmetic industry and the respective main reactions used during their manufacturing will be discussed briefly later. Before that, it may be useful to introduce the nomenclature of surfactants, which can become very complex and confusing.

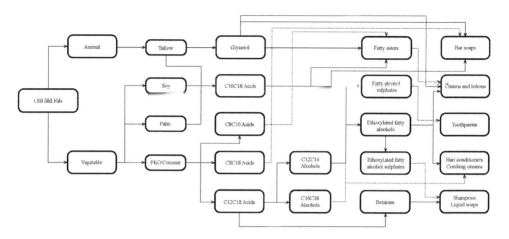

FIGURE 9.2   Chemical routes for the production of surfactants and related cosmetics.

For the purpose of labelling of cosmetics in accordance with U.S. regulations, the Cosmetics, Toiletry and Fragrance Association (now the Personal Care Products Council) created names for cosmetic ingredients. Rules for creating these names are included in the International Cosmetic Ingredient Dictionary (Cosmetic, Toiletry, and Fragrance Association, 1993). Some common names of the alkyl moiety of surfactants are listed in Table 9.1.

TABLE 9.1

Some 'Systematic' Common Names Used for the Fatty Moieties of Surfactants and Their Relation to IUPAC Nomenclature*

Carbon Chain Length	Number of Double Bonds	IUPAC Name	Common Name	Representative Examples
6	0	Hexyl	Caproyl	Caproic acid
7	0	Heptyl	Enanthyl	Heptyl alcohol
8	0	Octyl	Caprilyl	Caprylyl glycol
9	0	Nonyl	Pelargonyl	Pelargonic acid
10	0	Decyl	Capryl	Caprylic acid
11	0	Undecyl	Undecyl	Undecyl acetate
12	0	Dodecyl	Lauryl	Lauric acid
13		Tridecyl	Tridecyl	Trideceth-6
14	0	Tetradecyl	Myristyl	Myristyl amine oxide
15		Pentadecyl	Pentadecyl	Pentadecyl alcohol
16	0	Hexadecyl	Palmityl/cetyl	Cetyl alcohol
16	1	*cis*-Δ^9-Hexadecyl	Palmitoleyl	Palmitoleic acid
17		Heptadecyl	Margaryl	Margaryl alcohol
18	0	Octadecyl	Stearyl	Sodium stearate
18	1	*cis*-Δ^9-Octadecyl	Oleyl	Oleyl palmitate
18	2	*cis,cis*-$\Delta^{9,12}$-Octadecadienyl	Linoleyl	Linoleic acid
18	3	all*cis*-$\Delta^{9,12,15}$-Octadecatrienyl	Linolenyl	Linolenic acid
20	0	Eicosanyl	Arachidyl	Arachidic acid
20	4	all*cis*-$\Delta^{9,12,15}$-Eicosatetrienyl	Arachidonyl	Arachnodic acid
22	0	Docosanyl	Behenyl/erucyl	Behentrimonium chloride

Source: International Union of Pure and Applied Chemistry, 2013, *Nomenclature of Organic Chemistry: IUPAC Recommendations and Preferred Names* (A. H. Favre and H. W. Powell, Eds.), Cambridge: Royal Society of Chemistry.

* The IUPAC nomenclature of organic chemistry is a systematic method of naming organic chemical compounds as recommended by the International Union of Pure and Applied Chemistry (IUPAC).

Reactions and Specifications for Surfactants of Cosmetic Uses

To meet the cosmetic final product requirements, that is, functionality, low or no colour, odour and taste, mildness, and safety, surfactants must be manufactured under certain rigorous specifications, which in turn can only be achieved using specific production processes and controls. The raw materials of surfactants are, in general, non-polar molecules or contain the non-polar part that will constitute the lipophilic group of the surfactants. Specific reactions are used to functionalize these molecules, making them surfactants by 'introducing' polar moieties into them. This can be done by a series of different reactions and some auxiliary processes (Tables 9.2 and 9.3).

As can be seen, the production of surfactants is not a simple process. Besides the chemical reactions (some of which were previously mentioned), there are several auxiliary processes that allow the surfactants to be obtained within the (rigorous) specification ranges and limits required by cosmetics. A product specification is a set of characteristics that meet legal, quality, toxicological, environmental, safety, sensory, market and, above all, functional requirements. Thus, a product specification contemplates which items are minimally important for the commercialization of the product and which ranges, or limits, guarantee the fulfilment of the mentioned requirements. Some general correlations between specification requirements and marketing claims are shown in Table 9.4.

TABLE 9.2

Some Representative Chemical Reactions Used to Obtain Surfactants

Reactions	Definition/Importance	Some Representative Products That Make Use of This Technology
Ethoxylation	This reaction is regarded as a polyaddition, but it is a nucleophilic substitution, where the weak nucleophilic agent has increased its nucleophilicity by a previous reaction with a strong base. Thus, the nucleophile is the catalysed substrate and the electrophile is the ethylene oxide (EO). This reaction introduces a polyether moiety into the molecule, which is responsible for its solubility in water. It is used to create non-ionic surfactants.	Ethoxylated fatty alcohols Ethoxylated fatty acids Ethoxylated fatty esters Ethoxylated fatty amides Ethoxylated fatty amines
Propoxylation	Same as above except for the use of propylene oxide as an electrophile. This polyether in this case is less soluble in water.	Propoxylated alcohols
Alkoxylation	Generic name of the above reactions.	
Hydrogenation	Unsaturated fatty chains are prone to undergo oxidation and form oxides and peroxides, which cause rancidity and yellowing.	Hydrogenated oil ethoxylates Hydrogenated fatty acids Hydrogenated fatty esters
Sulphation	A reaction that introduces a sulphate group into an organic molecule, making the surfactant anionic.	Fatty alcohols (ethoxylated or not) sulphated
Sulphonation	A reaction that introduces a sulphonate group into an organic molecule, making the surfactant anionic.	Alkyl sulphonic acids
Esterification	Functionalization reaction of the molecule, making it an ester of variable solubility in water. It is used to make non-ionic biodegradable surfactants, normally used as emulsifiers and emollients.	Fatty esters (ethoxylated or not)
Amide formation	Functionalization reaction of the molecule. The amide bond is strong and resists better to hydrolysis when compared to ester bonds. It is used to make non-ionic surfactants.	Fatty alkanolamides
Quaternization	Reaction of a tertiary amine with sodium monochloroacetate, methyl chloride, etc.	Alkyl and alkylamido betaines Alkyl quaternary ammonium salts

TABLE 9.3

Some Representative Auxiliary Processes Used to Obtain Surfactants

Reactions	Definition/Importance	Products That Make Use of This Technology
Clarification	Chemical process that employs oxidizing agents to 'decompose' coloured by-products. It is used to improve the colour of surfactants.	Ethoxylated castor oil Alkyl ether sulphates Alkyl sulphosuccinates Fatty esters (ethoxylated or not)
Deodorization	Removal of volatile products that cause odour or other normally unwanted effects on products (dioxane, for example).	Alkoxylated products
Filtration	Removal of dust or any other particulate that may affect product appearance.	Liquid products
Distillation	Physical separation of products of different boiling points by means of temperature and/or pressure gradients. It helps to improve product odour and toxicity.	Esterification, amide formation reactions, base drying in the catalytic phase of the alkoxylation (it makes use of the stripping technique, which is the removal of water by vacuum and temperature, generally higher than 100°C when possible)
Blending	Operation to incorporate additives to surfactants or to blend surfactants to form dispersed, emulsified concentrates or simple solutions. Some concentrates may be powders. It makes it possible to handle viscous materials and enhances productivity. Water and hydrotropes can be added to some products to avoid their crystallization at low temperatures, for example.	Surfactant blends, concentrated surfactant solutions, emulsion concentrates
Drying	Drying process using spray driers or vacuum to allow for preparation of powder surfactants.	Powder surfactants, such as sodium dodecyl sulphate
Pelletizing	Obtaining products in the form of pellets, which facilitate handling and fractionation by cosmetic producers.	Solid products which tend to agglomerate, melt, form dust or absorb water at room temperature
Preservation	Addition of preservatives to avoid microbial contamination of surfactants and finished products. The antimicrobial additives must be permitted in cosmetics.	In general, surfactants which contain more than 30% of water must be preserved

Types of Surfactants

The most common way to classify surfactants is based on their structure. Both the polar head and the non-polar tail can be used as a criterion for structural classification, the first one being more common.

The classification based on the non-polar group refers to the type and number of hydrocarbon chains in the surfactant. With regard to the hydrocarbon chain, surfactants can be classified as linear or branched chain, the former being more suitable for cosmetic uses because of their higher biodegradation rate. In general, branched alkyl chain surfactants are of petrochemical origin, whereas linear surfactants are normally natural derivatives. The number of non-polar chains within a surfactant is not a classification criterion. Surfactants containing two or more non-polar chains and two or more polar heads are known as gemini surfactants (Menger and Keiper, 2000).

As described earlier, surfactants can be classified as anionic, cationic, non-ionic and amphoteric based on their polar head group. This classification is more popular and will be discussed in detail in the next section.

Anionic Surfactants

Anionic surfactants or anionics possess a negative charge in their hydrophilic portion when dissolved in water (Surfactant Science Series, 1981b). Representatives of anionics used in cosmetics include fatty acid soaps, alkyl sulphates, alkyl ether sulphates and alkyl sulphosuccinates, although many others can be

TABLE 9.4

Specification and Properties Required for Surfactants of Cosmetic Use

Specifications and Properties	Claims/Requirements	Comments
Physical form	Ease of handling	Solid and viscous liquid surfactants are difficult to handle. Different forms and special packaging minimize these difficulties. The use of additives facilitates the handling of some products, improving fluidity and/or avoiding solidification at low temperatures.
Colour, odour, taste	Acceptable organoleptic properties	Surfactants with no or low colour, odour and taste make it possible to obtain simpler finished product formulations without the use of masking agents or correctives (in cosmetics, for example, the use of dyes and fragrances constitute the major causes of allergies and irritations).
Content of contaminants as residual ethylene oxide, dioxane, nitrosamines	Low toxicity	Some ethoxylated surfactants are produced, for instance, from the reaction of long chain alcohols with ethylene oxide. In this reaction, a by-product called 1,4-dioxane can be formed. The higher the degree of ethoxylation, the greater the likelihood of 1,4-dioxane occurrence. It may be removed from the ethoxylated compounds by the addition of a vacuum line at the end of the polymerization process. This process is called deodorization. Skin absorption studies demonstrate that 1,4-dioxane has high permeability in human skin (U.S. Environmental Protection Agency, 1993). This substance is possibly carcinogenic to humans (Agency for Toxic Substances and Disease Registry, 2012).
Volatile content, especially volatile organic compounds (VOCs)	Low human and environmental toxicity	Some preparations may contain solvents and materials toxic to humans and the environment, and therefore are controlled in relation to their concentration, so that tolerable exposure limits are observed.
Water or moisture content	Proper functionality	Water is a very reactive substance and its presence in certain products poses a risk to the formulations of which they are part. Water may inactivate some catalysts, initiators, inhibitors or retarders of polymerization reactions, damaging the surfactant polymerization process. It can also compete in parallel reactions. Water represents problems in the ethoxylation, since it can generate glycols when reacting with ethylene oxide. The final products are very hygroscopic because they are ethoxylated, and this requires, in addition to moisture control, the use of special packaging.
Solid and/or active content	Proper functionality	Customers relate the content of solids and/or active matters to what they are buying/paying for. In general, diluted products are not valued.
Cloud point, turbidity, Krafft point, set point	Proper functionality	Such characteristics relate to the solubility of the surfactant in water. Surfactants of limited solubility represent limitation of use, as well as those of high solubility represent waste of use because they have high cmc.
Counts of microorganisms	Proper functionality, safety of use and legislation	These characteristics are related to the integrity of the product and, therefore, to its functionality, safety of use and compliance with current legislation.

more important for specific applications (e.g. acyl isethionates and acyl sarcosinates). As a rule, anionic surfactants are excellent foamers, detergents and wetting agents when compared to the other classes of surfactants. They are relatively low cost and widely used in personal care cleansing preparations. Some representative examples of anionics are listed in Table 9.5.

Alkaline fatty acid soaps are widely used in making bar soaps because their performance is satisfactory and their cost is low. These products are normally poor foamers and susceptible to water hardness. Their performance can be improved by the addition of chelating agents in the formulation or their replacement for synthetic surfactants, such as acyl isethionates (used mainly in syndet bars), which are

TABLE 9.5

Common Anionic Surfactants of Cosmetic Use

Anionic Surfactants	Manufacturing Processes
Acyl isethionates: sodium cocoyl isethionate	Esterification of a fatty acid with sodium isethionate.
Acyl sarcosinates: sodium lauryl sarcosinate	Reaction of a fatty acid chloride (most often in the coconut range) with sodium sarcosinate or *N*-methyl glycine.
Alkyl sulphosuccinates: sodium mono-, dialkylsulfosuccinates	Esterification of maleic acid, followed by sulphonation.
Alkyl sulphates and alkyl ether sulphates: sodium lauryl sulphate, sodium lauryl ether sulphates, ammonium lauryl sulphate	Alkyl sulphates and alkyl ether sulphates are commonly prepared using either sulphur trioxide and air or chlorosulphonic acid.

not sensitive to hard water. Hard water contains concentrations of salts of various di- and trivalent ions, notably calcium and magnesium. It is usually given in units of parts per million of calcium carbonate. In the United States, soft water is usually defined to be <60 ppm $CaCO_3$, while hard water is usually defined to be >120 ppm (The Soap and Detergent Association, 1987).

The alkyl sulphates and alkyl ether sulphates are the most commonly used products as foamy surfactants in hair and skin cleansers. Among the alkyl sulphates, the most important are the lauryl sulphates that have wide application in the cosmetic industry. Lauryl sulphates are obtained from the sulphonation of lauryl alcohol (from natural sources) with sulphonating agents (chlorosulphonic acid or sulphur trioxide), followed by neutralization with soda, triethanolamine, ammonia or monoethanolamine. The main characteristics of the alkyl sulphates are their high foaming power, easy to thicken, good solubility in water, pleasant odour and complete biodegradability. In the alkyl ether sulphates class, the most important products are the lauryl ether sulphates because of their excellent properties. They are more hydrophilic than their non-ethoxylated counterparts, have much lower irritability to eyes and skin, low set point, easy control of viscosity from addition of electrolytes and greater tolerance to water hardness. Lauryl ether sulphates are used in liquid soaps, shower gels, foam baths and shampoos. A 2 mol of ethylene oxide grade is generally preferable if a rich and stable foam is required or if viscosity is desirable in the final products. The 3 mol of ethylene oxide grade is more suitable when highly aerated foams must be generated. In general, increasing the degree of ethoxylation reduces skin and eye irritation.

Although alkyl sulphosuccinates produce little foam, have low detergent power and their formulations are rather difficult to thicken, they are very mild to the eyes and skin, have low toxicity and have excellent wetting power. They have low solubility in water, but this feature can be improved by alkoxylation with two or three moles of ethylene oxide. Sulphosuccinates are preferably used in baby and children's shampoos and other mild formulations, generally associated with a lauryl ether sulphate to improve their foam, detergency and thickening properties. The sulphosuccinate is produced by the condensation of maleic anhydride (or maleic acid) with the hydroxyl group of ethoxylated lauryl alcohol, giving the maleic mono lauryl ether acid, which is then reacted with sodium bisulphite to give the sodium mono lauryl ether sulphosuccinate.

Amino acid derivatives are a class of naturally derived anionic surfactants especially useful in applications where mildness is highly desired, such as facial cleansers, baby products and oral care products.

Cationic Surfactants

Cationic surfactants or cationics are characterized by having a positively charged hydrophilic group on a nitrogen atom attached to the hydrophobic fatty chain, usually known as the quaternary group (or simply quat) (Surfactant Science Series, 1990). They are often used for their substantivity (adsorption on substrates, for instance, hair and skin). Some important properties of these surfactants make them useful as bactericides in liquid soaps and mouthwashes, and antistatic agents in hair products. Conditioners are products used after hair washing, therefore their formulations do not need to contain anionic surfactants, since most of which are incompatible. This makes it possible to use cationic surfactants as antistatic and conditioning agents for hair fibres in formulations of conditioners, masks and combing creams.

TABLE 9.6

Common Cationic Surfactants of Cosmetic Use

Cationic Surfactants	Manufacturing Processes
Based on fatty nitrile: primary, secondary and tertiary amines, quaternary ammonium salts, alkyl diamines, ethoxylated amines quats	Reaction of fatty acids and ammonia. Conversion of dialkylamines to quaternaries.
Derived from fats and oils or their fatty acids: amidoamines, imidazolines, and their quaternized and/or ethoxylated derivatives	Amidoamines and imidazolines: reaction of fatty acids or tallow with various amines. The resulting amidoamines and imidazolines can be ethoxylated and/or quaternized with methyl chloride or methyl sulphate. Ester quats: reaction of tallow fatty acid with methyl diethanolamine or triethanolamine. The resulting aminodiester is then quaternized with methyl chloride or methyl sulphate.
alpha-Olefins or detergent alcohols derivatives: alkyl dimethylamines, dialkyl methylamines, dialkyl dimethylamine quaternaries, quaternary ethosulphate	Conversion to alkyl dimethylamines and further to fatty amine oxides or quaternaries. Some alkyl dimethylamine cationic surfactants are converted to alkyl betaine amphoteric surfactants.

The change in the lipophilic nature of the hair surface due to the adsorption of the cationic agent allows other ingredients of the formulation to be more compatible with hair. Thus, the cationic agents enable the deposition of fatty materials and emollients (oily components) on the hair surface, substances that contribute synergistically to the conditioning effects of the cationic agents.

The three groups consist of products derived from fatty nitrile intermediates, directly from fats and oils or their fatty acid derivatives, from alpha-olefins or fatty alcohols. Indeed, some individual product types may be derived from more than one source. Cationic surfactants are mainly based on long carbon chains (C16, C18, C22, etc.). Representative examples of cationics are listed in Table 9.6.

Non-Ionic Surfactants

Non-ionic surfactants or non-ionics are characterized by having non-charged hydrophilic groups attached to the fatty chain (Surfactant Science Series, 1998).

Non-ionics can be divided into ethers (the largest group), esters, amides and other miscellaneous surfactants. The ethers can be further subdivided into linear alcohol ethoxylates or propoxylates and others. Some representative examples of non-ionics are listed in Table 9.7.

Non-ionics are much less hard-water sensitive than anionics and normally generate less foam but are not very good detergents. They are very compatible with most raw materials used in cosmetics, have low irritancy to the skin and eyes and reduce surface tension efficiently, among other properties. These characteristics allow these surfactants to be used as emulsifiers.

Among the non-ionic surfactants, fatty alkanolamides are very popular in shampoos, because they present good thickening properties, contribute to stabilization of foam, enhance the solubilization of fatty esters, glycols, alcohols, essential oils, and so forth, and their conditioning effect and low detergency reduce the dryness caused by the anionic surfactants. Cocamide MEA, cocamide MIPA and cocamide DEA are very well-known alkanolamides, but the latter is being currently replaced due to toxicological concerns.

Other components of the non-ionic class are surfactants derived from polyols, such as glycerol esters. Glyceryl monostearate is most commonly used in lotions, creams and lipsticks. Glycol esters are used as emulsifying agents, dispersants, consistency agents, opacifiers and pearlizers. Many non-ionic surfactants are used as emollients, acting in the prevention and relief of dryness of the skin, as well as in its protection. They are substances that give softness and flexibility to the skin. They act by retaining water in the stratum corneum by forming a water-in-oil emulsion. The emollients also enhance the spreadability of the formulation, increase the penetration of active ingredients into the skin, assist in the dispersion of pigments, and act as emulsifiers and co-solvents.

Some non-ionic surfactants are used as solubilizers of fragrances, for example, derivatives of lauryl alcohol ethoxylated with two or three moles of ethylene oxide.

TABLE 9.7

Common Non-Ionic Surfactants of Cosmetic Use

Non-Ionic Surfactants	Manufacturing Processes
Alcohol ethoxylates	Ethoxylation of aliphatic alcohols with ethylene oxide. Alcohol ethoxylates can subsequently be converted to alkyl ether sulphates. Secondary alcohol ethoxylates are produced by ethoxylation detergent-range linear secondary alcohols in the presence of an acidic catalyst. The product is then further ethoxylated in the presence of an alkaline catalyst.
Alkyl polyglucosides and alkyl oligoglucosides: lauryl polyglucoside, decyl polyglucoside, decyl maltoside	Reaction of anhydrous glucose or n-butyl glucoside (from glucose and n-butanol) with fatty alcohols or fatty acids.
Alkyl dimethylamine oxides: lauryl dimethylamine oxide, stearyl dimethylamine oxide	Reaction of tertiary amines (e.g. fatty alkyl dimethylamine or fatty amines doubly ethoxylated at the nitrogen atom with hydrogen peroxide. Ethoxylated amine oxides.
Ethoxylated and propoxylated products: ethoxylated C12-C14 alcohols	Ethoxylation and propoxylation of detergent alcohols/
Fatty acid esters, fatty acid ethoxylates: monoglycerol fatty acid esters, ethoxylated sorbitan and sorbitol fatty acid esters, sucrose fatty acid esters, polyethylene or propyleneglycol fatty acid esters	Alkoxylation of fatty acids with ethylene oxide. Esterification of fatty acids with polyethylene glycols. Ethoxylated sorbitan and sorbitol fatty acid esters are produced by esterification of anhydrous sorbitol or sorbitol with fatty acids, followed by ethoxylation of the esters. Ethoxylated castor oil and ethoxylated hydrogenated castor oil are produced by reacting castor oil or hydrogenated castor oil with ethylene oxide.
Fatty alkanolamides (mono- and dialkanolamides)	Condensation of fatty acids or their methyl esters with alkanolamines or direct reaction of the oils with alkanolamines. Ethoxylated fatty alkanolamides (which have the characteristics of cationic surfactants) are prepared by reacting fatty alkanolamides with ethylene oxide.

Non-ionic surfactants can also be used as agents of consistency, especially ethoxylated cetostearyl alcohol with 20 moles of ethylene oxide, normally used in conjunction with the starting material of its synthesis (non-ethoxylated cetostearyl alcohol) in hair conditioners and creams.

Alkyl polyglucosides are a relatively new family of surfactants. They are derived from plant sugars and fatty alcohols (Hill, von Rybinsky, and Stoll, 2008). The resulting molecules are non-ionic surfactants with good water solubility due to their hydroxyl groups. They have a much higher tolerance to electrolytes than other non-ionic surfactants based on ethylene oxide. In addition, they are also good detergents and have a very high degree of biodegradability.

Amphoteric Surfactants

Amphoteric surfactants or amphoterics (or zwitterionic surfactants or zwitterionics) are characterized by having positive and negative sites in the same molecule in their polar portion (Surfactant Science Series, 1981a). The positive group is usually represented by a quaternary nitrogen atom and the negative group by a carboxylate or sulphonate group. Properties such as solubility, detergency, foaming and wetting power of amphoteric surfactants are dependent on the pH of the formulation and the length of the constituent fatty chain.

Amphoterics are generally relative mild, which makes them ideal for use in many personal care products. They are compatible with all types of surfactants. Amphoterics have excellent synergy with anionics, improving foam, mildness, detergency and thickening properties of the formulations containing them.

Amphoteric surfactants used in the cosmetic industry are imidazoline derivatives, betaines, amidobetaines and sulphobetaines, the most widely used being imidazoline derivatives and betaines.

The imidazoline derivatives are obtained from the condensation of coconut fatty acid with monoethyl ethanolamine, resulting in an imidazoline, which reacts with sodium monochloroacetate (one or two moles) to produce the amphoteric imidazoline. These compounds are commonly referred to as amphoacetates or propionates when the reaction is carried out with sodium monochloroproprionate.

Betaines are obtained by the reaction of cocodimethylamine with sodium monochloroacetate to obtain the cocobetaine or by condensation of the coconut fatty acid with dimethylaminopropylamine to obtain the cocamidopropylamine, which is subsequently reacted with sodium monochloroacetate to result in the cocamidopropyl betaine. Sulphobetaines are much less popular than alkyl betaines and amidobetaines, but they present good detergency and mildness, which make them useful in facial cleansers and make-up removers.

Some common amphoteric surfactants of cosmetic uses are listed in Table 9.8.

Surfactant Properties

Adsorption and Reduction of Surface and Interfacial Tension

Water is the most abundant substance on Earth, with a total volume of approximately 1,360,000,000 km^3, with 97.2% in oceans, 1.8% as ice, 0.9% in ground water, 0.02% in fresh water and 0.001% as water vapour. Water is a polar molecule due to difference of electronegativity between its atoms in an angular bond. The partially positively charged hydrogen atoms are at an angle of 104.5 degrees from each other with respect to the partially negatively charged oxygen. Consequently, the charge distribution and molecular geometry produce a considerably high dipole moment of 1.85 D. This polar structure is responsible for water's exceptionally strong self-attraction, resulting in a high cohesion and numerous unusual properties (Stillwell, 2013).

Water has an important ability to readily form multiple hydrogen bonds with itself and other polar substances. The water molecules at the air–water interface experience a net force of attraction that pulls them away from the interface and back into the bulk of the liquid. As a consequence, the entire liquid tends to take on a shape with the smallest possible surface contact area with air (or oil), resulting in a high surface tension (72–73 mN/m at 20°C) (Meyers, 2006). The intermolecular force of adhesion between water and a lipid substance (which can be dirt to be removed from skin or an active to be solubilized in a solution or an oil and water emulsion) is much lower than the intramolecular force of cohesion between

TABLE 9.8

Common Amphoteric Surfactants of Cosmetic Use

Amphoteric Surfactants	Manufacturing Processes
Alkyl betaines: alkyl dimethyl betaine, cetyl dimethyl betaine, lauryl–myristyl betaine	Reaction of tertiary amines (usually the fatty alkyl dimethylamines) with sodium monochloroacetate.
Alkylamido betaines: cocamidopropyl betaine, alkyl amidopropyl betaine, lauramidopropyl betaine	Reaction of fatty acids with diamines (usually dimethylaminopropylamine) to yield an amidoamine, which is then reacted with sodium monochloroacetate to yield the alkylamido betaine.
Amphoacetates (glycines, imidazoline): sodium lauroamphoacetate, sodium cocoamphoacetate	Reaction of sodium monochloroacetate or acrylic acid and long-chain substituted imidazolines or their precursors. Imidazoline is a reaction product of a fatty acid and aminoethylamine.
Sulphobetaines (amidopropyl hydroxysultaines, hydroxysultaines, or sultaines): sulphobetaines, cocamidopropyl hydroxysultaine, lauryl hydroxysultaine	Reaction of amidoamine intermediate, epichlorohydrin and sodium bisulphite. Alkyl amphopropyl sulfonate is prepared from epichlorohydrin, sodium bisulphite, and the same imidazoline/ amidoamine intermediate used to prepare the imidazoline or glycine amphoteric surfactants.
Amphopropionates: sodium lauriminodipropionate, sodium capryl amphopropionate	Reaction of alkylamines with methyl acrylate, followed by hydrolysis to the corresponding beta-alkyl aminopropionic acid.

the water molecules. The system minimizes its surface contact area by separating water from the lipid substances (Tanford, 1978; Stillwell, 2013).

As previously mentioned, surfactants accumulate at the air–water interface where their non-polar tails protrude into the air or lipid phase, avoiding unfavourable interactions with water, while the polar end of the molecule interacts favourably with water. The surfactants replace some of the surface molecules of water, decreasing the net surface tension. The most important methods for measuring interfacial properties in equilibrium conditions include the du Noüy ring method, Wilhelmy plate method, capillary rise method and spinning drop method (Surfactant Science Series, 1999).

The relationship between the molecular structure, the rate and degree of adsorption at the interface or surface, under certain conditions, differentiates the various types of surfactants and determines their use where the reduction of surface tension and related phenomena are required. The molecular characteristics required for a material to act as a surfactant in aqueous solution have been extensively discussed in the literature (Meyers, 2006; Rosen, 2004). However, it may be useful to discuss the basic functions of the molecular groups of the surfactants to understand their effects on interfacial and surface tension. The application of a surfactant greatly depends on its solubility in water, which ultimately can be adjusted by selecting the proper hydrophilic and hydrophobic groups. Often such a choice can be determined empirically with the help of specific tools (HLB system, for instance), discussed later. Fortunately, for cosmetic applications, there is a large 'menu' of surfactant options, which will fulfil the formulation and application requirements, and we can always adjust the formulation parameters and its composition to make it work. The 'net' solubility of a surfactant molecule leads to its preferential adsorption at the interfaces and changes its interaction energy with the interface of the liquid and the molecules in contact with it. Each of these functions are mainly related to the chemical nature of the hydrophobic group and, in some minor cases, to the hydrophilic group. When neither the proper choice of a surfactant nor the formulation adjustment is possible, then we can always 'think' of developing new surfactant molecules. Indeed, the cosmetic industry's special needs have always 'raised the bar' in the surfactants business.

The location of surfactant molecules at air–liquid, liquid–liquid and liquid–solid interfaces is of great interest to many applications in cosmetics, including the removal of undesirable materials from skin and hair (skin cleansers and shampoos), the change in wetting characteristics of a surface (waterproof products, products that should be spread on the skin and hair, hair conditioners), foams, and stabilization of solids or liquids finely divided in a liquid where there is no stability (stabilized dispersed or emulsified systems). In these and in numerous surfactant applications, the ability of the surfactant molecule to produce the desired effect is controlled by the chemical nature of the components of the system.

Wetting Properties

The ease and intensity of adsorption of surfactants at the solid–liquid interface dictate the application of the surfactants in cosmetics and are controlled by the chemical nature of the surfactant (including the nature of the hydrophilic and hydrophobic groups), the nature of the solid surface on which the surfactant is being adsorbed and the nature of the liquid. A small change in one of these three factors can significantly affect the adsorption process, and the sensitivity of the process results from the different types and magnitude of the operant adsorption mechanisms, which depend on the mechanisms that occur in all the, for example, intermolecular, electrostatic, dipole and van der Waals interactions, in addition to the hydrophobic effect (more specific to the area of surface activity), acid–base interaction, adsorption by electron polarization, adsorption by dispersion forces and hydrophobic bonding. The interactions range from very weak to stronger, more specific interactions between surfactant molecules and surfaces of opposing electric charges. The effect of the hydrocarbon chain within a homologous series of ionic surfactants shows that many of their characteristics, including surface tension and adsorption, vary regularly with the hydrophobic chain length.

The wetting power is the ability of a liquid substance to moisten or wet a solid surface, promoting its rapid spread over the surface (Figure 9.3 shows three different representations of wetting). A good surfactant is one which promotes the least contact time required for a liquid to moisten a solid, thereby providing and rapid wetting. Perhaps the most widely used method for evaluating the wetting power of surfactants is the Draves test (American Society for Testing and Materials, 1968) in which a piece of

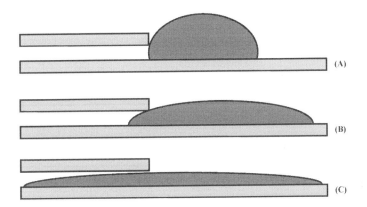

FIGURE 9.3 Representation of (a) poor wetting, (b) intermediate wetting and (c) very good wetting. (From Daltin, D., 2011, *Tensoativos: Química, Propriedades e Aplicações*, São Paulo: Edgard Blücher.)

cotton cloth tied to appropriate weights is placed on the surface of a surfactant solution and the time for complete wetting of the sample at a given concentration of surfactant, temperature, electrolyte concentration and so forth is determined. The standard conditions for the test are low ionic strength water, temperature of 25°C and a surfactant concentration of 0.1% by weight. Typical wetting times for various common surfactants are in the range of few seconds to several minutes (the wetting power of pure water is about 8 hours).

A method of correlating the structure of the surfactant with the wetting characteristics is through the use of hydrophile–lipophile balance value (HLB; discussed later). As a general rule, surfactants having HLB in the range of 7 to 9 exhibit better wetting characteristics for aqueous solutions on most solid surfaces.

Classically, for cosmetic purposes, we distinguish three categories of wetting phenomena, although in principle the tendency is to imagine wetting as a simple phenomenon of covering a surface by a liquid, being classified as adhesive wetting (film formation on hair and skin), spreading wetting (rubbing of creams and lotions) and immersive wetting (hair and skin cleansing) (Rosen, 2004).

Aggregation

The molecular structure of the hydrophobic chain, the polarity of the hydrophilic group and the 'environment' (composition, pH, temperature, presence of co-solutes, etc.) of the solution dictate surfactant solubility. While many surfactants have appreciable solubility in water, such a characteristic can change significantly with changes in the hydrophobic group. In general, the aqueous solubility of ionic substances increases as the water temperature rises, dependent on the energy of their crystalline reticulum and its heat of hydration. Ionic surfactants frequently undergo a sudden and discontinuous increase in their solubility at a specific temperature, which is known as Krafft temperature, TK or Krafft point. Below the Krafft point, the solubility of the surfactant is determined by the crystalline energy and heat of hydration of the system, in other words, it is a thermodynamic-driven process. The concentration of the surfactant monomeric species in solution (an individual dissolved surfactant molecule) is limited to an equilibrium value determined by such properties. Above the Krafft temperature the solubility of the monomer increases and reaches a point at which another sudden change takes place with the formation of aggregate species.

The aggregates may be viewed as arrangements structurally similar to a solid crystal or crystalline hydrate. The energy change when passing from the crystal to the aggregate is less than the change to the monomeric species in solution. The formation of aggregates favours an increase in solubility. The concentration of the surfactants in monomeric form may increase or decrease very gradually at higher concentrations and constant temperature, but the aggregates predominate above a critical concentration of surfactant. The total solubility of the surfactant depends not only on the solubility of the monomeric material, but also on the 'solubility' of the aggregates. A schematic representation of the temperature/solubility relationship of ionic surfactants is shown in Figure 9.4.

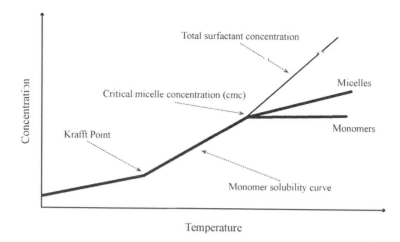

FIGURE 9.4 Temperature–solubility relationship for typical ionic surfactants. (From Meyers, D., 2006, *Surfactant Science and Technology*, 3rd ed., Hoboken, NJ: John Wiley & Sons.)

As water is added to the crystalline surfactant, the structure of the system undergoes a transition from a highly ordered crystalline state to a less organized state, usually referred as a mesophase, while a liquid phase is developed. These phases are characterized by having some physical properties of both fluid and crystalline structures, which means they have at least one dimension which is highly ordered, and the other dimensions correspond to phases that behave as unstructured fluids.

The liquid crystal structures or mesophases of surfactants are thermotropic (their structure and properties are determined by the temperature of the system) and lyotropic (their structure is determined by the specific interactions of the surfactant molecules with the solvent molecules).

One of several possible aggregates of surfactant in solution is called micelles. Micelles are the simplest aggregate form of surfactants. A complete understanding of surfactant systems, including correlations between chemical structures and surface properties, requires knowledge of the full spectrum of possible surfactant states (Laughlin, 1994). As shown in Figure 9.5, the aggregation possibilities range from highly ordered crystalline phases to dilute monomer solutions, which, while not completely unstructured, have an ordering only at the molecular level. Among these extremes are a variety of phases whose nature depends intimately on the chemical structure of the surfactant, the total composition of the continuous phase and the 'environment' of the system, as mentioned before.

Micelles and other aggregates can constitute good delivery systems for many kinds of cosmetics. A delivery system is a vehicle that makes an 'active' available to a target site. An effective delivery system is one that reaches the target and creates a high concentration reservoir for the active (Rosen, 2006).

Micelles

At the beginning of the 20th century, it was found that aqueous solutions containing surfactants did not behave according to the expected pattern for most solutions when the solute concentration is increased. In 1913, it was suggested that the unusual behaviour of surfactants could be attributed to an arrangement of individual molecules in solution aggregates above a well-defined concentration. (McBain, 1913; McBain and Martin, 1914). In 1920, it was reported that the osmotic activity of potassium stearate solutions indicated the presence of a considerable degree of association and suggested that the aggregate species be called micelles. To explain the changes in conductivity and osmotic activity, the formation of two distinct types of micelles was proposed: spherical species composed of ionic molecules of salts and non-ionic aggregates of lamellar structure. Subsequently, Hartley interpreted the results of such studies in terms of a single type of structure and in his model the micelles are essentially spherical with diameter approximately equal to twice the length of the hydrocarbon chain. Hartley suggested that the structure of micelles was composed of 50 to 100 molecules and that the association occurred in a narrow range

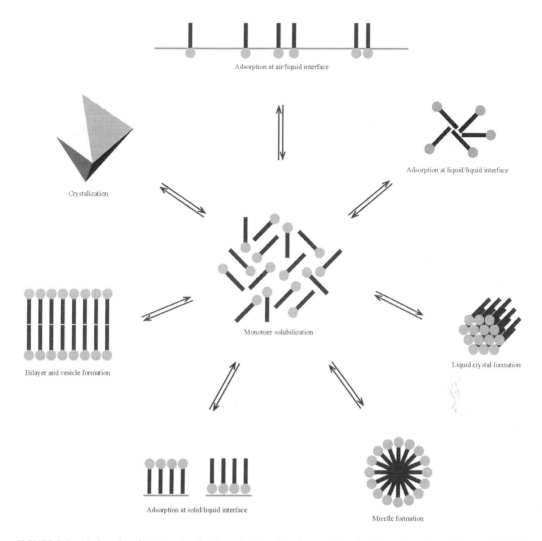

Adsorption at air/liquid interface

Crystalization

Adsorption at liquid/liquid interface

Monomer solubilization

Bilayer and vesicle formation

Liquid crystal formation

Adsorption at solid/liquid interface

Micelle formation

FIGURE 9.5 Modes of surfactant action for the reduction of surface and interfacial energies. (From Meyers, D., 2006, *Surfactant Science and Technology*, 3rd ed., Hoboken, NJ: John Wiley & Sons.)

of concentration. The interior of the micelle was described as being essentially hydrocarbon while the surface consisted of hydrophilic groups. Other proposed structures included spherical, lamellar, inverted (or reversed), disk and cylindrical or rod-like micelles, shown in Figure 9.6 (Meyers, 2006).

It is generally accepted that most surfactant molecules in aqueous solution may aggregate into structures containing 30–200 monomers in which the hydrophobic moieties of the molecules are associated and mutually protected from extensive contact with the aqueous phase (Meyers, 2006).

The formation of micelles has been explained by (1) the mass action model, in which the micelles and the monomeric species are in chemical equilibrium, and (2) the phase separation model in which the micelles constitute a new phase formed in the system above a certain concentration of surfactant (the critical micellar concentration, cmc). In each case, classical thermodynamic approximations are used to describe the micellization process (Rosen, 2004).

Early studies of the properties of surfactant solutions showed that small changes in concentration could cause very significant variations in the characteristics of these systems. The measurements of several properties show a very abrupt change as the surfactant concentration is increased, indicating some change in the solution, serving as evidence of the formation of aggregates or micelles in surfactant solutions in well-defined concentrations, as shown in Figure 9.7.

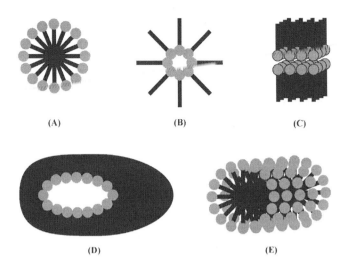

(A)   (B)   (C)

(D)   (E)

FIGURE 9.6 Five of the proposed micelle shapes, as interpreted from experimental data: (A) spherical; (B) lamellar; (C) inverted (or reversed); (D) disk; (E) cylindrical or rod-like. (From Meyers, D., 2006, *Surfactant Science and Technology*, 3rd ed., Hoboken, NJ: John Wiley & Sons.)

When a molecule containing both a hydrophilic group and a hydrophobic group is introduced into water, a distortion of the water structure occurs to accommodate the solute molecules, resulting in an increase in the free energy of the system. The physical result of this is the tendency of the solute to adsorb at the solution interface where preferred molecular orientations may reduce the free energy of the solution or the formation of molecular aggregates with their hydrophobic moieties directed into the core of the micelle. Micelle formation, therefore, is an alternative mechanism for reducing the free energy of the solution to minimize the distortion of the water structure.

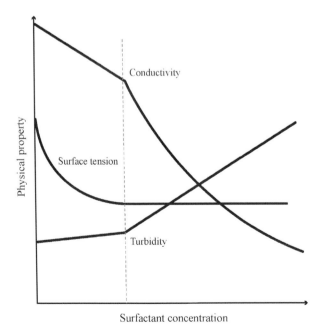

FIGURE 9.7 Representative effects of micelle formation on the physico-chemical properties of surfactant solutions. (From Meyers, D., 2006, *Surfactant Science and Technology*, 3rd ed., Hoboken, NJ: John Wiley & Sons.)

The micellization in a given surfactant system and the concentration at which micelles are formed are determined by the relative balance of the forces that favour and retard the molecular aggregation process. Since the magnitude of the opposing forces is determined by the chemical compositions of the surfactant molecules, where all other aspects (temperature, solvent pressure, etc.) are kept constant, it is the chemical structure of the surfactant molecules that ultimately controls the phenomenon. It is then possible to make reasonable generalizations about the characteristics of the surfactant micellization process and their chemical structure. In aqueous solutions it is generally found that the greater the length of the hydrophobic chain in a homologous series of surfactants, the greater the aggregation.

Many other factors affect the critical micelle concentration of surfactant systems, besides the nature of the hydrophobic chain. The type of hydrophilic group, including the counterions and the role of external factors not directly related to the chemical nature of the surfactant, is also important.

The length of the hydrophobic chain of the surfactant is taken as the main determinant factor of the cmc. Figure 9.8 shows how the cmc decreases as the number of carbon atoms in the hydrophobic chain of a homologous series of surfactants increases.

The effect of the hydrophilic group on the critical micelle concentration of a series of surfactants with the same hydrophobic chain may also vary considerably. In aqueous solution, for example, the critical micelle concentration for a C12 hydrophobic chain with an ionic group is close to 0.001 mol/L, whereas for a non-ionic of the same hydrophobic chain it is in the range of 0.0001 mol/L (Meyers, 2006).

Most, if not all, industrial surfactant applications involve the presence of co-solutes and other additives in aqueous surfactant solution. These co-solutes can potentially affect the micelle formation process through specific interactions with the surfactant molecules (thus altering the efficiency of the surfactant in solution) or by altering the thermodynamics of the micelle formation process by changing the nature of the solvent or of the various interactions leading, or opposing, to the formation of micelles. Examples of specific interactions between molecules of surfactants and co-solutes are common when the system contains polymeric materials. Commonly encountered changes in solution include the presence of a neutral electrolyte, pH changes, the addition of organic materials and the effect of temperature. When considering the possible reasons for the temperature effects on the critical micelle concentration, one of the key points is the degree of hydration of the hydrophilic group, since the structuring of the water molecules is very sensitive to changes in temperature. Ethoxylated surfactants that rely exclusively on hydrogen bonds for solubilization in aqueous solution commonly exhibit a phenomenon of inverse solubility, known as the cloud point. The cloud point is the temperature at which the solubility of the surfactant is such that its solution goes from clear to cloudy with phase separation. The dependence of the cmc of the

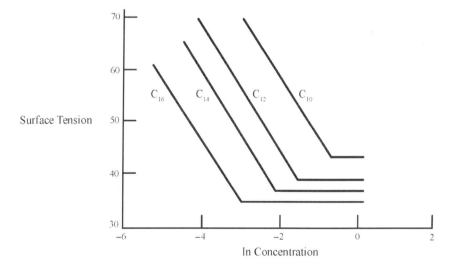

FIGURE 9.8 A schematic representation of the effects of hydrophobe chain length on surfactant cmc and surface tension reduction. (From Meyers, D., 2006, *Surfactant Science and Technology*, 3rd ed., Hoboken, NJ: John Wiley & Sons; and Rosen, M. J., 2004, *Surfactants and Interfacial Phenomena*, 3rd ed., Hoboken, NJ: John Wiley & Sons.)

ethoxylated non-ionic surfactants with temperature is especially important, since the interaction of the hydrophilic group of the surfactant with water molecules is essentially through hydrogen bonds.

Cmc determinations of binary mixtures of well-defined surfactants show that the greater the difference in cmc of the components of the blend, the greater is the change in the cmc of the component having the largest hydrophilic group, as the length of the hydrophobic chain is increased. The observed differences are explained as relatively small changes in the mole fraction of the lower carbon component due to the preferential aggregation of the more hydrophobic material and the difficulty in including the longer chain materials in the micelles relative to the components of shorter chain. The cmc of ethoxylated non-ionic surfactant mixtures is of particular interest, since in the synthesis of these compounds a mixture of oligomers is always produced.

Many mixtures of surfactants, especially ionic and non-ionic surfactants, have application properties significantly superior to those obtained from the individual components. Such synergistic effects greatly enhance many technological applications in the fields of emulsion formulation, emulsion polymerization and surface tension reduction, all of which are useful for cosmetics and personal care products. The use of mixed systems should always be considered as a method of achieving optimal performance for any practical application. Although mixtures of anionic and cationic surfactants in aqueous systems frequently result in precipitation of stoichiometric amounts of the two materials due to the ionic pairing of the hydrophilic portions of the two molecules, there are exceptions (Joensson et al., 1991).

The micelle formation process can be represented by Figure 9.9 and summarised as follows. As the amount of surfactant is increased, it adsorbs to all surfaces and interfaces of the system (air–liquid surface and container surface) until it is completely saturated. After this point, if more surfactant is added to the system, it will be molecularly dispersed in the aqueous solution. The molecular dispersion of the surfactant in water creates favourable interactions between its polar head and water, but also creates unfavourable (more numerous) interactions between its non-polar tail and water. The net result of the dispersion of the surfactant in water is an increase of the free energy of the system and is therefore thermodynamically unfavourable. The self-association of the surfactants in a micellar arrangement minimizes tail–water interactions and reduces the free energy of the system, since in the micelle the tails of the surfactant molecules are in contact with each other in favourable interactions, forming an essentially hydrophilic–polar topography from the polar heads of the surfactant molecules. The formation of micelles is a spontaneous and reversible process, and thus the micelles are thermodynamically stable, coexisting in equilibrium with the monomeric molecules of the surfactant in water. Substances essentially non-polar, such as essential oils and cosmetic actives of a lipid nature, can be 'dissolved' in the micelles core, resulting in relatively stable solubilized, emulsified or microemulsified systems.

Detergency

Detergency is a surface and colloidal phenomenon that reflects the physicochemical behaviour of matter at interfaces. Cleaning a solid substrate involves the removal of undesirable foreign matter from a given surface. The term 'detergency' is restricted to systems where the cleaning process is conducted in a liquid medium, a result of interfacial interactions between dirt, substrate and solvent system, and the process is not the solubilization of the dirt in the solvent, although this may contribute to the overall process.

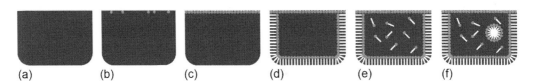

(a)   (b)   (c)   (d)   (e)   (f)

FIGURE 9.9 Representation of micelle formation in its various stages: (a) initial adsorption of surfactant molecules to air–water surface and consequent surface tension reduction; (b) additional adsorption of surfactant molecules to air–water surface and incremental surface tension reduction; (c) maximum adsorption of surfactant to the surfaces of the system and consequent maximum surface tension reduction; (d) 'solubilization' of surfactant molecules in aqueous solution and no additional surface tension reduction; (e) saturation of surfactant molecules in aqueous solution and no additional surface tension reduction; (f) micelle formation and no addition reduction on surface tension.

The essential difference between detergents and other surfactants lies in their ability to 'drag' dirt from a solid surface, that is, during the washing process. The greater or lesser capacity that a detergent has for removing dirt is closely linked to wetting and the reduction of surface tension. One way to measure the detergency of surfactants is basically to measure the amount of grease that a solution can remove from a given substrate within a given time. In detergency, the interaction between solid substrates and dispersed or dissolved materials is of fundamental importance.

Surfactants, being a class of substances that preferentially adsorbs at several types of interfaces due to their amphiphilic structure, are very useful in the detergency process. In most detergent-related adsorption processes it is the interaction of the hydrophobic moiety of the surfactant molecule with dirt particles that produces the detergent action. Such adsorption changes the chemical, electrical and mechanical properties at the various interfaces and depends strongly on the nature of each component. In cleaning the hair with anionic surfactants, for example, the adsorption of the surfactant onto the hair fibre and onto the dirt particles introduces repulsive electrostatic interactions that tend to reduce adhesion between the dirt and the hair, suspending the dirt and preventing its redeposition. The process is illustrated schematically in Figure 9.10. With non-ionic surfactants the mechanism is less clear. However, the steric repulsion between layers of adsorbed surfactants and solubilization are extremely important.

In general, there are two types of dirt found in detergent systems: liquid and oily materials, and particulate solid materials. Solid soils may consist of proteins, clays, carbon (soot) of various surface characteristics, metal oxides and so forth resulting from cellular metabolism, scaling of the skin and scalp, environmental pollution, and cosmetics. Liquid dirt may contain skin lipids, alcohols and fatty acids, mineral and vegetable oils, synthetic oils, and liquid components of creams and cosmetics. The three main types of lipids found in the stratum corneum are fatty acids, ceramides and sterols. In addition to the epidermal barrier lipids, a thin film of surface lipids covers the skin. These lipids are known as sebum, which is produced by the sebaceous glands. Sebum is composed of 60% triglycerides, 25% wax esters, 12% squalene, and 5% sterols and their esters (Downing et al., 1983).

The interfacial interactions of each of these with the solid substrate are usually very different and the dirt removal mechanisms may be similarly different. The removal of lipids from hair and skin is mainly via micellar solubilization and/or emulsification. Solubilization of oily materials in surfactant micelles is probably the most important mechanism for removal of oily dirt from hair and skin. Significant detergency only occurs above the cmc, reaching its maximum well above it. Since the adsorption of surfactants at interfaces involves the monomers, and the solubilization involves only the micellar form, the results seem to indicate that the solubilization is more important than the adsorption mechanisms in the general cleaning process (Meyers, 2006).

The degree of solubilization of the oily soil depends on the chemical structure of the surfactant, its concentration in the process and the temperature. At concentrations well above the critical micellar concentration (10- to 100-fold), larger micellar structures, which have a greater solubilizing capacity, can be found or some mechanism related to microemulsion formation may predominate. For a more efficient solubilization to occur, the contact between the surfactant in solution and the dirt should be maximized. The main mechanism for such a process is the 'rollback mechanism' (Figure 9.10), in which the adsorption of the surfactant at the dirt–solution and solid–solution interfaces causes the displacement of dirt by the cleaning solution. Once the oil droplet is formed in the solution, the increased interfacial area

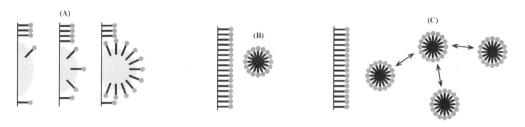

FIGURE 9.10 Rollback mechanism for detergency with (A) adsorption of the surfactant onto surfaces and dirt, (B) removal of the dirt by micelles, and (C) anti-redeposition of the dirt due to electrostatic or steric repulsion. (From Daltin, D., 2011, *Tensoativos: Química, Propriedades e Aplicações*, São Paulo: Edgard Blücher.)

accelerates the solubilization process. Alternatively, the oil droplet may be stabilized in the form of an emulsion by adsorption of more surfactant.

As a rule, surfactants with 12 and 14 atoms of carbon in the hydrophobic chain are the best detergents, because they have good solubility in water, form micelles quickly and adsorb very rapidly onto the soils, weakening their adhesion onto the skin and hair surface. Their ionic head groups prevent the 'micellized' or dispersed dirt from being redeposited onto the recent cleaned surface by electrostatic repulsion (or steric repulsion in the case of non-ionic surfactants).

Emulsion Formation

Emulsions are known to be relatively stable mixtures of water and oily or fatty components in the presence of a surfactant. Emulsions are very useful in cosmetics since they allow the formulation of a variety of creams and lotions with different actives for protection and treatment of skin. The different sensorial aspects required for each type of skin (dry, oily or mixed) and type of product (sunscreens, moisturizing lotions, anti-wrinkle creams, body products, facial products, etc.) can also be catered to. Water is the most commonly used ingredient in cosmetic emulsions. As not all ingredients of cosmetic emulsions are miscible with water, it is necessary to suspend heterogeneous mixtures of water with other raw materials such as paraffins, vegetable oils, natural and synthetic waxes, and inorganic pigments through emulsification.

Immiscible substances have low cohesive forces and therefore high interfacial tension, as previously discussed. Water and oil are mutually insoluble components, so that the emulsions are composed of separate phases stabilized by components which have affinity for both oil and water. To the human eye, the emulsion presents as a homogeneous phase. Emulsions are, however, heterogeneous systems, consisting of at least one immiscible liquid dispersed in another in the form of droplets, the diameter of which is generally greater than 0.1 μm. The interface increase involved in the emulsification process of 1 mL of two immiscible liquids in a test tube of 1 cm^2 cross section is 6×10^4 cm^2 (Surfactant Science Series, 1999). Such systems are naturally unstable due to increased surface area (and surface tension), but they are rendered stable using interfacial surfactants mixtures, known as the emulsifying system, or, using finely divided solids as in Pickering emulsions (Becher, 1965).

Within the two-phase liquid emulsion system, the liquid divided into droplets is known as the dispersed, discontinuous or internal phase. The surrounding liquid is the continuous or external phase. Cosmetic emulsions are mostly composed of oil phase, aqueous phase, emulsifiers and other specific additives. Emulsions are classified into four types according to the particle size of their dispersed or internal phase (Rosen, 2004):

Macroemulsions – Opaque, thermodynamically unstable emulsions with particles larger than 400 nm (0.4 μm), easily visible under a microscope.

Microemulsions – Thermodynamically stable, isotropic, transparent dispersions with droplets smaller than 100 nm (0.1 μm) and low viscosity (Newtonian).

Miniemulsions – An intermediate type between the first two, with particle size ranging from 100 to 400 nm (0.1 to 0.4 μm) and bluish-white appearance.

Multiple emulsions – The dispersed particles are emulsions thereof. In reality they are drops of a liquid dispersed in larger droplets of another liquid, which is then dispersed in a continuous phase, which is generally of the same nature as the first liquid.

The lower the interfacial tension between two immiscible liquids (water and oil) the greater is the ease of emulsifying them. The higher the interfacial tension between them, the greater is the mechanical energy that must be provided to the system for emulsification. As the interfacial tension between two pure immiscible liquids is always greater than zero, the dispersion from one liquid into another produces an increase in the interfacial area, resulting in a corresponding increase in the interfacial free energy of the system. The emulsion produced is therefore thermodynamically unstable, tending to phase separation to achieve a minimum of interfacial area. It is for this reason that immiscible liquids, when pure, do not form an emulsion and will separate in a process known as coalescence. Thus, the presence of a third component having affinity with each phase and the properties of migrating, adsorbing and accumulating at the interface are necessary to reduce the interfacial tension between the phases and facilitate the formation of the emulsion. This component is a surfactant and because in this case it is responsible for the emulsification of the phases, it is known as an emulsifier or emulsifying agent.

The balance between the philic and phobic portions of a surfactant molecule dictates how (and if) the surfactant will behave as an emulsifier. The HLB (hydrophile–lipophile balance value) is an indication of its emulsification behaviour. In the HLB system, surfactants are classified according to their hydrophilicity.

The nature and characteristics of the emulsion are largely determined by the nature of the emulsifier or combination of emulsifiers used by the formulator. The predominance of hydrophilic emulsifiers favours the formation of the oil-in-water (O/W) type emulsion. The lipophilic fraction of the surfactant molecule adsorbs to the oil droplets, while the hydrophilic or polar fraction of the molecule remains hydrated by water. In this way the oil droplets are as if they are 'entangled' by the molecules of surfactant that form a molecular film at the O/W interface. A common classification of emulsifiers is in the function of their molecular structure and considers the nature of their polar groups.

The self-emulsifying bases or waxes are composed of mixtures of water-soluble surfactants of high HLB and oil phase consistency agents (fatty thickeners such as cetostearyl alcohol and glyceryl monostearate) of low HLB, incapable of forming emulsions spontaneously. The advantage of these bases is the ease of preparation of the cream, but as a general rule they accept up to 10% of emollients. Above this they require additional emulsifiers.

The term 'stability', when applied to emulsions, usually refers to the resistance of the emulsions to the coalescence of their dispersed droplets. Flocculation, creaming and sedimentation represent a state in which the drops touch but do not combine to form a single drop. Thus, the most important factor for emulsion instability is coalescence that leads to the emulsion breaking, although visually flocculation, creaming and sedimentation are not desired. Figure 9.11 shows some instability forms.

Any discussion of emulsion stability should be approached not only on the stabilization mechanism, but also on the stability time and emulsion preparation conditions. The main factors related to coalescence are the physical nature of the interfacial film, the existence of electrical and steric barriers against the coalescence of the dispersed phase, the formation of liquid crystals in the interfacial film, the viscosity of the continuous phase, the volume ratio between phases and the temperature. Surfactants are, of course, crucial for the stabilization of emulsions. The droplets of dispersed liquid within the emulsion are constantly moving and colliding. Thus, it is very important to form an interfacial film of densely packed molecules of surfactant adsorbed around the droplets of the emulsion, with properties of strong intermolecular forces between them and the aqueous solution and high elasticity of the film, aiming at greater mechanical stability of the emulsion. For better packing of the surfactant molecules, it is important to use mixtures of at least two different surfactants in the emulsion system: one preferentially soluble in the oil phase, with long hydrophobic groups, preferably linear alkyl chains, and the other preferentially soluble in the aqueous phase, with stronger polar hydrophilic groups. The presence of electric charges in the interfacial films around the emulsion droplets constitutes a barrier to the approach of these droplets, a very important factor for O/W emulsions. The electric charge source is generally the surfactant layer adsorbed onto the droplets with their ionic hydrophilic group oriented towards the aqueous phase. The negative or positive charge created around the oil droplets delays the coalescence of the droplets in the same way that anionic surfactants in detergents keep the dirt in suspension, by the formation of a double electric layer, which acts against the force of attraction between the emulsified droplets.

Emulsions processed with non-ionic surfactants, especially those with large hydrophilic groups, such as ethoxylates, stabilize against coalescence through the formation of steric barriers of ethoxylated hydrophilic groups projecting into the aqueous phase and hydrated via hydrogen bonds, preventing the particles from colliding with each other. An increase in temperature disrupts the hydrogen bonds and dehydrates the surfactant's hydrophilic head group, favouring coalescence and leading to phase separation.

Lyotropic crystalline liquid crystals are binary or multicomponent systems of organic substances dispersed in a solvent, whose formation depends on concentration. They are formed mainly by amphiphilic

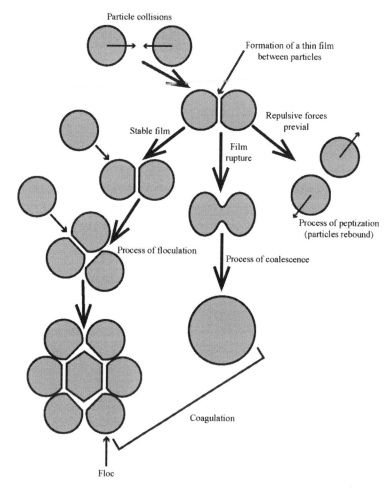

FIGURE 9.11 Types of emulsion instability. (From Surfactant Science Series, 1999, *Handbook of Detergents – Part A: Properties*, G. Broze, Ed., New York: Marcel Dekker.)

molecules, such as surfactants, phospholipids or glycerides and are common in cosmetics. Advantages of using this type of structure include the protection of lipophilic materials from thermal degradation and photodegradation, and the slower controlled transport to the stratum corneum of the components that are in the crystalline liquid phase. The formation of liquid crystals can then stabilize emulsions. As with most emulsions, stability against coalescence is determined by the nature of the liquid film separating the droplets. Interfacial films made of liquid crystals in O/W emulsions behave as a region of higher, rigid viscosity that resists the coalescence of the droplets and acts as a steric barrier preventing the approach of attracted particles by van der Waals forces. Some organic compounds used in emulsions, such as long carbon fatty alcohols (cetostearyl), alkylpolyglucosides and PPG-15 stearyl ether, favour the formation of liquid crystals. In the O/W emulsion these structures appear as globules dispersed in the aqueous phase, which aid in the hydration of the skin by releasing water slowly because intermolecular water is less available to evaporate.

An emulsion with smaller droplets that are more uniform in distribution is more likely to be stable. The use of correct emulsifiers improves the dispersivity of the droplets. As the volume of the dispersed phase increases, the interfacial film expands, reducing the stability of the system. Also, when the size and number of the dispersed phase droplets reduces the distance between the particles, there is an increase in coalescence. Thus, the selection and appropriate concentration of the emulsifier are crucial to stability.

Changes in temperature affect the stability of the emulsion, which can cause inversion or breakage. Temperature variability causes changes in the interfacial tension between the phases, the nature and viscosity of the interfacial film, the relative solubility of the emulsifiers in each phase, the vapour pressure and viscosity of the liquid phases, and the thermal agitation of the dispersed droplets.

Three basic rules form the basis for many methods of selecting emulsifiers and reducing development work (all methods aim to obtain more stable and cheaper emulsions):

- Emulsifiers preferentially soluble in oil, form water-in-oil (W/O) emulsions.
- Mixtures of emulsifiers – one soluble in oil and the other in water – produce more stable emulsions.
- For a more polar oil phase, use a more hydrophilic emulsifier, and for a non-polar oil, use a more lipophilic emulsifier.

The HLB Method of the Emulsifier

The HLB method was the first attempt to organize and standardize the choice of the emulsifying system as an alternative to the trial-and-error system. It is based on the partial solubility of the surfactant in the oil and aqueous phase, as a function of its amphiphilic structure, represented by the presence of non-polar and polar groups in the molecule. The system was developed by Griffin in 1947 and was named hydrophilic–lipophilic balance or lipophilic–hydrophilic equilibrium (HLB) (Griffin, 1949).

Griffin assigned a number for each surfactant, ranging from 0 to 20, the highest values representing more hydrophilic products. The HLB can be calculated from the molecular structure of the surfactant, the emulsification experimental data, the solubility in water by the determination of cloud point, and from the determination of the critical micelle concentration of the surfactant. Table 9.9 gives the HLB values for some representative emulsifiers.

TABLE 9.9

HLB Values for Selected Surfactants

Surfactant/Surfactant Blend	HLB
Sorbitan monolaurate	8.6
Sorbitan monopalmitate	6.7
Sorbitan monostearate	4.7
Sorbitan monooleate	4.3
Polyoxyethylene sorbitan monolaurate (polysorbate 20)	16.7
Polyoxyethylene sorbitan monopalmitate (polysorbate 40)	15.6
Polyoxyethylene sorbitan monostearate (polysorbate 60)	14.9
Polyoxyethylene sorbitan monooleate (polysorbate 80)	15.0
Span 80/polysorbate 80 (83/17 wt%)	6.0
Span 80/polysorbate 80 (65/35 wt%)	8.0
Span 80/polysorbate 80 (46/54 wt%)	10.0
Span 80/polysorbate 80 (28/72 wt%)	12.0
Span 80/polysorbate 80 (9/91 wt%)	14.0
Polyoxyethylene (4) lauryl ether	9.0
Polyoxyethylene (2) cetyl ether	5.0
Polyoxyethylene (10) cetyl ether	12.0
Polyoxyethylene (20) cetyl ether	16.0
Polyoxyethylene (100) stearyl ether	18.0
Polyoxyethylene (2) oleyl ether	4.0
Polyoxyethylene (10) oleyl ether	12.0
Polyoxyethylene (20) oleyl ether	15.3
Polyoxyethylene (5) nonylphenyl ether	9.8
Polyoxyethylene (9) nonylphenyl ether	13.4
Polyoxyethylene (12) nonylphenyl ether	14.0

The Required HLB Method

In addition to the HLB of the emulsifier, it is necessary to know the optimum HLB for the emollient or oil phase (Table 9.10) in order to make both compatible and ensure a more stable emulsion. When the required HLB of the oil phase is not known, it can be determined by experimental testing, which consists of making several emulsions of different HLBs, employing a mixture of known HLB emulsifiers. The HLB value in which the best emulsion is obtained will be considered the required HLB of the oil. In practice, it is common to use a mixture of oils and/or emollients. For the combinations, the required HLB will be a weighted average of the HLBs of the emollients, such that

HLB required to emulsify the oil phase = $\sum$ (Fraction of the emollient in the oil phase × Required HLB of the emollient)

Likewise, the HLB of the emulsifier system to be used to emulsify an oil phase (mixture of emollients) is calculated as the weighted average amount of emulsifier multiplied by the HLB of each. In this way, the HLB of the oil phase mixture can be made compatible with that of the emulsifier system. As an example, if it is desired to emulsify a paraffin with an HLB of 10, a mixture of emulsifiers of which HLB is 9 to 11 will be required. In addition, the rules of chemical compatibility and amount of emulsifier should be observed to obtain a stable emulsion. From the practice of emulsion processing, it has been established that the amount of emulsifier to be used represents 20 to 25% by weight of the oil phase. If the resulting product has good stability, the amount of emulsifier may be reduced. With an insufficient dose of emulsifier, emulsion breakdown and phase separation will occur. With an excessive amount of emulsifier, an excess of micelles occurs and that favours the cutaneous absorption of active vehicles and may reduce the viscosity of the emulsion.

The surfactant must be chemically compatible with other components of the formulation. A formulation may employ non-ionic and amphoteric mixed anionic agents, but not with cationic surfactants or salts. The final pH of the system should also be considered. Emulsifying agents that have an ester bond should not be used in strongly acid or alkaline media due to hydrolysis. With regard to the type of emulsifier that is compatible with an oil phase, the rule 'like attracts like' is applicable. For example, if the phase has constituents with unsaturated or branched carbon chains, the choice of emulsifiers should be based on oleates or isoesters.

Foam

Although the consumer usually associates foam with detergency, these terms are not synonymous because there are some efficient detergents that do not make too much foam. Foams have found many

TABLE 9.10

HLB Values for Oils and Waxes of Cosmetic Uses

Constituent	Emulsion Type	
	O/W	W/O
Liquid paraffin	12	4
Solid paraffin	10	4
Stearic acid	16	—
Beeswax	12	5
Castor oil	14	—
Fractionated coconut oil	15	—
Silicone oils	8–12	—
Vegetable oils	6–7	—
Emollient esters	12	—
Fatty acids/alcohols	14–16	—
Cottonseed oil	9	—

applications in the fields of firefighting, polymer insulation, rubber, concrete, ore extraction and topical pharmaceuticals, and have aesthetic utility in many personal care products and detergents, although their presence often does not contribute to the overall efficiency of the process. Foam is a psychological and functionally important factor with some surfactants, especially those used for the production of shampoos, skin cleansers, toothpastes and shave products.

All foams are unstable, but the degree of instability can be modulated using selected surfactants. The most widely used method for measuring foam formation and stability is the Ross-Miles test. (American Society for Testing and Materials, 1953). In the procedure suggested by Ross and Miles, the foam-forming solution is presented in a tall, cylindrical tempered receiver vessel with standard dimensions. A second quantity of the same solution is introduced from above by means of another standardized reservoir. This passes through the column and forms foam as a result of the turbulence on mixing with the receiving phase. The foam height is measured as soon as the reservoir is empty and also after 1, 2, 5, 10 and 20 minutes to evaluate stability. The dimensions of the standardized receiver and reservoir lead to a height of the setup greater than 1 meter. The foam height measurement is taken at the top of the foam column only.

In cosmetics, foams are produced by the introduction of air or other gas into water, increasing the interaction between two dissimilar substances, which means an increase in the surface tension. The formation of bubbles is not a favourable process. Once formed, the bubbles are encapsulated in a film of the liquid, the volume of gas being much larger than the volume of liquid. In many foams the gas bubbles are initially spherical and have no colloidal size, rapidly adopting a packed structure in which the individual bubbles are separated by a thin film of colloidal thickness. The thin film separating two or more gas bubbles is referred to as a layered structure (laminated structure) having essentially identical interfaces in close proximity.

Foams can undergo a series of perturbations leading to coalescence of the bubbles, ultimately resulting from the loss of the liquid from the lamellar layer until a critical thickness is reached and the film is no longer able to withstand the pressure of the gas in the bubble. The loss of liquid or its drainage from the lamella can be affected by factors such as a high viscosity of the liquid, which delays the drainage process and, in some cases, is opposed to the effects of disturbance; and the presence of electrical and steric interactions at the interface that may interfere with drainage. The addition of surfactants can alter any or all the characteristics of the system and hence increase or reduce the stability of the foam. The foam does not occur in pure liquids because there is no mechanism that delays the drainage of the lamellar film or no interfacial stabilization. When surfactant substances are present, however, their adsorption at the liquid–gas interface serves to retard the loss of liquid from the lamella and, in some cases, produces a mechanically more stable system. Surfactants also have the effect of lowering the surface tension of the system, thereby reducing the work required for the initial formation of the foam.

As the lamellar film between the adjacent bubbles is stretched as a result of gravity, agitation, drainage and so on, a new surface is formed in the regions where the film has a smaller instantaneous surfactant concentration, and a local surface tension increase occurs. A gradient of surface tension throughout the film is produced, causing the liquid to flow from the regions of lower surface tension towards the newly stretched surface, thus opposing the thinning of the film. Additional stabilizing action results from the fact that the diffusion of new surfactant molecules to the surface must also involve the transport of associated solvent in the surface area, again as opposed to the process of decreasing the interfacial film thickness by the drainage of the liquid. The mechanism can be seen as a 'healing' effect at the thinning site (Rosen, 2004; Meyers, 2006).

The amount of foam produced by a surfactant under a predetermined set of conditions increases with its concentration in the solution to a maximum which occurs near the cmc. Apparently, then, the critical micellar concentration of a surfactant can be used as a parameter to predict the foaming power of a material, but not necessarily the persistence of the foam. Any structural modification leading to lowering the cmc of a class of surfactants is expected to increase its efficiency as a foaming agent, such as an increase in its hydrophobic chain. In contrast, the branching of the hydrophobic chain, which increases the cmc, results in a lower foaming efficiency.

Anionic surfactants are those with the highest foaming power. However, some anionic surfactants used in cosmetics are sensitive to water hardness (soaps, lauryl sulphates, etc.) and do not foam in the presence of calcium and magnesium ions. Thus, complexing agents may act synergistically in the formation and stability of the foam.

The molecules of anionics suffering from mutual repulsion allow, in a simplified way, air to escape through the molecules of surfactants. Although various additives may enhance foam formation and stability, the best way to do this is by using other surfactants such as alkanolamides, betaines or any other surfactants (including other anionics) that interpose between the molecules of the anionic surfactants, reducing their repulsion (Figure 9.12).

The structure of the polar group obviously can positively or negatively impact the foaming properties of surfactants. When comparing the characteristics of the foam produced by sodium lauryl sulphate (Figure 9.13A) with the one formed by sodium lauryl ether sulphate (Figure 9.13B), an increase in foam stability is found for the latter. Because of its ethoxylated polar head group, sodium lauryl ether sulphate is able to make hydrogen bonds with water in the liquid bubble film, retarding its drainage to a greater extent than is observed with its non-ethoxylated analogue.

Thickening and Rheology

The structure of the surfactant, as well as the nature of the additives normally present in shampoos and liquid soap formulations, determine their viscosity at a given concentration. Linear alkyl sulphates, for

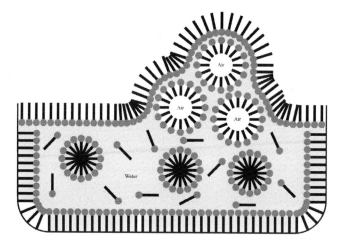

FIGURE 9.12   Behaviour of surfactants in a foam system. (From Daltin, D., 2011, *Tensoativos: Química, Propriedades e Aplicações*, São Paulo: Edgard Blücher.)

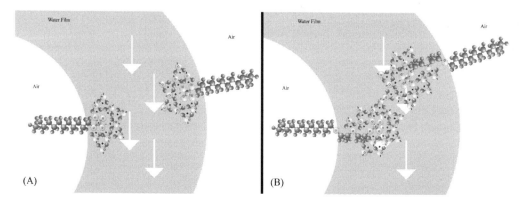

FIGURE 9.13   (A) Foam liquid film formed by sodium lauryl sulphate. (B) Foam liquid film formed by sodium lauryl ether sulphate. (From Daltin, D., 2011, *Tensoativos: Química, Propriedades e Aplicações*, São Paulo: Edgard Blücher.)

example, increase the viscosity of a shampoo compared to the branched analogues because of their tendency for higher packing and lower cmc, leading to the formation of the micellar isotropic phases more rapidly. The same explanations apply to the thickening effects of alkanolamides and amidobetaines. The alkyl amidobetaines, because they contain positive charges, decrease the repulsion of the anionic groups and lead to a tighter packing.

The viscosity of surfactant solutions/formulations may also be increased using salt or by an increase in the anionic surfactant concentration, such effects being greater in the presence of alkanolamides and alkyl amidobetaines. Excess salt may, however, lead to a decrease in viscosity after reaching a maximum. An explanation for the salt effect is due to the compression of the double electric layer between two charged micelle surfaces, which leads to the reduction of its effective charge and smaller inter-micellar repulsive forces. The micelle no longer restricted to its spherical shape can now 'grow' and change into a cylindrical and other shapes by the inclusion of additional monomer molecules. Spheres can move freely due to the reduced packing density, but the cylinders have a narrower translational and lateral movement, resulting in a higher viscosity. This viscosity behaviour can be investigated by the salt curve (Figure 9.14) (Rosen, 2004; Meyers, 2006).

Antistatic Properties and Conditioning

Washing hair with shampoos normally causes the hair fibres to become negatively charged because the pH of the washing solution is higher than the isoelectric point of keratin, which is 3.67 (Dias, 2015). This leaves hair difficult to comb when wet and, when combed after drying, static electricity causes the hair to be even more charged, increasing frizz and friction. Shampoos contain anionic detergent surfactants, which remove lipids from the hair, which, to some extent, condition and lubricate hair fibres. In addition, anionic surfactants can cause keratin denaturation (Harrold, 1959). These factors combined, make the hair surface even more negatively charged and consequently damaged.

Hair conditioners are products that are used after washing and can be formulated with cationic surfactants. The adsorption of a cationic surfactant molecule onto the hair surface results in an antistatic effect, which together with the lubricity imparted by the adsorption of the long hydrophobic chains of the cationic surfactant condition the hair fibres, make them easier to comb under both wet and dry conditions (Kulkarni and Shaw, 2016).

In addition, the change in the lipophilic nature of the hair surface due to the adsorption of the cationic agent allows other ingredients of the formulation to be more compatible with the hair. The cationic agents increase the deposition of fatty materials and emollients on hair. Cetyltrimethylammonium chloride, stearyl amidopropyl dimethylamine quats, esterquats, behentrimonium chloride and behentrimonium methosulphate are the most widely used cationic surfactants in hair conditioners.

Antimicrobial Properties

Cationic surfactants are very useful for their antimicrobial properties. Due to their high substantivity, cationic surfactants adsorb onto the membrane of the microbial cell, making cytoplasmic exchanges, essential to their metabolism, difficult, thus leading to their death. Some products also act on the

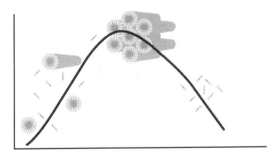

FIGURE 9.14   Salt curve. Viscosity behaviour of surfactant systems in the presence of different electrolyte concentrations.

cell's ribosome, interfering in the process of protein synthesis, also essential for the survival of the microorganism.

Biological activity seems to be characteristic of almost all types of cationic surfactants and small structural changes do not affect such activity. Among the various quaternary alkyltrimethylammonium compounds, the hexadecyl and tallow homologs generally have the best microbicidal action. The microbicidal activity of benzalkonium chlorides is usually higher in the presence of ethoxylated nonylphenols and amine oxides. Cetylpyridinium chloride is used in mouthwashes; toothpastes; and throat, breath and nasal sprays. It is an antiseptic that kills bacteria and other microorganisms. It has been shown to be effective in preventing dental plaque and reducing gingivitis (Asadorian and Williams, 2008; Haps et al., 2008; Meyers, 2006).

Mildness

Surfactants are evaluated for local toxicity such as skin irritation, eye irritation and systemic toxicity as determined by an oral lethal dose (LD50). Both *in vivo* and *in vitro* tests are used.

Most surfactants used in cosmetics are not potentially irritating to skin, although several irritative responses may occur. Cationic surfactants are considered the most aggressive but, because they are applied in low concentrations and usually in rinse-off products, they do not cause concern.

Eye irritation is also a very important factor since many surfactants are used in preparations for the hair and, therefore, reach the eyes with relative ease. The irritation is caused primarily by the monomeric surfactant molecules, due to its higher interaction and penetration through the skin. An increase in the hydrophobic chain length leads to a decrease in the degree of irritation.

The cationic surfactants have an oral LD50 in the range of 0.2 g/kg in mice. Other surfactants used in cosmetics have LD50s about 10- to 20-fold higher than the cationics. In general, the degree of irritability decreases in the following order: cationics > anionics > amphoterics > non-ionics (Meyers, 2006).

Biodegradability

The use of surfactants worldwide is growing at a speed exceeding the rate of population growth due to improved living conditions and availability of materials in the industrially least developed countries. Parallel to the greater use of surfactants is the difficulty of their disposal and impact on the environment.

While it is physically and chemically possible to remove surfactants from industrial effluents, such an operation is very costly, with costs prohibitive for industries. The way to solve the problem is to use surfactants with a ready and complete biodegradation.

The biodegradation of surfactants and other organic products is the result of bacterial action. In this degradation, the organic molecule undergoes several intermediary stages (primary biodegradation), until complete conversion into carbon dioxide, water and inorganic salts.

Microorganisms are able to degrade a wide variety of surfactants, using them for nutrition and as an energy source. The biodegradability of a product can be measured by the formation of by-products or the oxygen consumption of the medium when in the presence of microorganisms.

The chemical structure of the hydrophobic group is the most important factor in biodegradation, with higher degrees of branching leading to the lower rates of biodegradation. Some cationic surfactants, due to their characteristic high substantivity (and consequent microbiological properties), have lower rates of biodegradation and, therefore, their amounts are limited in several formulations.

Synergy of Surfactants and Polymers, and Optimization of Surfactant and Cosmetic Properties

Optimization of surfactant properties results in an improvement of the efficiency and efficacy of a cosmetic product or may even impart new functionalities.

The optimization of surfactant properties requires a complete understanding of how it interacts with the other ingredients of a given formulation, how the environmental conditions influence such interactions and the mechanism of the particular process (shampooing, hair conditioning, foam formation and stabilization, thickening, irritancy mitigation, etc.). A surfactant can interact with the vehicle and other ingredients by hydrogen bonding, hydrophobic interactions, polar and electrostatic interactions and so forth, which are dependent on pH, ionic strength (osmotic pressure), temperature and other factors.

In formulations containing two or more surfactants which are related to each other and, especially, those with the same charge, the performance properties are governed by the most highly surface active surfactant present in the system. On the other hand, if a given surfactant is mixed with a surfactant of a different charge (e.g. an anionic with a cationic, or a cationic with a zwitterionic), the resulting mixture might show better properties than those of the individual surfactants. If the mixture shows better properties than either component of the mixture by itself at the same concentration in the solution, the system is said to exhibit synergism. Mixing different charge types of surfactants that may exhibit synergism is an important method of enhancing performance properties. When synergism is present less mixed surfactant can be used to obtain the same level of performance, and the level of performance of the synergistic mixture (foaming, emulsifying ability) may be even greater than obtainable with either component by itself (Rosen and Dahanayake, 2000).

Synergy in mixtures of surfactants depends on an attractive molecular interaction between the two different surfactants. The existence of synergism in the system also depends on the values of the desired property in the two different surfactants. When reducing the critical micelle concentration (or increasing the adsorption efficiency at an interface), the greater the difference between the cmc values of the two surfactants, the stronger must be the attractive molecular interaction between them for synergism to exist. An increase in the length of the alkyl chains of the two surfactants and equal lengths of the two alkyl chains of the two surfactants (especially at interfaces) increase the attractive interaction between two different types of surfactants. An increase in the ionic strength of the solution generally decreases the attractive interaction of ionic surfactants. A decrease in the pH of the solution increases the attractive interaction between anionic surfactants and zwitterionics capable of accepting a proton, whereas an increase in the pH decreases it (Rosen and Dahanayake, 2000).

Synergism in the fundamental properties of surfactants is the basis for synergism in performance properties such as foaming, wetting, emulsification and detergency. Mixed monolayer formation at an interface can produce synergism in surface (or interfacial) tension reduction efficiency and effectiveness. Synergism in surface (or interfacial) tension reduction efficiency exists when a given surface (or interfacial) tension value can be obtained at a lower surfactant concentration of the mixture than of either of the two components of the mixture by itself. Surface (or interfacial) tension reduction effectiveness exists when the mixture can reach a lower surface (or interfacial) tension than can either of the two components of the mixture by themselves. If the critical micelle concentration of a given mixture is smaller than that of either component of the mixture by itself, that is, the mixture has a greater tendency to form micelles than either component alone, synergy occurs. Increased wetting and increased initial foaming by aqueous surfactant mixtures are correlated to synergism in surface tension reduction effectiveness. Synergism in interfacial tension reduction effectiveness in oil/water systems is correlated with increased detergency (Rosen and Dahanayake, 2000).

In shampoos, shower gels and liquid soaps, sodium lauryl ether sulphate is the preferred primary surfactant. The optimization of sodium lauryl ether sulphate properties is obtained by combining it with cocamidopropyl betaine, coconut alkanolamides, and lauryl and myristyl amine oxides at specific rates and conditions. The wetting, foaming, thickening and detergency properties of shampoos, shower gels and liquid soaps are increased by using such mixtures.

Surfactants can also exhibit synergy with other formulation ingredients, mainly polymers. For instance, conditioning shampoos were developed thanks to the understanding of the synergy between surfactants and polymers. They were developed to repair damage to the hair, including breakage or the removal of structural components or parts of hair that weaken or make it more vulnerable to chemicals or mechanical collapse, such as washing with shampoo and combing every day. Sunlight, pool water and cosmetic products, such as permanent products, whitening, straightening and some hair dyes, chemically alter the hair, increasing its propensity for further chemical and mechanical damage, as evidenced by increased sensitivity to abrasion, cuticle erosion and breakage of fibres.

Shampoos can damage hair in different ways. They can damage hair by abrasion/erosion both during the shampooing process itself, when hair is wiped against each other and stretched repeatedly while soaping, combing or drying with a towel or hair dryer. Shampoos can also slowly dissolve or remove structural and superficial lipids from the hair. The superficial lipids give lubricity to the hair, which protects during the act of combing and brushing.

The origin of conditioning shampoos can be attributed to the balm shampoo of the 1960s, followed by the introduction of polyquaternium-10 and the innovative work and scientific vision of Des Goddard in the 1970s and 1980s, who introduced the concept of the polymeric–surfactant complex coacervate, in which phases separate and deposit on the hair fibres during rinsing. During the 1980s and 1990s, complex coacervates were used as vehicles for the deposition of silicone on hair as conditioning shampoos evolved. This trend continues today and the two original polyelectrolytes – polyquaternium-10 and guar hydroxypropyltrimonium chloride – continue to be the dominant 'polymeric actives' in conditioning shampoos (Cosmetic and Toiletries, 2008).

The presence of a polyquaternium polymer in an anionic-based shampoo brings benefits not only for the application itself (hair conditioning) but also for the formulation. Polyquaternium polymers can interact with anionic surfactant aggregates to increase thickening (Figure 9.15) and foaming (Figure 9.16) properties. Once the mutual repulsion of anionic heads of surfactant molecules is reduced by the

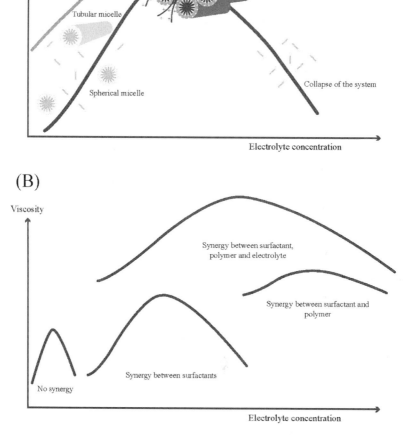

FIGURE 9.15   (A, B) Salt curves. Viscosity behaviours of synergistic surfactant–polymer systems in the presence of different electrolyte concentrations.

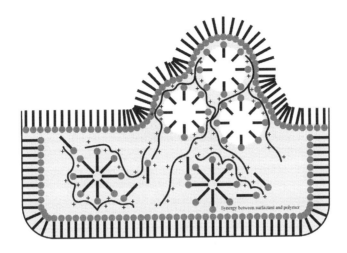

FIGURE 9.16   Foam behaviour of a synergistic surfactant–polymer system.

interaction with the polyquaternium polymer, the surfactant molecules can get closer and pack in different and denser arrangements. In addition, once the monomeric surfactant concentration is reduced, the mildness of the formulation is increased. Thus, the synergy between surfactants and polymers can enhance functional and structural properties of the product, positively influencing consumer acceptance and formulation stability.

Acknowledgements

Jessica Jaconetti Ydi provided kind support with the drawings.

REFERENCES

Adamson, A. W., and Gast, A. P. (1967). *Physical Chemistry of Surfaces*. New York: John Wiley & Sons.

Agency for Toxic Substances and Disease Registry. (2012). *Toxicological Profile for 1,4-Dioxane*. Atlanta: U.S. Department of Health and Human Services, Public Health Service.

American Society for Testing and Materials. (1953). Standard Test Method for Foaming Properties of Surface-Active Agents (ASTM D1173-53).

American Society for Testing and Materials. (1968). Draves Wetting Test Method (ASTM D2281-68).

Asadorian, J., and Williams, K. (2008). Cetylpyridinium chloride mouth rinse on gingivitis and plaque. *J. Dent. Hyg.*, 82.

Becher, P. (1965). *Emulsions: Theory and Practice* (2nd ed.). New York: Reinhold Publishing.

Cosmetics and Toiletries. (2008). *Hair Care: From Physiology to Formulation* (1st ed.) (A. C. Kozlowski, Ed.). Carol Stream, IL: Allured Publishing.

Cosmetic, Toiletry, and Fragrance Association. (1993). *International Cosmetic Ingredient Dictionary* (5th ed.) (J. Wenninger and G. McEwen Jr., Eds.). Washington, DC: Cosmetic, Toiletry, and Fragrance Association.

Daltin, D. (2011). *Tensoativos: Química, Propriedades e Aplicações*. São Paulo: Edgard Blücher.

Dias, M. F. (2015). Hair cosmetics: An overview. *Int. J. Trichology*, 7, 2–15.

Downing, D. T., Stewart, M. E., Wertz, P. W., Strauss, J. S., and Colton VI, S. W. (1983, April 27). Skin lipids. *Comp. Biochem. Physiol.*, 76B, 673–678.

Falbe, J. (1987). *Surfactants in Consumer Products: Theory, Technology and Application*. Berlin: Springer-Verlag.

Griffin, W. (1949). Classification of surface-active agents by HLB. *J. Soc. Cosmet. Chem.*, 1, 311–326.

Haps, S., Slot, D. E., Berchier, C. E., and Van Der Weijden, G. A. (2008). The effect of cetylpyridinium chloride-containing mouth rinses as adjuncts to toothbrushing on plaque and parameters of gingival inflammation: A systematic review. *Int. J. Dent. Hyg.*, 6, 290–303.

Harrold, S. (1959). Denaturation of epidermal keratin by surface active agents. *J. Invest. Dermatol.*, 32, 581–588.

Hayashida, M. (2010, July 12). The regulation of cosmetics in Japan. Retrieved from Pharmaceutical Affairs Law: http://www.yakujihou.com/img/CosRegulationJapan.pdf.

Hill, K., von Rybinsky, W., and Stoll, G. (2008). *Alkyl Polyglycosides*. Weinheim: Wiley, VCH.

International Union of Pure and Applied Chemistry. (2013). *Nomenclature of Organic Chemistry: IUPAC Recommendations and Preferred Names* (A. H. Favre and H. W. Powell, Eds.). Cambridge: Royal Society of Chemistry.

Joensson, B., Jokela, P., Kahn, A., Lindman, B., and Sadaghiani, A. (1991). Catanionic surfactants: Phase behavior and microemulsions. *Langmuir*, 7, 889–895.

Kogawa, A. C., Cernic, B. G., Domingos do Couto, L. G., and Salgado, H. R. (2017). Synthetic detergents: 100 years of history. *Saudi Pharm. J.*, 25, 934–938.

Kulkarni, V. S., and Shaw, C. (2016). *Essential Chemistry for Formulators of Semisolid and Liquid Dosages*. Amsterdam: Elsevier/Academic Press.

Laughlin, R. (1994). *The Aqueous Phase Behavior of Surfactants*. London: Academic Press.

McBain, J. (1913). Mobility of highly-charged micelles. *Trans. Faraday Soc.*, 9, 99–101.

McBain, J., and Martin, H. (1914). Constitution of soap solutions: Alkalinity and degree of hydrolysis of soap solutions. *J. Chem. Soc.*, 105, 954–977.

Meyers, D. (1988). *Surfactant Science and Technology*. New York: VCH.

Meyers, D. (2006). *Surfactant Science and Technology* (3rd ed.). Hoboken, NJ: John Wiley & Sons.

Official Journal of the European Union. (2018). *Council Directive 76/768/EEC of 27 July 1976 on the Approximation of the Laws of the Member States Relating to Cosmetic Products*. Retrieved from Eur-Lex: https://eur-lex.europa.eu/LexUriServ/LexUriServ.do?uri=CELEX:31976L0768:EN:HTML.

Rosen, M. J. (2004). *Surfactants and Interfacial Phenomena* (3rd ed.). Hoboken, NJ: John Wiley & Sons.

Rosen, M. R. (2006). *Delivery System Handbook for Personal Care and Cosmetic Products: Technology, Applications, and Formulations*. Norwich, NY: William Andrew Publishing.

Rosen, M. J., and Dahanayake, M. (2000). *Industrial Utilization of Surfactants: Principles and Practice*. Urbana, IL: AOCS Press.

Stillwell, W. (2013). *An Introduction to Biological Membranes: From Bilayers to Rafts*. New York: Elsevier.

Surfactant Science Series. (1981a). *Amphoteric Surfactants* (1st ed.). New York: Marcel Dekker.

Surfactant Science Series. (1981b). *Anionic Surfactants: Physical Chemistry of Surfactant Action* (Vol. 11) (E. Lucassen-Reynders, Ed.). New York: Marcel Dekker.

Surfactant Science Series. (1990). *Cationic Surfactants: Organic Chemistry* (J. Richmond, Ed.). New York: Marcel Dekker.

Surfactant Science Series. (1998). *Non-Ionic Surfactants: Organic Chemistry* (Vol. 72) (N. M. Van Os, Ed.). New York: Marcel Dekker.

Surfactant Science Series. (1999). *Handbook of Detergents – Part A: Properties* (G. Broze, Ed.). New York: Marcel Dekker.

Tanford, C. (1978). The hydrophobic effect and the organization of living matter. *Science*, 200, 1012–1018.

The Soap and Detergent Association. (1987). *A Handbook of Industry Terms*. Washington, DC: The Soap and Detergent Association.

U.S. Department of Health and Human Services. (2018). *Cosmetics & U.S. Law*. Retrieved from U.S. Food & Drug Administration: https://www.fda.gov/cosmetics/guidanceregulation/lawsregulations/ucm2005209.htm#U.S._Law.

U.S. Environmental Protection Agency. (1993). *Integrated Risk Information System on 1,4-Dioxane*. Cincinnati, OH: Office of Health and Environmental Assessment.

Willcox, M. (2000). *Pourcher's Perfumes, Cosmetics and Soaps* (10th ed.). Dordrecht: Kluwer Academic Publishers.

10

Oils

Fabricio Almeida de Sousa and Vânia Rodrigues Leite-Silva

CONTENTS

Introduction

'Oil' is a term that describes a material that is both hydrophobic (water repelling) and lipophilic (oil-loving). The term 'oil' includes hydrocarbons, triglycerides, esters, fatty alcohols and oil-soluble silicones (alkyl dimethicone). Dating back to antiquity, fats and oils have played an important role in the composition of cosmetics, providing emolliency and moisturisation, and acting as solvents and vehicles to carry other agents. The use of materials extracted from natural sources was standard practice from the time of the Egyptians (Poucher, 1959).

Oils can be classified by the source from which they are derived such as petrochemical or natural, animal or vegetable. Oils can be solid (waxy/butters) or liquid. The degree of unsaturation and branching in the molecule will increase liquidity.

Oil components for an important part of the formulation in many cosmetic products. All oils are emollients, but based on their different chemical structure they have very different additional properties.

Composition and Structure of Oils

Oils and Fats

Oils and fats are based on two simple building blocks: glycerol and fatty acids (Figure 10.1). Whilst there is only one form of glycerol, the fatty acid component can vary widely in structure and properties. Glycerol has three alcohol groups to which the fatty acids can be attached. The resulting products are monoglycerides (one fatty acid), diglycerides (two fatty acids) or triglycerides (three fatty acids). The fatty acids attached to the glycerol may or may not be identical, so a large number of different glycerides can be obtained from a limited number of building blocks.

Carboxylic acids with chain lengths of 6–24 carbon atoms are traditionally called fatty acids. In general, only the straight, even-numbered chains are present in formulations, although animal fats can contain both odd-numbered and branched-chain fatty acids. The most common simple fatty acids and their characteristics are shown in Table 10.1.

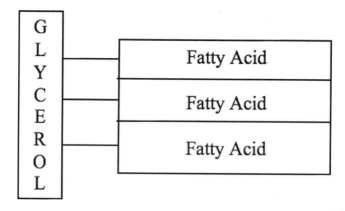

FIGURE 10.1   Example of a triglyceride.

TABLE 10.1

Typical Fatty Acid Composition of Oils and Fats

Carbon Atoms: Double Bonds	Common Name	Formula	Melting Point (°C)
Saturated Fatty Acids			
8:0	Caprylic acid	$C_8H_{12}O_2$	16.5
8:0	Caprylic acid	$C_8H_{12}O_2$	16.5
10:0	Capric acid	$C_{10}H_{20}O_2$	32
12:0	Lauric acid	$C_{12}H_{24}O_2$	43
14:0	Myristic acid	$C_{14}H_{28}O_2$	54
16:0	Palmitic acid	$C_{16}H_{32}O_2$	62
18:0	Stearic acid	$C_{18}H_{36}O_2$	79
20:0	Arachidic acid	$C_{20}H_{40}O_2$	76
Unsaturated Fatty Acids			
16:1	Palmitoleic acid	$C_{16}H_{30}O_2$	0
18:1	Oleic acid	$C_{18}H_{34}O_2$	13
18:2	Linoleic acid	$C_{18}H_{32}O_2$	-9
18:3	Linolenic acid	$C_{18}H_{30}O_2$	-17
20:4	Arachidonic acid	$C_{20}H_{32}O_2$	-50

Sources: Timberlake KC, *General Organic and Biological Chemistry: Structures of Life*, 6th ed., Pearson, 2018; and Weast RC, Astle MJ, Beyer WH (eds.), *CRC Handbook of Chemistry and Physics: A Ready-Reference Book of Chemical and Physical Data*, 66th ed., CRC Press, 1978.

Oils and fats are water-insoluble substances of plant or animal origin that consist predominantly of triglycerides, minor lipids (monoglycerides and diglycerides) and biologically active ingredients. Those that are solid or semi-solid at room temperature are normally called fats, while those that are liquid at room temperature are called oils.

Nature provides a great deal of variation in the composition of oils. Cold and temperate climates yield oils that are liquid at room temperature and mainly contain unsaturated C_{18} fatty acids, such as oleic, linoleic and linolenic acids. In tropical zones, fats with higher melting points and high content of saturated fatty acids are typical. Table 10.2 shows the typical fatty acid composition of oils and fats commonly used in cosmetics.

The oxidation stability of oils and fats is determined by their degree of unsaturation, the presence of natural or synthetic antioxidants and pro-oxidants such as metals, and the availability of oxygen. Oxidation of oils and fats normally takes place at the double bonds, resulting in the formation of hydroperoxides (primary oxidation products) that then lead to formation of ketones and aldehydes (secondary oxidation products).

Oils and fats are commonly used in personal care formulations such as soaps, wash lotions, shower gels and bath oils shaving creams, oral hygiene products, cleansing wipes and tissues, and antiperspirants and deodorants. The appropriate choice of an oil or fat system in a product is based on characteristics such as viscosity, stability, water dispersibility, and their emolliency and relipidising properties.

Oils are well suited to bath and shower formulations where good softness, moisturising and relipidising properties are required. Such properties are also of particular importance to dry and sensitive skin, where soap and detergents may cause drying and skin irritation. Oils rich in oleic acid, such as high oleic sunflower seed oil and high oleic rapeseed oil, are ideal for formulating creamy shower gels and bath oils, as they combine moisturisation with high stability.

Alcohols

Alcohols are classified as chemical compounds that have the general formula ROH, where R represents an alkyl group and –OH a hydroxyl group. Depending on how the –OH group is positioned on the alkyl group, alcohols are classified as primary, secondary or tertiary. For example, butanol has three structural isomers – *n*-butanol (1-butanol or butyl alcohol), 2-butanol (secondary butanol or isobutyl alcohol) and t-butanol (or tert-butyl alcohol: a tert-butanol with a hindered hydroxyl on the same carbon with three methyl groups) (Figure 10.2).

Fatty alcohols are oleochemicals derived from vegetable oils. These refined vegetable oils are first converted to a methyl ester or fatty acid, and the intermediate (methyl ester/fatty acid) then fractionated and hydrogenated to produce the fatty alcohol. In general, alcohols used in formulation are normal alcohols from natural fats and oils, having an even number of carbon atoms, and can be saturated or unsaturated.

TABLE 10.2

Typical Fatty Acid Composition of Oils and Fats Commonly Used in Cosmetics

Alcohols from Natural Fats/Oils	Number of Carbon Atoms	Alcohols from Synthetic Origin
Capryl alcohol (1-octanol)	8	2-ethyl hexanol (oxo, branched)
Pelargonyl alcohol (1-nonanol)	9	Isononanol (iso, branched)
Capric alcohol (1-decanol)	10	Isodecyl alcohol (branched)
Lauryl alcohol (1-dodecanol)	12	Dodecyl/tridecyl alcohol (mixed branched)
	13	Tridecyl alcohol (iso, branched)
Myristyl alcohol (1-tetradecanol)	14	
Cetyl alcohol (1-hexadecanol)	16	Isocetyl alcohol (branched)
Stearyl alcohol (1-octadecanol)	18	Isostearyl alcohol (branched)
Oleyl alcohol (1-octadecenol)	18	
Arachidyl alcohol (1-eicosanol)	20	
Behenyl alcohol (1-docosanol)	22	

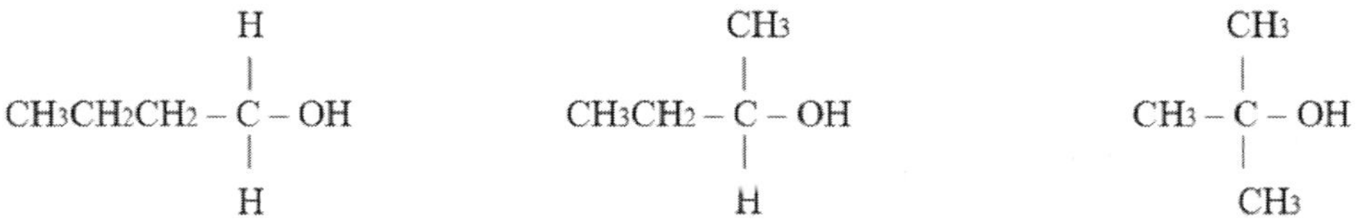

$$CH_3CH_2CH_2 - \overset{\displaystyle H}{\underset{\displaystyle H}{\overset{|}{\underset{|}{C}}}} - OH \qquad CH_3CH_2 - \overset{\displaystyle CH_3}{\underset{\displaystyle H}{\overset{|}{\underset{|}{C}}}} - OH \qquad CH_3 - \overset{\displaystyle CH_3}{\underset{\displaystyle CH_3}{\overset{|}{\underset{|}{C}}}} - OH$$

1 – butanol   2-butanol   Methyl-2-propanol

(butyl alcohol)   (s-butyl alcohol)   (t-butyl alcohol)

(a primary alcohol)   (a secondary alcohol)   (a tertiary alcohol)

FIGURE 10.2   Esters of butanol.

Synthetic alcohols are also used in formulation. Oxo alcohols are branched-chain alcohols, also termed synthetic higher alcohols. They can be mono-methyl branched or multi-carbon chained on the side at any or a specific interior carbon of the main carbon chain. Guerbet alcohols were first synthesised in the 1980s by Marcel Guerbet by aldol condensation at high temperatures in the presence of alkaline catalysts. They are primary, alpha-branched dimeric alcohols and are 100% defined branched at the second carbon position. Oxo alcohols and iso-alcohols are alpha-olefin based and are approximately 50% branched at the second carbon position. Oxo alcohols are about 50% linear and iso-alcohols are 100% multiple methyl branched.

Melting points or pour points are much lower for branched/Guerbet alcohols than for their linear counterparts of the same number of carbon chains. Linear unsaturated alcohols are liquid and suffer from poor heat stability due to unsaturation. Saturated Guerbet alcohols or branched iso-alcohols offer fluidity and also thermal stability and oxidation stability. These differentiating physico-chemical properties of branched chain alcohols make them immensely important in the synthesis and derivatization into cosmetics and personal care emollients.

Fatty alcohols are used as emulsifiers, emollients and thickeners in cosmetic formulations. (O'Lenick Jr., 2010). Table 10.3 shows the common names, carbon numbers and synthetic-branched counterparts of alcohols commonly used in cosmetic products.

Mineral Oil

The term 'mineral oil' refers to very highly refined liquid hydrocarbons derived from petroleum distillates, which are used in medicine, pharmaceuticals, cosmetics, food packaging, food contact applications

TABLE 10.3

Common Names, Carbon Numbers and Synthetic Branched Counterparts of Alcohols Commonly Used in Cosmetic Products

Alcohols from Natural Fats/Oils	Number of Carbons	Alcohols from Synthetic Origin
Capryl alcohol (1-octanol)	8	2-ethyl hexanol (oxo, branched)
Pelargonyl alcohol (1-nonanol)	9	Isononanol (iso, branched)
Capric alcohol (1-decanol)	10	Isodecyl alcohol (branched)
Lauryl alcohol (1-dodecanol)	12	Dodecyl/tridecyl alcohol (mixed branched)
	13	Tridecyl alcohol (iso, branched)
Myristyl alcohol (1-tetradecanol)	14	
Cetyl alcohol (1-hexadecanol)	16	Isocetyl alcohol (branched)
Stearyl alcohol (1-octadecanol)	18	Isostearyl alcohol (branched)
Oleyl alcohol (1-octadecenol)	18	
Arachidyl alcohol (1-eicosanol)	20	
Behenyl alcohol (1-docosanol)	22	

and food itself. Other terms often used interchangeably with mineral oil, include 'liquid petrolatum', 'liquid paraffin', 'paraffin oil', 'medicinal oil', 'white oil', 'white mineral oil', 'food grade oil', 'food grade white oil' and 'technical white oil'. Mineral oils (medium and low viscosity) are manufactured from crude mineral oils in various refining steps, such as distillation, extraction and crystallization, and are subsequently purified by acid treatment (oleum method) and/or hydrotreatment (catalytic hydrogenation). Mineral oils (medium and low viscosity) are mixtures of highly refined paraffinic and naphthenic liquid hydrocarbons with boiling points greater than 200°C. They are lightweight, inexpensive, odourless and tasteless.

Mineral oils are a common ingredient in baby lotions, cold creams, ointments and cosmetics. Examples are their use to prevent brittleness and breaking of eyelashes, in cold cream, and to remove make-up and temporary tattoos. A common concern regarding mineral oil is the presence in many lists of comedogenic (i.e. clogs skin pores) substances that were developed many years ago and are frequently quoted in the dermatological literature. However, more recently, highly refined and purified oils commonly used in cosmetics and skin care products are non-comedogenic.

Esters

Esters are the product of the reaction of any acid (usually organic) and alcohol. Common fatty acids are caprylic, capric, lauric, myristic, palmitic, stearic, oleic, linoleic and behenic acids, and dicarboxilic acids such as adipic acid. Common alcohols are isopropyl, *n*-butyl, ethylhexyl, myristyl or oleyl alcohol, as well as polyvalent alcohols such as ethylene glycol, polyglycerol, propylene glycol and glycerol. A wide range of combinations can be used to produce the large range of synthetic esters available to the cosmetic industry.

Carboxylic acids can be esterified by alcohols in the presence of a suitable acidic catalyst as illustrated in Scheme 10.1. The initial step is protonation of the acid to give an oxonium ion (1) that can then undergo an exchange reaction with an alcohol to give the intermediate (2), and that can lose a proton to become an ester (3). Each step in the process is reversible, but in the presence of a large excess of alcohol, the equilibrium point of the reaction is displaced so that esterification proceeds virtually to completion. However, in the presence of water, which is a stronger electron donor than are aliphatic alcohols, formation of the intermediate (2) is not favoured and esterification will not proceed fully.

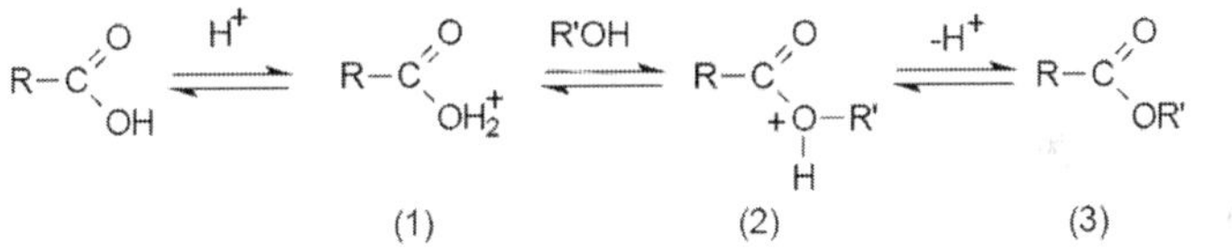

SCHEME 10.1 Acid-catalysed esterification of fatty acids.

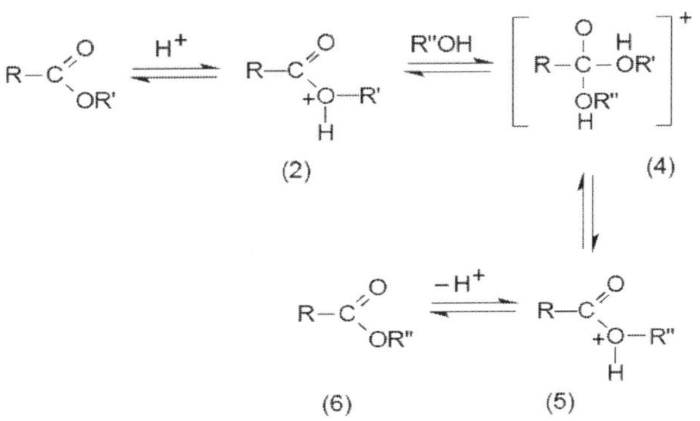

SCHEME 10.2 Acid-catalysed transesterification of lipids.

Ester exchange or transesterification occurs under similar conditions (Scheme 10.2). Initial protonation of the ester is followed by the addition of the exchanging alcohol to give the intermediate (4), which can be dissociated via the transitions state (5) to give the ester (6). Again each step is reversible and in the presence of a large excess of the alcohol, the equilibrium point of the reaction is displaced so that the product is almost entirely the required ester (6). Water must once more be excluded, as it would produce some hydrolysis by dissociation of an intermediate analogous to (4) (R"=H) to a free acid.

Esters are utilised as cosmetic ingredients to form semi-occlusive films for moisturising benefits, help maintain the skin's softness and plasticity, reduce the itching sensation often present in dry skin, and improve the appearance of the stratum corneum. They act mainly as moisturisers, plasticisers and tactile modifiers when applied to skin (Gorcea, 2008).

Silicones

'Silicones' is a generic name for many classes of organo-silicone polymer that consist of an inorganic siloxane (Si-O) backbone with pendant organic groups (usually methyl). It is this structure that gives silicones their unique combination of properties and, in particular, their surface properties (Figure 10.3).

Silicones have been used in skin care products since the 1950s. The first applications involved basic silicone fluids (INCI name: dimethicone), which are linear polymers with a wide range of molecular weights. Dimethicones remain important for their emollient properties and their ability to improve the skin feel of many types of skin care formulations. Cyclomethicones were introduced in the late 1970s. They are volatile, low-viscosity silicone fluids that act as cosmetic solvents. They are particularly suited for use with other silicones and as delivery vehicles for a variety of active ingredients.

The main reason silicones are used in all types of skin care products is because of their sensory properties. Studies on the emollient properties of various materials have shown that silicones deliver greater emolliency values than many commonly used cosmetics ingredients both during and after application. They are described as smooth, velvety, nongreasy or nonoily, and can impart these sensory properties to cosmetics and toiletry formulations, improving the overall product by masking the negative feel associated with other ingredients (Girboux and Courbon, 2008).

Volatile silicones provide transient effects characterised by slight lubricity, a light texture, fast spreading, and good distribution of the product application, while leaving no residual effects. They are often included in formulations to remove the greasy or oily feel of hydrocarbon-based emollients and are the basis for 'oil free'–type claims (De Backer and Ghirardi, 1993). Silicone elastomers are used to give a dry, powdery feel to skin care formulations (Van Reeth and Starch, 2003).

Characterisation Methods

Unfortunately there is no easy general formula to choose the ideal oil for cosmetic products, but identifying the key product parameters and knowledge of the ingredient characteristics can guide the steps of the

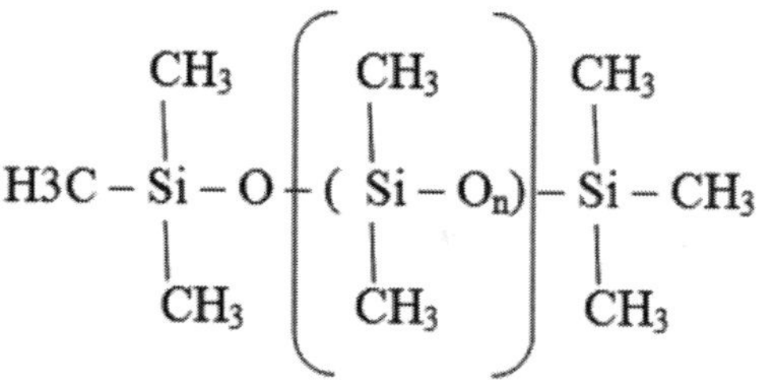

$$
H_3C - \underset{\underset{CH_3}{\overset{CH_3}{\vert}}}{Si} - O \left(\underset{\underset{CH_3}{\overset{CH_3}{\vert}}}{Si} - O_n \right) \underset{\underset{CH_3}{\overset{CH_3}{\vert}}}{Si} - CH_3
$$

FIGURE 10.3 Chemical structure of silicones.

development of new and improved cosmetic products. Thus effective characterisation of oil ingredients is an important aid to cosmetic formulation design and development.

The first step in performing a general characterisation of an oil sample is to establish its identity. We need to establish if the sample is a single component fat, such as hydrogenated soybean oil, or a mixture of two or more components (either by design or possible contamination with a minor oil). Having established the identity of the oil, the product quality should be assessed to determine if it will have an appropriate shelf life and functional properties for the intended application. In order to establish this information, a number of chemical, physical and sensory analytical methods are available.

Chemical Methods

Traditionally, simple wet chemistry methods, such as iodine value and hydroxyl value, were used to characterise oils. Table 10.4 provides a summary of the most commonly used indices, including references to the appropriate standard methods for their determination. When used in combination, these indices give a rough idea of the identity of the sample. However, a typical fat may contain tens or hundreds of chemical components; therefore, these methods cannot provide definitive identification.

In the past few decades, advances in chromatography, spectrometric methods and mass spectrometry have revolutionized the identity assessment of oils. These methods can be used to accurately identify complex oils and fats.

Physical Methods

A routine characterisation of an oil-based formulation covers determination of solid fat content (SFC), polymorphic stability, crystallization kinetics and rheology. Additional techniques can be used for troubleshooting and when formulating novel or unique products. For example, oil migration can be studied by rheology and X-ray diffraction in conjunction with chemical analysis to monitor the migration of tracer components. Methods for characterisation of the properties are summarised next.

TABLE 10.4

Summary of Chemical Methods for Identity Assessment

Index	Purpose/Information	Standard Method
Iodine value	Measures degree of insaturation by adding iodine to unsaturated double bonds in the fatty acids.	IUPAC 2.205, AOCS Cd 1-25
Hydroxyl value	Measures the degree of free hydroxyl groups in the sample giving indirect information on the glyceride composition in terms of mono-, di- and triglycerides.	IUPAC 2.241, AOCS Cd 4-40
Saponification value	Measures the relative proportions between fatty acids and alcohol in the sample and gives information on the average molecular weight. High values mean low molecular weight, and low values mean high molecular weight.	IUPAC 2.202, AOCS Cd 3-25
Refractive Index	Another measure of the degree of unsaturation of the sample	IUPAC 2.102, AOCS Cc 7-25
Acid value	Content of free fatty acids, hydrolytic breakdown	IUPAC 2.201, AOCS Cd 3d-63
Peroxide value	Primary oxidation products	IUPAC 2.501, AOCS Cd 8b-90
Colour	Oxidation, coloured contaminants	IUPAC 2.103, AOCS Cc 13e-92
Fatty acid composition	Identification	IUPAC 2.301+2.304, AOCS Ce 1e-91

Melting Properties

The melting point of an oil is normally a poor descriptor of its properties. Oils with the same melting point may have completely different appearances at room temperature: one oil may be a brittle solid and another a plastic semi-solid. This behaviour is more closely reflected in the solid fat content (SFC) profile, therefore it is SFC rather than melting point that is used to describe oils.

The amount of solid fat in a system can be determined by several techniques. A direct method utilises low-resolution pulsed NMR techniques. The relaxation times of hydrogen nuclei differ depending on whether they are in a liquid or solid state, therefore measurement of the relaxation time after a suitable pulse sequence can be correlated directly to the solid fat content. The oil requires a suitable tempering procedure in order to obtain the correct crystal form and equilibrated solids content. The official method for this determination is IUPAC 2.150, AOCS Cd 16-81, Cd 16b-93.

Differential scanning calorimetry (DSC) is another convenient method for determining oil crystallization and melting behaviour. The solid content at different temperatures can be measured by partial integration of the heat flow curve generated by the DSC. The measurement is indirect and should be corrected for the actual melting enthalpy of a sample that is completely solidified in the correct crystal form. DSC provides a complete melting and solidification profile in one measurement, making it useful for characterising the overall behaviour of an oil.

SFC determination can also be used to assess the degree of compatibility between two or more oils. The oils are blended in different proportions and subjected to a suitable tempering regime. The solid fat content is measured and plotted as a function of the ratio between the different fats. Fat incompatibility is seen as a non-linear relationship between the mixing ratios and SFC at low temperatures.

Polymorphism

Triacylglycerols and other oils can crystallize in different crystal forms that have, on average, the same chemical composition. Each polymorphic form will have a characteristic melting temperature, powder X-ray diffraction pattern, infrared spectral signature or Raman spectra. At least three main polymorphic forms have been identified for triglycerols, namely alpha (α) and beta (β) forms, the latter subcategorized and listed in order of increasing stability (β' being the most stable) (Sato and Ueno, 2005). The polymorphic form can change during processing and storage, affecting the characteristics of the oil, and potentially relating to changes in the sensory and textural properties of the product in storage. The polymorphic stability of oils or oil mixtures is most commonly determined by X-ray diffraction methods, which provide unequivocal information about the polymorphic form. Measurements of this type are primarily applied to the total oil phase, but studies of single components are also important in order to trace interaction effects. In well-defined oil systems, DSC can also be used to measure polymorphism, as melting behaviour may be correlated to crystals present in the oils.

Crystallization Rates

Crystallization kinetics are determined by DSC, nuclear magnetic resonance spectroscopy (NMR) or rheological analysis. It is used to characterise the dynamic crystallization behaviour of the full formulation, the total oil phase or the oil alone. Dynamic DSC is used for screening critical temperature ranges and identifying polymorphic events during crystallization. When the critical time and temperature ranges are identified, isothermal DSC can be used to study the crystallization process in detail. In this case, the sample is rapidly cooled from the melted state, and the appearance of an exothermal crystallization peak is registered. Crystallization peak time as a function of temperature can then be compared in different oil systems.

Where the influence of shear conditions during the crystallization process is important, these can be identified by rheological measurements of macroscopic samples of formulated systems. Knowledge of the crystallization kinetics, and their dependence on temperature, shear, and presence or absence of additives and minor components are important for understanding how the formulation will alter during processing and manufacturing.

Consistency and Rheology

Consistency and rheological behaviour are important determinants and thus descriptors of the textural, sensory and migration-related aspects of the formulation. Techniques employed include rotational and oscillatory rheological measurements at low deformations (applied as a function of the composition and temperature) and texture analysis at large deformations (penetrometer measurements). Rheological analysis also gives an insight into a product's microstructure, which influences shelf life, sensory and processing properties.

Formulating Skin Care Products

Oils are components of a wide range of skin care formulations, such as emulsions/lotions, emulsions/creams, ointments and anhydrous products. The choice of oils for cosmetic formulations may be governed by aspects such as raw material origin, biological functionality claims, stability considerations, consistency and sensory properties. Formulating with oils that are solid at room temperature may lead to increased body and consistency of the final product. Considering the crystallization properties of different vegetables fats while optimizing processing parameters may affect formulation characteristics such as consistency, texture and hardness.

Emulsions and Microemulsions

Emulsions are dispersions of liquid droplets in a liquid or semi-solid matrix. If the droplets consist of oil dispersed in an aqueous liquid, they are called oil-in-water (o/w) emulsions, while water-in-oil (w/o) emulsions are water disperse in an oil matrix. Elegant emulsion-based cosmetic formulations can be prepared by incorporating oils using a balanced surfactant system based on non-ionic surfactants, such as polyglycerol esters, fatty alcohol ethoxylates, monoglycerides, sucrose and sorbitan esters, polymers or combinations of emulsifiers and stabilisers. The surfactant system associates primarily with either the oil or aqueous phase but has sufficiently balanced polarity to reside at the interface between the two phases. Thus the surfactant system stabilises the interface and allows two phases to exist as a stable emulsion. This is achieved by stabilising the aqueous (hydrophilic) and oil (lipophilic) components together as an emulsion formulation using the hydrophilic–lipophilic balance (HLB) system (Griffin, 1949, 1954; Davies, 1957).

The required hydrophilic–lipophilic balance (rHLB) value for common vegetable oils is 7–10, and 5 for emulsification into w/o and o/w emulsions, respectively. If replacing other ingredients such as mineral oil or silicones, the estimated rHLB value of the ingredient should be considered and the surfactant system matched appropriately (see Table 10.5 for rHLB values).

In most cases the emulsification of liquid oil is straightforward. Excellent stability and sensory properties can be achieved using medium-chain monoglycerides, fatty alcohol ethoxylates, unsaturated long-chain monoglycerides and polyglycol esters of fatty acids. After proper homogenization, emulsions with a small particle size and good stability are obtained.

However, if the oil is partially crystalline, the formulation may be more complex, and the crystallization properties of the oil must be considered. The type of application is also crucial. Ideally the fat used should crystallize rapidly in small crystals without going through extensive polymorphic transitions after solidification. Large crystals and crystals that transform from one polymorph to another often cause instability by weakening the droplet surfaces or inducing flocculation (process of particles/droplets agglomerating). The degree of crystallinity may be used to adjust the consistency and textural properties of the emulsions and microemulsions.

The study investigated the formulation effects of laurocapram- and iminosulfurane-derived penetration modifiers on human stratum corneum using thermal and spectral analyses. First, formulations of penetration modifiers were assessed as enhancers/retardants using the model permeant diethyl-m-toluamide followed by investigation of their mechanisms of action using DSC and attenuated total reflectance FTIR spectroscopy.

TABLE 10.5

Required Hydrophilic–Lipophilic Balance (rHLB) Values for Common Oils in Cosmetic Formulations

INCI Name	rHLB ± 1	INCI Name	rHLB ± 1
Aleurites Moluccana Seed Oil	7	Grape (*Vitis vinifera*) Seed Oil	7
Almond Oil NF	6	Hybrid Safflower (*Carthamus tinctorius*) Oil	9
Anhydrous Lanolin USP	10	Isopropyl Myristate	11.5
Avocado (*Persea gratissima*) Oil	7	Isopropyl Palmitate	11.5
Avocado (*Persea gratissima*) Oil	7	Jojoba (*Buxus chinensis*) Oil	6,5
Babassu Oil	8	Jojoba (*Buxus chinensis*) Oil	10
Beeswax	12	Macadamia (Ternifolia) Nut Oil	7
Borage (*Borago officinalis*) Seed Oil	7	Mangifera Indica (Mango) Seed Butter	8
Brazil Nut Oil	8	Mineral Oil	10.5
C12-15 Alkyl Benzoate	13	Myristyl Myristate	8.5
Cannabis Sativa Seed Oil	7	Olive (*Olea europaea*) Oil	7
Canola Oil	7	Oryza Sativa (Rice Bran) Oil	7
Caprylic/Capric Triglyceride	5	Peanut Oil NF	6
Carrot (*Daucus carota sativa*) Seed Oil	6	Petrolatum	7
Castor (*Ricinus communis*) Oil	14	PPG-15 Stearyl Ether	7
Ceresin	6	Retinyl Palmitate	6
Cetearyl Alcohol	15.5	Safflower (*Carthamus tinctorius*) Oil	8
Cetyl Alcohol	15.5	Sesame (*Sesamum indicum*) Oil	7
Cetyl Esters	10	Shea Butter (*Butyrospermum parkii*)	8
Cetyl Palmitate	10	Soybean (*Glycine soja*) Oil	7
Coconut Oil	8	Stearic acid	15
Daucus Carota Sativa (Carrot) Root Extract	6	Stearyl Alcohol	15.5
Diisopropyl Adipate	9	Sunflower (*Helianthus annus*) Oil	7
Dimethicone	5	Sweet Almond (*Prunus amygdalus dulcis*) Oil	7
Dog Rose (*Rosa canina*) Hips Oil	7	Theobroma Cacao (Cocoa) Seed Butter	6

Source: www.lotioncrafter.com/pdf/Emulsions_&_HLB_System.pdf.

Conclusion

Oils are an important component of many cosmetic products. A thorough understanding of their properties and the properties they impart to a cosmetic product, as individual ingredients and in combination with other oils and components, is critical to the successful formulation. Consideration of aesthetics, sensory properties, cosmetic activities such as emolliency and water-proofing, and stability are all influenced by the appropriate choice of oil and oil combinations.

Acknowledgements

Thanks to Jari Alander for information about characterisation methods.

REFERENCES

Blank IH. Action of emollient creams and their additives. *J Am Med Assoc.* 1957, 164(4): 412–415.
Conti A, Rogers J, Verdejo P, Harding CR, Rawlings AV. Seasonal influences on stratum corneum ceramides 1 fatty acids and the influence of topical essential fatty acids. *Int J Cosmet Sci.* 1996, 18(1): 1–12.
Chlebarov S. *Notabene Medici*, vols. 2 and 3. 1989, 74–76 and 124–127.

De Backer G, Ghirardi D. Silicones: Utilization in fat-free cosmetics. *Parfums Cosmet Aromes*. 1993, 114: 61–64.

Davies JT. A quantitative kinetic theory of emulsion type, I. Physical chemistry of the emulsifying agent (PDF). *Gas/Liquid and Liquid/Liquid Interface* (Proceedings of the International Congress of Surface Activity). 1957: 426–438.

De Graef V, Vereecken J, Smith KW, Bhaggan K, Dewettinck K. Effect of TAG composition on the solid fat content profile, microstructure and hardness of model fat blends with the same saturated fatty acid content. *Eur J Lipid Sci Technol*. 2012, 114(5): 592–601.

DiNardo JC. Is mineral oil comedogenic? *J Cosmet Dermatol*. 2005, 4: 2–3.

General Electric (GE). Silicones: Silicone Specialties For Personal Care Materials of Applications. Waterford, NY, 1999. P1-16 (CDS 4711(3/99).

Girboux AL, Courbon E. Enhancing the feel of vegetable oils with silicone. *Cosmet Toiletries*. 2008, 123(7): 49–56.

Gorcea M. Efficient strategies for skin moisturization. *NYSCC Cosmetiscope*. 2008, 14(2): 6, 7.

Griffin WC. Classification of surface-active agents by 'HLB' (PDF). *J Soc Cosmet Chem*. 1949, 1(5): 311–326.

Griffin WC. Calculation of HLB values of non-ionic surfactants (PDF). *J Soc Cosmet Chem*. 1954, 5(4): 249–256.

Himawan C, Starov VM, Stapley AGF. Thermodynamic and kinetic aspects of fat crystallization. *Adv Colloid Interface Sci*. 2006, 122(1–3): 3–33.

Johnson DH. *Hair and Hair Care*. Cosmetics Science and Technology, 17th ed. Marcel Dekker, 1997.

Lanzet M. Comedogenic effects of cosmetics raw materials. *Cosmet Toiletries*. 1986, 101: 63–72.

Lochhed RY. Formulating conditioning shampoos. *Cosmet Toiletries*. 2001, 116: 11.

Lodén M, Maibach HI (eds.). *Dry Skin and Moisturizers, Chemistry and Function*. CRC Press, 2000.

Morrison R, Boyd R. *Química Orgânica*, 6th ed. Fundacao Calouste Gulbekian, 1978.

Nissim Garti KS. *Crystallization and Polymorphism of Fats and Fatty Acids*. Marcel Dekker, 1998.

O'Lenick Jr AJ. Silicones. Comparatively Speaking: Essential Fatty Acid vs. Trans Fatty Acid. *Cosmet Toiletries*. December 8, 2010.

Poucher WA. *Perfumes, Cosmetics and Soaps*, vol. 2., 7th ed. Chapman & Hall, 1959, p. 3.

Rousset P, Rappaz M, Minner E. Polymorphism and solidification kinetics of binary systems POS-SOS. *J Am Oil Chem Soc*. 1998, 76(7): 857–864.

Sato K, Ueno S. Polymorphism in fats and oils. In: *Bailey's Industrial Fat and Oil Products*. Wiley, 2005, pp. 77–120.

Smith K, Bhaggan K, Talbot G, van Malssen KF. Crystallization of fats: Influence of minor components and additives. *J Am Oil Chem Soc*. 2011, 88(8): 1085–1101.

Sonwai S, Mackley MR. The effect of shear on the crystallization of cocoa butter. *J Am Oil Chem Soc*. 2006, 83(7): 583–596.

Sonwai S, Rousseau D. Structure evolution and bloom formation in tempered cocoa butter during long-term storage. *Eur J Lipid Sci Technol*. 2006, 108(9): 735–745.

Spector AA. Essentiality of fatty acids. *Lipids*. 1999, 34(Suppl): S1–S3.

Timberlake KC. *General Organic and Biological Chemistry: Structures of Life*, 6th ed. Pearson, 2018.

Van Malsen K, Peschar R, Schenk H. Real-time X-ray powder diffraction investigations on cocoa butter. I. temperature-dependent crystallization behavior. *J Am Oil Chem Soc*. 1996, 73(10): 1209–1215.

Van Reeth I, Starch M. Novel silicone thickening technologies: Delivery the appropriate rheology profile to optimize formulation performance. *J Appl Cosmetol*. 2003, 21: 97–107.

Weast RC, Astle MJ, Beyer WH (eds.). *CRC Handbook of Chemistry and Physics: A Ready-Reference Book of Chemical and Physical Data*, 66th ed. CRC Press, 1978.

Wiechers J. Relative performance testing: Introducing a tool to facilitate cosmetic ingredient selection. *Cosmet Toiletries*. 1997, 112(9): 79–84.

Zocchi G. Skin feel agents. In: Paye M, Barel A, Maibach HI. (eds.), *Handbook of Cosmetic Science and Technology*, 2nd ed. CRC Press, 2006, pp. 248–260.

11

Moisturizers

Catherine Tolomei Fabbron Appas, Ariane Dalanda Silva Ladeira, Thamires
Batello Freire, Patricia Santos Lopes and Vânia Rodrigues Leite-Silva

CONTENTS

Introduction

Moisturizers are one of the most important and commonly used product groups for skin care. They are generally applied daily to facilitate skin elasticity and flexibility, making the skin feel healthier, improving appearance and preventing premature aging. Whilst they may comprise a combination of ingredients, their basic function is to promote, restore and maintain the hydration balance of the skin. Moisturizing skin care products are generally formulated and marketed for specific body regions such as the face and neck, body, hands and feet. They often contain specific ingredients for particular roles such as anti-aging, sun protection, skin whitening, tanning. Agents included for these specific roles are covered in other chapters, whilst the focus of this chapter is the formulation of skin care products for moisturization. A moisturizer usually refers to a formulated product containing an emollient and/or humectant that in combination with water in the formulation provides direct hydration of the stratum corneum. These agents promote and prolong the moisturizing effect of water.

Skin Hydration

The functional and aesthetic properties of the skin are directly related to skin hydration. Although the stratum corneum has relatively low hydration compared to the viable epidermal regions, the presence of water has an important function in giving the skin surface its elasticity and softness. The amount of water in the stratum corneum also regulates the activity of specific hydrolytic enzymes that are important for the normal desquamation process from the skin surface (Wu and Polefka, 2008; Qassem and Kyriacou, 2013; Crowther, 2016). As reviewed in Chapter 3, the stratum corneum provides the main skin barrier function (Kelleher et al., 2015), protecting against loss of water, invasion by microorganisms and penetration of toxic agents (Wu and Polefka, 2008; Falcone et al., 2015; du Plessis et al., 2013; Darlenski and Fluhr, 2016). The corneocytes and lipids of the stratum corneum form the physical barrier, while the slightly acid hydrolipid film on the skin surface serves as an additional biochemical barrier. Skin barrier function and appearance are maintained by the interaction between the transepidermal water loss (TEWL), stratum corneum hydration, sebum levels on the skin and the pH of the skin surface (du Plessis et al., 2013). The epidermis is waterproof due to the presence of intercellular lipids such as cholesterol esters and ceramides in the stratum corneum (Iobst et al., 2006) that also provide the primary barrier to both percutaneous absorption and TEWL (Bouwstra et al., 2003). The water in the stratum corneum enables the maturation of corneocytes by means of enzymatic reactions (Iobst et al., 2006; Beny, 2000; Rawlings and Harding, 2004). The lipids of the skin surface, which are derived from the sebaceous glands, contribute to the epicutaneous emulsion that has an important role in hydrating the skin (Iobst et al., 2006; Inoue, 2006; Rossi and Vergnanini, 1997).

The natural moisturizing factor (NMF) present in the stratum corneum is important for maintaining proper moisture levels within the region. NMF is a degradation product of filaggrin, a histidine-rich protein that acts on the keratin filaments within the cells of the stratum granulosum, aggregating them to form the keratin bundles that are responsible for the rigid structure of the stratum corneum cells. Filaggrin is degraded within the stratum corneum into a number of hygroscopic molecules such as urea and pyrrolidone carboxylic acid, glutamic acid and other individual amino acids, which are collectively called the NMF of the skin. Filaggrin degradation to NMF is reported to be dependent on ambient humidity levels, with an optimum humidity range for NMF production of 80–95%. The role of NMF and its optimum levels *in vivo* are not well known, although it has been observed experimentally that filaggrin levels are lower in the skin of patients with atopic dermatitis, even in areas that appear unaffected (Jungersted et al., 2010). The quantity and quality of NMF components within the stratum corneum is important in the regulation of TEWL, and in prevention of dry skin (Robinson, 2009; Stamatas et al., 2011). Robinson et al. (2010) showed that even short-term exposure to water alone (such as washing or bathing) reduces skin hydration, and this remained low over a 4-hour post-exposure period. In addition, significant increases in skin pH were observed in the immediate post-soak period, although this recovered over the following 4 hours. Topical application of NMF was shown to reverse the skin dehydrating effects of soaking. Rawlings and Harding (2004) also showed that the pH of the stratum corneum surface increases when the skin is completely occluded and thereby exposed to increased humidity. While the repair of the barrier after acute disorders usually proceeds at acidic pH, skin regeneration is delayed at neutral pH, which prevents the corresponding discharge process of newly secreted polar lipids (glucosylceramides) towards maturation of lamellar bilayer ceramides. Furthermore, an increase in proteolytic activity observed under conditions of high pH triggered an increase in corneodesmolysis aberrations and cohesion of the corneocytes.

Reduced skin hydration can provide characteristic clinical symptoms of dry skin or xerosis, such as flaking, itching, opacity, redness, fissures/cracks and tightness that is noticeable both visually and by feel (Rossi and Vergnanini, 1997; Loden, 2003). In dry skin, the envelopes of the corneocytes are fragile due to the diminished action of the enzyme transglutaminase (Rawlings and Harding, 2004). Maintenance of skin hydration and the capacity of cell renewal of the body are vital in the preservation of health, softness, flexibility, elasticity and skin youthfulness (Rossi and Vergnanini, 1997; Leonardi, 2004).

The barrier function and the maintenance of skin hydration are influenced by the intrinsic characteristics of the skin layers. The stratum basale is nourished and hydrated by the capillaries of the dermis (Beny, 2000). Water is retained by the corneocytes due to the action of glycosaminoglycans (Beny, 2000; MacMary et al., 2006; Girard et al., 2000). The stratum spinosum cells are formed of lamellar bodies responsible for the formation of the hydrolipidic mantle and the keratohyalin granules (Harris, 2003). In the stratum granulosum there is also the presence of keratin filled with granules, which after cell maturation, take on the flatness of keratinocytes, and these keratin-forming plates offer resistance to mechanical stress. The stratum corneum cells, which are flattened and without nuclei, are composed mostly of the fibrous protein keratin (Figure 11.1). The thickness of the stratum corneum varies with body site and can change whether or not the skin is moisturized. Egawa (2007) showed that after 1.5 hours of moisturization, the thickness can double.

Skin Moisturization

Given our knowledge of the structure and physiology of the skin (Chapter 2), it is vital to understand the site of action of moisturizing agents, so as to ensure that a cosmetic product will have high efficacy by targeting to its site of action within the skin and low systemic toxicity by not reaching the bloodstream (Loden, 2003; Leonardi, 2004). A number of different hydration mechanisms can be used to protect the hydrolipidic barrier, including occlusion, emollience, barrier enhancement, wetting and active hydration.

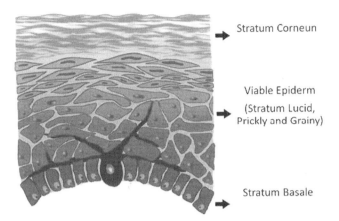

Stratum Corneun

Viable Epiderm
(Stratum Lucid,
Prickly and Grainy)

Stratum Basale

FIGURE 11.1   Skin layers of the epidermis.

Occlusion achieved through the use of formulation components that produce a thin layer of oil on the skin, favouring the formation of the hydrolipidic mantle and retarding the evaporation of water is illustrated in Figure 11.2. This is most effective when applied to the skin after bathing, with a view to retaining the water in the stratum corneum, and give softness and a velvety appearance to the skin. Examples of components that confer occlusive properties to a product are petrolatum, lanolin, isopropyl myristate and mineral oil (Table 11.1) (Yokota and Maibach, 2006).

Emollients are a broad class of ingredients that are among the most important in skin care and cosmetic formulations (Douguet et al., 2017). Lipophilic emollients can influence the sensory characteristics of the products in which they are included, offering a wide variety of sensations such as greasy, oily, dry, sticky or velvety when applied to the skin (Parente et al., 2005). Emollients can soften the skin and replenish the natural skin lipids in the lipid domains between the corneocytes in the stratum corneum, to create a smoother skin surface and provide cohesion and levelling of the corrugated edges of the corneocytes (Evangelista et al., 2014). Additional benefits of emollient application include a partial occlusive function to attenuate TEWL and increase hydration (Figure 11.3) (Polaskova et al., 2015).

Water absorption can be promoted by the action of humectants, which attract and retain water from both the environment and the dermis (Figure 11.4) (Rawlings et al., 2004). Examples of humectants are sorbitol, glycerol, propylene glycol, glycosaminoglycan, elastin and collagen (Table 11.1). There is some debate about the use of humectants, because in dry environments they can absorb water from the skin, thereby increasing skin dryness. Humectants are therefore most suitable in wet environments (Sethi et al., 2016).

Substances that can mimic the action of the constituents of the NMF by promoting intracellular hydration are said to provide hydration therapy. They are often recommended for the treatment of xerosis or dry skin because they are structurally similar to natural substances in the stratum corneum, resulting in increased compatibility with the epithelial cells and consequent improvement of the therapeutic and aesthetic results (Rossi and Vergnanini, 1997; Padamwar et al., 2005). Free water in the stratum corneum facilitated by the NMF favors the biochemical reactions of proteases, that in turn contribute to the development of NMF constituents through the proteolysis of filaggrin (Rawlings and Harding, 2004).

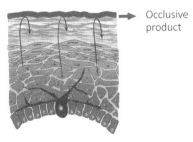

Occlusive
product

FIGURE 11.2   Film formed on skin when an occlusive product is applied. The arrows indicate that water loss from the skin is prevented by occlusion.

TABLE 11.1

Classification and Examples of Moisturizing Agents

Mechanism of Action	Example	Reference
Occlusive Physically block TEWL, induce expression of key barrier differentiation markers (filaggrin and loricrin), increase stratum corneum thickness, and reduce T-cell infiltrates in the normal-appearing or non-lesional AD skin	Petrolatum	Ghadially et al., 1992; Flynn et al., 2001; Czarnowicki et al., 2016
	Paraffin	Flynn et al., 2001; Darlenski and Fluhr, 2011
	Lanolin	Ghadially et al., 1992; Flynn et al., 2001
	Mineral oil	Flynn et al., 2001
	Silicones	Douguet et al., 2017
	Zinc oxide	Brandrup et al., 1990
	Squalene and wool alcohol	Flynn et al., 2001
	Fatty acids	Flynn et al., 2001
	Silk sericin	Padamwar et al., 2005
Humectants Hydrophilic compounds that hydrate the stratum corneum Attract water from the skin and the environment into the stratum corneum (transepidermal)	Glycerin	Loden, 2003; Loden, 2003; Shao et al., 2015; Kikuchi and Tagami, 2008; Short et al., 2007
	Urea	Loden, 2003; Flynn et al., 2001; Kikuchi and Tagami, 2008
	Alpha hydroxy acids	Loden, 2003; Flynn et al., 2001; Kikuchi and Tagami, 2008
	Sugars	Shao et al., 2015
	Sorbitol (derivated sugar)	Loden, 2003; Flynn et al., 2001
	Propyleneglycol	Flynn et al., 2001
	Ethyleneglycol	Robinson et al., 2008
	Free amino acid	Hoffman et al., 2014
	PCA (2-pyrolidone-5-carboxylic acid)	Loden, 2003; Flynn et al., 2001
	Sodium lactate	Loden, 2003; Flynn et al., 2001
Emollients Hydrophobic compounds that provide an occlusive barrier to prevent water loss Smooths skin by filling spaces between skin flakes with droplets of oil	Mineral oil	Douguet et al., 2017; Parente et al., 2005; Jackson, 1992
	Esther	Douguet et al., 2017; Flynn et al., 2001; Robinson et al., 2008
	Fatty acids	Nasrollahi et al., 2018; Jung et al., 2013
	Cholesterol	Jung et al., 2013
	Squalene	Parente et al., 2005
	Structural lipids (ceramides)	Jung et al., 2013–51
	NTP-CE (Panthopenol-containing emollient)	Stettler et al., 2017
	Silicones fluid (cyclomethicone)	Parente et al., 2005
	DFD-01	Jackson et al., 2017
	Mucilage	Hadi et al., 2015
Protein rejuvenators Claimed to rejuvenate skin by replenishing essential proteins in skin	Collagen	Hou et al., 2012, 56
	Keratin	Short et al., 2007; Kraft and Lynde, 2005
	Elastin	Kraft and Lynde, 2005
	Tretinoin	Gilchrest, 1996
Stimulating aquaporines Lead to a five-fold increase in water permeability, which subsequently increases stratum corneum hydration	AQP3	Hadi et al., 2015

Source: Adapted from Lynde CW, *Skin Therapy Lett.* 2001;6(13):3–5.

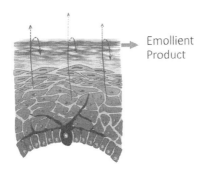

FIGURE 11.3 Emollient products intercalate in the lipid domains between corneocytes. Partial occlusion by emollients retards water loss from the skin surface, as indicated by the dotted lines.

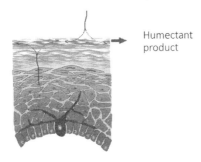

FIGURE 11.4 Humectants attract and retain water in the stratum corneum from both the environment and deeper skin layers, as illustrated by the arrows.

There is a correlation in the decline of NMF with age, due to both reduced profilaggrin synthesis compared to young skin and decreased barrier function. In addition, modulation of the profilaggrin–filaggrin system and the water present in the stratum corneum alters the regulation of desquamation (Rawlings and Harding, 2004). NMF extracellular components that can be found in corneocytes include sugars, hyaluronic acid, urea, and lactate, thus these agents may be useful moisturizer active ingredients.

Moisturizing products may include a single active ingredient or multiple, complimentary ingredients and approaches. Benson et al. (2010) demonstrated the permeation of urea into *in vitro* human skin and the increase in skin hydration in human volunteer skin by optical coherence tomography (OCT). Both occlusion and a novel magnetophoretic array device increased the permeation of urea and skin hydration, with the latter enhancement technique showing a greater effect.

Active ingredients used as moisturizers have favorable safety and toxicity properties. Even when used on large areas of the body, and for a long period of time, they are rarely associated with adverse effects (Halvarsson and Loden, 2007). However, Buraczewska et al. (2007) stated that there is a lack of knowledge about the long-term effects of moisturizing on skin barrier function. They suggested that some moisturizers may weaken the function of the barrier, while others may strengthen it, so given that moisturizers are often part of an individual's daily skin care routine it is important to understand their long-term effects on skin properties. To test the efficacy of skin moisturizers, sensitive and reliable methods are required (Agner et al., 2000).

Table 11.1 provides a classification and examples of commonly used moisturizers. It should be noted that some agents can act by more than one mechanism and can therefore be classified into more than one category, for example mineral oil can act as an emollient and/or by occlusion.

Equally important to the raw materials is the vehicle in which they are incorporated. Indeed it has been shown that the moisturization of the skin can vary for an active substance like urea if the formulation in which it is applied is different (Chaves et al., 2014).

The sensorial properties (discussed in Chapter 24) of different vehicles can vary markedly and are directly related to the body site of intended application. As an example, we can see two different

TABLE 11.2

Examples of Formulations Containing Urea

Raw Material (INCI)	Body Lotion (% w/w)	Foot Cream (% w/w)	Function
Urea	5.0	5.0	Active
Ceteareth 20	2.0	1.5	Surfactant
Cetearyl alcohol	3.0	3.5	Donor consistency (wax)
Triethanolamine	pH 5.5–6.5	×	pH adjustment
Carbomer	0.15	×	Donor consistency (gum)
Lanolin	2.0	5.0	Emollient
Caprilic/capric triglyceride (CCT)	5.0	2.0	Emollient
Glycerin	4.0	3.0	Humectant
Phenoxyethanol	0.7	0.7	Preservant
Glyceryl stearate	×	3.0	Surfactant and emulsifying agent
Petrolatum	×	4.0	Emollient
Mineral oil		1.5	Emollient
Dimeticone	×	0.5	Emollient
BHT	0.05	0.05	Antioxidant
EDTA	0.1	0.1	Chelating agents
Acqua	q.s. 100	q.s. 100	Vehicle

formulations with urea (Table 11.2). Both are for the same purpose to 'moisturize the skin', but the site of application is different, in this case the body and foot. Moisturizing lotion intended for the body is very fluid, as it is designed to be easy to apply to a large area, dry quickly and is not greasy. The moisturizer for the foot is more viscous and greasier, containing a higher proportion of oil in the vehicle. The formulations in Table 11.2 are 'typical formulas' to aid in teaching practical skills to students and do not have stability, safety or efficacy testing. The authors recommend that any products being developed are tested in advance to verify their suitability for application to the skin.

FIGURE 11.5   Confocal Raman microspectroscopy. (From http://mvascientificconsultants.com/capabilities.)

Assessment of Skin Hydration

Skin barrier function is evaluated using non-invasive biophysical methods such as transepidermal water loss (du Plessis et al., 2013; Gardien et al., 2016), capacitance and conductance. Whilst these methods are relatively quick, convenient and inexpensive to perform, they share some characteristics that can influence the measurements, such as sensitivity to environmental changes, the amount of electrolyte present, the area of contact and applied pressure. However, these instruments remain the mainstay of routine hydration assessment of topical products. Confocal Raman microspectroscopy (CRS; Figure 11.5) can directly measure the molecular composition and structure of the skin and has proven to be a powerful technique for biomolecular analysis (Falcone et al., 2015). CRS and infrared spectrometry can be used to map the relative concentration and distribution of endogenous molecules in the skin, including water and concentration gradients of the NMF constituents (Flach and Moore, 2013).

There is a continuous increase in the both the development and use of spectroscopic imaging methods to monitor biological and biophysical issues in skin science and dermatology. The application of these assessment approaches to monitor hydration and moisturization actives is described next.

Electrical Current Methods

Dry stratum corneum is a dielectric medium and its electrical properties change with the change in moisture content (Chaves et al., 2014). As the electrical properties of the skin are directly related to the water content of the stratum corneum, this can be used to monitor skin hydration and the moisturizing effect of skin care products (Clarys and Barel, 2016).

The most widely used non-invasive approach to measure skin moisture content is by measuring its ability to conduct electrical current. Currently there are four widely used instruments: the Skicon® (I.B.S. Company, Japan), Corneometer® (Courage & Khazaka, Germany), MoistureMeter® (Sephora, Hong Kong) and the Nova Dermal Phase meter® (Nova Technology Corp. USA), based on measuring conductivity, capacitance and impedance, that provide indirect assessment of skin hydration (Wu and Polefka, 2008). These devices all use rigid probes that must be in contact with the skin (Figure 11.6) (Ezerskaia et al., 2016).

The measurement of total electrical resistance (impedance) of the skin of an alternating current frequency (F) is the most widely used method to assess the hydration state of the skin surface. A number of physical factors can play an important role in the impedance measurements, such as the design of the electronic oscillator circuit and the frequency of the electric current, the geometry of the electrode applied to the skin (distance between the electrodes), the depth measurement of the electric field in the skin, and the direct or indirect galvanic contact with the skin surface (Clarys and Barel, 2016).

In addition to water content, other factors can also influence the electric current flow of the skin including orientation of the dipole moments of skin, applied product constituents such as keratin, and the ionic

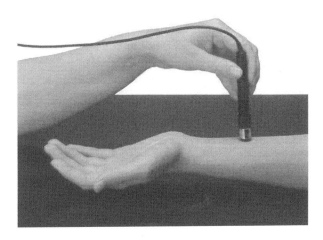

FIGURE 11.6 Probe in contact with skin. (From www.courage-khazaka.de/index.php/en/products/scientific/55-corneometer.)

movement of salts and other electrolytes. Salts are common components of skin care products such as moisturizers and can also be delivered to the skin through sweat (Wu and Polefka, 2008). Other factors that can influence electric flow measurements are the skin contact area with the probe, the pressure applied, ambient temperature and humidity (Ezerskaia et al., 2016). Another limitation of the electrical methods is that they do not provide information about the depth without the simultaneous use of several instruments (Wu and Polefka, 2008). They are also not suitable for measuring changes in moisture levels over time and to visualize the spatial distribution and heterogeneity of the hydration capacity of the whole face skin (Ezerskaia et al., 2016).

The Corneometer CM 825® (Courage & Khazaka Electronic GmbH, Germany) measures skin hydration based on the dielectric constant of water (Chaves et al., 2014). The measuring capacitor shows changes in capacitance of the skin surface in arbitrary units (corneometrics units) (Polaskova et al., 2015; Clarys and Barel, 2016), ranging from 20 (very dry) to 120 units (quite hydrated). The depth of penetration of electric current into the skin is about 45 mm (Clarys and Barel, 2016). This instrument has a flat probe with an interposition grid covered by gold electrodes. The interposition grid is coated by a vitrified dielectric material, therefore there is no direct galvanic contact between the electrode surface and the skin during measurement. The application pressure of the measurement probe is less than 1 N as ensured by a spring system triggering capacitance measurements. The frequency changes from 0.95 MHz in a hydrated environment to 1.15 MHz in a dry environment (Clarys and Barel, 2016).

Transepidermal Water Loss (TEWL)

The stratum corneum's ability to prevent uncontrolled evaporation of water from the epidermal layers (permeability barrier) is reflected by the parameter TEWL. Low TEWL is a characteristic of intact and healthy skin, and there is generally a correlation between TEWL and hydration of the stratum corneum (healthy and unchanged skin). However, there is poor correlation in a number of conditions such as the hyperhydration state immediately after the application of surfactants, on palm skin and measurements taken at a skin location with a strongly perturbed epidermal barrier (Leite e Silva et al., 2009). TEWL measurement is commonly used to evaluate skin barrier function and can be useful in the evaluation of skin conditions such as atopic dermatitis. TEWL is a non-invasive *in vivo* measurement of water loss through the stratum corneum that is considered one of the most important parameters for the skin barrier function (Kelleher et al., 2015; Gardien et al., 2016; Kottner et al., 2013; Fluhr and Darlenski, 2014). Several TEWL measuring instruments with different technologies are commercially available, and are routinely used throughout the world in research and development and dermatology practice (Kottner et al., 2013).

TEWL is a density measure of condensed water diffusion (or flux) from the highly hydrated deeper layers of the dermis and epidermis to the skin surface, and is generally expressed in grams per square meter per hour ($g/m^2/h$) (Ezerskaia et al., 2016; Kottner et al., 2013). TEWL values are affected by the condition and function of the stratum corneum. There is convincing evidence that a TEWL increase is associated with dysfunction of a normal skin barrier, and that a decreased TEWL is an indicator of an intact or retrieved skin barrier. However, there is no precise definition of the normal TEWL. TEWL is affected by a number of factors including anatomical location, weather conditions (summer, winter, air conditioning/central heating, sun exposure, etc.), and measuring devices, including calibration and accuracy (Kottner et al., 2013).

As TEWL is a very sensitive parameter, there are a number of variables that can influence measurement accuracy and reliability. The study protocol should ensure that the measurements are performed, whenever possible, in an air-conditioned room of 18–21°C and relative humidity of 40–60%. The study volunteers should acclimatize in air-conditioning for 20–30 min before the first measurement. Measurements should be carried out on the same day and season, avoiding measurements during the summer, ambient flow of air or direct light (influences temperature) at the test site. The remaining steam in the probe should be dried after each measurement (seconds) and during repeated measurements to ensure no carry-over (Fluhr and Darlenski, 2014).

A decrease in TEWL, parallel to increase in hydration of the stratum corneum, is observed after the application of occlusive substances (oils, petrolatum) on the skin. A high TEWL is recorded directly (10–15 min) after application of moisturizing agents on the skin surface. This effect is not due to increased hydration of the stratum corneum, but reflects the evaporation of the water incorporated in the cosmetic product itself. When an occlusive cover is applied on the skin, the TEWL is inhibited. After removal of the occlusion, the water that has accumulated in the porosity of the skin during occlusion evaporates and diffuses from the outside showing a TEWL rate higher than the base value and the time function (Gioia and Celleno, 2002).

There are different methods to measure TEWL: closed chamber, vented chamber and open chamber method. The closed chamber potentially obstructs the skin and therefore is unable to perform continuous measurement. The disadvantage of the vented chamber method is the dry or moistened carrier gas can interfere with the microenvironment on the surface of the skin. The open chamber method, which does not interfere with microclimate and does not clog the skin, is a reliable and useful tool for both individual and continuous measurements of water loss by evaporation (Fluhr and Darlenski, 2014).

Therefore, the TEWL may be measured *in vivo* using a closed or an open chamber method. The measurement method with an open chamber is widely used in clinical and experimental research and is the standard against which newer closed chamber methods are developed (du Plessis et al., 2013).

There are a number of TEWL meters in common usage in clinical, industry and research environments. The ServoMed evaporimeter (Seromed AB, Stockholm, Sweden) is based on the open chamber evaporation method. The fundamental principle of measurement is based on the fact that the pressure gradient of the water vapour immediately above the surface of the skin is proportional to the difference between the pressure measured at two different perpendicular heights above the skin on the diffusion zone. This device consists of a movable measuring probe connected to a signal processing unit. The Teflon capsule has a cylindrical measuring chamber that is open at both ends, incorporated with paired sensors (hygrosensors and thermosensors) for determining the relative humidity and temperature at two different levels above the skin surface.

The Tewameter TM 300® (Courage & Khazaka Electronics GmbH, Germany) indirectly measures the density gradient of water evaporation from the skin by two pairs of sensors (temperature and humidity) within the hollow cylinder, determining the TEWL in grams per square meter per hour ($g/m^2/h$) (Polaskova et al., 2015) (Figure 11.7). The Tewameter consists of a movable measuring probe connected to a processor unit, which can be connected to a computer for data management. It has a cylindrical and opened measurement chamber connected to the probe, with two sensor units positioned at 3 and 8 mm from the skin surface. A clamp secures the probe holder at a fixed position at the moment of measuring the skin surface (Ferreira, 2014).

Gardien et al. (2016) examined the reliability of the Tewameter TM 300 for repeat measurements in burn scars, with intra-class correlation coefficient (ICC) values between 0.85 and 0.94. Mean TEWL values on burn scars were significantly higher than healthy skin ($p < 0.001$). Indeed, significant correlations were found between TEWL values of hypertrophic scars and erythema affected skin ($r = 0.60$, $p = 0.001$). TEWL values at the burn scar skin changed over time as the skin healed with significantly different TEWL values for 3 to 6 months and 3 to 12 months for old scars (respectively, $p = 0.021$ and $p = 0.002$). In

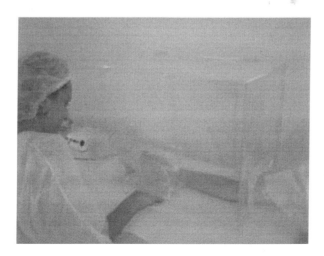

FIGURE 11.7 TEWL measurement using a Tewameter in an environmental chamber. (From Leite e Silva VR Desenvolvimento de formulações cosméticas hidratantes e avaliação da eficácia por métodos biofísicos (tese), University of São Paulo, Brazil, 2009.)

total or partial burns, the function of the stratum corneum is damaged, resulting in a deviation of TEWL in scar tissue compared to normal skin.

The two most commonly used closed chamber TEWL instruments are the AquaFlux (Diox Systems Ltd., United Kingdom) and VapoMeter (Delfin Technologies Ltd., Finland).

The VapoMeter is a portable and battery-operated instrument with a humidity sensor inside a cylindrical measurement chamber. The AquaFlux consists of a measurement chamber with a condenser that removes water vapour originating from the skin surface, thus maintaining low humidity within the chamber. Sensors within the chamber determine the relative humidity and temperature, and the humidity gradient is converted to flux density using the same diffusion gradient principle as the open chamber system. The manufacturers claim that the waisted design of the measuring chamber provides reduced sensitivity to the probe angle and reduces perturbation of the sample surface microclimate. In both cases the chamber is closed by the skin or other measurement surface during the measurement period and is thereby unaffected by ambient airflows. The sensor monitors the increase of relative humidity inside the chamber during the measurement. Imhof et al. (2014) who developed the AquaFlux system, showed advantages in the accuracy, sensitivity and repeatability performance characteristics of the AquaFlux, whilst the VapoMeter benefits from speed and mobility.

The use of TEWL as a single parameter to characterize the direct hydration of the skin should be applied cautiously. A combination of the measurement of TEWL with other non-invasive methods is recommended as the most reliable and accurate approach (Leite e Silva, 2009).

Optical Methods

Optical methods based on light absorption or dispersion, such as optical coherence tomography (OCT), infrared spectroscopy (such as attenuated total reflectance Fourier transform infrared [ATR-FTIR] spectroscopy) and confocal Raman spectroscopy have been used for non-invasive *in vivo* evaluation of skin properties (Falcone et al., 2015).

Confocal Raman spectroscopy (CRS) has been used to evaluate changes in the stratum corneum (Simpson et al., 2013), including the investigation of individual cells, such as the characterisation of stem cells (Falcone et al., 2015). This technique has potential as a non-destructive tool for many skin

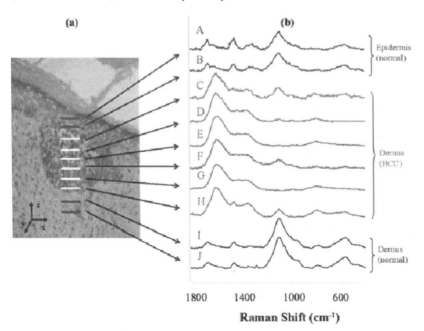

FIGURE 11.8 Confocal Raman profile of human epidermis and dermis. (From Choi J, et al., *Biopolymers*, 2005;77(5):264–72.)

applications, such as monitoring the distribution of physiological components, diagnosis of pathological conditions, measurement of topically applied actives and excipients, and assessing skin hydration (Hoppel et al., 2016). The composition of the skin and the changes in hydration by depth can be determined in a non-invasive way (Figure 11.8) (Wu and Polefka, 2008; Egawa and Sato, 2015).

CRS has been employed in the *in vivo* study of the structural components of the epidermal barrier (lipids, proteins, NMFs and water gradient), including changes in the integrity and protective function of the epidermal barrier. The CRS microscope combines the Raman scattering spectra of compositions, such as water, natural moisturizing factor and urea in human skin with the techniques of confocal laser microscopy using a small orifice to permit the measurement of water content at a specific depth from the skin surface. The depth of measurement is limited by the depth range of the laser, therefore the dermis may not be reached by the CRS microscope (Sato et al., 2015). CRS can be used in combination with spectroscopic techniques such as infrared spectroscopy to better elucidate the skin's micro composition. For example, after evaluating the compounds of interest measured in the Raman spectrum, Darlenski and Flurh identified additional substances from the initial spectrum (Darlenski and Fluhr, 2016).

CRS and IR spectrometry are based on determination of the vibrational modes of the chemical bonds of a molecule (Figure 11.9) (Takaoka et al., 2010). Compared to other analytical methods they have the advantage of low interference from skin fluorescence/luminescence and the presence of water, and therefore have received considerable interest in the evaluation of biological molecules within the skin. CRS requires very small amounts of samples and is not destructive, so the material can be recovered (Baby et al., 2006).

CRS and IR do have some limitations. The strong absorption of infrared radiation by water (Raman spectroscopy Fourier transform) limits the penetration depth of light to a few micrometers, so this technique reflects the hydration only in the most external layers of the stratum corneum. The use of hydrating substances on the skin can potentially interfere with the electromagnetic emission and thus influence the optical properties of the pretreated skin (Darlenski and Fluhr, 2011).

The physical basis of the CRS technique is the scattering of inelastic light by different molecules: Raman spectra are obtained for the specific chemical structure of the molecules. Considering the intensities of the Raman bands at certain changes, the ratio of water to protein can be calculated (Darlenski and Fluhr, 2011). This technique, which uses a highly advanced spectrophotometer equipped with a fiber optic probe to obtain spectra near infrared reflectance (NIR) spectra of skin, is used because it can accurately determine water within the skin using a combination of OH bonds and HOH water occurring in the

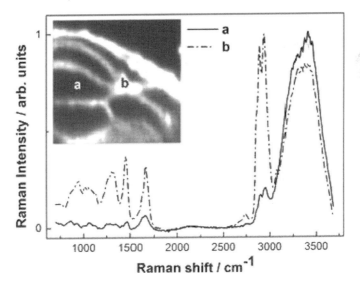

FIGURE 11.9 Spectra extracted from the water inclusions (trace a) and the surrounding lipid and protein rich regions (trace b) in a skin sample hydrated with neat PBS for 24 h. Spectra were normalized with respect to the maximum intensity for clarification. (From Alber C et al., *Biochim Biophys Acta* 2013;1828(11):2470–8. With permission.)

NIR region, which are good indicators of the level of skin hydration and water content. NIR spectra can also provide information about other components of the skin, including lipids and proteins, which can differentiate between different types of water on the skin, and its instrumentation can easily be equipped with fiber optic probes for in vivo measurements (Qassem and Kyriacou, 2013).

To show the feasibility of the method for measuring skin hydration and surface lipids in natural conditions, and their changes due to external stimuli, infrared optics using infrared spectroscopy absorption were developed to facilitate the quantitative and simultaneous measurement of hydration levels of skin and sebum. The method uses detection between three different wavelengths: 1720, 1750, 1770 nm. These correspond to the vibrational bands of lipids, which are the prominent water absorption bands in order to evaluate the balance between these factors related to the integrity of skin, and to select the treatment and appropriate products for skin care (Ezerskaia et al., 2016). This method was useful in showing the hydration of normal skin and skin treated with formulations, and good correlation with other biophysical measurements.

Whilst these spectroscopic methods are excellent analytical tools, allowing a more direct measurement of water as well as the potential to measure other ingredients, they are more expensive than the electrical techniques and are far more complex to use. Therefore, although they have a place in the research and development process, they have not replaced conventional electrical methods in the routine monitoring of skin hydration in clinical trials (Crowther, 2016).

Caloric Loss

Calorimetry is a technique for measuring the thermal properties of materials to establish a connection between temperature absorbed or emitted during controlled heating, and the specific physical properties of substances. It is the only method for the direct determination of enthalpy associated with a process of interest. Calorimeters are often used in chemistry, biochemistry, cell biology, biotechnology, pharmacology and recently in nanoscience to measure the thermodynamic properties of biomolecules and materials at the nanoscale. Among the many kinds of calorimeters, differential scanning calorimetry (DSC) is particularly popular in formulation development and assessment. DSC is a thermal analysis apparatus for measuring how the physical properties of a sample change with temperature as a function of time. In other words, the device is a thermal analysis instrument that determines the temperature and heat flow associated with transitions in materials as they are heated in a controlled way over time. A DSC measures the quantity of heat that is radiated or absorbed by the sample, based on the temperature difference between the sample and reference material (Gill et al., 2010). It is particularly applied to monitor phase transition changes (Gill et al., 2010). DSC is a widely used method for the assessment of structural changes of the stratum corneum, allowing the assessment of the effects that compounds applied to the skin have on its structure. It is used in the pharmaceutical and polymer industries to characterize the thermal behaviour of drug components (Takaoka et al., 2010). Leite-Silva (2009) used DSC to evaluate and compare in vitro changes in the stratum corneum caused by moisturizing products when applied to the skin.

Snake skin has been used in *in vitro* skin studies as an alternative to human skin, as it provides similar barrier qualities as the human stratum corneum and does not require the sacrifice of laboratory animals (skin is replaced approximately every 2–3 months in adult snakes). The membrane is devoid of hair follicles, easy to store and does not undergo microbial degradation (Takaoka et al., 2010; Baby et al., 2006).

Conclusions/Future Directions

Given the importance of hydration of the skin, new raw materials are in constant development. This is a very dynamic area in cosmetics and personal care product development, thus there is a clear need to continue to develop and improve the evaluation techniques that are so important to develop moisturizers with excellent efficacy and safety. Hydration is essential for good skin function as well as good sensorial aspects and maintaining a youthful appearance.

Acknowledgements

The authors thank Bruna Dallabrida for the artwork, Luiz Gustavo Matheus and Alberto Kurebayashi for the formulations, Dr. Jeffrey Grice for the wise advice, Alessandra Andrei for assisting with organization of the table, and Pierre-Yves Libouban for assisting with organization of the references.

REFERENCES

Agner T, Held E, West W, Gray J. Evaluation of an experimental patch test model for the detection of irritant skin reactions to moisturisers. *Skin Res Technol.* 2000;6(4):250–4.

Alber C, Brandner BD, Bjorklund S, Billsten P, Corkery RW, Engblom J. Effects of water gradients and use of urea on skin ultrastructure evaluated by confocal Raman microspectroscopy. *Biochim Biophys Acta.* 2013;1828(11):2470–8.

Baby AR, Lacerda AC, Velasco MV, Lopes PS, Kawano Y, Kaneko TM. Spectroscopic studies of stratum corneum model membrane from Bothrops jararaca treated with cationic surfactant. *Colloids surf B Biointerfaces.* 2006;50(1):61–65.

Benson HA, Krishnan G, Edwards J, Liew YM, Wallace VP. Enhanced skin permeation and hydration by magnetic field array: Preliminary in-vitro and in-vivo assessment. *J Pharm Pharmacol.* 2010;62(6):696–701.

Beny MG. Fisiologia da pele. *Cosmet Toiletries.* 2000;12:44–50.

Bouwstra JA, Honeywell-Nguyen PL, Gooris GS, Ponec M. Structure of the skin barrier and its modulation by vesicular formulations. *Prog Lipid Res.* 2003;42(1):1–36.

Brandrup F, Menne T, Agren MS, Stromberg HE, Holst R, Frisen M. A randomized trial of two occlusive dressings in the treatment of leg ulcers. *Acta Derm Venereol.* 1990;70(3):231–5.

Buraczewska I, Berne B, Lindberg M, Torma H, Loden M. Changes in skin barrier function following long-term treatment with moisturizers, a randomized controlled trial. *Br J Dermatol.* 2007;156(3):492–8.

Chaves CAC, D'Agostino MC, Baby AR, Leite-Silva VR. Biometrological methods to evaluate in vivo the skin hydratation of different commercial moisturizers containing 10.0% urea as the main claim. *Biomed Biophar Res.* 2014;11(1):101–10.

Choi J, Choo J, Chung H, Gweon DG, Park J, Kim HJ, et al. Direct observation of spectral differences between normal and basal cell carcinoma (BCC) tissues using confocal Raman microscopy. *Biopolymers.* 2005;77(5):264–72.

Clarys P, Barel AO. Measurement of skin surface hydration. In: Humbert P, Maibach H, Fanian F, Agache P, editors. *Measuring the Skin.* Cham: Springer International Publishing; 2016, pp. 1–5.

Crowther JM. Understanding effects of topical ingredients on electrical measurement of skin hydration. *Int J Cosmet Sci.* 2016;38(6):589–98.

Czarnowicki T, Malajian D, Khattri S, Correa da Rosa J, Dutt R, Finney R, et al. Petrolatum: Barrier repair and antimicrobial responses underlying this "inert" moisturizer. *J Allergy Clin Immunol.* 2016;137(4):1091–1102.e7.

Darlenski R, Fluhr JW. In vivo Raman confocal spectroscopy in the investigation of the skin barrier. *Curr Probl Dermatol.* 2016;49:71–79.

Darlenski R, Fluhr JW. Moisturizers and emollients. In: Fluhr JW, editor. *Practical Aspects of Cosmetic Testing: How to Set up a Scientific Study in Skin Physiology.* Heidelberg: Springer; 2011, pp. 123–41.

Douguet M, Picard C, Savary G, Merlaud F, Loubat-Bouleuc N, Grisel M. Spreading properties of cosmetic emollients: Use of synthetic skin surface to elucidate structural effect. *Colloids Surf B Biointerfaces.* 2017;154:307–14.

Egawa M, Hirao T, Takahashi M. In vivo estimation of stratum corneum thickness from water concentration profiles obtained with Raman spectroscopy. *Acta Derm-Venereol.* 2007;87(1):4–8.

Egawa M, Sato Y. In vivo evaluation of two forms of urea in the skin by Raman spectroscopy after application of urea-containing cream. *Skin Res Technol.* 2015;21(3):259–64.

Evangelista MT, Abad-Casintahan F, Lopez-Villafuerte L. The effect of topical virgin coconut oil on SCORAD index, transepidermal water loss, and skin capacitance in mild to moderate pediatric atopic dermatitis: A randomized, double-blind, clinical trial. *Int J Dermatol.* 2014;53(1):100–8.

Ezerskaia A, Pereira SF, Urbach HP, Verhagen R, Varghese B. Quantitative and simultaneous non-invasive measurement of skin hydration and sebum levels. *Biomed Opt Express*. 2016;7(6):2311–20.

Falcone D, Uzunbajakava NE, Varghese B, de Aquino Santos GR, Richters RJ, van de Kerkhof PC, et al. Microspectroscopic confocal Raman and Macroscopic biophysical measurements in the in vivo assessment of the skin barrier: Perspective for dermatology and cosmetic sciences. *Skin Pharmacol Physiol*. 2015;28(6):307–17.

Ferreira JMEdFG. Metodologia dinâmica para avaliação da hidratação cutânea. Universidade Lusófona de Humanidades e Tecnologias, Lisbon, Portugal; 2014.

Flach CR, Moore DJ. Infrared and Raman imaging spectroscopy of ex vivo skin. *Int J Cosmet Sci*. 2013;35(2):125–35.

Fluhr JW, Darlenski R. Transepidermal water loss (TEWL). In: Berardesca E, Maibach HI, Wilhelm K-P, editors. *Non Invasive Diagnostic Techniques in Clinical Dermatology*. Heidelberg: Springer; 2014, pp. 353–6.

Flynn TC, Petros J, Clark RE, Viehman GE. Dry skin and moisturizers. *Clin Dermatol*. 2001;19(4):387–92.

Gardien KL, Baas DC, de Vet HC, Middelkoop E. Transepidermal water loss measured with the Tewameter TM300 in burn scars. *Burns*. 2016;42(7):1455–62.

Ghadially R, Halkier-Sorensen L, Elias PM. Effects of petrolatum on stratum corneum structure and function. *J Am Acad Dermatol*. 1992;26(3 Pt 2):387–96.

Gilchrest BA. A review of skin ageing and its medical therapy. *Br J Dermatol*. 1996;135(6):867–75.

Gill P, Moghadam TT, Ranjbar B. Differential scanning calorimetry techniques: Applications in biology and nanoscience. *J Biomol Tech*. 2010;21(4):167–93.

Gioia F, Celleno L. The dynamics of transepidermal water loss (TEWL) from hydrated skin. *Skin Res Technol*. 2002;8(3):178–86.

Girard P, Beraud A, Sirvent A. Study of three complementary techniques for measuring cutaneous hydration in vivo in human subjects: NMR spectroscopy, transient thermal transfer and corneometry – Application to xerotic skin and cosmetics. *Skin Res Technol*. 2000;6(4):205–13.

Hadi H, Razali SN, Awadh AI. A comprehensive review of the cosmeceutical benefits of vanda species (Orchidaceae). *Nat Prod Commun*. 2015;10(8):1483–8.

Halvarsson K, Loden M. Increasing quality of life by improving the quality of skin in patients with atopic dermatitis. *Int J Cosmet Sci*. 2007;29(2):69–83.

Harris MINC. Pele: Estrutura, Propriedades e Envelhecimento. SENAC; 2003.

Hoffman DR, Kroll LM, Basehoar A, Reece B, Cunningham CT, Koenig DW. Immediate and extended effects of sodium lauryl sulphate exposure on stratum corneum natural moisturizing factor. *Int J Cosmet Sci*. 2014;36(1):93–101.

Hoppel M, Kwizda K, Baurecht D, Caneri M, Valenta C. The effect of a damaged skin barrier on percutaneous absorption of SDS and skin hydration investigated by confocal Raman spectroscopy. *Exp Dermatol*. 2016;25(5):390–2.

Hou H, Li B, Zhang Z, Xue C, Yu G, Wang J, et al. Moisture absorption and retention properties, and activity in alleviating skin photodamage of collagen polypeptide from marine fish skin. *Food Chem*. 2012;135(3):1432–9.

Imhof B, Xiao P, Angelova-Fischer I. TEWL, Closed-chamber methods: AquaFlux and VapoMeter. In: Berardesca E, Maibach HI, Wilhelm K-P, editors. *Non Invasive Diagnostic Techniques in Clinical Dermatology*. Heidelberg: Springer; 2014, pp. 345–52.

Inoue S. Biotechnology in skin care (II): Moisturization. In: Lad R, editor. *Biotechnology in Personal Care*. Boca Raton, FL: CRC Press; 2006.

Iobst S, Santhanam U, Weinkauf RL. Biotechnology in skin care (I): Overview. In: Lad R, editor. *Biotechnology in Personal Care*, vol. 29. New York: Informa Healthcare; 2006.

Jackson EM. Moisturizers: What's in them? How do they work? *Dermatitis*. 1992;3(4):162–8.

Jackson JM, Grove GL, Allenby K, Houser T. DFD-01 reduces transepidermal water loss and improves skin hydration and flexibility. *Dermatol Ther (Heidelb)*. 2017;7(4):507–14.

Jung JY, Nam EH, Park SH, Han SH, Hwang CY. Clinical use of a ceramide-based moisturizer for treating dogs with atopic dermatitis. *J Vet Sci*. 2013;14(2):199–205.

Jungersted JM, Scheer H, Mempel M, Baurecht H, Cifuentes L, Hogh JK, et al. Stratum corneum lipids, skin barrier function and filaggrin mutations in patients with atopic eczema. *Allergy*. 2010;65(7):911–8.

Kelleher M, Dunn-Galvin A, Hourihane JO, Murray D, Campbell LE, McLean WH, et al. Skin barrier dysfunction measured by transepidermal water loss at 2 days and 2 months predates and predicts atopic dermatitis at 1 year. *J Allergy Clin Immunol.* 2015;135(4):930–5.e1.

Kikuchi K, Tagami H, Japanese Cosmetic Scientist Task Force for Skin Care of Atopic D. Noninvasive biophysical assessments of the efficacy of a moisturizing cosmetic cream base for patients with atopic dermatitis during different seasons. *Br J Dermatol.* 2008;158(5):969–78.

Kottner J, Lichterfeld A, Blume-Peytavi U. Transepidermal water loss in young and aged healthy humans: A systematic review and meta-analysis. *Arch Dermatol Res.* 2013;305(4):315–23.

Kraft JN, Lynde CW. Moisturizers: What they are and a practical approach to product selection. *Skin Therapy Lett.* 2005;10(5):1–8.

Leite e Silva VR. Desenvolvimento de formulações cosméticas hidratantes e avaliação da eficácia por métodos biofísicos (tese). University of São Paulo, Brazil; 2009.

Leite e Silva VR, Schulman MA, Ferelli C, Gimenis JM, Ruas GW, Baby AR, et al. Hydrating effects of moisturizer active compounds incorporated into hydrogels: In vivo assessment and comparison between devices. *J Cosmet Dermatol.* 2009;8(1):32–39.

Leonardi GR. Cosmetologia aplicada. São Paulo: Medfarma. 2004;51.

Loden M. Role of topical emollients and moisturizers in the treatment of dry skin barrier disorders. *Am J Clin Dermatol.* 2003;4(11):771–88.

Loden MM, H. Dry Skin and Moisturizers: Chemistry and Function. Cosmetics & Toiletries. 1999.

Lynde CW. Moisturizers: What they are and how they work. *Skin Therapy Lett.* 2001;6(13):3–5.

Mac-Mary S, Creidi P, Marsaut D, Courderot-Masuyer C, Cochet V, Gharbi T, et al. Assessment of effects of an additional dietary natural mineral water uptake on skin hydration in healthy subjects by dynamic barrier function measurements and clinic scoring. *Skin Res Technol.* 2006;12(3):199–205.

Nasrollahi SA, Ayatollahi A, Yazdanparast T, Samadi A, Hosseini H, Shamsipour M, et al. Comparison of linoleic acid-containing water-in-oil emulsion with urea-containing water-in-oil emulsion in the treatment of atopic dermatitis: A randomized clinical trial. *Clin Cosmet Investig Dermatol.* 2018;11:21–28.

Padamwar MN, Pawar AP, Daithankar AV, Mahadik KR. Silk sericin as a moisturizer: An in vivo study. *J Cosmet Dermatol.* 2005;4(4):250–7.

Parente ME, Gambaro A, Solana G. Study of sensory properties of emollients used in cosmetics and their correlation with physicochemical properties. *J Cosmet Sci.* 2005;56(3):175–82.

du Plessis J, Stefaniak A, Eloff F, John S, Agner T, Chou TC, et al. International guidelines for the in vivo assessment of skin properties in non-clinical settings: Part 2. Transepidermal water loss and skin hydration. *Skin Res Technol.* 2013;19(3):265–78.

Polaskova J, Pavlackova J, Egner P. Effect of vehicle on the performance of active moisturizing substances. *Skin Res Technol.* 2015;21(4):403–12.

Qassem M, Kyriacou PA. In vivo optical investigation of short term skin water contact and moisturizer application using NIR spectroscopy. *Conf Proc IEEE Eng Med Biol Soc.* 2013;2013:2392–5.

Rawlings AV, Canestrari DA, Dobkowski B. Moisturizer technology versus clinical performance. *Dermatol Ther.* 2004;17 Suppl 1:49–56.

Rawlings AV, Harding CR. Moisturization and skin barrier function. *Dermatol Ther.* 2004;17 Suppl 1:43–48.

Robinson C, Hartman RF, Rose SD. Emollient, humectant, and fluorescent alpha,beta-unsaturated thiol esters for long-acting skin applications. *Bioorg Chem.* 2008;36(6):265–70.

Robinson M, Visscher M, Laruffa A, Wickett R. Natural moisturizing factors (NMF) in the stratum corneum (SC). II. Regeneration of NMF over time after soaking. *J Cosmet Sci.* 2010;61(1):23–29.

Robinson MH. The Natural Moisturizing Factor of the Skin: Effects of Barrier Perturbation and Anatomical Location and Relation to Biophysical Measurements. Dissertation, University of Cincinnati; 2009.

Rossi ABR, Vergnanini AL. Mecanismos de hidratação da pele. *Cosmet Toiletries.* 1997;9:33–37.

Sato S, Maruyama Y, Kamata H, Watanabe S, Kita R, Shinyashiki N, et al. Evaluation of water measurement techniques for human skin by dielectric spectroscopy and confocal Raman spectroscopy. *Trans Mater Res Soc Jpn.* 2015;40(2):133–6.

Sethi A, Kaur T, Malhotra SK, Gambhir ML. Moisturizers: The slippery road. *Indian J Dermatol.* 2016;61(3):279–87.

Shao P, Shao J, Han L, Lv R, Sun P. Separation, preliminary characterization, and moisture-preserving activity of polysaccharides from *Ulva fasciata*. *Int J Biol Macromol.* 2015;72:924–30.

Short RW, Chan JL, Choi JM, Egbert BM, Rehmus WE, Kimball AB. Effects of moisturization on epidermal homeostasis and differentiation. *Clin Exp Dermatol.* 2007;32(1):88–90.

Simpson E, Bohling A, Bielfeldt S, Bosc C, Kerrouche N. Improvement of skin barrier function in atopic dermatitis patients with a new moisturizer containing a ceramide precursor. *J Dermatol Treat.* 2013;24(2):122–5.

Stamatas GN, Nikolovski J, Mack MC, Kollias N. Infant skin physiology and development during the first years of life: A review of recent findings based on in vivo studies. *Int J Cosmet Sci.* 2011;33(1):17–24.

Stettler H, Kurka P, Wagner C, Sznurkowska K, Czernicka O, Bohling A, et al. A new topical panthenol-containing emollient: Skin-moisturizing effect following single and prolonged usage in healthy adults, and tolerability in healthy infants. *J Dermatol Treat.* 2017;28(3):251–7.

Takaoka A, Baby AR, Prestes C, Pinto V, Silva M, Velasco Y, et al. Study of the interaction of moisturizers with a Crotalus durissus biomembrane by differential scanning calorimetry and RAMAN spectroscopy. *J Basic Appl Pharm Sci.* 2010;31(1):53–58.

Wu J, Polefka TG. Confocal Raman microspectroscopy of stratum corneum: A pre-clinical validation study. *Int J Cosmet Sci.* 2008;30(1):47–56.

Yokota M, Maibach HI. Moisturizer effect on irritant dermatitis: An overview. *Contact Dermatitis.* 2006;55(2):65–72.

12

Preservation and Preservatives

Dene Godfrey

CONTENTS

Introduction

Most cosmetic products contain water, and water is essential for microbial growth. The availability of nutrients will enhance the growth of microbes, and most cosmetic products contain ingredients that can act as a nutrient source. Emulsion-based products, such as moisturisers, tend to contain higher levels of nutrient sources than most other product types and, as a consequence, tend to require most robust protection from microbial growth. This chapter will discuss the reasons why antimicrobial preservation is required; the possible effects of no, or inadequate, preservation; and highlight the most commonly used preservatives available to solve the problem. There will also be a discussion on preservative blends, so-called secondary preservatives and the position on 'preservative-free' claims. Finally, both human and environmental safety are considered.

The Consequences of Microbial Growth in Cosmetics

There are several possible adverse effects that may occur in the absence of a sufficiently robust preservation system, and these may be broadly divided into aesthetics and safety issues:

1. Visible growth is one of the more obvious signs of microbial contamination, whether it be a black mould (e.g. *Aspergillus brasiliensis* – formerly *A. niger*) or other discolourations from various different bacteria (e.g. *Pseudomonas aeruginosa* can produce a blue/green, yellow/green or a red/brown colour). The metabolites produced by microbes are often acidic, and the reduction in pH may lead to breakdown of the emulsion. Additionally, the emulsifiers themselves may be used as a nutrient source, and this may also lead to phase separation. Microbial growth can also produce unpleasant odours.
2. There are also clear safety issues with microbial contamination. The presence of pathogenic organisms in a product being applied to skin, especially if the skin is damaged, may lead to infection. Staphylococcal infections are particularly easily spread in this manner. Far more seriously but, fortunately, far less common is the possibility of contaminated product entering the eye. For example, sufficient numbers of *Pseudomonas aeruginosa* in the eye will cause blindness.[1]

It is essential, therefore, from the dual considerations of product integrity/aesthetics and consumer safety that cosmetic products are sufficiently well-preserved to withstand any possible microbial contamination from reasonably foreseeable conditions of use.

The regulation of cosmetics around the world varies considerably, but no cosmetic product should be unsafe for use, either from a toxicological or a microbiological point of view. It is, therefore, essential that steps be taken to ensure that the product is proven to be adequately preserved.

Sources of Microbial Contamination

There are three principal sources of microbial contamination in cosmetics:

1. *During the manufacturing process*. Unless the product is manufactured in totally sterile conditions, there will be contamination from the atmosphere and from the skin and clothes of the operators. This is unavoidable but should be kept to a minimum by using the best manufacturing practice with an emphasis on plant hygiene.
2. *Raw materials*. Some raw materials may have levels of microbial contamination, especially if they are natural in origin. This should be controlled by the supplier to ensure that any contamination is kept to acceptably low levels.
3. *Consumer use*. This source of contamination is almost entirely unavoidable, as the consumer will inevitably have some contact with the product unless it is supplied in packs designed to eliminate contact or in single-use packs.

What Is Preservation?

Preservation may be defined as the art of protecting products against microbial attack during their shelf life.

It should not be used to compensate for unhygienic manufacturing conditions, but only to allow for the contamination introduced during consumer use.

What Is a Preservative?

A preservative is any substance used to kill or prevent the growth of microorganisms which, by their growth, will spoil or contaminate a raw material or product.

Microbial Challenge Testing

The most effective and consistent means of establishing the efficacy of a preservation system is to subject the product to a microbial challenge test. There are many variations on this test but, in its most simple form, the methods prescribed by various pharmacopoeia are a good basis, e.g. United States Pharmacopoeia (USP) and Pharmacopoeia Europa (Ph. Eur.). In the Ph. Eur. method, the product is inoculated with cultures of organisms representative of the four main subgroups of likely contaminants:

Gram-negative bacteria – *Pseudomonas aeruginosa*

Gram-positive bacteria – *Staphylococcus aureus*

Yeasts – *Candida albicans*

Moulds – *Aspergillus brasiliensis*

These cultures are inoculated in separate samples, and then checked for surviving colonies at specified time points. There are different criteria for success for bacteria than for fungi; both are measured on the degree of logarithmic reduction in colony numbers.

In order to pass the criteria for bacteria, the colony count must be reduced by at least log 2 within 2 days, and by at least log 3 within 7 days, with no increase in numbers thereafter (being tested also after 14 and 28 days).

In order to pass the criteria for fungi, the colony count must be reduced by at least log 2 within 7 days, with no increase thereafter, again, up to 28 days.

The criteria described here, from the Ph. Eur., are the 'A' criteria. There are also 'B' criteria, which are less stringent and are applied only in situations where the possibility of microbial contamination is greatly reduced, usually due to packaging (e.g. single-use packs).

The USP criteria include the Gram-negative bacterium *Escherichia coli* and are less stringent than those of the Ph. Eur., requiring only that there is no increase in the initial numbers of *A. brasiliensis* and *C. albicans* up to 28 days, and only a 1 log reduction of bacteria by 7 days; then a 3 log reduction by 14 days, with no increase at 28 days.

Minimum Inhibitory Concentration Values

When evaluating preservatives and preservative combinations in the development stage of the process, it is common practice to determine the minimum inhibitory concentration (mic) value. This is the lowest concentration at which the test substance prevents further growth of the test organism. This value is specific to both the substance and the organism tested. The concentration required to kill the organism is usually much higher than the mic value. Manufacturers of preservatives usually offer mic data in support of their products, but this information should be treated with a little caution. There is no standard method for the mic test and the results are subject to wide variability, as there are several factors that have a significant effect on the result. The most important variables are the inoculum count employed and the actual strain of the test organism. The lower the inoculum count, the lower the concentration of test substance required to prevent cells from growing, and it is possible to manipulate the data by using a low inoculum count to produce low mic figures, thereby making the test substance appear unduly active. Microbial cells from all species evolve and adapt to their surroundings, and microbial resistance, or tolerance, can develop in laboratory strains, leading to incorrect mic results. There is also the potential for huge variability in response between the different strains of the same species. (A strain is a subset of a bacterial species differing from other bacteria of that species by some minor but identifiable difference.)

The result of these potential variations within the mic test is that direct comparison of mic data should only be made for the data on one specific substance (i.e. comparing the mic of the substance for all the organisms tested at the same time) and should not be 'read across' to other substances tested separately, other than with a high degree of caution.

For the aforementioned reasons, there are no mic data quoted for the preservatives being discussed, only general statements for guidance on the types of organisms against which they are effective.

Criteria for Preservatives/Preservation Systems

In order to determine the usefulness of a preservative, or a preservation system, it is important to understand the attributes required for successful preservation of a product.

The most important criterion is broad-spectrum preservation, that is, the system must be effective against bacteria (normally divided into Gram-negative bacteria and Gram-positive bacteria) and fungi (normally divided into yeasts and moulds). These subdivisions are important, as some preservatives are only effective in one or two of these areas. If the proposed preservation system for any given product is not effective against any of these subgroups, or even just one specific common microorganism, it is not suitable for use without another substance being included that will control the gap in the activity spectrum. For example, *Pseudomonas aeruginosa* is a common Gram-negative bacterium that is relatively difficult to control. If the preservation system controlled all other microorganisms, but allowed *P. aeruginosa* to survive, these bacteria would grow even faster than normal due to the absence of competitive organisms. Partial preservation is barely more effective than the complete absence of preservation. Broad-spectrum antimicrobial protection is the key to a safe and stable product.

The components of the preservation system must be compatible with the other ingredients used in the formulation. Any interaction may result in reduced activity; this would become apparent in the results of the microbial challenge test, unless the interaction develops over a more prolonged period.

The preservation system should be effective at the target pH of the product. Not all preservatives are equally effective across the entire pH range typically seen in moisturisers.

The preservation system should not be prone to discolouration, especially if being considered for use in non-coloured products, and should not impart any detectable/undesirable odour. It may be possible to mask any odour, but this is not ideal.

The preservation system must be appropriate for the intended use of the final product – a substance permitted only in rinse-off products is clearly not suitable for a moisturising lotion.

The preservation system must be as safe as possible. It is inherent amongst biologically active substances to have some potential for skin irritation and, albeit a lower potential, for sensitisation. The choice of preservative is a balance between efficacy and risk of a skin response. As a general rule, the more potent the antimicrobial activity, the higher the risk of a skin response (in relative terms), but this is usually mitigated by the use of lower concentrations due to the more powerful activity.

Commonly Used Preservatives

There are 57 entries on Annex VI of the EU Cosmetics Directive (list of permitted preservatives), some covering multiple substances, and other antimicrobial substances are also employed in the EU and in other markets. However, the number of commonly used preservatives is relatively low, and the majority of these are discussed here.

Parabens

INCI names:

Methylparaben
Ethylparaben
Propylparaben
Butylparaben

'Paraben' is a contraction of 'esters of *para*hydroxy*benz*oic acid'. The methyl-, ethyl-, propyl- and butyl- esters are the ones in general use as preservatives, along with their respective sodium salts (INCI names – sodium methylparaben, etc.; Figure 12.1).

Parabens have been used as preservatives in personal care products since the late 1920s and have a long and successful history of use, being amongst the preservatives with the lowest rates of irritation and sensitisation, with sensitisation rates amongst dermatological patients as low as 1.2%.[2]

Parabens are primarily active against fungi, but they also generally have moderate activity against bacteria.

Methylparaben is the least active of the group, and the antimicrobial activity increases with the increasing carbon chain length of the ester group. The activities of ethylparaben and propylparaben lie between those of methylparaben and butylparaben.

All the parabens have relatively low water solubility, and this decreases with the increasing carbon chain length of the ester group; methylparaben being soluble up to 0.25% in water (25°C), butylparaben being only 0.02% soluble.[3,4]

Frequently, combinations of parabens are employed in order to maximise the total concentration available in the aqueous phase and also, therefore, the antimicrobial activity. It is relatively unusual to use a single paraben ester in a preservation system. Methylparaben is typically used at 0.1–0.25% and propylparaben, typically at 0.1–0.15% (as a combination), with the other parabens being mostly used as components of more complex blends, rather than being added as individual ingredients. The typical total parabens concentration is 0.2–0.4%.

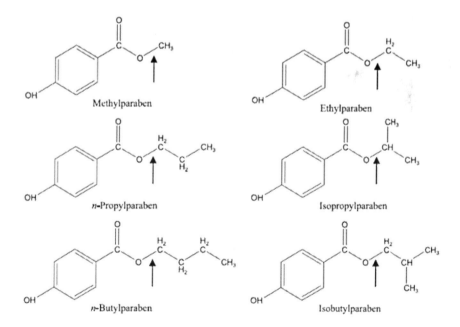

FIGURE 12.1 Paraben structure: hydrolysis at the point indicated by the arrow leads to the generation of the parent *p*-hydroxybenzoic acid.

In a hot process (c. 80°C), the parabens may be added to either phase prior to emulsification. It is sometimes recommended to split the parabens addition between the two phases, but this is of limited or no value as the parabens will partition between the phases during the emulsification/mixing process.

For cold processes, given the difficulty of dissolving the parabens at low temperatures, the sodium parabens may be used. These are highly water-soluble (<50%) but will exert a very high pH, which will require adjustment. A 0.1% aqueous solution of sodium methylparaben is approximately pH 9, so an adjustment is important.

The parabens are compatible with most commonly used cosmetic ingredients, but they may be inactivated by some non-ionic species. This effect is strongly concentration-dependent, as the paraben ester may be encapsulated within any micelles in the system. Below the critical micelle concentration (CMC), therefore, there is no inactivation, with a progressive reduction in activity above the CMC. There is little effect on the activity of parabens from the pH, within the range pH 4–8.

Parabens are permitted globally for use in personal care products but with differing maximum permitted concentrations:

European Union
Methylparaben – 0.4%
Ethylparaben – 0.4%
Propylparaben + butylparaben – total 0.14%
Maximum total parabens concentration – 0.8%
(These concentration figures refer to the 4-hydroxybenzoic acid equivalent; for example, the *actual* total of propylparaben and butylparaben = 0.19%, approx.)

United States
Safe as used.

Japan
Total maximum parabens – 1.0% (Japanese name: parahydroxybenzoate esters)

Imidazolidinyl Urea

Imidazolidinyl urea is primarily an antibacterial preservative, with very little antifungal activity. It has been widely used in many markets since its introduction in the 1980s.

Imidazolidinyl urea is highly water-soluble and should be added to the product at temperatures below 40°C to avoid decomposition. It retains its activity over a broad pH range, from pH 3 to 9. Typical use concentrations are 0.2–0.5%. The potential for imidazolidinyl urea to release very low levels of formaldehyde is perceived by some to be a disadvantage (see 'General Note on Formaldehyde Donors').

Imidazolidinyl urea is permitted globally, with the following restrictions:

European Union
Maximum 0.6%.

United States
Safe as used.

Japan
In rinse-off products only, with a maximum of 0.075% (expressed as free formaldehyde) with special labelling requirements: 'should not be used by infants, or by people who are hypersensitive to formaldehyde'.

Diazolidinyl Urea

Diazolidinyl urea (Figure 12.2) is primarily an antibacterial preservative, with activity against moulds but fairly weak anti-yeast activity. It has been widely used in many markets since its introduction in the 1980s. It is classified as a formaldehyde donor (see 'General Note on Formaldehyde Donors').

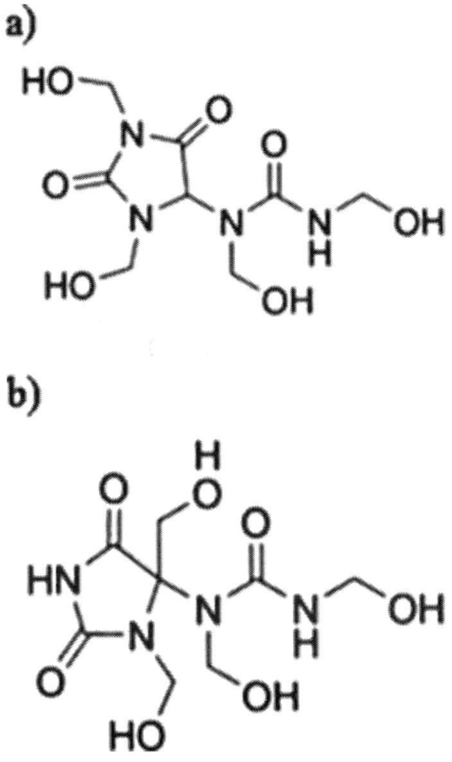

FIGURE 12.2 Diazolidinyl urea. (a) Original structure, (b) Hoeck's revised structure.

Diazolidinyl urea is highly water-soluble and should be added to the product at temperatures below 40°C to avoid decomposition. It retains its activity over a broad pH range, from pH 3 to 9. Typical use concentrations are 0.1–0.3%.

Diazolidinyl urea is permitted in many territories, with the following restrictions:

European Union
Maximum 0.5%.

United States
Maximum 0.5%.

Japan
Not permitted.

DMDM Hydantoin

DMDM hydantoin (Figure 12.3a) is a broad-spectrum preservative but has better activity against bacteria than against fungi. It is classified as a formaldehyde donor (see 'General Note on Formaldehyde Donors').

DMDM hydantoin is highly water-soluble (it is usually supplied as a 55% aqueous solution) and should be added to the product at temperatures below 40°C to avoid decomposition. It retains its activity over a broad pH range, from pH 3 to 9. Typical use concentrations are 0.15–0.4%.

DMDM hydantoin is permitted in many territories, with the following restrictions:

European Union
Maximum 0.6%.

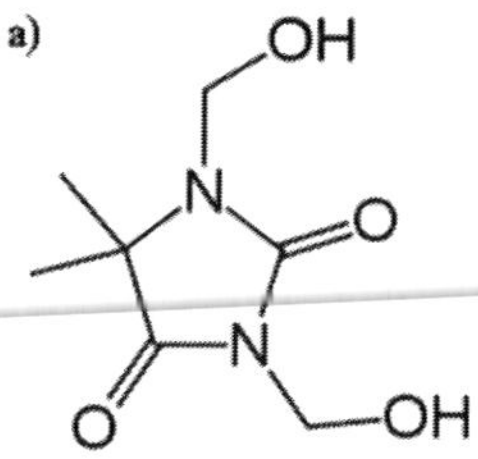

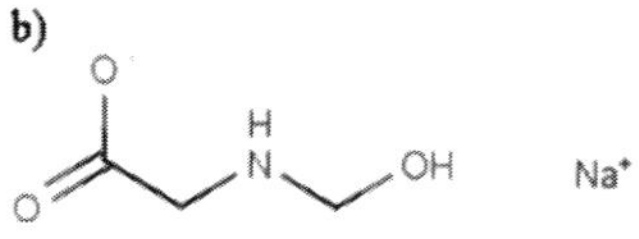

FIGURE 12.3 (a) DMDM hydantoin (1,3-dimethylol-5,5-dimethylhydantoin), (b) sodium hydroxymethylglycinate.

United States
Safe as used.

Japan
In rinse-off products only, with a maximum of 0.075% (expressed as free formaldehyde), with special labelling requirements: 'should not be used by infants, or by people who are hypersensitive to formaldehyde'.

Sodium Hydroxymethylglycinate

Sodium hydroxymethylglycinate (sodium HMG) is a broad-spectrum preservative Figure 12.3b) but has slightly better activity against bacteria than against fungi. It is classified as a formaldehyde donor (see section 'General Note on Formaldehyde Donors').

Sodium HMG is highly water-soluble and should be added to the product at temperatures below 40°C to avoid decomposition. It retains its activity over a broad pH range, from pH 3 to 9. Typical use concentrations are 0.4–0.8%.

Sodium HMG is permitted in many territories, with the following restrictions:

European Union
Maximum 0.5%.

United States
Safe as used.

Japan
Not permitted.

General Note on Formaldehyde Donors

Formaldehyde is classified as a category 1 carcinogen (i.e. a proven human carcinogen), and this has led to concerns over the safety of substances having the potential to release formaldehyde. There is no proof, however, that the use of formaldehyde donors in cosmetics results in sufficient exposure to formaldehyde to be of concern. Historically, available formaldehyde levels in cosmetic products have been exaggerated due to a flaw in the standard methods used to detect formaldehyde.[5,6] In the case of formaldehyde donors, the donor molecule exists in equilibrium with free formaldehyde and the other reaction product. The standard method for determining formaldehyde requires a derivatisation step which disturbs the equilibrium, thereby producing more free formaldehyde, that is, formaldehyde that is not actually present in the product prior to the analytical method being performed.

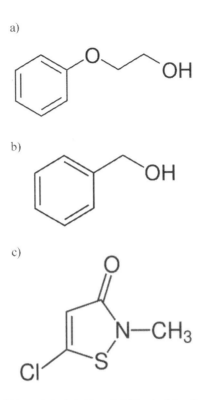

FIGURE 12.4 (a) Phenoxyethanol, (b) benzyl alcohol, (c) methylchloroisothiazolinone.

Phenoxyethanol

Phenoxyethanol is one of the most widely used preservatives Figure 12.4a). It is a broad-spectrum preservative, but it is slightly weaker against Gram-positive bacteria than the other species types.

Phenoxyethanol is slightly water-soluble (approximately 2.4%) and is preferably added to the product at temperatures below 40°C to reduce the possibility of evaporative loss, although this is only likely to be an issue should the manufactured batch be held at a high temperature (>80°C) for a prolonged period. It retains its activity over a broad pH range, from pH 3 to 9. Typical use concentrations are 0.4–1.0%. Phenoxyethanol is more commonly used in combination with other preservatives and is rarely used alone.

Phenoxyethanol is permitted in most territories, with the following restrictions:

European Union
Maximum 1.0%.

United States
Safe as used.

Japan
Maximum 1.0%.

Methylchloroisothiazolinone/Methylisothiazolinone

Methylchloroisothiazolinone/methylisothiazolinone (MCI/MI) (Figure 12.4c) is one of the most widely used preservatives. It is a broad-spectrum preservative system, effective at extremely low concentrations. It is usually supplied as a 1.5% active aqueous solution, stabilised with high concentrations of magnesium salts to prevent degradation of the MCI component.

Methylisothiazolinone (MI) is also available as a single component.

MCI/MI retains its activity over a broad pH range, from pH 3 to 7.5, but MCI breaks down rapidly above pH 8. MI does not contribute towards any antimicrobial activity in this combination product. Typical use concentrations are 7.5–15 ppm (0.00075–0.0015%) of active MCI/MI.

MI is highly stable and retains its activity from pH 3 to 9. Typical use concentrations are 0.005–0.01%. The MCI/MI combination is permitted in most territories, with the following restrictions:

European Union
Maximum 0.0015% (active MCI/MI).

United States
Maximum 0.0015% (active MCI/MI).

Japan
Maximum 0.0015% (active MCI/MI) – rinse off products only.
Methylisothiazolinone is permitted in most territories, with the following restrictions:

European Union
Maximum 0.01% (active MI).

United States
Maximum 0.01% (active MI).

Japan
Maximum 0.01% (active MI) – rinse-off and leave-on products.

At the time of this writing, Cosmetics Europe recommends the cessation of use of MI in leave-on products due to reported high frequency of allergic reactions to this material.

Benzyl Alcohol

Benzyl alcohol (Figure 12.4b) is a widely used preservative. It is broad spectrum in its range of activity, but it is slightly weak against Gram-positive bacteria.

Benzyl alcohol is slightly water-soluble (approximately 2%) and is preferably added to the product at temperatures below 40°C to reduce the possibility of evaporative loss, although this is only likely to be an issue should the manufactured batch be held at a high temperature (>80°C) for a prolonged period. It retains its activity over a broad pH range, from pH 3 to 8.5. Typical use concentrations are 0.4–1.0%. Benzyl alcohol is more commonly used in combination with other preservatives and is rarely used alone.

Benzyl alcohol is permitted in most territories, with the following restrictions:

European Union
Maximum 1.0%. Benzyl alcohol is one of the 26 designated fragrance allergens, and products containing it must be labelled accordingly.

United States
Safe as used.

Japan
Maximum 5.0%.

Sorbic Acid/Potassium Sorbate

Sorbic acid and potassium sorbate (Figure 12.5a) are food-approved preservatives. They are broad-spectrum preservatives, but they are slightly weak against bacteria.

Sorbic acid is very slightly water soluble, but the potassium salt is highly soluble. They retain their activity over a narrow pH range, from pH 3 to 6. Typical use concentrations are 0.3–0.5%. (See 'General Note on Organic Acids'.)

Sorbic acid/potassium sorbate is permitted in most territories, with the following restrictions:

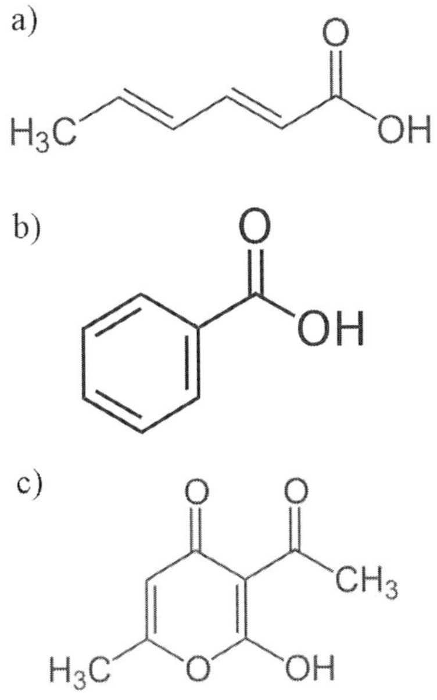

a)

b)

c)

FIGURE 12.5 (a) Sorbic acid, (b) benzoic acid, (c) dehydroacetic acid.

European Union
Maximum 0.6% (as the acid).

United States
Safe as used.

Japan
Maximum 0.5%.

Benzoic Acid/Sodium Benzoate

Benzoic acid and sodium benzoate (Figure 12.5b) are food-approved preservatives. They are broad-spectrum preservatives, but they are slightly weak against bacteria.

Benzoic acid is very slightly water-soluble, but the sodium salt is highly soluble. They retain their activity over a narrow pH range, from pH 3 to 5. Typical use concentrations are 0.2–0.4%. (See 'General Note on Organic Acids'.)

Benzoic acid/sodium benzoate is permitted in most territories, with the following restrictions:

European Union
Rinse-off products, except oral care products: 2.5 % (as acid).
Oral care products: 1.7 % (as acid).
Leave-on products: 0.5 % (as acid).

United States
Safe as used.

Japan
Maximum 0.2%.

Dehydroacetic Acid/Sodium Dehydroacetate

Dehydroacetic acid and sodium dehydroacetate (Figure 12.5c) are broad-spectrum preservatives, but they are slightly weak against bacteria.

Dehydroacetic acid is very slightly water soluble, but the sodium salt is highly soluble. They retain their activity over a narrow pH range, from pH 3 to 6.5. Typical use concentrations are 0.3–0.5%. (See 'General Note on Organic Acids'.)

Dehydroacetic acid/dehydroacetate is permitted in most territories, with the following restrictions:

European Union
Maximum 0.6% (as the acid).

United States
Safe as used.

Japan
Maximum 0.5%.

General Note on Organic Acids

The organic acids described here are only active in their undissociated form – the ionic species have zero antimicrobial activity, hence the restrictive pH range over which they retain their activity. All three acids are 100% active at pH3, because they are completely undissociated at this pH, but, as the pH increases, the concentration of ionic species increases, and the activity drops rapidly. For example, benzoic acid is approximately twice as active at pH 5.0 than at pH 5.5. For this reason, it is vital that any finished product being submitted for microbial challenge testing has the pH adjusted to the highest end of the pH specification for that product. This is then testing the worst-case scenario, i.e. the pH at which the antimicrobial protection is at its weakest. If the product passes challenge at the highest pH within the specification range, it will easily pass at any other pH within the specification.

Chlorphenesin

Chlorphenesin (Figure 12.6a) has broad-spectrum antimicrobial activity against bacteria and fungi, being also effective against *Pseudomonas* spp. and other problematic Gram-negative bacteria.

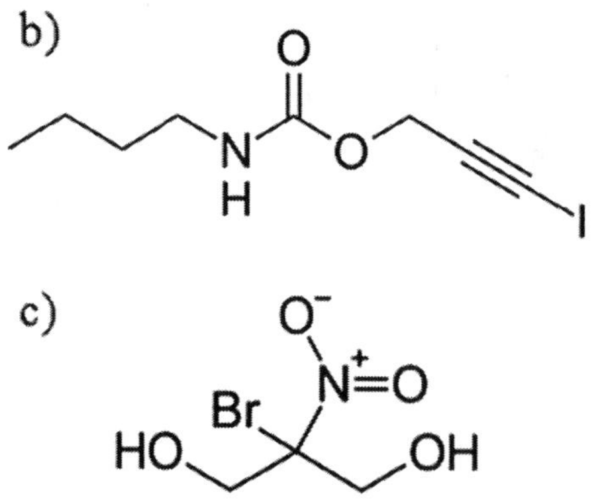

FIGURE 12.6   (a) Chlorphenesin (3-(4-chlorophenoxy)-1,2-propanediol), (b) iodopropynyl butylcarbamate, (c) bronopol (2-bromo-2-nitropropane-1,3-diol).

Summary

Cosmetic products susceptible to microbial growth must be preserved for aesthetic reasons, but also, more important, for consumer safety.

The combination of preservatives used must endow broad-spectrum protection against bacteria and fungi.

Microbial challenge testing is vital to ensure that products are adequately protected against microbial growth.

Preservative blends offer more robust protection than most single substances.

Preservatives must be used carefully and at the lowest concentration required to achieve sufficient antimicrobial protection to ensure safety in use, and to reduce the low risk of skin response even further, as much as possible.

Preservatives are tightly regulated and it is important to be aware of the restrictions in target markets, and equally important to be aware of changes to regulations as they occur.

'Preservative-free' claims are contentious and may be best avoided.

There is no evidence of any adverse environmental impact resulting from the use of preservatives in cosmetics.

NOTES

1. Engel LS, Hill JM, Moreau JM, Green LC, Hobden JA, O'Callaghan RJ. Pseudomonas aeruginosa protease IV produces corneal damage and contributes to bacterial virulence. *Invest. Ophthalmol. Vis. Sci.* 39, (3), 662–665, (1998).

2. Steinberg D. Frequency of preservative use. *Cosmet. Toiletries* 125, (November), 46–51, (2010).

3. Azelis product literature.

4. Clariant product literature.

5. Tallon M, Merianos JJ, Subramanian S. Non-destructive method for determining the actual concentration of free formaldehyde in personal care formulations containing formaldehyde donors. *SOFW-J.* 135, (5), 22–32, (2009).

6. Winkelman JGM, Voorwinde OK, Ottens M, Beenackers AACM, Janssen LPBM. Kinetics and chemical equilibrium of the hydration of formaldehyde. *Chem. Eng. Sci.* 57, 4067–4076, (2002).

7. http://personalcaretruth.com/2010/11/parabens-in-perspective-an-introduction/.

8. http://personalcaretruth.com/2010/11/parabens-in-perspective-part-i/.

9. http://personalcaretruth.com/2010/11/parabens-in-perspective-part-ii/.

10. http://personalcaretruth.com/2010/11/parabens-in-perspective-part-iii/.

11. http://personalcaretruth.com/2010/11/parabens-in-perspective-part-iv/.

12. http://personalcaretruth.com/2010/12/parabens-in-perspective-part-v/.

13. http://personalcaretruth.com/2010/12/parabens-in-perspective-part-vi/.

14. http://personalcaretruth.com/2010/12/parabens-in-perspective-part-vii/.

15. http://personalcaretruth.com/2010/12/parabens-in-perspective-part-viii/.

16. http://personalcaretruth.com/2011/01/parabens-in-perspective-part-ix/.

17. http://personalcaretruth.com/2011/01/parabens-in-perspective-part-x/.

18. http://eur-lex.europa.eu/legal-content/EN/TXT/?uri=CELEX:32009R1223.

19. Gottschalck TE, Bailey JE (Eds.). *International Cosmetic Ingredient Dictionary and Handbook.* Personal Care Products Council, Inc.

20. http://ec.europa.eu/health/scientific_committees/consumer_safety/index_en.htm.

21. http://ec.europa.eu/health/scientific_committees/videos/videos/video_committees_en.htm.

22. http://ec.europa.eu/health/scientific_committees/consumer_safety/opinions/index_en.htm.

23. Lee HB, Peart TE, Svoboda ML. Determination of endocrine-disrupting phenols, acidic pharmaceuticals, and personal-care products in sewage by solid-phase extraction and gas chromatography-mass spectrometry. *J. Chromatogr. A* 1094, 122–129, (2005).

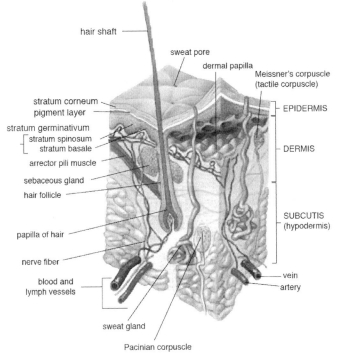

FIGURE 2.1 Diagram of full thickness human skin showing the major strata, subcutis, and various other features. The skin has two principle layers, the epidermis and the dermis, which, over most of the body, is thicker than 1 mm. The epidermis (approximately 100 μm thick) is further split into two layers, the stratum corneum and the stratum germinativum (or viable epidermis). A number of appendages are also shown and these include the hair follicle and shaft, the associated sebaceous glands and erector muscles, and sweat ducts. The dermis contains a dense network of blood vessels, which provides constant support for the germinative tissue and important body temperature regulation. There are also nociceptors (pain receptors) and tactile nerve endings (Meissner's corpuscles). (From upload.wikimedia.org/Wikipedia/commons/2/27/skin.png.)

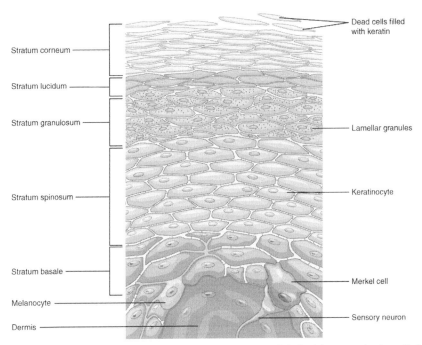

FIGURE 2.2 Cell strata of the epidermis are arranged in layers that are all derived from germinative cells in the basal layer (the stratum basale that forms the barrier between the dermis and the epidermis). This layer contains the skin's stem cells that generate the daughter cells that are forced outwards from the basal layer. As these cells mature they build protein bridges between them and this generates cell–cell communication and gives the cells spiny contours (the stratum spinosum). They also start to synthesise keratins. Further outwards the cells create packets of lipids, which give the cells a granular appearance (the stratum granulosum). The lipid granules are eventually exocytosed and the lipid contents reform to create the intercellular lipid lamellae. The cells subsequently lose their nuclei and become clear (the stratum lucidum) and the cells are no longer considered viable as they form the stratum corneum. Other cells contained in the epidermis are the pigment-producing melanocytes and the sensory Merkel cells. (From opentextbc.ca.)

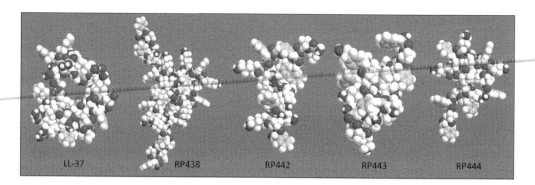

FIGURE 15.2 Structural space filling representation of human cathelicidin LL-37, LLGDFFRKSKEKIGKEFK RIVQRIKDFLRNLVPRTES, in addition to the novel dAMPs RP438 (CPGFAKKFAKKFKKFAKKFAKFAFAF)-disulphide-dimer); RP442, FAFAFAFKKAFKKFKKAFKKAF; PR443, FAFAFOAFOOAFOOFOOAFOOAF; and RP444, FAOOFAOOFOOFAOOFAOFAFAF. In the space filling models, red represents oxygen, blue represents nitrogen, grey is carbon and hydrogen is white.

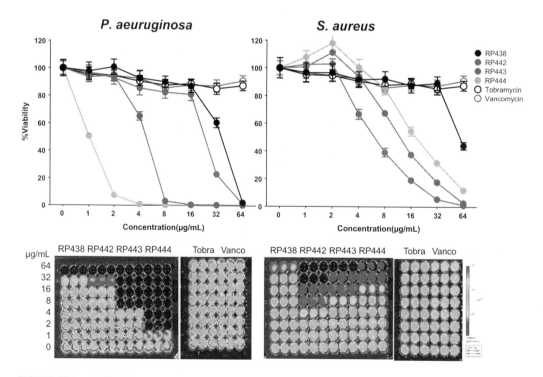

FIGURE 15.3 dAMPs kill *P. aeruginosa* and *S. aureus* within 30 minutes of exposure. RP438, RP442, RP443, RP444, tobramycin and vancomycin were applied to bioluminescent *P. aeruginosa* 19660 and *S. aureus* 49525, and bactericidal effectiveness assessed following incubation for 30 minutes. The bioluminescence of viable bacteria was evaluated non-invasively with an IVIS Lumina bioimaging system. Data represent the mean ± SE of triplicate replicates from two independent experiments. For some points, the error bars are shorter than the height of the symbols.

13

Powders in Cosmetic Formulations

Newton Andréo-Filho, Heather A.E. Benson,
Vânia Rodrigues Leite-Silva and Gislaine Ricci Leonardi

CONTENTS

Introduction

Powders are important as both cosmetic ingredient components and as cosmetic products, therefore it is important to understand their physical and chemical properties and how these can influence cosmetic product manufacture and performance. Powder-based cosmetic products include make-up tablets (eye shadow, blusher, etc.), eyeliner and cosmetic talcum powder. However, powders are also used as raw ingredients in many cosmetic and personal care products, and may subsequently be presented within the final product dissolved as solution or suspended as particles, or both. Particle shape, size and size distribution will determine flowability, both during the manufacturing process and in the organoleptic characteristics such as texture, odour, skin feel and adherence of solid finished cosmetic products. Solubility, and properties that influence solubility such as polymorphic forms, hydrates and so forth, will influence the choice of other formulation excipients, the manufacturing process and the organoleptic properties of the final product. Clearly, understanding the properties of powder raw materials is essential to the formulation of a cosmetic product that is effective, stable, safe and has suitable organoleptic characteristics that will determine consumer preference. This chapter provides an overview of the forms in which powders are used as cosmetic products; the physicochemical properties of powder raw materials; and how these properties influence the formulation design, manufacturing process and final product performance.

Cosmetic Formulations in Solid Form

Solid formulations have better stability compared to liquid and semi-solid formulations and are generally less prone to microbial spoilage and temperature-related degradation. Consequently storage and transportation requirements can be less stringent and therefore less expensive. Examples of solid cosmetic formulations are:

Stick – Formulation constituents are shaped, either by fusion and later solidification (lipsticks) or by compression (eyeliners), usually creating cylindrical systems, that undergo wear when applied to skin, resulting in material transfer.

Soap – Presented in a range of shapes, depending on the mold used during manufacturing. Their base raw materials are surfactants, saponified from the chemical reaction between a strong alkali and grease, animal or vegetable oils.

Film – Consists of a thin, elongated layer containing a single dose of one or more active compounds.

Powders – Contain one or more dry active compounds, with reduced particle size, with or without excipients, as detailed next.

Cosmetic Formulations in Powder Form

Cosmetic powder formulations generally consist of more than one dry powder ingredient mixed to ensure homogeneity. Most common examples are make-up products such as facial powders and eye shadows, and talcum powders designed to provide a pleasant body odour (Figure 13.1, part I). Powder-based formulations provide a number of advantages, including the enhanced stability outlined earlier. Powders can provide very effective skin coverage, important for effective make-up performance particularly in covering skin blemishes, birthmarks and providing sun protection. Examples of powder-based make-up products are described next.

Facial powders are utilized to provide local coverage to the application area, giving a uniform aspect to the skin. They can also decrease skin oiliness and reduce shine, a property that is particularly important during photography. They are formulated with finely milled and pulverized substances, and are designed to be applied with a brush or sponge.

Pressed facial powders have similar composition to facial powders, however, an agglutinant is added during the pressing process. Pressed powders are used to retouch, utilizing a sponge.

Eye shadows are presented in powder, pressed powder and solid cream forms, or are packed in pencil form. They are applied to colour the eyelids. Pressed eye shadows are the most common product. To be effective they must spread evenly with a light texture and adhere on the eyelids.

Foundations are generally presented as emulsions, with pigments in suspension, but there are also powder versions in the form of mineral make-up. They colour the skin and give it a uniform appearance, camouflaging skin imperfections. They are generally used beneath facial powders and combine to give the perfect skin 'illusion'.

Blushes are used to colour the cheekbones and provide facial shape definition. They must have a smooth texture for easy application and uniform coverage. They can be presented as powder, pressed powder, liquid and solid cream forms.

Powder Cosmetic Raw Materials

In addition to solid form cosmetic products, powder raw materials are incorporated as both active compounds and formulation excipients in a wide range of cosmetic products such as semi-solid creams, gels, suspensions and aerosols (Figure 13.1, part II). Where the powder is incorporated within a liquid formulation, it may remain dispersed as solid particles (e.g. pigments), dispersed in colloidal form (e.g. gelling agents), or be completely dissolved (e.g. antimicrobial preservatives) or partially dissolved, depending on the formulation and manufacturing processes involved in producing the final product. The physicochemical properties of the solid raw material will determine all aspects of its performance within the formulation and manufacturing process. For example, an alteration in the raw material particle size will change the flow properties of a powder mix, potentially changing homogeneity, flow in manufacturing

Part I

Part II

FIGURE 13.1 Examples of commercial products with powders as an essential raw material. Part I – Some make-up products for facial care: (A) makeup palette; (B) pencil eyeliner; (C) lipstick; (D) stick foundation; (E) pencil for eyes; (F) liquid foundation. Part II – Other products with powder as an essential raw material: (A) facial mask; (B) aerosol antiperspirant; (C) nail polish; (D) sunscreen lotion; (E) exfoliating soap; (F) stick deodorant.

equipment and powder compression, and the performance of the final product on the skin. It can also alter sedimentation time in a suspension formulation, such as make-up foundation. Thus, knowledge of the raw material's physicochemical properties is an important tool that allows the formulator to effectively choose the different techniques and equipment for development and manufacture of more stable and effective formulations, with great consumer acceptance.

Physicochemical Aspects of Powders That Influence Cosmetic Formulations

The way in which physicochemical properties influence the quality and manufacturing processes of cosmetic formulations is the same as many other human use products including drug and food products. Consequently there is a large body of knowledge derived from multiple disciplines. With the exception of hearing, all our other senses can be stimulated through the use of a cosmetic product. Thus whilst the efficacy and safety of any product are of paramount importance, in cosmetic formulations the organoleptic characteristics, including within the final packaging and in use are very important considerations.

For a facial foundation containing suspended particulate material (e.g. clay), it is essential that when applied to facial skin it provides uniform coverage and acts as a base for the application of subsequent make-up steps. It must also continue to provide uniform coverage for a prolonged period and in varied environmental conditions. In addition, during storage within the packaging and in use, the product should be a uniform system that does not separate or require extensive shaking prior to use. The uniformity of the suspension formulation will be determined by the particle size and homogeneity of particulate materials. Thus a formulator may need to reduce particle size to reduce sedimentation during storage and ensure uniformity within the formulation during application. It is important to be mindful that in cosmetic formulation the aim is to retain the particles on the skin surface to ensure effective coverage with minimal absorption, therefore an optimal particle size that achieves this outcome should be used.

Clearly it is knowledge of the physical properties of the particulate raw materials that forms the basis for the design and development of cosmetic formulations that are stable, safe and have good sensorial products. Apart from the composition of the solid material, the most important properties that influence particle performance are particle size, particle shape and particle surface characteristics. These properties related to the shape and crystal structure not only dictate the properties of the particles in the raw material form, but also affect the way in which the particles can be processed during manufacturing. The presence of polymorphic forms can affect the choice of manufacturing conditions and stability of a cosmetic formulation, so it is important to identify polymorphs early in the design phase. These properties can influence all aspects of cosmetic formulation throughout the manufacturing process, in the final product both in storage and in use. This is discussed in detail in the following section.

Particle Shape and Crystal Structure

Molecules in solid particles are held together by intermolecular forces. The strength of the interaction between the molecules is dependent on the individual atoms within the molecular structure and the intermolecular forces of attraction. These can be primary valence bonds, such as ionic attraction that hold the ions; organic or inorganic; tightly together; or weaker forces of attraction such as van der Waals forces, dipolar interaction and hydrogen bonds for organic molecules. Solids can be crystalline (highly ordered lattice structure of repeated units) or amorphous (disordered structure) as shown in Figure 13.2a. Amorphous solids tend to have higher solubility and better compression characteristics, but also tend to be less stable than crystals. Examples are proteins, peptides, some sugars, fatty compounds and polymers. Crystalline solids have a definite shape and contain a highly ordered array of molecules, ions and atoms held together by non-covalent molecules interactions. Unlike amorphous solids, crystalline solids have distinct melting points, easily verified by thermal analysis such as differential scanning calorimetry (DSC). The difference between crystalline structure and amorphous can be observed by powder X-ray diffraction (XRD) analysis. Figure 13.2b shows examples of thermograms and diffractograms of solid raw material used in cosmetic products.

The internal packing of the atoms, molecules or ions is known as the crystal structure. This internal arrangement will influence the way in which the crystal grows, and can determine the external shape of particles formed by precipitation or solidification. The overall shape of the crystal is termed the crystal habit, for example acicular (needle), prismatic, pyramidal, tabular, equant, columnar and lamellar types (Figure 13.3). Whilst these crystal habits have different external shapes, they might have the same internal structure as can be seen from their XRD patterns. The crystal habit acquired depends on the conditions of crystallisation such as solvent used, temperature, concentration and presence of impurities. Thus, it is very important to control these conditions during the preparation of crystalline raw materials to ensure batch-to-batch consistency in crystal habit. The crystal habit can influence the flow properties and compressibility of a powder material thereby affecting many aspects of the powder performance in both the manufacturing processes and final product.

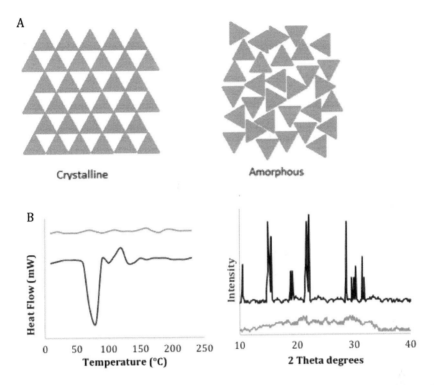

FIGURE 13.2 Schematic representation of variations in crystalline and amorphous solid raw materials. (A) Triangles represent molecules of solid raw material within a rigid and organized structure as a crystal lattice (left) or a fragile and disorganized amorphous powder structure (right). (B) Typical profile of crystalline (black line) and amorphous (grey line) in differential scanning calorimetry (left) and X-ray diffraction (right).

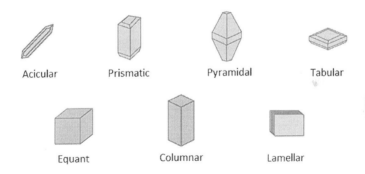

FIGURE 13.3 Typical crystal habits of crystalline raw materials.

Crystal shapes with more regular dimensions, such as cubic and rhombohedral, tend to have better flow characteristics when compared to acicular and planar shapes, where one or two crystal dimensions predominate. Where the powder does not flow efficiently, mixing processes require greater time to achieve formulation homogeneity, and manufacturing processes must be designed accordingly. If there is a change in the crystal habit, such as due to altered conditions during production of the raw material, this can lead to poor mixing and product failure.

Product performance will also vary depending on the raw material crystal shape. For example, acicular (needle)-shaped particles may be more irritating when in contact with the skin compared to cubic and rhombohedral particles. They may also not provide effective skin coverage.

Clearly it is important that the formulator critically evaluate the different crystal shapes a raw material can exhibit and choose the most appropriate based on the requirements of formulation manufacturing and the desired performance of the final product.

Polymorphism

The crystallization process can also alter the arrangement of molecules within the crystal structure, forming different polymorphs. When polymorphism occurs, the molecules are arranged differently within the crystal lattice structure, leading to differences in their spectroscopy patterns, like in powder XRD analysis, providing a technique for identifying the presence of polymorphs.

Polymorphs have different physical and chemical properties, such as different melting points and solubilities. They generally also exist in different crystal habits. Thus, polymorphs are different crystal structures from the same chemical entity, with different physical and chemical properties that can profoundly alter their performance in the manufacture, stability and performance of cosmetic products.

The formulator must consider the appropriate polymorphic form to use. For example, if the formulator chooses the lower melting point polymorph that exhibits a higher dissolution rate, they may expect that the raw material will be soluble in formulation, and that the time and energy (temperature and agitation) required solubilizing the material will be lower. On the other hand, choosing the lower melting point polymorph is generally also the least stable and might be a problem if we consider the incorporation of this raw material in a solid cosmetic form, as it may compromise the stability and thus performance of the product. It is also possible that an unstable low melting point polymorph can revert to a more stable polymorphic form during storage, thus potentially changing critical physicochemical properties that alter the product manufacturing and performance.

Cocoa butter, for example, is a fatty raw material with six different polymorphs, each one with a specific melting point (Table 13.1). This raw material is used as stick for lip protection, and emollient and moisturizer in lotions (Figure 13.4). Generally, the manufacturing processes of products with cocoa butter involve warming up to melting followed by mixing with other compounds. The heating, without rigorous temperature control, might eliminate any crystal structure of cocoa butter due to overheating, leading to metastable crystal formation during the cooling step. This behaviour can be problematic as metastable crystals have a lower melting point than regular crystals. It may directly impact the manufacturing process because it takes more time to solidify and may not form a consistent product, particularly if it contains powders that sediment such as a lipstick with inconsistent pigmentation.

Metastable polymorphs in cocoa butter, formed due to overheating, can impact product performance. Products compounded with raw materials whose polymorphs can change in viscosity during storage due to transition to a more stable crystalline form can lead to modifications on application of products like creams and lotions, causing problems on spreading and touch, impairing the sensorial features of the product.

Scientists have used the transition of crystalline forms to control the release of active compounds from nanostructures like solid lipids nanoparticles and nanostructured lipid carriers. When the fatty materials that form the colloidal carriers are more fluid they can incorporate and retain more active compound in their inner core. As the fatty materials change to form a more organized structure with crystals predominating, the accommodation of the active ingredient in the lipid matrix reduces, causing it to be released from the nanostructure.

Clearly, a thorough understanding of the chemical composition, crystalline structure and crystal habit of all powder raw materials is critical to the design of cosmetic product formulations. This information should be sought from raw material suppliers who are required to deliver the materials in their established specifications, an essential factor for effective manufacturing process validation; for inter- and intra-batch product uniformity; and allowing the manufacture of safe, effective, stable cosmetic products with suitable organoleptic characteristics.

TABLE 13.1

Melting Points (mp) and X-Ray Diffractograms of Cocoa Butter Polymorph Crystals

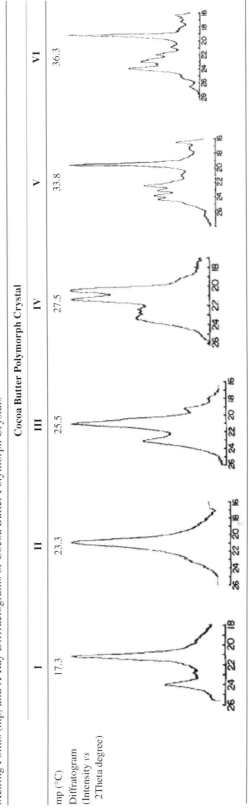

	I	II	III	IV	V	VI
			Cocoa Butter Polymorph Crystal			
mp (°C)	17.3	23.3	25.5	27.5	33.8	36.3
Diffratogram (Intensity vs 2Theta degree)						

Source: Data from Wille, R. L.; Lutton, E. S., *J. Am. Oil Chem. Soc.* 43, 491–496, 1966.

FIGURE 13.4   Cocoa butter raw material and uses in lipstick and body lotion.

Manufacturing Area for Powder Manipulation

To manufacture powders for cosmetic products, it is necessary to segregate the powder manipulation area inside the manufacturing area to avoid cross-contamination. This is primarily important when using fine particulate raw materials, which are easily spread in the air, causing potential risk of cross-contamination and to the health and safety of personnel. An exhaust system with hoods should be installed near the manufacturing equipment to ensure that powders suspended in the air do not reach the surface of other equipment, utilities or tools inside the room, or leave by doors during transit between areas. In addition, areas can have different air pressure with powder processing carried out in an area of relatively low pressure, reducing the risk of airborne powder materials escaping against a pressure gradient to contaminate other areas where contamination of products, equipment or personnel without adequate personal protective equipment (PPE). The latter is an important health and safety consideration, as personnel involved in powder manipulation will be aware of the respiratory risks and be required to use appropriate PPE, but personnel elsewhere on site will not be protected from fine particles suspended in the air.

Cross-contamination of cosmetic products can have significant consequences for cost and brand. Many fine materials are pigments or colourants so cross-contamination can change the colour of products, impairing the maintenance of one pattern. Hypoallergenic products and preservative-free products are produced in multipurpose plant factories, where potentially sensitizing compounds and preservatives may be used to compound a product batch before the manufacture of a hypoallergenic product. Cross-contamination of these products could result in sensitization of consumers.

In addition to these physical plant measures, production standard operational procedures (SOPs), validated cleaning procedures and rigorous quality control are essential. These are in keeping with the principles and practice of good manufacturing practices (GMPs) in every aspect of the production process from raw materials to the finished cosmetic product.

Important Machinery for Powder Manipulation

There is a wide variety of equipment used in powder manipulation and production. This includes particle size reduction (grinding) of individual powders and mixing of several powders to ensure a homogeneous dispersion (e.g. colourants in an eye shadow or pigments in a lipstick formulation). Examples of the manufacturing process for two powder-based cosmetic products are summarised in Figures 13.5 and 13.6. In both cases it is vital to generate a homogeneously dispersed mixture with appropriate properties to flow and adhere to their site of application.

In the eye shadow product (Figure 13.5) the first step is the effective mixing of the colourant and other powders, followed by mixing with the binder, which facilitates compaction. The press machine is essential for molding the powder into a suitable form and texture. It is clear that the effective mixing and molding will depend on the powder raw materials having appropriate properties such as, density, agglutination, average particle size and granulometric size distribution. In addition, process variables such as the rate of addition of binder, mixing speed and time, design of mixing chamber, velocity, and force

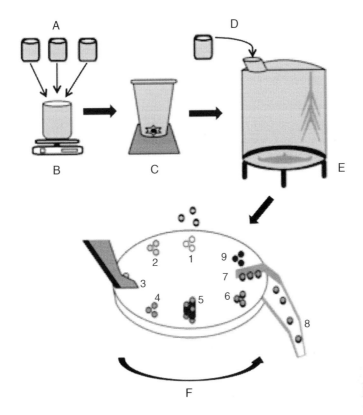

FIGURE 13.5 Schematic illustration of eye shadow manufacturing process. (A) Separation of powder pigments and raw materials in warehouse. (B) Weighing and fractionation. (C) Mixing by high-speed mixer. (D) Adding an oil binder to promote stickiness of the particles. (E) Intense homogenization by high-shear mixer. (F) Compression table at press machine: (1) empty compression chambers; the molds are outside of press machine; (2) the molds are put in compression chambers; (3) the filler dispenses powder mixture over the molds; (4) while the compression table spins, the filler takes off the excess powder; (5) superior punches come down and compress the powder inside molds; (6) superior punches lift out the compression chamber, and at the same time inferior punches eject molds containing compacted eye shadow powder; (7) a fin drives the molds off the compression table; (8) molds are gathered in a bin for the next step of the process; (9) inferior punches are in the upper position up to where molds were put off. Afterwards, inferior punches assume the lower position to a new circle.

of compression must be considered. SOPs must be accurately described and precisely followed to avoid errors during the manufacturing process. Process quality control will ensure that good manufacturing practice is followed.

In the second example (Figure 13.6) of a lipstick manufacturing process, the mixing step is the same as for eye shadow, but in this case, the colourants are mixed with melted fatty material to form a homogeneous colourant dispersion. This uses a different mixer type with a warmed chamber that can produce a mixture of the colourants, fatty material and other adjuvants with the right consistency to process through a three roll mill to facilitate the production of a homogeneously mixed mass that is then transferred to molds and cooled. Once it is in a solid form, the formulation is retrieved from the mold and packaged.

During the process (Figure 13.6), there are several steps that are critical to obtain a cosmetic product of suitable quality. The mixing and grinding steps are reliant on the mixing tank design, mix capability, warming diffusion, and ability to eliminate lumps or aggregates (steps C to E). Steps G and H are related to transfer of the formulation to the mold, with each mold cavity forming a lipstick. The temperature of formulation, agitation to maintain a homogeneous mixture throughout processing, transfer rate and cooling rate might all influence the quality of lipstick obtained. For example, while an adequate temperature is necessary to maintain the formulation in fluid form necessary for the transfer into mold cavities, over-warming might lead to an excessive drop in viscosity that could result in sedimentation of the pigment

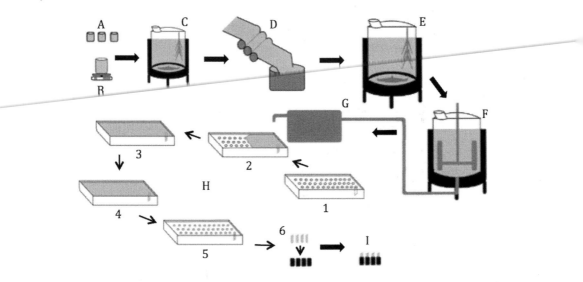

FIGURE 13.6 Schematic illustration of lipstick manufacturing process. (A) Separation of powder and other oil raw materials in warehouse. (B) Weighing and fractionation. (C) Warming and homogenization of oil raw material inside a jacketed reactor and dispersion of pigments. (D) Grinding lumps of pigments in formulation for colour homogenization by three roll mill and smoothing. (E) Addition of other formulation materials and blending inside a jacketed reactor. (F) Formulation maintained under soft agitation and warming for filling the molds. (G) Filler dispenses formulation over the molds to fill whole mold. (H) Molds in different steps of lipstick formation: (1) empty molds before to receive formulation; (2) flooding the mold cavities to form the lipstick, note the excess of formulation dispensing over the mold; (3) mold totally filling; (4) mold under cooling to solidify lipstick formulation; (5) mold filled, which the formulation excess was previously withdrawn after solidification; (6) lipsticks retrieved from mold and put in package. (i) Lipstick packaged.

inside the cavities. Providing sufficient agitation to maintain a homogeneous mixture throughout the process is essential to ensure that the mixture transferred into the mold forms a quality product. The rate of transfer is also important for a good quality of lipstick. A low transfer rate can result in a lipstick with parts solidified at different times creating fragile points, whilst a high transfer rate can trap air pockets in the cavity causing shape failure.

Examples of Powder Form Substances Employed in Cosmetics

Many active compounds employed in cosmetics are presented in powder form and can have different origins, including vegetable (e.g. horse chestnut, guarana), animal (e.g. collagen, amino acids) and mineral (e.g. fluor, calcium and clay). Many other powders are used as excipients or adjuvants in cosmetic formulation (e.g. talcum powder, silica). Some powders can be used both as actives and excipients, such as the clays. Coloured clays are aluminium silicate with changing iron, magnesium, manganese oxides and titanium dioxide concentrations in their composition, resulting in many colours. Natural or modified clays are often employed as active substances in cosmetics, such as masks, due to their high adsorption capacity, including oiliness produced by sebaceous glands. Clays can also be employed to adjust rheological properties and to stabilize formulations including emulsions and suspensions.

Future Trends

Even during the global financial crisis the beauty and cosmetic market continued to grow. This is driven in part by the substantial innovation in all aspects of the cosmetics industry, including in the use of powders and powder-based cosmetic products. We now have lipsticks in pencil form; 100% natural origin mineral pigments (from coloured clays); and make-up products with added attributes, such as moistening, anti-aging and sun protection.

Novel and distinguished textures, nanotechnology, and the union of biodiversity and packaging sophistication concepts are being embraced by cosmetic manufacturers. Cosmetic products are increasingly being developed to disguise imperfections and provide a uniform skin appearance. The trend is to obtain make-up cosmetic products that create a natural appearance, to enhance the features rather than cover them up. The powder presence in formulations can grant a perfect finish to make-up products and provide good sensorial properties. Small-sized particles such as natural silica microspheres give a matte and dry effect to hide skin imperfections. These mineral powders allow a matte make-up finish, and play an important role in the product sensorial properties (softness, smoothness), but they show some opacity that can lead to lack of luminosity in some concentrations and may also exhibit poor colour stability. Coating techniques, such as air fluid bed coating, can help to avoid pigment degradation and improve resistance to interaction with naturally present skin oiliness, extending product efficacy.

There is also an increasing development of microspheres with a soft-focus effect (to disguise optical wrinkles and expression lines), and coated and micronized pigments that help to provide good skin coverage.

Final Considerations

The role of powders as and in cosmetics has been established since the beginning of the use of products to enhance beauty. This has developed from crudely pulverized natural powders to highly sophisticated nanomaterials, with powder manipulation processes developing in their complexity and sophistication. There are dermato-cosmetic bases, obtained from an emulsion then dehydrated, that when in contact with water again are capable of being reconstituted to obtain an oil-in-water emulsion.

A substantial advantage of powders is the many stability-related benefits they bring to the final product and the better performance in regard to skin coverage power, resulting in better skin appearance and hiding of skin imperfections. Makeup foundation used to require application of a moisturizer followed by facial powder to provide coverage to complexion imperfections, oiliness control and to increase skin softness. Nowadays liquid foundations are suspensions that provide skin moistening and texture improvement in one multifunctional product. However, where total skin coverage is desired and skin oiliness is problematic, applying a powder foundation provides the most effective outcome.

There is no doubt that technical scientific knowledge brings advantages to the formulator in this increasingly competitive and technological area, and who wants to continue to drive innovation in the use of powders in cosmetic formulation.

Acknowledgements

We thank Taís Souza Barboza for her valuable help with the text.

BIBLIOGRAPHY

Aburjai, T.; Natsheh, F. M. Plants used in cosmetics. *Phytother. Res.* 17, 987–1000, 2003.

Buschmann, H. J.; Schollmeyer, E. Applications of cyclodextrins in cosmetic products: A review. *J. Cosmet. Sci.* 53, 185–191, 2002.

Carretero, M. I. Clay minerals and their beneficial effects upon human health. *Appl. Clay Sci.* 21, 155–163, 2001.

Carretero, M. I.; Pozo, M. Clay and non-clay minerals in the pharmaceutical and cosmetic industries. Part II. Active ingredients. *Appl. Clay Sci.* 47, 171–181, 2010.

Draelos, Z. D. Colored facial cosmetics. *Dermatol. Clin.* 18 (4), 621–631, 2000.

Gao, Y.; Muzzio, F. J.; Ierapetritou, M. G. Scale-up strategy for continuous powder blending process. *Powder Technol.* 235, 55–69, 2013.

Keck, C. M.; Kovačević, A.; Müller, R. H.; Savić, S.; Vuleta, G.; Milić, J. Formulation of solid lipid nanoparticles (SLN): The value of different alkyl polyglucoside surfactants. *Int. J. Pharm.* 474 (1–2), 33–41, 2014.

Kuo, C. L.; Wang, C. L.; Ko, H. H.; Hwang, W. S.; Chang, K. M.; Li, W. L.; Huang, H. H.; et al. Synthesis of zinc oxide nanocrystalline powders for cosmetic applications. *Ceram. Int.* 36 (2), 693–698, 2010.

Luckewicz, W.; Saccaro, R. Determination of ascorbyl dipalmitate in cosmetic whitening powders by differential scanning calorimetry. *J. Soc. Cosmet. Chem. Detroit.* 41, 359–367, 1990.

Luckham, P. F.; Rossi, S. The colloidal and rheological properties of bentonite suspensions. *Adv. Colloid Interface Sci.* 82 (1–3), 43–92, 1999.

Silva, E. C. D.; Paola, M. V. R. V. D.; Matos, J. D. R. Análise térmica aplicada à cosmetologia. *Br. J. Pharm. Sci.* 43 (3), 347–356, 2007.

Su, C. Y.; Tang, H. Z.; Chu, K.; Lin, C. K. Cosmetic properties of TiO$_2$/mica-BN composite powder prepared by spray drying. *Ceram. Int.* 40 (5), 6903–6911, 2014.

Teixeira Neto, E.; Teixeira Neto, A. A. Modificação química de argilas: desafios científicos e tecnológicos para obtenção de novos produtos com maior valor agregado. *Quím. Nova* São Paulo, 32 (3), 809–817, 2009.

Wille, R. L.; Lutton, E. S. Polymorphism of cocoa butter. *J. Am. Oil Chem. Soc.* 43, 491–496, 1966.

Zague, V.; Santos, D. de A.; Baby, A. R.; Velasco, M. V. R. Argilas: Natureza das máscaras faciais. *Cosmet. Toiletries* 19, 64, 2007.

14

Natural Products and Stem Cells and Their Commercial Aspects in Cosmetics

Sonia Trehan, Rose Soskind, Jemima Moraes, Vinam Puri and Bozena Michniak-Kohn

CONTENTS

Introduction

Natural products form a large and diverse group of compounds derived from sources encompassing the plant and animal kingdoms examples of which include among others, insects, marine algae and microbes. Commonly used ingredients comprise herbs, minerals, vitamins, antioxidants, essential oils, enzymes and hormones that have become increasingly more popular in cosmetic and personal care products due to the increasing number of reports concerning the potential harmful effects of some synthetic ingredients on the human body. Stem cells are another natural source that is gaining enhanced popularity in the cosmetic field. Stem cells carry inherent implausible capabilities for applications in skin care for both dermaceuticals and cosmeceuticals. Skin is the largest organ of the human body, and skin care is not only related to skin health but it also has a great influence on a person's overall feeling of self-worth, personality, confidence and lifestyle. This is one reason why customers are currently increasingly demanding cosmetics produced with naturally derived ingredients, containing the smallest possible number of synthetically produced compounds (Ligeza et al., 2016). This chapter reviews natural ingredients that are most often used in cosmetic formulations and discusses properties and qualities as well as the regulatory aspects of these compounds.

Natural Ingredients in Cosmetic Formulations

Plant-Based Materials

Humans have relied for a long time on nature to provide their basic needs such as food, clothing, fertilizers, flavors, fragrances and medicines. Plants and other natural sources have formed the basis of traditional medications that have been in existence for thousands of years providing people with remedies for diverse treatments (Gurib-Fakim, 2006). The preparations derived from herbs, roots, stems and other materials of plant origin are often referred to as 'botanicals'. These are widely used in cosmetic products in the form of extracts obtained from fresh plants. The use of plants dates back to historical times and in the coming years, the market will see many more additional products containing 'naturals' to provide the population with both medicinal and cosmetic formulation options. Natural raw materials such as plant extract from seeds, herbs, flowers and fruits have always been part of research and development studies as evidenced by the large number of cosmetic products with botanicals being commercialized. In addition, the naturally occurring mixtures of active compounds in plants are often more effective than individual molecules and combinations of synthetic molecules (Lee et al., 2017). There are a large variety of natural ingredients for cosmetic applications known in the scientific literature. A summary of the most common ingredients is provided next.

Mango (*Mangifera indica*) is used in cosmetic products since the kernel of the fruit contains an emollient oil rich in oleic, stearic acids and triglycerides, which has also lately been reported to release drugs at a remarkably greater rate than the standard paraffin-based ointment formulation (Nahar et al., 2017).

Green tea (*Camellia sinensis*) contains more than 500 individual chemical compounds, which include tannins, flavonoids, amino acids, vitamins, caffeine and polysaccharides. The flavonoids (polyphenols) have been shown to possess anti-inflammatory, antioxidant, antiallergic, antibacterial and antiviral effects, while the tea tannins have antiseptic and antioxidant effects (Katiyar and Elmets, 2001). The polyphenol epicatechins found in green tea extract is an important antioxidant used in a variety of cosmetic formulations. It can affect the cellular stress system within the skin in two ways: directing effects on signal transduction and modifying connections in the dominant antioxidative status in the cells (Kerscher and Buntrock, 2017).

French sea pine (*Pinus pinaster* ssp. *atlantica*) contains an antioxidant known as *pycnogenol* that has been claimed to be an antiaging compound, due to its capacity to neutralize most naturally occurring oxygen radical species. This is the reason that pycnogenol has been considered to be significantly more active as an antioxidant when compared to coenzyme Q10, alpha-lipoic acid and grape seed extract. In a clinical trial, 30 women diagnosed with melasma showed a significant decrease in the average surface area of melasma when they were supplemented with 25 mg of pycnogenol as part of meals three times a day. Also pycnogenol is often included in sunscreens and other skin care products (Ni et al., 2002; Kerscher and Buntrock, 2017).

Red clover (*Trifolium pretense*) has proven anti-inflammatory effects and is recommended for treatment of diverse skin changes as often occur in psoriasis, eczema and acne. In a published study, hairless mice were subjected to a topical application of isoflavones from red clover flowers prior to ultraviolet (UV) exposure, and following this UV exposure, the results showed photoprotective effects, a reduction in the inflammatory edema and suppression of contact hypersensitivity induced by moderate doses of solar-simulated UV radiation (Aburjai and Natsheh, 2003). Some cosmetic lotions contain 'equol' (an isoflavone from red clover) and it is thought to protect the immune system from photosuppression in advance to any sunburn reaction, even when applied after exposure. Due to this fact it is considered as a cosmetic photoprotective agent and most recently as an antiaging substance (Kerscher and Buntrock, 2017).

Oat (*Avena sativa*) has proven skin protection against damage from environmental effects such as UVA/UVB irradiation, pollution, smoke, bacteria and free radicals, and also reduces discomfort, irritation and inflammation of the skin (Aburjai and Natsheh, 2003). In addition, oats also play an important role as moisturizing agents and emollients in cosmetic formulations and most recently they have been used as rheology modifiers (Cizauskaite et al., 2017). For example, oatmeal is one of the important ingredients in the Johnson & Johnson highly successful Aveeno brand series of products.

Pomegranate (*Punica granatum*) is a fruit that contains a significant amount of polyphenols, which provide a rich source of antioxidant, anti-inflammatory and anticarcinogenic actives (Kerscher and Buntrock, 2017).

Soy (*Glycine max*) contains two inhibitors known as trypsin and Bowman-Birk, and both have demonstrated significant improvements in mottled skin pigmentation, blotchiness, dullness, fine lines and skin tone. In addition, several preliminary *in vivo* human studies have demonstrated the skin-lightening effect of non-denatured soy extracts (Lee et al., 2017).

Artichoke (*Cynara scolymus*) contains a polyphenolic flavonoid known as silybin or silibin which has antioxidant, anti-inflammatory, and anticarcinogenic properties. Several studies have shown skin photoprotection effects for the compound when silibin was topically applied prior to, or immediately after, UV exposure. In addition, silibin is widely applied to skin from various antiaging and sun-protective skin care products. Silymarin/silibinin is found in many moisturizers that are used to prevent cutaneous oxidative damage and photoaging (Lee et al., 2017).

Turmeric (*Curcuma longa*) contains the active compound named curcumin. It is applied in natural cosmetic products due to various benefits such as moisture retention, antiaging and antioxidant activity (Prasad et al., 2014). In addition, turmeric has been shown to enhance the bioavailability and increase antioxidant activity as part of nanodelivery carriers such as solid lipid nanoparticles, nanoliposomes, nanoniosomes and nanoemulsions (Golubovic-Liakopoulos et al., 2011; Tavano et al., 2014).

Ginseng (*Panax*) is one of the safest and most potent natural antiaging agents for the skin. Several published studies have demonstrated that ginsenosides play a role as skin protectants and anti-wrinkle compounds as well as provide protection against harmful solar radiation. In addition, the ginsenosides known as Rd and compound K have received special attention in recent years for their natural antiaging effects, and these could be potential novel cosmetic ingredients (Lee et al., 2017).

Ginkgo biloba (*Ginkgo biloba* L.) contains unique polyphenols including terpenoids (ginkgolides, bilobalides), flavonoids and flavonol glycosides with anti-inflammatory effects. The flavonoid fractions quercetin, kaempferol, sciadopitysin, ginkgetin and isoginkgetin underwent *in vitro* studies which demonstrated the compounds' ability to promote fibroblast proliferation in human skin. In addition, ginkgo extracts are widely applied in many cosmetic products as known antioxidants and collagen synthesis promoters (Lee et al., 2017).

Aloe vera (*Aloe vera*) contains C-glycosylated chromone known as aloesin and it plays a role as a competitive tyrosinase inhibitor, reducing both the hydroxylation of tyrosine to dihydroxyphenylalanine and oxidation of dihydroxyphenylalanine to dopachinone. Furthermore, in a published study, aloesin was topically administered four times a day for 15 days to treat hyperpigmentation in human skin following UV radiation, and the results showed that aloes at 100 mg/g vehicle suppressed skin pigmentation by 34% when compared to the vehicle control (polyethylene glycol PEG 8000; Sigma, USA) ($n = 15$; $P < 0.05$) (Choi et al., 2002; Lee et al., 2017).

Camphor tree (*Cinnamomum camphora*) extracts contain the compound termed camphor, which is widely applied in repellent products and as preservative in cosmetic products. Recently, a study has shown other advantages of camphor for skin health such as the promotion of fibroblast proliferation, maintaining or recovering collagen and elastin in the skin that had been exposed to UV radiation, and preventing thickening of the epidermis and subcutaneous fat layer (Tran et al., 2015).

The natural *essential oils* are extracted from flowers, grass, herbs, fruit, peel of citrus fruits, seeds, leaves, bark, rhizomes and roots. They are recognized as important natural ingredients for formulations and have such unique effects that they are classified within their own group. The plants and their essential oils have been used in cosmetics, as incenses, perfumes, spices, for nutrition and also in medications. The essential oils have been used for over a thousand years to relieve pain, treat tension and fatigue, invigorate the body, and to produce a sense of relaxation. The properties of essential oils can change depending on their origin and composition; some oils have medicinal properties such as antioxidant activity against free radicals, as well as anti-inflammatory and antimicrobial activities (Innocenti et al., 2010; Mangena and Muyima, 1999). Some common constituents present in these essential oils and their claimed properties are the following:

Terpenes – Inhibit the accumulation of toxins and are said to help discharge existing toxins from the human organism.

Esters – Antifungal, calming and relaxing effects.

Aldehydes – Anti-infectious with a sedative effect on the central nervous system; when applied on the skin they may be irritating, but may have a deep calming effect when inhaled.

Ketones – Stimulate cell regeneration, promote the formation of tissue, and liquefy mucous.

Alcohols – Commonly recognized for their antiseptic and antiviral activities.

Phenols – Responsible for the fragrance of the oil, and have antiseptic, antibacterial and antioxidant properties.

Oxides – Principal compounds of eucalyptus oil and play a role as anesthetics and antiseptics (Higley and Alan, 2005).

Most of the natural ingredients in the perfumery industry come from essential oils, and their main application is in cosmetic toiletries (including personal care products, fine fragrances, cosmetics, bath products, deodorants, hair products) and household products (including air fresheners, laundry products, washing liquids, surface cleaners and disinfectants). The cosmetic industry includes natural oils in their research and development, and use the compounds to offer a pleasant aroma, or provide conditioning in a hair care product, or emolliency or improvement of elasticity of the skin (Ligeza et al., 2016; Aburjai and Natsheh, 2003).

Natural essential oils can provide other benefits, such as physiological (anti-inflammatory effects, bactericidal effects, antiviral, antioxidant, anticancer activities, brain stimulation, anxiety-relieving, sedation, antidepressant activities, as well as increasing cerebral blood flow) and psychological (the traditional aromatherapy use aromatic oils considering the terms yin [passive and calming] and yang [active and simulating]). Some additional studies have shown effects of fragrances on memory and mood, once the fragrances are absorbed through inhalation and are able to cross the blood–brain barrier and interact with receptors in the central nervous system), and cosmetic advantages (emolliency, pleasant scent, shine effects, cooling and long-lasting refreshing feeling to the skin) (Ligeza et al., 2016). Essential oils and their constituents (terpenes, terpenoids) have been widely investigated as safe and suitable skin penetration enhancers for both hydrophilic and hydrophobic actives. Terpenes derived from essential oils have been shown to be efficient in delivering many actives through the skin including nicardipine hydrochloride, caffeine, hydrocortisone, triamcinolone acetonide and propranolol hydrochloride, but the mechanism of their action is not yet fully comprehended (Herman and Herman, 2015).

The essential oil derived from sweet basil (*Ocimum basilicum*) showed potential as a skin permeation enhancer, and improved the percutaneous absorption of indomethacin by increasing active partitioning into the stratum corneum (SC) and by disrupting the skin morphology (Avetisyan et al., 2017). Another study with peppermint oil showed that low concentrations of the oil had the ability to reduce the percutaneous penetration of benzoic acid (retarding action), while at higher concentrations it was shown to decrease the integrity of the dermal barrier (Nielsen, 2006). Furthermore, studies showed that turpentine, eucalyptus and peppermint oils improved permeation of ketoconazole by modification of the skin barrier but without any change of their structure. In addition, *Alpinia oxyphylla* oil showed a higher affinity for the lipophilic SC and consequently reduced the polarity of the SC, thereby enhancing the permeation of the lipophilic indomethacin into the dorsal rat skin (Fang et al., 2004).

The technology of microencapsulation to promote increased stability of essential oils with controlled release has been a subject of considerable research in the recent years. The encapsulation of essential oils has been investigated for various reasons, for example, protection from oxidative decomposition and evaporation, odour masking, or to provide controlled release in the formulation. Several microencapsulation methods have been developed that can be adapted to different types of active agents including essential oils, generating particles with a variable range of sizes, shell thicknesses and permeability, and providing a tool to modulate the release rate of the active compounds. The essential oils that were subjected to microencapsulation and reported in the literature are lemon oil, thyme oil, citronella oil, vanilla oil, clove oil, peppermint oil, mint oil, orange oil, eucalyptus oils, rosemary oil, rose oil and cinnamon oil (Martins et al., 2014). Some common essential oils known in the literature for cosmetic applications are described in Table 14.1.

Waxes and Butters

Natural waxes and butters can be sources of emollients, thickeners and film formers, and often silicone derivatives are available (O'Lenick et al., 2008).

Waxes are esters comprised of long-chain fatty acids that have been reacted with higher alcohols. Wax esters have two fatty groups: one in the alcohol portion and one in the acid portion (O'Lenick et al., 2008). Carnauba wax is a high melting point wax derived from the leaves of carnauba palm trees (*Copernicia cerifera*). Although the tree grows throughout the world, only the variety found in northeastern Brazil produces the wax used in the cosmetic industry. Carnauba wax can used to harden softer waxes for reduced elasticity and crystallization, and can be used in a wide variety of make-up products (Corbeil et al., 2000). Candelilla wax is taken from the outer surface of the *Euphorbia cerifera* plant found in northern Mexico. The wax is light brown in colour, and is hard and shiny. It can be found in creams, lipsticks and other make-up products. The wax has a high melting point and can be used in many products that require resistance to heat. Jojoba oil is a waxy ester that is liquid at room temperature and comes from *Simmondsia chinensis* evergreen shrubs found in the southwestern United States and in Mexico (O'Lenick et al., 2008). Japan wax, or sumac wax, is not a true wax. Instead, it is a fat extracted

TABLE 14.1

Plant-Derived Essential Oils and Their Cosmetic Applications

Essential Oil	Plant Source	Application
Coconut oil	*Cocos nucifera*	Emollient in cosmetics and useful for protection and prevents drying of skin.
Castor oil	*Ricinus communis*	Smoothing and moisturizing qualities essential for conditioning of skin, also castor oil and/or its esters, are useful as vehicles, emollients or solubilizers for toiletry, cosmetic, hair and skin care formulations.
Olive oil	*Olea europaea*	Known especially for moisturization of dry skin, applied in lipsticks, shampoos, hand lotions, soaps, massage oils and dandruff treatment products.
Lavender oil	*Lavendula angustifolia, latifolia, stoechas and intermedia*	Widely employed in soaps, lotions and perfumes for fragrance effects.
Grape seed oil	*Vitis vinifera*	Useful in anti-aging and skin-lightening cosmetic creams and lotions due to antioxidant effect.
Melaleuca oil	*Melaleuca alternifolia*	Employed in cosmetics such as hair care formulations and skin care creams.
Basil oil	*Ocimum Sanctum*	Applied in skin care products for acne treatment due to antibacterial, anti-inflammatory and antiulcer activities.
Vetiver oil	*Chrysopogon zizanioides*	Cosmetic ingredient used for its pleasant aroma and antimicrobial activity.
Chamomile oil	*Matricaria chamomilla*	Ability to reduce inflammation due to the presence of flavonoids. Safe for skin care, and credited with a gentle analgesic effect. Anti-inflammatory, anti-erythema and antipruritic, at the same time as being gentle, soothing and antiseptic. May help in whitening age spots, minor wounds, burns or insect bites; or applied for dry skin, windburn, sunburn, or even chronic skin conditions such as acne and psoriasis.
Geranium oil	*Pelargonium graveolens*	Widely used in the fragrance industry due to its strong rose-like fragrance and also in aromatherapy. It is considered as a cleansing, toning and sharpening oil, and is useful in treating skin conditions such as greasy, over-oily, acne, congested skin and eczema.
Black cumin seed oil	*Nigella sativa*	Derived from seeds with demonstrated radical-scavenging, antibacterial, antifungal, anticarcinogenic, analgesic and anti-inflammatory properties. Moreover, the low toxicity of *N. sativa* oil, evidenced by high LD50 values, suggests a wide margin of safety.
Sweet almond oil	*Prunus dulcis*	Used as emollient and emulsifier for chapped hands, in lotions (both moisturizing and night skin care formulations), gels, blushes, make-up bases, lipstick, soaps, skin cleansing, creams, and as an ointment base. Some studies showed weak antibacterial effect.
Anise oil	*Pimpinella anisum*	Possesses weak antibacterial effect and is applied in cosmetic products such as toothpastes, perfumes, soaps, creams and lotions.
Carrot seed oil	*Daucus carota*	Displays antimicrobial and antifungal activity. Applied in cosmetic products such as soaps, creams, lotions and perfumes. Used in certain sunscreen formulations and as a source of beta-carotene and vitamin A.
Copaiba oil	*Copaifera officinalis*	This oil has shown antibacterial and anti-inflammatory activity. It is applied in cosmetics products such as soaps, bubble baths, creams, lotions and perfumes.
Rosewood oil	*Aniba roseaeodora*	It is recognized as a rich linalool natural source, which is widely used in perfumery. Applied in other cosmetics such as soaps, creams and lotions.

Sources: Kerscher, Martina, and Heike Buntrock, 2017, Cosmetics and cosmeceuticals, in *Cosmetic Medicine and Surgery*, edited by Pierre Andre, Eckart Haneke, Leonardo Marini, and Christopher Rowland Payne, London: CRC Press; Burger, Pauline, et al., 2017, *Medicines Medicines* 4 (2):41; Avetisyan, Arpi, et al., 2017, *BMC Complementary and Alternative Medicine* 17:60; Ligeza, M., et al., 2016, *Family Medicine and Primary Care Review* 18 (4):443–7; Aburjai, Talal, and Feda M. Natsheh, 2003, *Phytotherapy Research* 17 (9):987–1000; Higley, Connie and Alan Higley, 2005, Essential oils constituents, in *Reference Guide for Essential Oils*, Olathe, KS: Abundant Health; Khan, A. I., and A.E. Abourashed, 2011, Natural ingredients, in *Encyclopedia of Common Natural Ingredients Used in Food, Drugs and Cosmetics*, Hoboken, NJ: John Wiley & Sons.

from the fruit of the *Rhus succedanes* plant and can be used as a plant-based alternative to beeswax (Corbeil et al., 2000).

Butters are triglycerides that generally have a titer point between 20°C and 40.5°C, at which the mixture of fatty acids solidifies. *Cocoa butter* (*Theobroma cacao*) is used as an emollient in topical cosmetic formulations, and recently has been recognized as a good source of natural antioxidants (Fankem et al., 2017). South America and the Brazilian rainforest offer various plants with common butters used in the industry that include cocoa butter extracted from the cocoa bean (*Theobroma cacao*), cupuaçu butter from the related *Theobrama grandiflorum* tree and murumuru butter from the murumuru palm tree (*Astrocaryum murumuru*). India is another source of many butters used in cosmetic products, including kokum butter extracted from the seeds of the *Garcinia indica* tree, mango butter from the *Mangifera indica* tree and shea butter derived from the *Butyrospermum parkii* tree. Interestingly, shea butter is considered an antioxidant and is believed to include the compound allantoin that contributes to shea butter's medicinal properties including wound healing as well as reduction of rheumatic pains and inflammation (O'Lenick et al., 2008).

Balms, Gums and Resins

Plant-based balms, gums and resins have been used for thousands of years. One example of their use is for embalming the dead, a technique used extensively by the ancient Egyptians (Corbeil et al., 2000). Balms are perfumed, tree-based resins. One example used in the cosmetic industry is Balsam of Peru from the *Myroxylon pereirae* tree that grows in Central America. The balsam can be used in perfumes, and the essential oil form is known as cinnamein. Balsam of Peru must be used with caution since it is an allergen for many people (Corbeil et al., 2000). Some plants produce polysaccharide gums that can be used in various cosmetic products as thickeners and emulsifiers, as well as bases for masks and foundations. Many gums are also edible. While several gums are derived from marine-based plants and are discussed in the 'Marine-Based Materials' section of this chapter, there are also gums from land plants that are commonly used in the cosmetic industry. However, some of these gums are being replaced by more stable synthetic ingredients that can often be produced with a higher yield than through natural sources (O'Lenick et al., 2008). Cellulose is a major part of plant fibres, so many cellulose-based gums are of natural origin and may be sourced as by-products of the wood pulp process used in making paper. Starches from grains and tubers can also be used as thickening agents, and may assist with the viscosity of various formulations. The size and shape of the starch granules is dependent on the plant source (Dweck, 2011). Acacia trees grown in Africa are the source of water-soluble acacia gum, also known as arabic gum. Acacia gum reduces irritation of mucous membranes, though it is highly tacky and may be blended with other gums to reduce tackiness. Acacia gum is stable in a range of pHs, though electrolytes and salts can decrease acacia gum viscosity (Dweck, 2011). Tragacanth gum comes from the *Astragalus gummifer* plant that grows in the Middle East. The gum can be used to achieve high viscosities that can be stable over a range of heat and pH values (Dweck, 2011). Carob bean gum, or locust bean gum, is from carob trees (*Ceratonia siliqua*) found in the Mediterranean and Central America. Carob bean is not as commonly used in the industry as the other plant-based gums since carob bean gum involves an expensive preparation. Guar gum is from the Indian cluster bean (*Cyamopsis tetragonolobus*), and is ionic and a good thickening agent that easily swells in cold water. Only a small amount of guar gum is needed to produce excellent thickening and stabilizing properties (Dweck, 2011). Karaya gum comes from the dried resin of the *Sterculia urens* tree grown in India. Better quality gums have a transparent white colour, whereas lower grades tend to be yellow or pinkish-brown. The gum can be pulverized into a water-insoluble powder that is heat-sensitive and should thus be prepared in cold water. The presence of electrolytes and acidity decrease the viscosity of the gum. In cosmetics, Karaya gum is mainly used in fixatives (Corbeil et al., 2000). Benzoin gum comes from the resin of trees of the *Styrax* genus that are found in Thailand and Indonesia, and are named based on the place of origin, usually either Sumatra or Siam. Skin lotions with benzoin gum may stimulate cutaneous blood circulation and provide a skin tightening effect. The gum may also be used in perfumes as a fixative and is considered to have a pleasant odour when burned (Corbeil et al., 2000). Elemi, dammara and sandarac plant–based gums can be used in nail polishes to provide gloss and increase adhesion to nails (Corbeil et al., 2000). Myrrh is a resin from the *Commiphora myrhha* tree grown in Africa and the Middle East. It has a strong odour that can be used in perfumes and incense, and is considered to have astringent and antiseptic properties (Corbeil et al., 2000).

Colourants

Various colourants used in the cosmetic industry can be derived from plant sources. Henna has been used since ancient times and is the most commonly used hair colourant in the Middle East. Henna is also used in many countries, including India, for creating decorative patterns on the skin. Henna comes from the *Lawsonia alba* bush of North Africa and India, and provides an orange colour. Henna contains lawsone, which is a chemical compound that provides protection from the sun. It is important to note that henna's colouring effects usually require acidic conditions. Boiled chamomile flowers can also be used as hair dyes by providing a yellow colour. Chlorophyll is a green photosynthetic pigment that commercially can be extracted from nettle and spinach. It can also have deodorizing properties and is thus used in some deodorants and toothpastes. Xanthophyll is a yellow photosynthetic pigment that may also be used in the industry. Indigo is a blue pigment that comes from the *Indigofera* genus of flowers and is the oldest pigment known. Indigo is now often derived from synthetic processes. Saffron is a dark orange pigment from the *Crocus sativus* flower that has a strong tinting power due to the presence of carotene, a precursor of vitamin A. The dye is difficult to extract, and as a result is expensive. Saffron can also be used as an essential oil. Several sources of red pigment exist that include madder that has been used since ancient times and is from the *Rubia tinctorium* root; carthamin derived from the safflower (*Carthamus tinctorius*) bush; annatto, also known as bixin, from the seeds of the *Bixa orellana* plant; and alkanet reddish-brown pigment from the *Alkanna tinctoria* husk (Corbeil et al., 2000).

Scrubs

Many scrubs and abrasives are derived from a variety of plants, usually in the form of powders or flours. Examples of sources include argan kernel (*Argania spinosa*), murumuru seed (*Astrocaryum murumuru*), oat bran or kernel (*Avena sativa*), cinnamon bark (*Cinnamomum cassia*), bitter orange peel (*Citrus aurantium*), coconut (*Cocos nucifera*), coffee bean (*Coffea arabica*), fennel seed (*Foeniculum vulgare*), soybean seed (*Glycine soja*), barley seed (*Hordeum vulgare*), avocado seed (*Persea gratissima*), almond (*Prunus amygdalus*), cocoa (*Theobroma cacao*), and vanilla seed (*Vanilla* genus). The loofah plant (*Luffa cylindrical*) produces a cylindrical fruit that can be dried and used as a scrubbing sponge. In addition to plant-based materials, soils, minerals and clays can also be used as natural sources of scrubs (Dweck, 2011).

Clay Minerals

Clay minerals have been the focus of many studies including those in the pharmaceutical sciences, cosmetology, geology, materials science, medicine, food science and biotechnology. In particular, clay minerals have been increasingly utilized in cosmetics, and major advances have been achieved in the research and innovation related to these materials. In the past, the unique properties of clay minerals have attracted great interest in the industry, especially because these materials are used in pharmaceutical compounding of formulations and these clay minerals are abundant in the nature and not expensive (López-Galindo and Viseras, 2004; Viseras et al., 2007; Auerbach et al., 2004). Clay minerals are natural crystalline earthy materials with a fine grain size (particle size of less than 2 μm) composed of hydrated aluminium silicates, with magnesium, iron, calcium, potassium or sodium present as essential constituents, organized in different structures often as superimposed alternating layers. In addition to minerals, clays may also contain organic compounds, such as soluble salts, quartz particles, pyrite, calcite, other non-clay minerals and amorphous components (Auerbach et al., 2004).

Clays are easily found in nature and can be sourced globally. They occur in nature in different colours (Stepkowska and Jefferis, 1992) such as white, red, beige, yellow and brown, and may be used for many purposes (Murray, 2007), among which are those related to personal care or health care when they are directly applied on the skin or mucosae or when added to cosmetic or pharmaceutical products for topical application. Each clay mineral has a unique cosmetic or therapeutic function, and can be used for wound healing, skin lightening, for colour embellishment, sebum absorption, antisepsis, moisturization, vascularization and toxin elimination (Gomes and Silva, 2007). Before they can be utilized in the manufacturing process of dermaceuticals, clay minerals extracted from natural sources need to be processed in order to attain maximum purity and ideal grain size (desiccation, pulverization, sieving and wet separation of the clay fraction, sterilization by heat, etc.) (Murray, 2007). In some cases, the clay minerals may be subjected to a chemical process to enhance some specific property (as it is the case with the so-called homoionic clay minerals) or even change their properties (such as with the organoclay minerals) (de Paiva et al., 2008). The main properties attributed to clays used for cosmetic applications include the surface properties (surface area, cation exchange capacity or CEC, layer charge [either neutral or charged], sorption, dispersibility), rheological properties (thixotropy, rheopexy, viscosity, plasticity), and other physical and mechanical properties (particle size, shape, colour, softness, opacity, reflectance, iridescence and so on) (Moraes et al., 2017). Currently, one can identify the presence of clay minerals in a large variety of cosmetic products, such as facial creams, sunscreen, skin cleansers, shampoos and make-up (liquid and powder foundations, eye shadow, facial masks, lipsticks, etc.). The ever-increasing use of clay minerals in cosmetic products either as dermatological active ingredients or as excipients is due to the versatility of these materials, and to their particular physical and chemical properties (López-Galindo et al., 2007).

Nowadays, a great deal of scientific information about clay minerals is available and the pharmaceutical and cosmetic industries have an increasing interest in incorporating this natural resource into their products (Viseras et al., 2005; Cerezo et al., 2005; Ghadiri et al., 2015). Even before the industry started to incorporate clay minerals in cosmetic products, these materials were already used in geotherapy and fangotherapy ('mudding' at aesthetic clinics and spas). Several skin conditions many with an inflammatory component such as ulcers, comedones, acne and seborrheic dermatitis, can be treated efficiently with clay masks, due to the ability of some clay minerals to adsorb dirt, oil and toxins (Carretero and Pozo, 2009; Clijsen et al., 2008). Three clay-based cosmetic products with potential therapeutic properties have been under development by the Gomes group (Gomes and Silva, 2007). One of them contains a blend of bentonite from the Porto Santo Island (in the Madeira Archipelago) with biogenic carbonate sand from Porto Santo. The bentonite acts as an excipient, promoting the adhesion of the active component (in this case, the sand) to the skin. The biogenic carbonate sand contains calcium, magnesium and strontium, which are released into the skin in a controlled manner until they reach the extracellular matrix, where cellular metabolic and catabolic chemicals also circulate. Moreover, another product has been developed by the same group (Gomes and Silva, 2007) which contains bentonite gel and numerous chemical elements present in the biogenic carbonate sand of Porto Santo, intended for topical use in the treatment of certain rheumatic diseases. The same authors also mentioned a third product that they were developing based on a blend of several ingredients, among them biogenic silica (the active component), a refined bentonite from Porto Santo and sodium hydroxide (pH conditioner). According to the scientists, the silicon provided by the biogenic silica is essential for the synthesis of collagen fibres and therefore helps treat osteo-articular and muscular-skeletal conditions. Considering the diversity of the chemical structures in clay minerals, the Pasbakhs research group (Pasbakhsh et al., 2013) conducted a scientific study and showed that the structure of such clay minerals makes them ideal carriers for actives, as the clays occur as nanotubes that can accommodate active molecules in their interior. The physical and chemical properties of halloysites (such as the level of impurities and the volume of the internal cavity of the nanotubes) vary greatly among samples obtained from different sources, to the extent that some of them may be suitable for use as carriers but others may not. Another study, conducted by the Valenti group (Valenti et al., 2012), showed that the topical application of kaolinite clay masks in rats promoted an increase in the amount of collagen fibres in their skin. The group of rats that received skin treatment with clay masks for a period of 7 days displayed a significant increase in the percent area of collagen fibres ($51.74 \pm 1.28\%$) in comparison to the control group ($43.39 \pm 1.79\%$). A third group of rats, which received skin treatment with retinoic acid for the same period of time, displayed a statistically insignificant increase in the area of collagen fibre ($45.66 \pm 1.10\%$) in comparison to the control group. The authors conclude that the clay could be useful for treating skin aging, as corroborated by their findings.

Clay minerals also have been included in formulations of sunscreens due to their excellent properties, acting as a barrier that blocks solar radiation and thus protects the cellular DNA against potentially serious damage. In order to be useful for sunscreen formulations, the clay minerals must have high index of refraction and optimal light dispersion properties. The bentonite clay and the hectorite organoclay have been shown to meet the required specifications and are already being applied in sunscreen products (Ghadiri et al., 2015; Mattioli et al., 2016). Some clay minerals utilized in cosmetic products and their main properties are described next:

> *Bentonite (smectite or montmorillonite clay mineral)* – Bentonite is the most widely used in the cosmetic industry for stabilization of emulsions. It can also modify the rheological behaviour of the emulsion, resulting in a final product with features that are both acceptable and desirable to consumers (Viseras et al., 2007; Soleymani et al., 2016).

> *Opaque effect clay minerals* – Palygorskite, sepiolite, kaolinite, smectite and talc all have opaque effects and high sorption capacity, which is why they are widely used in such cosmetic formulations such as creams, powders and emulsions. These clay minerals ensure the opacity of the formulations, help remove skin oiliness together with toxins, and also help cover up skin maculae and patches (Viseras et al., 2007).

> *Illite* – This clay mineral has high reflectance and iridescence, which makes it the most recommended for make-up products such as eye shadows and lipsticks. Recently, micas of the muscovite type have been used in the production of moisturizing creams in order to create a brightening effect on the skin (Carretero and Pozo, 2010).

Ingredients from Land Animals and Insects

Proteins and Related Ingredients

Proteins have been used for cosmetic purposes since ancient times. One such protein, collagen, is found in all multicellular organisms and is the main protein found in skin. Collagen is also present in connective tissues, tendons and bones. With skin aging, the molecular structure of collagen is modified and causes skin to look wrinkled and dry and have reduced elasticity. Several types of collagen are marketed that are extracted from animal waste produced after preparing animals for dietary consumption. Soluble collagen, also known as native collagen, is generally the type used in skin and hair care products. This collagen can be hydrolyzed with free amino acids, resulting in no residue on skin and hair. Collagen is known for its moisturizing properties and can be used in creams to increase the skin penetration of formulation ingredients. However, there is some controversy over the effectiveness of collagen for use in hair products. Collagen can also be hydrolyzed and denatured to obtain gelatin. Elastin is a protein that is often found in the body together with collagen and, as the name implies, elastin contributes to the skin's elasticity. This protein is used in creams, lotions, shampoos and conditioners, face masks, and other cosmetic formulations. Non-animal–derived options are available for collagen and elastin, including plant and synthetic sources (Corbeil et al., 2000).

Keratin forms a set of protein filaments of significance in the human body. In its soft form, it is one of the primary proteins of the skin's upper layer, the epidermis. Keratin is available in a hard form that is found in human hair and nails. Keratin is also found in animals, including in wool, claws, beaks and horns. Hydrolyzed keratin can be used for a long-term effect since it stays on the hair and can be used for conditioning, voluminizing effects and replacing cysteine lost during hair-damaging procedures. While there has been a rising trend in using hydrolyzed keratin in hair products, as with collagen, the effectiveness of keratin in hair products is questionable. Keratin can also be used in skin products, as it can help with skin strength and elasticity (Mokrejs et al., 2017). Further, studies on wool-derived keratin have shown that this type of keratin can stimulate skin cell migration and collagen expression (Barba et al., 2008). As with collagen and elastin, there are non-animal–based alternatives available for keratin (Corbeil et al., 2000).

Hyaluronic acid is another ingredient that is found in the skin, as a major component in the extracellular matrix. Hyaluronan is the term used to encompass hyaluronic acid and sodium hyaluronate. With aging, there is a loss of hyaluronic acid and a corresponding decrease in skin elasticity. An animal source of hyaluronic acid used in the cosmetic industry is the cockscomb. Topical application of high molecular weight hyaluronic acid solutions will result in formation of hydrating films that do not penetrate through the stratum corneum of the skin (Barel et al., 2014).

Animal Milk

Animal milk has been used for thousands of years as an ingredient of various cosmetic recipes and is still a source of proteins that are used in modern cosmetics. For example, Cleopatra bathed in milk to improve her skin appearance (Rajanala and Vashi, 2017). One such protein derived from milk is casein, which is rich in amino acids (Corbeil et al., 2000). Lactic acid is an α-hydroxy acid that is also derived from milk and is very common to find in cosmetic formulations. Lactic acid creates acidity and can be used in combination with its sodium salt, sodium lactate, to maintain the pH of cosmetic properties. Ethyl lactate may have an application in acne products by lowering the pH of sebaceous follicle ducts. Lactic acid can also be used as a humectant, which may improve skin smoothness and reduce photoaging, and can serve as an exfoliating agent in chemical peels (Smith, 1996).

Chicken Eggs

Chicken eggs also provide a source of protein and other types of ingredients in cosmetics. Albumin found in egg whites is sometimes used in face masks and provides astringent properties. Egg yolk powder can be used for hair conditioning and for face masks. Egg oil is similar to human lipids and is used as an emollient. While lecithin is found in animals and plants, it is often extracted from egg yolks and can also be used as an excellent emollient. Cholesterol is found in animal fats and eggs, and can be used as an emulsifier. Since cholesterol is found in the skin and hair, it is believed to have skin regenerating effects, as well as hair nourishing and growing properties. Cholesterol is also believed to control dandruff and stimulate keratin gland production. Hydrolyzed egg powder may be used to enhance the texture of various cosmetic formulations (Corbeil et al., 2000).

Silk

Silk is an ingredient that was discovered in China and has been used in various industries for thousands of years. Silk is a filament secreted by silkworms of *Bombyx mori* silk moths. Silk is comprised of about 25–30% sericin, a sticky material that surrounds the fibroin protein that is found in the center of silk. Traditionally, sericin had been degummed from silk, and its disposal caused environmental concerns, particularly water pollution. Sericin is primarily composed of the amino acid serine, and the protein has antibacterial, antioxidant, irritation-reducing and moisturizing properties (Barel et al., 2014). Ground silk proteins can be used in facial powders and foundations, and can also improve hair texture and shine (Corbeil et al., 2000).

Waxes, Fats and Miscellaneous Actives

Lanolin is produced by the sebaceous glands of sheep and can be extracted from sheep wool grease. Lanolin has similarities with human sebum and is primarily composed of a mixture of fatty acids and of esters from high molecular weight alcohols. This mixture provides sheep with protection from the outside environment. Lanolin is one of the most important animal-based ingredients in the cosmetic industry and has been used for cosmetic purposes for thousands of years. For example, at around 400 B.C., the ancient Greeks used lanolin as an antiwrinkle pomade (Corbeil et al., 2000). Lanolin can be used as a primary emulsifier for emulsions with an oil continuous phase, and as a secondary emulsifier or stabilizer for emulsions with a water continuous phase. Lanolin's branched fatty acids allow for ease of spreadability and provide a moisturizing, emollient effect when applied to the skin. Further, lanolin has low comedogenic effect on the skin. Modern, controlled esterification procedures have resulted in lanolin being a substance that is easy to handle that is also odourless and colourless. Various derivatives of lanolin are used in industry, including water-soluble ethoxylated forms, acetylated lanolin, isopropyl lanolate and lanolin cholesterol (Barel et al., 2014).

Tallow and animal fats are also used in the cosmetic industry. The fatty acids contained in tallow, in decreasing order, are oleic acid, palmitic acid, stearic acid, linoleic acid and myristic acid. These ingredients are commonly found in cosmetics including in creams. Extraction procedures can be used to isolate these fatty acids, including saponification. Hydrogenation can be used to convert oleic acid into stearic acid. Oleic acid is often used as an emulsifier and emollient, and can be used to control the viscosity and consistency of formulations. Oleic acid is a precursor to several non-ionic, stable emulsifiers that include oleate, dioleate, trioleate and sesquioleate sorbitan. As with many other animal-derived products mentioned in this chapter, the fatty acids found in tallow can also be found in the fats of plants (Corbeil et al., 2000).

Glycerin, also known as glycerol, is a viscous, sweet-tasting liquid that is the by-product of saponified animal fats and oils. It is believed that glycerin was discovered in 1779 by the Swedish chemist Carl Wilhelm Scheele (Corbeil et al., 2000). Glycerin derivatives are used in many industries, including in plastics or as nitroglycerin in explosives. Moreover, glycerin is now one of the primary ingredients in countless cosmetic products as a solvent and humectant. Glycerin is also considered an effective moisturizer due to its hygroscopic properties. However, large amounts of glycerin should be avoided in skin products since glycerin may be irritating. Nowadays, glycerin is often obtained as the by-product of biodiesel production and is not necessarily from animal sources (Fan et al., 2010). Researchers are also exploring conversion of glycerin to other chemicals that can be used in cosmetic products, including citric acid, lactic acid, 1,3-dihydroxyacetone (DHA), and 1,3-propanediol (Tan et al., 2013).

Bees digest honey to produce wax comprised of over 80 compounds that is used for constructing honeycombs. Beeswax is considered the most important natural wax used in cosmetics, and can be used to regulate the consistency of creams and ointments, as well as in stick formulations (Williams and Schmitt, 1992). Yellow beeswax is the crude material, while white beeswax is a bleached form and beeswax absolute is an alcohol extract. With the addition of sodium borate, also known as borax, yellow and white beeswax become stable emulsifiers and thickeners that are very useful in many cosmetic formulations. However, borax is banned in the European Union for use cosmetic products. Polyethylene glycol-20 sorbitan beeswax is a derivative product used as a surfactant in cosmetics at concentrations of up to 11%. Beeswax absolute can be used in fragrances (Leung et al., 2010). Myricyclic alcohol can be extracted from beeswax and used in pomades for its wound-healing properties. Oftentimes, synthetic beeswax substitutes are used nowadays in cosmetic formulations (Corbeil et al., 2000). Bees also produce other products, the most famous being honey, a sugary substance that is based on flower nectar. Honey components are able to penetrate the epidermis, stimulate skin cell activity, improve skin tone and heal wounds. It should be noted that honey contains some volatile compounds that are destroyed by heat. Honey is becoming more popular as an active ingredient in various cosmetic formulations, including in skin creams, facial masks and hair conditioners. In addition to honey, bees produce royal jelly that is fed to larvae and queen bees (Corbeil et al., 2000). Royal jelly contains vitamins, amino acids and hormones that are important for maintaining cell function, promoting hair growth, ensuring nail elasticity, and reducing hyperpigmentation and age spots (Ramadan and Al-Ghamdi, 2012). Notably, royal jelly contains vitamin B12 and can thus be used for oil and sensitive skin types (Corbeil et al., 2000). Propolis is a sticky substance composed of a mixture of tree resin and bee gland secretions that bees use for maintaining their hive and for mummifying any intruders of the hive. Propolis has antimicrobial, anti-inflammatory and antioxidant properties. Propolis can be found in cosmetic products, particularly in several European countries (Corbeil et al., 2000). Spermaceti is a wax extracted from the head of sperm whales. However, only synthetic forms of this substance are currently available due to restrictions on commercial whaling (Corbeil et al., 2000).

Perfumes

For centuries, musk has been used in perfumes and other fragrance products. Originally, musk specifically referred to a secretion from sexual glands in male musk deer (*Moschus moschiferus*). While female musk deer also have the gland, males at reproductive age produce the odorous cyclic ketone compound muscone used in industry. More commonly, the term 'musk' refers to any compound that has a similar smell as to that emitted by male musk deer. Musk compounds are amongst the most important substances used in fragrances due to their odour and their fixative properties that cause slowed release of volatile fragrance compounds. Nowadays, it is less common to find musk compounds extracted from animals due to animal concerns, limited availability of the natural ingredient, and expense (Sommer, 2004). Some plants also produce musky aromas, either through their flowers, roots or seeds. These include the muskflower (*Mimulus moschatus*) native to North America, the muskwood or silver-leaved musktree (*Eurybia argophylla* or *Guarea Swartzei*) that grows in Jamaica and Australia, ambrette seeds (*Hibiscus abelmoschus* or *Abelmoschus moschatus*) that are found in India, and musk thistle (*Carduus nutans*) that grows in Europe and Asia (Panda, 2003). However, while many musk compounds have natural origins, most are synthesized as nitro musks, polycyclic musks or macrocyclic musks. The earliest synthetic musk compound was a nitro musk synthesized by chance in 1888 by the German chemist Albert Baur when he was working on improving TNT explosives (Sommer, 2004). However, there has been some debate on the toxicity of synthetic musks from human exposure (Taylor et al., 2014).

As mentioned, musk originally referred to secretions, particularly during mating season, from the male musk deer found in East Asia. It has been used since ancient times and was brought to Europe by the crusaders (Sommer, 2004). There are various forms of musk that are available on the market that are named based on the source of the deer, including Tonkin musk (also known as Tibetan or Oriental musk) that is considered the best variety of musk and Kabardin musk (also known as Russian or Siberian musk) that is a cheaper variety. While not required to obtain the compound, generally, musk deer are killed and the glands are removed for use (Miller and Miller, 1990). Unfortunately, musk deer are currently an endangered species, and efforts are being taken to avoid using natural musk from these deer. Trade of musk from many Asian countries, including Afghanistan, India, Nepal and Pakistan, has been banned by the Convention on International Trade in Endangered Species of Wild Fauna and Flora (CITES). Further, the European Union has banned musk trade from Russia (Species+, 2018).

Ambergris, also known as ambra, is a metabolic product from the digestive system of sperm whales that is used for its fragrance and aphrodisiac properties. It is not to be confused with tree resins known as amber that are common in the Baltic region of Europe. Due to threatened whale populations, the International Whaling Commission banned commercial whaling since 1986, though Norway and Iceland continue to have commercial whaling operations (International Whaling Commission, 2018). Possession and trade of ambergris is forbidden in the United States by the Endangered Species Act of 1973 (United States House Committee on Merchant Marine and Fisheries, 2003). Industrial synthesis methods have been developed in order to use ambergris fragrance from non-animal sources (Panten et al., 2014).

Other animals have been used for deriving musky aromas. Castoreum is secreted from glands located near the anus of beavers. The beavers must be killed to isolate the compound. An odour similar to castoreum comes from hyraceum, obtained from the secretions of hyraxes, which are African mammals. Muskrat (*Ondatra zibethicus*) is a member of the beaver family that has glands that secrete a musky odour, with male muskrats producing more of the substance. Civet is an odour extracted from the civet cat (*Viverra civetta*) found primarily in North Africa and Asia. Male civet cats also secrete more of the odorous compound, which is of a better quality than that of female civet cats. The substance may be used in some Middle Eastern countries to tint and add shine to hair and eyebrows. Castoreum and civet are sometimes counterfeited (Corbeil et al., 2000).

Colourants

Use of animal-based colourants is limited due to expensive extraction and colour variations. Insects provide another source for cosmetic pigments. Cochineal, a red dye, is from the dried, pulverized bodies of cochineal beetle (*Coccus cacti*) native to Mexico and Central America. This dye was once used by the Aztecs as body paint, as a fabric dye and as medicine. It takes about 150,000 beetles to create one kilogram of dye. Thus, the pigment is relatively expensive. Carminic acid is a purified version of cochineal that can be used to create various shades of red, orange and yellow depending on pH and the addition of salts to make carmine dye. For example, the aluminium lake version is blue-red and has a decrease in bluish tint with a corresponding decrease in pH (Dweck, 2002).

Styling Aid

Shellac is one of the oldest styling aids and resin-like secretion produced by the *Kerria lacca* insect that lives in India. Shellac is a hydrophobic material that is used in hair sprays and some foundations (Corbeil et al., 2000).

Marine-Based Materials

The cosmetic industry is always searching for novel bioactives and other types of ingredients for formulations. There has been a growing trend of investigating marine-derived cosmetic ingredients since the oceans represent a reservoir of materials, many of which are yet to be explored. So far, more than 250,000 marine species are known, and thousands of bioactives have been identified (Mora et al., 2011).

Antiaging Actives

As mentioned in the animal ingredients section of this chapter, collagen has skin regenerative properties and is present in all multicellular organisms. This includes fish and jellyfish (Zhuang et al., 2009). Further, collagen-stimulating actives can be used in antiaging applications. One example is the extract of the *Chlorella vulgaris* microalga (Wang et al., 2015). Keratin is another protein that can be extracted from marine sources, specifically from fish scales (Corbeil et al., 2000). Matrix metalloproteinase (MMP) enzymes play a role in skin aging by degrading collagen and forming wrinkles. Thus, inhibitors of matrix metalloproteinases are being investigated as antiaging actives for cosmetic formulations (Zhang and Kim, 2009). Sources of inhibitors include peptides and proteins found in seahorses (Ryu et al., 2010) and Atlantic cod (Lodemel et al., 2004), as well as phlorotannins, which are polyphenol compounds mainly sourced from brown algae and not found terrestrially. Examples of phlorotannins being investigated are eckol and dieckol from *Ecklonia stolonifera* that inhibit MMP-1 (Min-Jeong et al., 2006), and 6,6′-bieckol from *Ecklonia cava* that inhibits MMP-2 and MMP-9 (Zhang et al., 2010).

The omega-3 fatty acids eicosapentaenoic acid and docosahexaenoic acid found in fish have antiaging, photoprotective and wound healing properties (McDaniel et al., 2008). It has been found that eicosapentaenoic acid inhibits UV-induced MMP-1 expression (Kim et al., 2005), and topical application enhances collagen and elastin expression in skin (Kim et al., 2006).

Many marine organisms produce carotenoid pigments that have antiaging, photoprotective and antioxidant effects. One example is β-carotene, a vitamin A precursor that is able to prevent reactive oxygen species formation. The green alga *Dunaliella salina* has been used to commercially produce β-carotene for several decades (Borowitzka and Borowitzka, 1989). Astaxanthin is a carotenoid that can be found in various marine algae, bacteria and yeasts. This includes the bacteria genera *Paracoccus* and *Agrobacterium*, and the yeast genera *Rhodotorula*, *Phaffia* and *Xanthophyllomyces* (Corinaldesi et al., 2017). Industrial-scale production of astaxanthin is available using the green alga *Haematococcus pluvialis* as a source (Wan et al., 2014). Although more of astaxanthin can be extracted from algae, yeasts can be grown more quickly and easily, and can be genetically modified to increase carotenoid production (Mata-Gómez et al., 2014). Fucoxanthin is another type of carotenoid that is mainly produced by seaweed (Peng et al., 2011).

Exopolysaccharides from bacteria living in extreme environments, including Antarctic waters and deep-sea hydrothermal vents, are another source of antiaging actives. In addition to being emulsifiers and thickeners, exopolysaccharides can be used for reduction of wrinkles. *Alteromonas macleodii* produces the polysaccharide HYD657 that is already available on the market in cosmetic products. The polysaccharide HE 800 found in *Vibrio diabolicus* is structurally similar to hyaluronic acid found in human skin and is believed to stimulate collagen (Corinaldesi et al., 2017).

Seawater has a variety of minerals that are able to provide hydration and benefit to the skin. Sea salts can also be used in skin care such as for skin lightening and cleansing effects (Kim, 2012). Likewise, sea muds are composed of various nutrients and minerals that are considered beneficial to the skin, including possibly providing antiaging effects and helping alleviate skin disorders, including acne, eczema and psoriasis (Matz et al., 2003). Of note is the mud and minerals from the Dead Sea that are found in many cosmetic formulations, and have various benefits to the skin including improving smoothness and texture (Ma'or et al., 1997). Although heavy metals have been detected in Dead Sea mud, these are of little concern since these compounds are below levels of toxicity (Ma'or et al., 2015).

Polysaccharide Thickening Agents and Film Formers

Chitin is one of the most abundant polysaccharides and is found in the exoskeleton of crustaceans. Chitosan is obtained after the alkaline deacetylation of chitin. In cosmetic applications, chitosan can be used in moisturizers, cleansers, and hair and skin fixatives, and is known to have antimicrobial activity (Li et al., 1992). Chitosan is considered the only natural cationic gum that increases in viscosity in acidic conditions. Chitooligosaccharide derivatives may have antioxidant properties (Kim, 2012). Further, chitosan is also being explored as a biomaterial for nanotechnology applications (Beck et al., 2014).

Polysaccharide-based gums obtained from marine sources are among the most commonly used thickening agents used in the cosmetic industry and can also be used as emollients. Algal cell walls contain many useful polysaccharides, including alginate and fucoidans from brown seaweed, and carrageenan and agar from red seaweed. Carrageenan usually comes from the *Chondrus crispus* ('Irish moss') species and *Gigartina* genus of red seaweed. There are three chemical forms of carrageenan: (1) the kappa type forms rigid gels, (2) the iota type forms soft gels and (3) the lambda type is a thickener that does not form gels. Carrageenan is composed of D-galactopyranose units, while agar is composed of both D-galactopyranose and L-galactopyranose. Alginates are the primary matrix polysaccharides of brown seaweed and are composed of D-mannuronic acid and L-guluronoic acid linkages. Fucoidans are sulfated polysaccharides that are commercially available as extracts from several brown seaweed species, including *Fucus vesiculosus* and *Undaria pinnatifida* (Kim, 2012).

Moisturizing Agents

Squalene is a fatty substance and natural emollient and antioxidant that is found in high concentrations in shark liver, though it can also be extracted from some vegetable oils. Squalane, the saturated form of squalene, chemically resembles sebum that is naturally found in the skin, and thus squalane can be well absorbed by the skin to provide nutrients and hydration effects. Squalane, especially with other ingredients such as vitamin E and hyaluronic acid, can be used for protecting against photoaging and brown spots, and for reducing fine lines and wrinkles (Barel et al., 2014). Linoleic acid and γ-linolenic acid are omega-6 polyunsaturated fatty acids that restore transepidermal water loss (TEWL) (Magdassi and Touitou, 1999). Sources of these fatty acids include seaweed of the *Laminaria* genus and microalgae of the *Nannochloropsis* genus (Bellou and Aggelis, 2013). Some halophile bacteria that reside in high salinity conditions, such as *Ectothiorhodospira halochloris*, produce a compound known as ectoine in response to osmotic stress. Ectoine is able to bind to water molecules, and topical administration is well-tolerated and can provide long-term hydration (Galinski et al., 1985). Collagen from fish and jellyfish can also be used as an emollient (Corbeil et al., 2000).

Antimicrobial Preservatives, Antioxidants and Photoprotectants

Seaweed, microalgae and halophyte plant extracts are being explored as antimicrobial preservatives. Examples of seaweed being investigated include *Himanthalia elongata* and the *Synechocystis* species (Plaza et al., 2010), and examples of microalgae are *Isochrysis galbana*, *Chlorella marina*, *Nannochloropsis oculata* and *Dunaliella salina* (Srinivasakumar and Rajashekhar, 2009). Halophyte extracts being explored include the falcarindiol compound of *Crithmum maritimum* leaves (Meot-Duros et al., 2010), as well as extracts from *Pistacia lentiscus* leaves and fruits (Lopes et al., 2016). Cationic peptides can also be used as antimicrobial agents since they are able to disrupt microbial membranes, and such peptides can be found in half-fin anchovy (*Setipinna taty*) (Song et al., 2012) and Atlantic mackerel (*Scomber scombrus*) (Ennaas et al., 2015). As mentioned earlier in this chapter, chitosan is also known to have antimicrobial properties (Corbeil et al., 2000).

Natural antioxidant compounds that are derived from seaweeds and are being investigated as safe alternatives to synthetic antioxidants include phlorotannins, polysaccharides, fucosterol and carotenoids. Fucosterol is a compound found in brown seaweed that, in addition to being an antioxidant, is believed to be able to reduce cholesterol and have antidiabetic activity. Carotenoid pigments from other types of marine organisms can also serve as antioxidants (Kim, 2012). Additionally, antioxidant phenolic compounds are produced by marine halophyte plants, with high levels noted in *Limonium wrightii* (Aniya et al., 2002) and *Salicornia ramosissima* (Surget et al., 2015), to name a couple species. Moreover, collagen from fish and jellyfish has antioxidant and photoprotectant activity (Zhuang et al., 2009).

Several marine organisms, primarily those that undergo photosynthesis, produce natural sunscreen compounds that provide protection by absorbing UV rays. Scytonemin is a pigment found in the extracellular sheath of some cyanobacteria (Rastogi et al., 2015). Another UV-absorbing compound is mycosporine-like amino acids (Derikvand et al., 2017).

Pigments

As previously discussed, carotenoids are the most common pigments found in nature. Many marine organisms, especially microalgae, produce carotenoid pigments that also have antioxidant properties, as well as potential anti-inflammatory and photoprotective effects (Kim, 2012).

Phycobiliproteins found in cyanobacteria and algae have fluorescent properties of varying colour shades that have drawn the interest of the cosmetic industry. Phycoerythrin is a red pigment that is primarily being extracted from the red microalga *Porphyridium cruentum*, whereas phycocyanin is a blue pigment that has antioxidant and antiradical properties and is mainly sourced from the cyanobacterium *Spirulina platensis* (de Jesus Raposo et al., 2013).

Another natural source of blue pigments is the diatom microalga *Haslea ostrearia*, which produces a compound referred to as marennine. Interestingly, because this organism is present in some oyster refining tanks, marennine may become trapped in oyster gills and provide a greenish colouring to the oysters housed in the tanks. Other diatoms belonging to the *Haslea* genus have recently been found to produce marennine-like compounds (Gastineau et al., 2014).

Microbial-Based Ingredients

The skin is home to a diverse set of microorganisms that form a microbiome dependent on endogenous factors from the human host and exogenous factors from the environment. As a result, promicrobial and antimicrobial methodologies are being explored for use in therapeutic and personal care products.

Probiotics

Probiotics are live microorganisms being explored for their health benefits. The scientist Elie Metchnikoff is considered the father of probiotics. While probiotics have primarily been used for several years in products targeting the gastrointestinal system, recently there has been a growing interest in probiotics for skin care products. Studies have indicated that probiotics may have health benefits that include producing antimicrobial compounds, stimulating the immune system, improving skin barrier function, restoring acidic skin pH and providing antiaging effects. It is important to note that some health benefit claims are based on preliminary evidence, so further studies on probiotics are required (Sharma et al., 2016). In the personal care industry, probiotics are being investigated as actives in products that can either be ingested or that can be applied topically. *Lactobacillus* strains are among the most commonly used probiotics, though newer strains are also being explored. Probiotics present formulation and manufacturing challenges that include difficulty in maintaining long-term viability and in achieving safe and effective concentrations. Further, some regulatory agencies including the U.S. Food and Drug Administration (FDA) have a safety limit for the amount of microorganisms that can be found in cosmetic products. Specifically, the FDA allows for 500 colony-forming units per gram in eye products, and 1000 colony-forming units per gram for cosmetics applied to other areas of the body (Huang and Tang, 2015).

Gums

Several gums are derived from microbial fermentation processes. Xanthan gum is derived from the fermentation of sugars by *Xanthomonas campestris* bacterium. The viscosity is maintained in varying temperatures and pH ranges. Xanthan gum can provide foam to many shower and bathing products, including shampoos and soaps. Gellan gum comes from the fermentation of the *Sphingomonas elodea* bacterium and can be used as a gelling agent even in very low concentrations. Sclerotium gum is derived from the fermentation of the *Scelrotium rolfsii* bacterium (Dweck, 2011).

Stem Cells

Stem cells are undifferentiated biological cells that have the remarkable potential to differentiate into many different types of cells and to divide to form more stem cells using mitosis. Due to this capability, these cells can be called the foundation of development in plants, animals and humans. In the recent years, stem cells have been amongst the top trending topics in the field of life sciences. This interest in these cells is because of the fact that they possess the potential of doing incredible jobs like human tissue replacement, cancer therapy and drug delivery, and extensive research is being conducted on these applications. Human skin is the organ that displays regeneration capabilities in the most visible manner that becomes lesser with age. This implies that the skin must also contain cells that are responsible for such regeneration (Lavker and Sun, 2000). Depending on the origin of the stem cells, they could be called plant or animal stem cells.

Stem cell research can revolutionize medicine, more than anything since antibiotics.

– Ronald Reagan

Plant Stem Cells

Plant stem cells represent the ultimate origin of most of the food, air and fuels consumed by humans and therefore, may be the most important cells for our well-being (Greb and Lohmann, 2016). Plant stem cell systems are differentiated into primary and secondary meristems, primary being the ones established during embryogenesis and secondary being the ones established post-embryonically. The stem cell systems are also divided based on whether they are involved in longitudinal or lateral growth into root apical meristem, shoot apical meristem and cambium (Figure 14.1). Plant cell culture technology consists of several techniques to grow plant cells, tissues or organs in microbe- and pollution-free environments all year round with consistency (Morus et al., 2014). The basis of this technology involves inducing slight mechanical damage on the plant causing the callus to appear. The callus is a group of undifferentiated stem cells. According to Dal Toso and Melandri (2011) these cells can be cultivated *in vitro* to allow derivation of plant products even from endangered or rare species. It is also mentioned that plan cell culture technology is advantageous over conventional methods in terms of environmental sustainability while concluding that the technique possesses the potential to likely help provide unlimited sources of natural products.

Animal Stem Cells

The two main types of stem cells from animals and humans, that have been worked on in the recent times are embryonic and non-embryonic stem cells. Embryonic stem cells are those that are derived from a 5-day preimplantation embryo. These are pluripotent cells that can grow into cells or tissues of any of the three germ layers – ectoderm, mesoderm and endoderm. These can be cultured *in vitro* to allow proliferation without differentiation for a long time. Non-embryonic stem cells, also known as somatic or adult stem cells are found in organs and differentiated tissues, and have shown to possess limited ability to differentiate and self-renew. Another type of stem cell is induced pluripotent stem cells (iPSCs) that are somatic cells reprogrammed to assume an embryonic stem cell–like state.

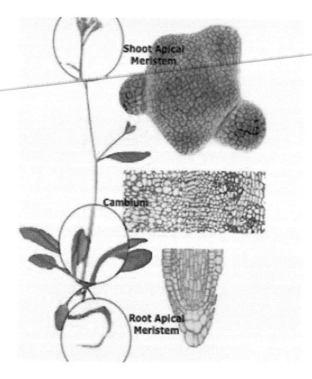

FIGURE 14.1 Representation of the three different types of plant stem cell systems: shoot apical meristem, cambium and root apical meristem. (From Greb, Thomas, and Jan U. Lohmann, 2016, *Current Biology* 26 (17):R816–21.)

Plant vs. Animal Stem Cells

Even though they are both undifferentiated cells with similar capabilities, there are certain differences in plant and animal stem cells. In both plants and animals, stems cells are maintained in specialized micro-environments called stem cell niches and are maintained by signals from specialized cells (Heidstra and Sabatini, 2014). The signals that control stem cells are more understood in animals than in plants so far. However, the astonishingly long life and robustness of plants have made plant stem cells a topic of inter-est in terms of their ability to act on animals. Use of animal- or human-based stem cells is also somewhat objectionable and not currently well received in our society (Bongso and Richards, 2004). The use of stem cells in the area of cosmetic research will be the focus of this chapter.

Applications of Stem Cells in Cosmetics

It is becoming a question asked more and more if the incredible regeneration potential, robustness and longevity of plant stem cells can be used to improve similar characteristics in humans. Gotlieb Haberlandt suggested in 1902 that an individual plant cell is capable of regenerating an entire plant (Byrne et al., 2003) and this was demonstrated in 1958 when a carrot was cloned by *in vitro* cultivation of carrot cells (Fehér, 2015). The peculiar characteristic by which plant cells are able to switch back and forth between differentiated somatic cells and undifferentiated stem cells is of specific interest in the cos-metic industry (Byrne et al., 2003). Apple tree is one of the many plant types that are being researched in order to explore the use of their stem cells in cosmetics including others such as grape, jasmine, common horehound and blue cornflower (Morus et al., 2014; Schmid et al., 2008) showed in a study that Uttwiler Spätlauber stem cell extract showed positive effects on longevity of skin and hair. They showed increase in viability of umbilical cord blood stem cells and lesser decrease of viable cells in the presence of UV light as a result of presence of Uttwiler Spätlauber stem cell extract.

Table 14.2 presents a list of some of the many products that claim to contain plant-based stem cells and are currently being sold in the market. Even though there are several products in the market those claims to use plant stem cells, the credibility of their claims is questionable. The Food and Drug Administration has issued warnings to several cosmetic companies from time to time stating violations including that these products in many cases contain 'unapproved drugs'. Other than regulation, another aspect that needs to be considered in cosmetic products containing stem cells is their nomenclature. It has been seen that most of the products contain extracts of stem cells and not actually live stem cells which cannot have the same cosmetic effects (Trehan et al., 2017). A study has also shown the effect of a novel antiaging laser therapy to be synergistic in the presence of a plant-based stem cell extract (Beri and Milgraum, 2015). Although there is further research needed to understand the mechanisms behind this, the authors were able to conclude that the combining fractionated Q-switched 1,064 nm with a topical plant stem cell extract and *N*-acetyl glucosamine showed excellent *in vivo* antiaging effects and that this opens up doorways for application to wound repair and scar healing. Another research group has established cloudberry cell culture and has developed an *in vitro* system for industrial scale production of cloudberry cells in bioreactors, leading to industrial scale application in commercial cosmetic preparations (Nohynek et al., 2014). Tito et al. (2011) developed a new cosmetic active ingredient from tomato-cultured stem cells, which showed high amounts of antioxidants and protection to the skin against heavy metals. In yet another research, extract of *Coffea bengalensis*, which contains minimal caffeine, was produced from cell cultures for beneficial cosmetic applications on the skin (Bimonte et al., 2011). Redefined ginger is another company with claims that they are using active plant cells from *Zingiber officinale*, which are extracted using biotechnology and cell culture. In their clinical study, they show that their product containing 0.5% of these active plant cells are able to reduce wrinkles, close the pores of the skin, reduce sebum production and also decrease the shininess and orange peel appearance of the skin; problems that all arise with skin aging (Naolys, 2016). Another company called Active Skin care Solution has a range of products under the brand Xtemcell which contains a patented technology that they mention allow organic stem cells to regenerate naturally via a process known as mimeticism. Using its technology, the company claimed it was able to deliver high concentrations of lipids, proteins, amino acids and phytoalexins to the outer layers of the skin resulting in increased cell renewal and reduced damage by other factors (Backman et al., 2014).

There is much research needed in this area, along with regulatory guidelines to be laid in order to have actual application of the technology seen in products to fully utilize the potential and to see considerable effects of using such products.

Not only do stem cells derived from animals also demonstrate potential in treating several different kinds of diseases, they are also being investigated for cosmetic applications. Soft tissue augmentation,

TABLE 14.2

A Non-Extensive List of Cosmetic Products Being Sold Which Claim to Contain Plant Stem Cells or Plant Stem Cell Extracts

Product Name	Product Description	Company
Stem Cellular ™ Anti-Wrinkle Moisturizer	Anti-aging moisturizer with plant-based stem cells for all skin types	Juice Beauty
Stem Cellular ™ Anti-Wrinkle Booster Serum	A blend of fruit stem cells boost radiance and enhance your skin's tone and texture	Juice Beauty
Stem Cellular ™ Anti-Wrinkle Eye Treatment	Organic, essential fatty acids from evening primrose, borage and linseed combat dark circles	Juice Beauty
Plant Stem Cell Science Nourishing Cleansing Cream	Softening cleanser for dry and sensitive skin	Intelligent Nutrients
Lift Serum Intense with Stem Cell Recovery Complex	A serum with powerful lifting and firming effects	Lancer
Plantscription Youth-Renewing Power Night Cream	Cellular repair with plant stem cell technology to restore skin's regenerative power to bring it back to its youth	Origins

Note: These are only references made to claims by the companies selling these products.

which is widely used in plastic and reconstructive surgery, has been shown to utilize human mesenchymal stem cells, and those constructs have shown to retain predefined shape and dimensions (Alhadlaq et al., 2005). There is no doubt that stem cell technology has a great range of possible applications from animals as well as from plants, and with continuous research being conducted on this technology, there will be a tremendous range of beneficial products that would be used in the field of cosmetics. However, currently because of the objectionable nature of human and animal-based stem cells and only partial use of plant-based stem cells, there needs to be a lot more work done on this research. Hopefully, in the near future, we will be able to see much more widely accepted, safe and ethical applications of this relatively novel technology for the benefit of humans.

Quality and Regulatory Aspects

Many natural products are used in cosmetics, and several others have a great potential for the development of new cosmetic and pharmaceutical products. These natural products have chemical constituents like amino acids, phenolics, flavonoids, tannins and vitamins. Thus, use of natural products in personal care cosmetics may impact the skin biological functions (Fonseca-Santos et al., 2015) in a beneficial way to treat skin disorders (De Wet et al., 2013) or negative impact leading to adverse effects. Notwithstanding the admired faith in the mildness of natural ingredients, numerous events of adverse reactions to natural or plant extracts have been reported, specifically topical side effects such as contact urticarial, allergic and irritant contact dermatitis, and phototoxic and photoallergic reactions (Ernst, 2000; Corazza et al., 2014).

Cosmetic ingredients with the alike constituent name may vary in quality depending on its source and purification process used by manufacturer. Even trace amounts of impurities may attribute to key dissimilarities in safety and quality. The type of impurities in an ingredient may depreciate product quality and hence causing cutaneous allergy and/or irritation. Elevated levels of impurities may also cause discolouration of product or worsening of smell on extended exposure to sunlight or high temperatures.

The FDA has the legal authority to inspect cosmetic establishments as well as cosmetics offered for import. Section 704 of the Federal Food, Drug, and Cosmetic Act (FD&C Act, April 2014) (FDA, 2014) authorizes the FDA to conduct inspections of cosmetic firms at reasonable times, in a reasonable manner, and without prior notice in order to ensure non-violation of laws and regulations, to regulate proper labelling and to ensure product safety. The FDA may warrant an inspection of a cosmetic establishment based on FDA surveillance and compliance initiatives, significance of consumer or trade complaints received, the type of products, the company's compliance history and various other factors. The FDA can issue warning letters to companies who have violated the laws and request necessary corrective actions to be taken (FDA, 2017).

Under U.S. law, cosmetic products and ingredients do not need FDA approval (except coal tar hair dyes) to enter the market. The FDA only regulates, but not approves, cosmetic products under the Federal Food, Drug and Cosmetic Act (FD&C, Title 21, Chapter VI). Nevertheless, U.S. law poses legal responsibility to the companies and individuals for the safety and proper labelling of their products. The FDA can take legal action against a cosmetic product on the market, based on public health priorities and available resources showing the product as 'adulterated' or 'misbranded'.

Under the FD&C Act, Section 601 [21 U.S.C. 361], a cosmetic shall be considered adulterated if the product or its container is composed, entirely or in a portion, of any poisonous, decomposed or deleterious substance or prepared and or held in unhygienic conditions, which can render the contents harmful to the consumer health. A cosmetic or personal care product is deemed misbranded according to the FD&C Act, Section 602 [21 U.S.C. 362], if the product label bears misleading, wrong, unclear or incomplete information. Even improper or misleading packaging design, device or any combination is also considered misbranded.

FDA (21 CFR 700.35) restricts the use of some terms such as 'sunscreen' or similar sun protection terms without relevant explanation in cosmetics that can, according to the claims, lead the product to be regulated as a drug or a drug/cosmetic. A grey area exists between drugs and cosmetics regulation and partially sheds light on the situation with stem cell–based cosmetics. Regulations in the European market are stringent for use of stem cells in cosmetics. However, the U.S. FD&C Act applies an 'intent' prerequisite for a product to be designated as cosmetic. Most industries in the United States implement an even-tempered way of qualifying their topical formulations containing stem cells, as 'cosmetic' by adjusting claims to bypass the tough drug regulations.

The FDA regulates the inclusion and concentration of some ingredients in cosmetics. Any ingredient that makes cosmetic product deleterious upon intended use is illegal. However, the FDA specifically prohibits use of some ingredients like bithinol, halogenated salicylanilides (di-, tri-, metabromsalan and tetrachlorosalicylanilide), chlorofluocarbon propellants for domestic consumption, vinyl chloride, zirconium-containing complexes and so forth that can cause cancer, skin sensitization, or serious lung or skin diseases (FDA, 2013). The FDA restricts or controls the limits of use or presence of certain substances like chloroform, methylene chloride and mercury compounds in cosmetics to a very limited area of use or in certain percentage, beyond which it may have harmful effects when used as intended. Many stem cell–containing topical creams are entering the U.S. market due to lack of regulatory approval and with no compulsory safety assessments for cosmetics. In lieu of public safety, it is important for regulatory bodies in the United States to enforce some mandatory safety evaluations for stem cell–containing cosmetics similar to what is done in Europe (Panis, 2011).

The market for natural and organic cosmetics and personal care products has rapidly grown globally during the last few years. The FDA regulates cosmetics under the authority of the FD&C Act and the Fair Packaging and Labeling Act (FPLA). The expression 'natural' is not regulated, but the term 'organic' is a regulated expression that can or should be used only if the product has been certified by an authorized certification program.

Organic Certification in the United States

The OASIS (Organic and Sustainable Industry Standards) certification program is established by the cosmetic industry in the United States, while NSF International developed NSF/ANSI 305 for personal care products containing organic ingredients in order to administer and coordinate the U.S. voluntary standardization and conformity assessment system. Another certification system, the National Organic Program (NOP) is regulated by the U.S. Department of Agriculture (USDA) to define and regulate the term 'organic' as it applies to agriculture products.

NSF/ANSI 305: Personal Care Products Containing Organic Ingredients

NSF/ANSI 305 voluntary standard is designed only for and allows the 'contains organic ingredients' designation on label for personal care products with organic content of 70% or more that comply with all other requirements of the standard. NSF standard allows for limited chemical processes that are standard for personal care products but would not be allowed for food products. NSF/ANSI 305 also requires companies to state the exact percentage of organic content based on the requirements of the standard.

USDA's National Organic Program

A personal care product may be eligible to be certified 'organic' under the NOP regulations if it is made up of agricultural ingredients, and can meet the USDA/NOP organic production, handling, processing and labelling standards. Under NOP organic labelling categories, 100% organic products must contain only organically produced ingredients, except water and salt. Products labeled as 'Made with organic ingredients' must contain at least 70% organic ingredients and product label can list up to three of the organic ingredients or 'food' groups on the principal display panel. The aforementioned products may display the USDA organic seal and must display the certifying agent's name and address. Products made with less than 70% organic ingredients may specify 'organically produced' for the particular USDA-certified constituents on the ingredients statement on the information panel but cannot use the term 'organic' on the principal display panel. Such products may not display the USDA organic seal or certifying agent's name and address (USDA, 2008).

Organic and Sustainable Industry Standards (OASIS)

OASIS was founded by more than 30 large global brands, smaller specialty brands, and private cosmetic players as a non-profit mutual benefit organization to generate a true industry standard. Acknowledging chemistry as an essential fragment of cosmetic production, OASIS certifies products that use safer, green chemistry principles and integrates organic feedstock.

The Organic Consumers Association (OCA) launched the Coming Clean Campaign in 2004 with a goal of eliminating products falsely labeled as 'organic' and limiting organic claims to personal care products that are certified to USDA organic standards (OCA, 2004).

Retailers are now increasingly hesitant to sell non-certified organic personal care products. For example, Whole Foods' new policy obligates that 'organic' or 'made with organic ingredients' claims must be certified under the USDA National Organic Program, similar to food. A more limited 'contains organic ingredients' claim for personal care may be certified under the NSF/ANSI 305 standard. Organic claims that are not certified, including 'organics' in branding, will no longer be accepted for sale (OCA, 2010).

Organic and Natural Cosmetic Certification in Other Countries

ECOCERT is a private organization in France that has established superior standards and respect for the environment throughout the manufacturing process of cosmetics to preserve a real augmented value for natural and organic cosmetics. In Germany, comprehensive guidelines of the BDIH qualifies cosmetic products as 'Certified Natural Cosmetics' that use natural raw materials such as plant waxes and fats, herbal extracts and essential oils, and aromatic materials specifically obtained from controlled biological cultivation or controlled biological wild collection. The Soil Association (UK) does annual inspections of manufacturing facilities and approves product and labels as per Soil Association Health and Beauty Standards. While in Canada, the IOS Cosmetics Standard, established by the private company Certech, is used for natural and organic certification. At least 95% of the constituents should be of natural origin to be certified as natural cosmetics, and products with certified organic ingredients, free of chemicals and preservatives can acquire organic certification. Packaging must be recyclable and the products and their individual ingredients must not have been tested on animals; must be virtually free of synthetic ingredients; and must not contain pesticides, harmful preservatives, artificial colours and fragrances (Certech 2008). Australian Certified Organic (ACO) is one of the most stringent and largest organic certifying arms that blends national and international relevant standard necessities for the public to have an easy choice for organic products. In Australia, the Cosmetics Raw Materials Scheme provides wide-ranging validation of natural components to be used in ACO- and COSMOS-certified products (ACO, n.d.).

Current and Future Trends and Approaches Going Forward

The natural cosmetics global market has a proficient upsurge from almost $7 billion in 2007 estimated to be $15 billion in 2017 (Statista, 2018) and to be appreciated at $22 billion by the end of 2024, listing a compound annual growth rate of 8.8% over the projection period 2016–2024 according to Persistence Market Research (PMR) estimates. Global organic and natural personal care business is predominantly trend-motivated. Hence, modern and budding trends of organic and natural products are encouraging new contenders to hit the market via acquisition by modern competitors. The evolution of the global natural and organic personal care product market is navigated by dynamics such as rising online customer access, extension of distribution channels and present retailers' swing towards offering premium personal care products. Other driving factors include new product development, marketing strategies, and consumers' preference to spend more on premium products leading to wide progression of the global natural and organic personal care products market. Talking about market share value, the skin care segment is predicted to emerge the supreme attractive segment with 30.9% share by 2024, trailed by other product segments like hair care, oral care and cosmetics (Persistence, 2016). The global skin care market is anticipated to observe substantial growth over the forecast period due to increasing health consciousness of consumers globally, preference of clean label products, demand of natural and organic personal care products, and online shopping. The hypermarket/supermarket segment is a major channel for sales of natural and organic personal care products, and was projected to register a compound annual growth rate of 9.6% between 2016 and 2024, followed closely by the specialty retailers segment. Amazon.com, Drugstore.com, Walgreens, Sephora.com and Vitacost, for example, are substantial distribution channels to empower product diffusion and easy consumer reach to enlarge the consumer circle. According to regional segmentation analysis, North America is the major global market for natural and organic personal care products, with Europe as another major market; and Japan, China and India being the major markets in Asia-Pacific. The European region is one of the leading markets, and the demand for superior quality and high-end perfumes, natural, organic and 'waterless' personal care products is increasing. This region has also a growing inclination for waterless personal care products, due to ecological alarms and water protection concerns, with consequential innovation of dry shampoos and beauty bars. Predicted optimistic future stance for global revenue is predicted to exceed $22 billion by 2024 for the organic and natural personal care products global market (Persistence, 2016).

L'Oréal SA, Burt's Bees, Estee Lauder, Coty Inc., Weleda AG, Avon Products Inc., Bare Escentuals Beauty Inc., Arbonne and Aveeno are vital companies in the organic and natural personal care market worldwide. Recently, after endorsing the rise in consumer interest in natural ingredients, Ulta, a large beauty retailer in the United States, has launched its own natural products line that includes a top-selling lip oil infused with green tea and avocado extracts (Harper, 2017). Kline & Company also made a similar prediction about the decline of the synthetic cosmetics sector and growth of the natural skin care segment over the next two years. The 'natural' personal care products market in the United States grew swifter than the personal care product market as a whole, by an estimated 7.5% in 2014. Dianne Feinstein, a Democratic senator from California, introduced the Personal Care Products Safety Act in April 2015, a bill to strengthen regulation of ingredients in personal care products. In her testimony to the Senate Health, Education, Labor and Pensions Committee, Feinstein mentioned, 'Our skin is our largest organ, and many ingredients contained in these products – whether it be lotion, shampoo, or deodorant – are quickly absorbed by the skin. … There is increasing evidence that certain ingredients in personal care products are linked to a range of health concerns, ranging from reproductive issues, such as fertility problems and miscarriage, to cancer' (Feinstein, 2015). Parabens and phthalates pose more risk of breast cancer due to endocrine effects as revealed by some recent studies too. A study at the University of California, Berkeley, demonstrated a significant drop in levels of hormone-disrupting chemicals by brief discontinuity in use of shampoos and lotions containing such chemical ingredients (Arnold, 2016). For the past few years, many big companies including Johnson & Johnson have taken action to reformulate their range of baby and adult products to eliminate and reduce trace amounts of ingredients with questionable safety. Successful measurements have been done on revamping prototypes even to completely omit certain ingredients that may generate trace quantities of carcinogenic compounds. Clean and Clear® facial cleanser is one of the most popular reformulated products of Johnson & Johnson (Kessler, 2015). Sephora offers a comprehensive category of natural products like oat milk dry shampoo, coconut cleansing oil, and kale and spinach age-prevention cream with the latest addition of the 100% plant-based skin care line Biossance that relies on the mega-moisture molecule squalane. They are consistently trying to acquire more brands to expand the range of cosmetics containing natural products. Sephora has also swayed other established brands like Estée Lauder to remove certain potentially harmful chemicals and add natural ingredients to their products. A trade group of manufacturers and retailers called the Natural Products Association has certified around 1500 ingredients and products that come from a 'renewable source found in nature'. Procter & Gamble, Johnson & Johnson, Avon, Colgate Palmolive and at least two dozen other companies have agreed to eliminate certain possible carcinogens from their products.

A big revolution driven by customers' needs and expectations towards natural, safe healthy and sustainable personal care products fosters a need of collaborative initiatives towards transparency of ingredients and for the FDA to execute more authority and regulations for cosmetics and personal care products. Personal care industries and consumer advocates are demanding better regulatory approaches for natural and personal care products to exclude harmful chemicals, ensure better safety and certification of chemicals in products, and transparency to consumers. Presently, the Personal Care Products Council, the Environmental Working Group, and several major companies and other advocacy groups support Feinstein and Senator Susan Collins's (R–ME) Personal Care Products Safety Act (Feinstein 2015). Under this bill, the FDA would provide regulations on good manufacturing practices for cosmetics and personal care products, and test the safety of at least five compounds of personal care products every year. The FDA would have authority to recall any unsafe products. This bill also enforces requirement of complete label information on ingredients, and warnings including ingredients not appropriate for children and those that should be professionally administered on product package as well as online due to trendy online purchasing. Under the bill companies would also be required to provide consumers with contact details and report serious adverse events to the FDA within 15 days. This act is the beginning of better regulation of natural products in personal care and cosmetics in the United States.

BIBLIOGRAPHY

Aburjai, T., and F. M. Natsheh. 2003. Plants used in cosmetics. *Phytotherapy Research* 17 (9):987–1000.
ACO. n.d. Natural Cosmetic Ingredients Assessment. Available from http://aco.net.au/resources/natural-cosmetic-ingredients-assessment/ (assessed on March 2018).
Alhadlaq, A., M. Tang, and J. J. Mao. 2005. Engineered adipose tissue from human mesenchymal stem cells maintains predefined shape and dimension: implications in soft tissue augmentation and reconstruction. *Tissue Engineering* 11 (3–4):556–66.
Aniya, Y., C. Miyagi, A. Nakandakari, S. Kamiya, N. Imaizumi, and T. Ichiba. 2002. Free radical scavenging action of the medicinal herb Limonium wrightii from the Okinawa islands. *Phytomedicine* 9 (3):239–44.
Arnold, C. 2016. Toward a better beauty regimen: reducing potential EDC exposures from personal care products. *Environmental Health Perspectives* 124:A188.
Auerbach, S. M., K. A. Carrado, and P. K. Dutta. 2004. *Handbook of Layered Materials.* 1st ed. London: Marcel Dekker.
Avetisyan, A., A. Markosian, M. Petrosyan, N. Sahakyan, A. Babayan, S. Aloyan, and A. Trchounian. 2017. Chemical composition and some biological activities of the essential oils from basil Ocimum different cultivars. *BMC Complementary and Alternative Medicine* 17:60.
Backman J., T. Isohanni, K. M. Oksman-Caldentey, L. Nohynek, H. Rischer, R. Puupponen-Pimiä. 2014. Cosmetic compositions containing cloudberry cell culture preparation. Patent WO2013124540.
Barba, C., S. Mendez, A. Roddick-Lanzilotta, R. Kelly, J. L. Parra, and L. Coderch. 2008. Cosmetic effectiveness of topically applied hydrolysed keratin peptides and lipids derived from wool. *Skin Research and Technology* 14 (2):243–8.
Barel, A. O., M. Paye, and H. I. Maibach. 2014. *Handbook of Cosmetic Science and Technology.* Boca Raton, FL: CRC Press.
Beck, R., S. Guterres, and A. Pohlmann. 2014. *Nanocosmetics and Nanomedicines New Approaches for Skin Care.* Berlin: Springer Berlin.
Bellou, S., and G. Aggelis. 2013. Biochemical activities in Chlorella sp. and Nannochloropsis salina during lipid and sugar synthesis in a lab-scale open pond simulating reactor. *Journal of Biotechnology* 164 (2):318–29.
Beri, K., and S. S. Milgraum. 2015. Neocollagenesis in deep and superficial dermis by combining fractionated Q-switched ND: YAG 1,064-nm with topical plant stem cell extract and N-acetyl glucosamine: open case series. *Journal of Drugs in Dermatology: JDD* 14 (11):1342.
Bimonte, M., C. Antonietta, T. Annalisa, B. Ani, C. Francesca, A. Fabio, C. Gabriella et al. 2011. *Coffea bengalensis* for antiwrinkle and skin toning applications. *Cosmetics & Toiletries* 126 (9):644.
Bongso, A., and M. Richards. 2004. History and perspective of stem cell research. *Best Practice & Research: Clinical Obstetrics & Gynaecology* 18 (6):827–42.
Borowitzka, L. J., and M. A. Borowitzka. 1989. β-Carotene (Provitamin A) production with algae. In *Biotechnology of Vitamins, Pigments and Growth Factors*, edited by E. J. Vandamme. Dordrecht: Springer Netherlands.
Burger, P., A. Landreau, M. Watson, L. Janci, V. Cassisa, M. Kempf, S. Azoulay et al. 2017. Vetiver essential oil in cosmetics: What is new? *Medicines Medicines* 4 (2):41.
Byrne, M. E., C. A. Kidner, and R. A. Martienssen. 2003. Plant stem cells: Divergent pathways and common themes in shoots and roots. *Current Opinion in Genetics & Development* 13 (5):551–7.

Carretero, M. I., and M. Pozo. 2009. Clay and non-clay minerals in the pharmaceutical industry. Part I. Excipients and medical applications. *Applied Clay Science* 46 (1):73–80.

Carretero, M. I., and M. Pozo. 2010. Clay and non-clay minerals in the pharmaceutical and cosmetic industries Part II. Active ingredients. *Applied Clay Science* 47 (3–4):171–81.

Cerezo, P., A. Garcés, M. Galindo, C. Aguzzi, C. Viseras, and A. López-Galindo. 2005. Estudio de capacidad de enfriamiento y extensibilidad de peloides usados en distintos balnearios. Paper read at Libro de Resúmenes de la XIX Reunión Científica de la Sociedad Española de Arcillas, at Salamanca, Spain.

Certech. 2008. *International Organic Standard – Natural and Natural Organic Cosmetic Certification.* Available from http://www.makingcosmetics.com/articles/Certech-IOS-Cosmetics-Standard.pdf.

Choi, S., Y. I. Park, S. K. Lee, J. E. Kim, and M. H. Chung. 2002. Aloesin inhibits hyperpigmentation induced by UV radiation. *Clinical and Experimental Dermatology* 27 (6):513–15.

Cizauskaite, U., R. Marksiene, V. Viliene, R. Gruzauskas, and J. Bernatoniene. 2017. New strategy of multiple emulsion formation based on the interactions between polymeric emulsifier and natural ingredients. *Colloids and Surfaces A: Physicochemical and Engineering Aspects* 515:22–33.

Clijsen, R., J. Taeymans, W. Duquet, A. Barel, and P. Clarys. 2008. Changes of skin characteristics during and after local Parafango therapy as used in physiotherapy. *Skin Research and Technology* 14 (2):237–42.

Corazza, M., A. Borghi, R. Gallo, D. Schena, P. Pigatto, M. Michela Lauriola, Fabrizio Guarneri et al. 2014. Topical botanically derived products: Use, skin reactions, and usefulness of patch tests. A multicentre Italian study. *Contact Dermatitis* 70 (2):90–97.

Corbeil, C., D. Mehran, and A. R. Mehran. 2000. *Nature in Cosmetics and Skin Care a Compendium of Ingredients Used in Cosmetic and Skin Care Chemistry.* Carol Stream, IL: Allured.

Corinaldesi, C., G. Barone, A. Dell'Anno, F. Marcellini, and R. Danovaro. 2017. Marine microbial-derived molecules and their potential use in cosmeceutical and cosmetic products. *Marine Drugs* 15 (4): 118.

Dal Toso, R., and F. Melandri. 2011. Sustainable sourcing of natural food ingredients by plant cell cultures. *AgroFOOD Industry Hi-Tech* 22 (2):30–32.

de Jesus R., M. Filomena, R. M. S. C. de Morais, and A. M. M. B. de Morais. 2013. Health applications of bioactive compounds from marine microalgae. *LFS Life Sciences* 93 (15):479–86.

de Paiva, L. B., A. R. Morales, and F. R. V. Diaz. 2008. Organoclays: properties, preparation and applications. *Applied Clay Science* 42 (1–2):8–24.

De Wet, H., S. Nciki, and S. F. van Vuuren. 2013. Medicinal plants used for the treatment of various skin disorders by a rural community in northern Maputaland, South Africa. *Journal of Ethnobiology and Ethnomedicine* 9 (1):51.

Derikvand, P., C. A. Llewellyn, and S. Purton. 2017. Cyanobacterial metabolites as a source of sunscreens and moisturizers: a comparison with current synthetic compounds. *European Journal of Phycology* 52 (1):43–56.

Dweck, A. C. 2002. Natural ingredients for colouring and styling. *International Journal of Cosmetic Science* 24 (5):287–302.

Dweck, A. C. 2011. *Formulating Natural Cosmetics.* Carol Stream, IL: Allured.

Ennaas, N., R. Hammami, L. Beaulieu, and I. Fliss. 2015. Purification and characterization of four antibacterial peptides from protamex hydrolysate of Atlantic mackerel (Scomber scombrus) by-products. *Biochemical and Biophysical Research Communications* 462 (3):195–200.

Ernst, E. 2000. Adverse effects of herbal drugs in dermatology. *British Journal of Dermatology* 143 (5):923–9.

Fan, X., R. Burton, and Y. Zhou. 2010. Glycerol (byproduct of biodiesel production) as a source for fuels and chemicals – Mini review. *Open Fuels Energy Science Journal* 3:17–22.

Fang, J. Y., Y. L. Leu, T. L. Hwang, and H. C. Cheng. 2004. Essential oils from sweet basil (*Ocimum basilicum*) as novel enhancers to accelerate transdermal drug delivery. *Biological & Pharmaceutical Bulletin* 27 (11):1819–25.

Fankem, P. M., S. N. Kwanga, M. L. Sameza, F. Tchoumbougnang, R. Tchabong, L. T. Ngouné, P. M. J. Dongmo. Antioxidant and Antifungal Activities of Cocoa Butter (*Theobroma cacao*), Essential Oil of *Syzygium aromaticum* and a Combination of Both Extracts against Three Dermatophytes. *ARSJETS Journal* 2017: 37, 255–272.

FDA. 2013. Cosmetic Good Manufacturing Practices; Draft Guidance for Industry. Available from www.fda.gov/CosmeticGuidances.

FDA. 2014. Guidance for Industry FDA Records Access Authority Under Sections 414 and 704 of the Federal Food, Drug, and Cosmetic Act. Available from www.fda.gov/downloads/Food/GuidanceRegulation/UCM292797.pdf (assessed on March 2018).

FDA. 2017. FDA Authority Over Cosmetics: How Cosmetics Are Not FDA-Approved, but Are FDA-Regulated. Available from www.fda.gov/Cosmetics/Labeling/Regulations/ucm126438.htm.

Fehér, A. 2015. Somatic embryogenesis – Stress-induced remodeling of plant cell fate. *Biochimica et Biophysica Acta*. 1849 (4):385–402.

Feinstein, D. 2015. Senators Introduce Bill to Strengthen Personal Care Product Oversight. Washington, DC: Office of Senator Dianne Feinstein.

Fonseca-Santos, B., M. A. Corrêa, and M. Chorilli. 2015. Sustainability, natural and organic cosmetics: consumer, products, efficacy, toxicological and regulatory considerations. *Brazilian Journal of Pharmaceutical Sciences* 51:17–26.

Galinski, E. A., H. P. Pfeiffer, and H. G. Trüper. 1985. 1,4,5,6-Tetrahydro-2-methyl-4-pyrimidinecarboxylic acid. A novel cyclic amino acid from halophilic phototrophic bacteria of the genus Ectothiorhodospira. *European Journal of Biochemistry* 149 (1):135–9.

Gastineau, R., F. Turcotte, J. B. Pouvreau, M. Morançais, J. Fleurence, E. Windarto, F. S. Prasetiya et al. 2014. Marennine, promising blue pigments from a widespread *Haslea* diatom species complex. *Marine Drugs* 12 (6):3161–89.

Ghadiri, M., W. Chrzanowski, and R. Rohanizadeh. 2015. Biomedical applications of cationic clay minerals. *RSC Advances* 5 (37):29467–81.

Golubovic-Liakopoulos, N., S. R. Simon, and B. Shah. 2011. Nanotechnology use with cosmeceuticals. *Seminars in Cutaneous Medicine and Surgery* 30 (3):176–180.

Gomes, C., and J. Silva. 2007. Minerals and clay minerals in medical geology. *Applied Clay Science* 36 (1–3):4–21.

Greb, T., and J. U. Lohmann. 2016. Plant stem cells. *Current Biology* 26 (17):R816–21.

Gurib-Fakim, A. 2006. Medicinal plants: traditions of yesterday and drugs of tomorrow. *Molecular Aspects of Medicine* 27 (1):1–93.

Harper, T. 2017, May. What's Driving the Billion-Dollar Natural Beauty Movement? Available from www.fastcompany.com/3068710/whats-driving-the-billion-dollar-natural-beauty-movement (assessed on January 30, 2018).

Heidstra, R., and S. Sabatini. 2014. Plant and animal stem cells: similar yet different. *Nature Reviews Molecular Cell Biology* 15 (5):301–12.

Herman, A., and A. P. Herman. 2015. Essential oils and their constituents as skin penetration enhancer for transdermal drug delivery: a review. *Journal of Pharmacy and Pharmacology* 67 (4):473–85.

Higley, C. and A. Higley. 2005. Essential oils constituents. In *Reference Guide for Essential Oils*. Olathe, KS: Abundant Health.

Huang, M-C. J., and J. Tang. 2015. Probiotics in personal care products. *Microbiology Discovery* 3 (1):5.

Innocenti, G., S. Dall'Acqua, G. Scialino, E. Banfi, S. Sosa, K. Gurung, M. Barbera et al. 2010. Chemical composition and biological properties of rhododendron anthopogon essential oil. *Molecules* 15 (4):2326–38.

International Whaling Commission. 2018. Commercial whaling. Available from https://iwc.int/commercial.

Katiyar, S. K., and C. A. Elmets. 2001. Green tea polyphenolic antioxidants and skin photoprotection (Review). *International Journal of Oncology* 18 (6):1307–13.

Kerscher, M., and H. Buntrock. 2017. Cosmetics and cosmeceuticals. In *Cosmetic Medicine and Surgery*, edited by Pierre Andre, Eckart Haneke, Leonardo Marini, and Christopher Rowland Payne. London: CRC Press.

Kessler, R. 2015. More than cosmetic changes: taking stock of personal care product safety. *Environmental Health Perspectives* 123 (5):A120–A127.

Khan, A. I., and A. E. Abourashed. 2011. Natural ingredients. In *Encyclopedia of Common Natural Ingredients Used in Food, Drugs and Cosmetics*. Hoboken, NJ: John Wiley & Sons.

Kim, H. H., S. Cho, S. Lee, K. H. Kim, K. H. Cho, H. C. Eun, and J. H. Chung. 2006. Photoprotective and anti-skin-aging effects of eicosapentaenoic acid in human skin in vivo. *Journal of Lipid Research* 47 (5):921–30.

Kim, H. H., C. M. Shin, C-H. Park, K. H. Kim, K. H. Cho, H. C. Eun, and J. H. Chung. 2005. Eicosapentaenoic acid inhibits UV-induced MMP-1 expression in human dermal fibroblasts. *Journal Lipid Research* 46 (8):1712–20.

Kim, S-K. 2012. *Marine Cosmesceuticals Trends and Prospects.* Boca Raton, FL: CRC Press.

Lavker, R. M., and T. T. Sun. 2000. Epidermal stem cells: Properties, markers, and location. *Proceedings of the National Academy of Science USA* 97 (25):13473–5.

Lee, C. H., G. C. Huang, and C. Y. Chen. 2017. Bioactive compounds from natural product extracts in Taiwan cosmeceuticals – Mini review. *Biomedical Research (India)* 28 (15):6561–6.

Leung, A. Y., I. A. Khan, and Ehab A. Abourashed. 2010. *Leung's Encyclopedia of Common Natural Ingredients Used in Food, Drugs and Cosmetics.* New York: John Wiley.

Li, Q., E. T. Dunn, E. W. Grandmaison, and M. F. A. Goosen. 1992. Applications and properties of chitosan. *Journal of Bioactive and Compatible Polymers* 7 (4):370–97.

Ligeza, M., D. Wygladacz, A. Tobiasz, K. Jaworecka, and A. Reich. 2016. Natural cold pressed oils as cosmetic products. *Family Medicine and Primary Care Review* 18 (4):443–7.

Lodemel, J. B., W. Egge-Jacobsen, and R. L. Olsen. 2004. Detection of TIMP-2-like protein in Atlantic cod (Gadus morhua) muscle using two-dimensional real-time reverse zymography. *Comparative Biochemistry and Physiology Part B Biochemistry and Molecular Biology* 139 (2):253–9.

Lopes, A., M. J. Rodrigues, C. Pereira, M. Oliveira, L. Barreira, J. Varela, F. Trampetti et al. 2016. Natural products from extreme marine environments: searching for potential industrial uses within extremophile plants. *Industrial Crops and Products* 94:299–307.

López-Galindo, A., C. Viseras, and P. Cerezo. 2007. Compositional, technical and safety specifications of clays to be used as pharmaceutical and cosmetic products. *Applied Clay Science* 36 (1):51–63.

López-Galindo, A., and C. Viseras. 2004. Pharmaceutical and cosmetic applications of clays. *Interface Science and Technology* 1:267–289.

Ma'or, Z., S. Yehuda, and W. Voss. 1997. Skin smoothing effects of Dead Sea minerals: comparative profilometric evaluation of skin surface. *International Journal of Cosmetic Science* 19 (3):105–10.

Ma'or, Z., L. Halicz, M. Portugal-Cohen, M. Z. Russo, F. Robino, T. Vanhaecke, and V. Rogiers. 2015. Safety evaluation of traces of nickel and chrome in cosmetics: the case of Dead Sea mud. *Regulatory Toxicology and Pharmacology Regulatory Toxicology and Pharmacology* 73 (3):797–801.

Magdassi, S., and E. Touitou. 1999. *Novel Cosmetic Delivery Systems.* New York: Marcel Dekker.

Mangena, T., and N. Y. O. Muyima. 1999. Comparative evaluation of the antimicrobial activities of essential oils of *Artemisia afra, Pteronia incana* and *Rosmarinus officinalis* on selected bacteria and yeast strains. *Letters in Applied Microbiology* 28 (4):291–6.

Martins, I. M., M. F. Barreiro, M. Coelho, and A. E. Rodrigues. 2014. Microencapsulation of essential oils with biodegradable polymeric carriers for cosmetic applications. *Chemical Engineering Journal* 245:191–200.

Mata-Gómez, L. C., J. C. Montañez, A. Méndez-Zavala, and C. N. Aguilar. 2014. Biotechnological production of carotenoids by yeasts: an overview. *Microbial Cell Factories* 13 (1):1–11.

Mattioli, M., L. Giardini, C. Roselli, and D. Desideri. 2016. Mineralogical characterization of commercial clays used in cosmetics and possible risk for health. *Applied Clay Science* 119:449–54.

Matz, H., E. Orion, and R. Wolf. 2003. Balneotherapy in dermatology. *Dermatologic Therapy* 16 (2):132–40.

McDaniel, J. C., M. Belury, K. Ahijevych, and W. Blakely. 2008. Omega-3 fatty acids effect on wound healing. *Wound Repair and Regeneration* 16 (3):337–45.

Meot-Duros, L., C. Magne, S. Cerantola, H. Talarmin, C. Le Meur, and G. Le Floch. 2010. New antibacterial and cytotoxic activities of falcarindiol isolated in Crithmum maritimum L. leaf extract. *Food and Chemical Toxicology* 48 (2):553–7.

Miller, R. A., and I. Miller. 1990. *The Magical and Ritual Use of Perfumes.* Rochester, VT: Destiny Books.

Min-Jeong, J., K. Su-Nam, and C. Hye-Young. 2006. The inhibitory effects of eckol and dieckol from Ecklonia stolonifera on the expression of matrix metalloproteinase-1 in human dermal fibroblasts. *Biological & Pharmaceutical Bulletin* 29 (8):1735–9.

Mokrejs, P., M. Hutta, J. Pavlackova, P. Egner, and L. Benicek. 2017. The cosmetic and dermatological potential of keratin hydrolysate. *Journal of Cosmetic Dermatology* 16 (4):e21–e27.

Mora, C., D. P. Tittensor, S. Adl, A. G. B. Simpson, B. Worm, and G. M. Mace. 2011. How many species are there on earth and in the ocean? *PLoS Biology* 9 (8):e1001127.

Moraes, J. D. D., S. R. A. Bertolino, S. L. Cuffini, D. F. Ducart, P. E. Bretzke, and G. R. Leonardi. 2017. Clay minerals: Properties and applications to dermocosmetic products and perspectives of natural raw materials for therapeutic purposes – A review. *International Journal of Pharmaceutics* 534 (1–2):213–9.

Morus, M., M. Baran, M. Rost-Roszkowska, and U. Skotnicka-Graca. 2014. Plant stem cells as innovation in cosmetics. *Acta Poloniae Pharmaceutica* 71 (5):701–7.

Murray, H. H. 2007. *Applied Clay Mineralogy: Occurrences, Processing, and Application of Kaolins, Bentonites, Palygorskite-Sepiolite, and Common Clays.* Amsterdam: Elsevier

Nahar, M. K., S. A. Lisa, K. Nada, and M. Begum. 2017. Characterization of seed kernel oil of Bangladeshi mango and its evaluation as cosmetic ingredient. *Bangladesh Journal of Scientific and Industrial Research* 52 (1):43–48.

Naolys. 2016. Naolys Active Cells: Refine Ginger Restores Skin Texture. Available from www.naolys.com/media/refine_ginger_en.pdf.

Ni, Z., Y. Mu, and O. Gulati. 2002. Treatment of melasma with Pycnogenol®. *Phytotherapy Research* 16 (6):567–71.

Nielsen, J. B. 2006. Natural oils affect the human skin integrity and the percutaneous penetration of benzoic acid dose-dependently. *Basic & Clinical Pharmacology & Toxicology* 98 (6):575–81.

Nohynek, L., M. Bailey, J. Tähtiharju, T. Seppänen-Laakso, H. Rischer, K-M. Oksman-Caldentey, and R. Puupponen-Pimiä. 2014. Cloudberry (Rubus chamaemorus) cell culture with bioactive substances: establishment and mass propagation for industrial use. *Engineering in Life Sciences* 14 (6):667–75.

O'Lenick, A. J., David C. Steinberg, K. Klein, and C. LaVay. 2008. *Oils of Nature.* Carol Stream, IL: Allured.

OCA. 2004. Organic Integrity in the Health and Beauty Care Aisle. Available from www.organicconsumers.org/campaigns/coming-clean (assessed on March 2018).

OCA. 2010. Whole Foods Imposes One-Year Deadline on Brands to Drop Bogus Organic Claims. Available from www.organicconsumers.org/press/whole-foods-imposes-one-year-deadline-brands-drop-bogus-organic-claims (assessed on March 2018).

Panda, H. 2003. *The Complete Technology Book on Herbal Perfumes & Cosmetics.* Delhi: National Institute of Industrial Research.

Panis, S. 2011. Stem Cells Intended for Cosmetic Use Only: Regulation in Belgium, Europe and the United States. Available from http://nrs.harvard.edu/urn-3:HUL.InstRepos:8784344 (accessed March 5, 2018).

Panten, J., H. Surburg, and B. Holscher. 2014. Recent results in the search for new molecules with ambergris odor. *Chemistry Biodiversity* 11 (10):1639–50.

Pasbakhsh, P., G. J. Churchman, and J. L. Keeling. 2013. Characterisation of properties of various halloysites relevant to their use as nanotubes and microfibre fillers. *Applied Clay Science* 74:47–57.

Peng, J., J. P. Yuan, C. F. Wu, and J. H. Wang. 2011. Fucoxanthin, a marine carotenoid present in brown seaweeds and diatoms: metabolism and bioactivities relevant to human health. *Marine Drugs* 9 (10):1806–28.

Persistence. 2016. Global Natural and Organic Personal Care Products Market is Expected to be Valued at US$ 21,776.9 Mn by the End of 2024. Available frHi Lenom www.persistencemarketresearch.com/market-research/natural-organic-personal-care-product-market.asp (assessed on January 30, 2018).

Plaza, M., S. Santoyo, L. Jaime, G. Garcia-Blairsy Reina, M. Herrero, F. J. Senorans, and E. Ibanez. 2010. Screening for bioactive compounds from algae. *Journal of Pharmaceutical and Biomedical Analysis* 51 (2):450–5.

Prasad, S., A. K. Tyagi, and B. B. Aggarwal. 2014. Recent developments in delivery, bioavailability, absorption and metabolism of curcumin: the golden pigment from golden spice. *Cancer Research and Treatment* 46 (1):2–18.

Rajanala, S., and N. A. Vashi. 2017. Cleopatra and sour milk – The ancient practice of chemical peeling. *JAMA Dermatology* 153 (10):1006.

Ramadan, M. F., and A. Al-Ghamdi. 2012. Bioactive compounds and health-promoting properties of royal jelly: a review. *Journal of Functional Foods* 4 (1):39–52.

Rastogi, R. P., R. R. Sonani, and D. Madamwar. 2015. Cyanobacterial sunscreen scytonemin: role in photoprotection and biomedical research. *Applied Biochemistry and Biotechnology* 176 (6):1551–63.

Ryu, B., Z-J. Qian, and S-K. Kim. 2010. SHP-1, a novel peptide isolated from seahorse inhibits collagen release through the suppression of collagenases 1 and 3, nitric oxide products regulated by NF-κB/p38 kinase. *Peptides* 31 (1):79–87.

Schmid, D., C. Schürch, P. Blum, E. Belser, and F. Züll. 2008. Plant stem cell extract for longevity of skin and hair. *SÖFW Journal* 134 (5):30–35.

Sharma, D., M. M. Kober, and W. P. Bowe. 2016. Anti-aging effects of probiotics. *Journal of Drugs in Dermatology* 15 (1):9–12.

Smith, W. P. 1996. Epidermal and dermal effects of topical lactic acid. *Journal of the American Academy of Dermatology: Part 1* 35 (3):388–91.

Soleymani, A. R., R. Chahardoli, and M. Kaykhaii. 2016. Development of UV/H2O2/TiO2-LECA hybrid process based on operating cost: Application of an effective fixed bed photo-catalytic recycled reactor. *Journal of Industrial and Engineering Chemistry* 44:90–98.

Sommer, C. 2004. The role of musk and musk compounds in the fragrance industry. *Handbook of Environmental Chemistry* 3:1–16.

Song, R., R. B. Wei, H. Y. Luo, and D. F. Wang. 2012. Isolation and characterization of an antibacterial peptide fraction from the pepsin hydrolysate of half-fin anchovy (Setipinna taty). *Molecules* 17 (3):2980–91.

Species+. 2018. *Physeter macrocephalus.* Available from www.speciesplus.net/-/taxon_concepts/10761/legal.

Srinivasakumar, K. P., and M. Rajashekhar. 2009. In vitro studies on bactericidal activity and sensitivity pattern of isolated marine microalgae against selective human bacterial pathogens. *Indian Journal of Science and Technology* 2 (8):16–23.

Statista. 2018. Global Market Value for Natural Cosmetics from 2007 to 2017 (in US billion dollars). Available from www.statista.com/statistics/673641/global-market-value-for-natural-cosmetics/ (accessed on January 30, 2018).

Stepkowska, E. T., and S. A. Jefferis. 1992. Influence of microstructure on firing colour of clays. *Applied Clay Science* 6 (4):319–42.

Surget, G., V. Stiger-Pouvreau, K. Le Lann, N. Kervarec, C. Couteau, L. J. M. Coiffard, F. Gaillard et al. 2015. Structural elucidation, in vitro antioxidant and photoprotective capacities of a purified polyphenolic-enriched fraction from a saltmarsh plant. *Journal of Photochemistry & Photobiology, B: Biology* 143:52–60.

Tan, H. W., A. R. Abdul Aziz, and M. K. Aroua. 2013. Glycerol production and its applications as a raw material: a review. *Renewable and Sustainable Energy Reviews* 27:118–27.

Tavano, L., R. Muzzalupo, N. Picci, and B. de Cindio. 2014. Co-encapsulation of lipophilic antioxidants into niosomal carriers: percutaneous permeation studies for cosmeceutical applications. *Colloids and Surfaces B: Biointerfaces* 114:144–9.

Taylor, K. M., M. Weisskopf, and J. Shine. 2014. Human exposure to nitro musks and the evaluation of their potential toxicity: An overview. *Environmental Health* 13 (1):14.

Tito, A., A. Carola, M. Bimonte, A. Barbulova, S. Arciello, F. de Laurentiis, I. Monoli et al. 2011. A tomato stem cell extract, containing antioxidant compounds and metal chelating factors, protects skin cells from heavy metal-induced damages. *International Journal of Cosmetic Science* 33 (6):543–52.

Tran, T. A., M. T. Ho, Y. W. Song, M. Cho, and S. K. Cho. 2015. Camphor induces proliferative and anti-senescence activities in human primary dermal fibroblasts and inhibits UV-induced wrinkle formation in mouse skin. *Phytotherapy Research* 29 (12):1917–25.

Trehan, S., B. Michniak-Kohn, and K. Beri. 2017. Plant stem cells in cosmetics: current trends and future directions. *Future Science OA* 3 (4):FSO226.

United States House Committee on Merchant Marine and Fisheries. 2003. Endangered Species Act of 1973, as amended through the 108th Congress.

USDA. 2008. Cosmetics, Body Care Products, and Personal Care Products. Available from www.ams.usda.gov/sites/default/files/media/OrganicCosmeticsFactSheet.pdf (assessed on March 2018).

Valenti, D. M. Z., J. Silva, W. R. Teodoro, A. P. Velosa, and S. B. V. Mello. 2012. Effect of topical clay application on the synthesis of collagen in skin: an experimental study. *Clinical & Experimental Dermatology* 37 (2):164–8.

Viseras, C., C. Aguzzi, P. Cerezo, and A. Lopez-Galindo. 2007. Uses of clay minerals in semisolid health care and therapeutic products. *Applied Clay Science* 36 (1):37–50.

Viseras, C., P. Cerezo, A. Garcés, C. Aguzzi, M. Setti, and A. López-Galindo. 2005. Composición mineral y características texturales de sólidos "madurados" en aguas mineromedicinales. Paper read at Libro de Resúmenes de la XIX Reunión Científica de la Sociedad Española de Arcillas, at Salamanca, Spain.

Wan, M., D. Hou, Y. Li, J. Fan, J. Huang, S. Liang, W. Wang et al. 2014. The effective photoinduction of *Haematococcus pluvialis* for accumulating astaxanthin with attached cultivation. *Bioresource Technology* 163:26–32.

Wang, H-M. D., C-C. Chen, P. Huynh, and J-S. Chang. 2015. Exploring the potential of using algae in cosmetics. *BITE Bioresource Technology* 184:355–62.

Williams, D. F., and W. H. Schmitt. 1992. *Chemistry and Technology of the Cosmetics and Toiletries Industry.* London: Blackie Academic & Professional.

Zhang, C., and S. K. Kim. 2009. Matrix metalloproteinase inhibitors (MMPIs) from marine natural products: the current situation and future prospects. *Marine Drugs* 7 (2):71–84.

Zhang, C., Y. Li, X. Shi, and S. K. Kim. 2010. Inhibition of the expression on MMP-2, 9 and morphological changes via human fibrosarcoma cell line by 6,6(-bieckol from marine alga *Ecklonia cava. BMB Reports* 43 (1):62–68.

Zhuang, Y., H. Hou, X. Zhao, Z. Zhang, and B. Li. 2009. Effects of collagen and collagen hydrolysate from jellyfish (Rhopilema esculentum) on mice skin photoaging induced by UV irradiation. *Journal of Food Science* 74 (6):H183–H188.

15

Designed Antimicrobial Peptides: A New Horizon

Kathryn W. Woodburn, Chia-Ming Chiang, Jesse Jaynes and L. Edward Clemens

CONTENTS

The use of peptides in topical cosmeceutical products has increased dramatically over the past several years (Schagen, 2017). Many studies show the wide range of possible topical applications of biologically active peptides for improving the skin and its properties (Zhang and Falla, 2009). A class of endogenously produced peptides called antimicrobial peptides (AMPs), often produced in skin, provide an innovative alternative for treating skin infections, as they provide rapid and complete microbicidal activity with selective targeting of prokaryotic cell membranes (Zasloff, 2002). Furthermore, in this age with the promiscuous use of antibiotics causing bacterial resistance, AMPs offer a prime advantage in that they exhibit limited selection for microbial resistance.

Antimicrobial Peptides

Antimicrobial peptides, also known as host defense peptides, are evolutionarily highly conserved components of the innate immune system and provide the first line of defense against invading pathogens in all multicellular organisms (Zasloff, 2002). Their significance in host defense is underscored in plants and insects by their ability to live in bacterial environments without lymphocytes and antibodies. Over the last 25 years, a great deal of information has been published describing naturally occurring peptides that possess antimicrobial activity (Blondelle et al., 1996; Kagan et al., 1994; Zasloff, 1987). AMPs exist in a range of sizes but are, in general, between 20 and 40 amino acids in length. AMPs have direct antimicrobial activities and are able to kill both Gram-negative and Gram-positive bacteria, including multidrug-resistant strains, mycobacteria (including *Mycobacterium tuberculosis*), enveloped viruses, parasites and fungi (Jenssen et al., 2006).

AMPs function by selectively differentiating between prokaryotic and eukaryotic membranes. The topological cationic amphipathic nature of AMPs endears them as potent biocidal agents by allowing direct electrostatic interaction with both Gram-positive and Gram-negative bacterial cell membrane walls which are rich in anionic phospholipids and negatively charged lipopolysaccharides with subsequent disruption of barrier function (Figure 15.1) (Gordon et al., 2005). In contrast, mammalian cells

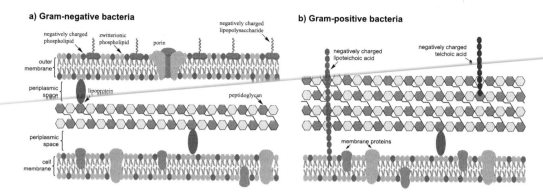

FIGURE 15.1 Cationic AMPs selectively target, and lyse, the negatively charged cell walls of both Gram-positive and Gram-negative bacteria. The electrostatic interaction between a cationic amphipathic peptide and the negatively charged outer bacterial cell wall, inherently due to negatively charged phospholipids, facilitate direct biocidal activity of the cationic AMP in disrupting the bacterial cell membrane wall with subsequent cell lysis. (a) Gram-negative bacteria possess an inner and outer membrane separated by periplasm and peptidoglycan. The outer membrane contains porins and negative-charged lipopolysaccharides. (b) Gram-positive bacteria possess a single cell membrane surrounded by a porous layer of peptidoglycan interspersed with negatively charged teichuronic and lipoteichoic acids. In contrast, eukaryotic cell membranes contain zwitterionic phosphatidylcholine lipids, stabilized with cholesterol, thereby generating a rigid bilayer preventing membrane disruption.

contain more zwitterionic phospholipids framed with cholesterol and cholesterol esters that act as membrane stabilizers (Gordon et al., 2005). The remarkable targeting and direct contact disruption of the bacterial membrane by AMPs, makes bacterial resistance less likely to develop, as AMP microbiocidal activity is a non-specific mechanism rather than by a specific protein target (Yeaman and Yount, 2003).

AMPs, in addition to direct antimicrobial activities, modulate immune responses (Hancock and Sahl, 2006). They affect innate immune cell functions including inducing and modulating chemokine and cytokine generation, angiogenesis production and wound repair (Hancock and Sahl, 2006). Endogenous AMPs prevent skin infections by inhibiting invading microorganisms and maintaining a balanced commensal flora. Once this biochemical barrier is broken, bacteria or bacterial factors gain access to epidermal keratinocytes and induce inflammation (Meyer-Hoffert et al., 2011). In human skin, AMPs are mainly produced by keratinocytes, neutrophils, sebocytes or sweat glands, and are expressed constitutively or following an inflammatory stimulus.

The importance of AMPs in skin biology is underscored in several human skin disease states where there is an inverse relationship between severity of the disease and the level of AMP production. AMP over-expression can lead to protection against skin infections as observed in patients with inflammatory skin diseases such as psoriasis and rosacea, in which the lesions generally never become super-infected. In skin infections, such as acne vulgaris, increased levels of AMPs can be found in inflamed tissues, indicating a role of these peptides in the immune reaction to infection. Decreased AMP levels in burn tissue predisposes the burn victim to bacterial colonization and wound infection (Bhat and Milner, 2007). This infection interplayed with an associated proinflammatory response and subsequent immunosuppression can lead to sepsis and eventual death. The broad-spectrum antimicrobial, antifungal, and antibiofilm activity with minimal possibility of bacterial resistance coupled with immunomodulatory potential make AMPs appropriate and effective candidates as topical anti-infective drugs in treating many infections including atopic dermatitis, infectious acne, acute bacterial skin and skin structure infections, and even combat-related bacterial and fungal wound infections (Schittek et al., 2008).

Human cathelicidin LL-37, a 37-amino acid amphipathic peptide whose chemical structure is presented in Figure 15.2, is a prime example of an AMP which has been evaluated for its pharmacologic utility. LL-37 is expressed endogenously in skin, the gastrointestinal and respiratory tracts, and in immune cells. Elevated levels of LL-37 have been found in wound healing and in angiogenesis and augments inflammatory responses, binding and neutralizing LPS, and promoting re-epithelialization and wound closure (Dürr et al., 2006). LL-37 is presently being evaluated as a potential treatment for polymicrobial-infected wounds (Duplantier and van Hoek, 2013).

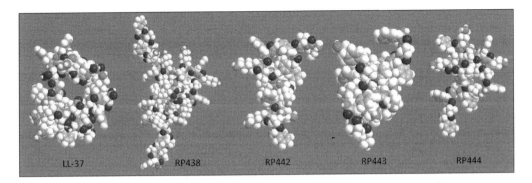

FIGURE 15.2   (See Colour Insert.) Structural space filling representation of human cathelicidin LL-37, LLGDFFRK SKEKIGKEFKRIVQRIKDFLRNLVPRTES, in addition to the novel dAMPs RP438 (CPGFAKKFAKKFKKFAKKF AKFAFAF)-disulphide-dimer); RP442, FAFAFAFKKAFKKFKKAFKKAF; PR443, FAFAFOAFOOAFOOFOOAFO OAF; and RP444, FAOOFAOOFOOFAOOFAOFAFAF. In the space-filling models, red represents oxygen, blue represents nitrogen, grey is carbon and hydrogen is white.

Despite much promise and the dire need for new anti-infectives, AMPs have not garnered appeal due to apparent high cost and concerns relating to toxicity, and when given systemically, poor bioavailability (Zasloff, 2002). Recent advances in solid phase peptide chemistry have reduced manufacturing costs and preliminary data has shown designed antimicrobial peptides to exhibit more favourable eukaryotic cytotoxicity profiles.

Designed Antimicrobial Peptides

Designed antimicrobial peptides (dAMPs) are laboratory-synthesized peptides which have been rationally chemically analogued from naturally occurring AMPs (Zasloff, 2002). AMPs have direct antimicrobial activities in addition to modulating immune responses (Hancock and Sahl, 2006). Based upon the functions and structure in naturally occurring AMPs, with numerous variations added to the structural repertoire by design, rationally designed, laboratory-created peptides have demonstrated increased potency, efficacy, safety, specificity and reduced toxicity in comparison to their natural templates (Blondelle et al., 1996; Gunshefski et al., 1994; Jaynes et al., 1988; Arrowood et al., 1991; Barr et al., 1995; Ballweber et al., 2002; Ma et al., 2002; Visser et al., 2012; Clemens et al., 2017).

The dAMPs described herein are derived from the defensin family of AMPs (Figure 15.2). Defensins, an AMP class, have demonstrated effects on innate immune cell functions, including inducing and modulating chemokine and cytokine generation, angiogenesis, and wound healing (Hancock and Sahl, 2006). Defensins are widely distributed in mammalian epithelial cells and phagocytes with known broad-spectrum activity against Gram-positive and Gram-negative bacteria, fungi and enveloped viruses (Ganz, 2003). Additionally, they possess immunomodulatory and anti-inflammatory properties, and are upregulated in response to challenges by infectious agents (Huang et al., 2007; Hancock and Sahl, 2006). Use of non-natural amino acids, not encoded by the Universal Genetic Code, the specific replacement of lysine (K) with ornithine (O), as was evident with RP443 and RP444, following several iterative evaluations has been shown to increase antibacterial activity while also enhancing proteolytic stability.

Topical administration is the preferred route for treating skin conditions such as burn wounds where the non-perfused tissue limits bioavailability of systemic antibiotics. Moreover, AMPs topically applied to infected skin in the form of an ointment, cream or gel can penetrate into the stratum corneum where they will kill the infectious pathogens.

In *in vitro* studies, the dAMPs RP438 and RP443 showed superior antimicrobial activity against methicillin-resistant *Staphylococcus aureus* (MRSA) ATCC 33591 and against *Klebsiella pneumoniae* ATCC 10031 when compared to LL-37 (Table 15.1).

TABLE 15.1

Minimum Bactericidal Concentration (μg/mL) of LL-37 and dAMPs

AMPs	Minimum Bactericidal Concentration (μg/ml)	
	MRSA (ATCC 33591)	***K. pneumonia* (ATCC 10031)**
LL-37	256	64
RP438	64	16
RP442	64	64
RP443	32	32

Note: The minimum bactericidal concentration (MBC) is the lowest concentration of an antibacterial agent required to kill a particular bacterium.

dAMPs Exhibit Broad-Spectrum Bactericidal Activity

dAMPs were screened for their bactericidal effectiveness against four strains of Gram-positive and Gram-negative bacteria using the minimum inhibitory concentration (MIC) assay (Table 15.2). The dAMPs showed robust activity against both Gram-positive and Gram-negative bacterial species including the methicillin-resistant *S. aureus* (MRSA) ATCC 33591.

dAMPs Are Active against Gram-Negative and Gram-Positive Biofilms

Innovation in antimicrobial agents and treatment is currently needed due to the rapid emergence of bacterial resistance which is made more problematic by pathogen resilience due to biofilm formation (Vazirani et al., 2015). Bacteria within biofilms are projected to be 20–1000 times less sensitive to antibiotics than their planktonic counterparts (Elder et al., 1995), as they are protected from antibiotics and the host's immune system, hence the imperative need to find efficacious agents that are able to encroach biofilms and eradicate the bacteria. In addition to inhibiting planktonic growth, the evaluated dAMPs were active against Gram-positive and Gram-negative biofilm bacteria (Table 15.3).

dAMPs Rapidly Eradicate Bacteria

dAMPs rapidly eradicate *Pseudomonas aeruginosa* and *S. aureus* in *in vitro* time-kill assays. The concentration-dependent bactericidal activity of the dAMPs and conventional antibiotics, tobramycin and vancomycin, against *P. aeruginosa* and *S. aureus* is shown in Figure 15.3 following 30 minutes incubation. The dAMPs rapidly kill both Gram-negative and Gram-positive bacteria in a concentration-dependent

TABLE 15.2

Minimum Inhibitory Concentrations (μg/mL) against Gram-Positive and Gram-Negative Bacteria

Gram-Positive Bacteria	**RP438**	**RP442**	**RP443**	**RP444**
E. faecium 700221	16	8	8	32
MRSA 33591	32	16	16	16
S. epidermis 51625	32	64	32	64
S. pneumoniae 49619	32	32	32	32
Gram-Negative Bacteria	**RP438**	**RP442**	**RP443**	**RP444**
E. aerogenes 13048	32	64	64	8
A. baumannii 17978D-5	16	2	4	2
P. aeruginosa 19660	32	64	64	32
P. aeruginosa 27853	32	64	16	8

Note: Minimum inhibitory concentration (MIC) is the lowest concentration that will inhibit in vitro growth of planktonic bacteria (10^4 CFU/well).

TABLE 15.3

Minimum Biofilm Eradication Concentration (μg/mL) Values against
Gram-Positive and Gram-Negative Biofilm Bacteria

Gram-Positive Bacteria	RP438	RP442	RP443	RP444
MRSA 33591	94	47	24	47
S. epidermis 51625	94	47	12	47
Gram-Negative Bacteria	**RP438**	**RP442**	**RP443**	**RP444**
A. baumannii 17978D-5	47	6	12	2
P. aeruginosa 27853	375	750	188	188

Note: Minimum biofilm eradication concentration (MBEC) is the lowest concentration needed to kill biofilm bacteria. Data represent the mean of three
replicates from three independent experiments.

manner, underscoring the bactericidal mechanism of action is via specific, and direct, membrane disruption and immediate lysis (Figure 15.3). For *P. aeruginosa*, the order of bactericidal dynamics was RP444
followed by RP443 and then RP442 and RP438, while for *S. aureus* it was PR442, RP443, RP444 and
then RP438. In contrast, tobramycin, an aminoglycoside which disrupts cell membranes and protein
synthesis in Gram-negative bacteria, and vancomycin, a bactericide that acts by inhibiting cell wall
synthesis in Gram-positive bacteria, exhibited no trace of antibacterial activity following 30 minutes of
exposure.

The concentration-dependent time-kill evaluation of the dAMPs and vancomycin and tobramycin
on *P. aeruginosa* 19660 is displayed in Figure 15.4. As anticipated, due to their respective modes of
action, vancomycin showed no evidence of bactericidal activity during the 2-hour evaluation period, with
tobramycin exhibiting bactericidal effects following 60 minutes of incubation. The striking immediate

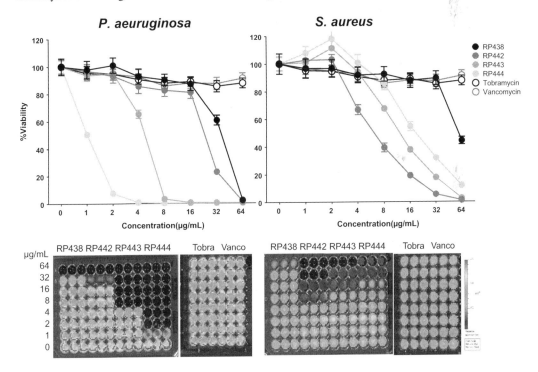

FIGURE 15.3 (See Colour Insert.) dAMPs kill *P. aeruginosa* and *S. aureus* within 30 minutes of exposure. RP438,
RP442, RP443, RP444, tobramycin and vancomycin were applied to bioluminescent *P. aeruginosa* 19660 and *S. aureus*
49525, and bactericidal effectiveness assessed following incubation for 30 minutes. The bioluminescence of viable bacteria was evaluated non-invasively with an IVIS Lumina bioimaging system. Data represent the mean ± SE of triplicate
replicates from two independent experiments. For some points, the error bars are shorter than the height of the symbols.

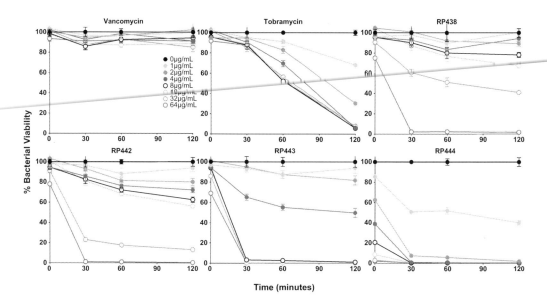

FIGURE 15.4   dAMPs rapidly eradicate *P. aeruginosa* in an increasing dose-dependent manner. The bactericidal effectiveness of vancomycin, tobramycin and the dAMPs were evaluated against bioluminescent *P. aeruginosa* 19660 through 2 hours of incubation. The bioluminescence of viable cells was quantitated non-invasively with an IVIS Lumina bioimaging system. Data represent the mean ± SE of triplicate replicates from two independent experiments. For some points, the error bars are shorter than the height of the symbols.

bactericidal activity of the dAMPs relates to direct electrostatic interaction with the bacterial cell membranes. *P. aeruginosa* was killed by RP444 within 30 minutes with as little as 1 µg/mL peptide.

dAMPs Selectively Target Bacterial Cell Membranes

dAMPs selectively disrupt bacterial cells in concentrations much lower compared to effects on eukaryotic cells. An example is presented in Figure 15.5 with human corneal epithelial cells. RP444 exhibits limited cytotoxicity to human corneal epithelial cells following 24-hour incubation. In contrast, complete killing of *P. aeruginosa* occurs within 30 minutes with as little as 1 µg/mL RP444 with higher concentrations of 20 µg/mL needed to kill 50% of the *S. aureus*.

Pathogens Do Not Develop Resistance against dAMPs

The increasing emergence of antibiotic resistance also highlights the need for novel alternatives that provide rapid and complete microbicidal activity with minimal safety-related effects, while exhibiting limited susceptibility to mechanisms of microbial resistance. *P. aeruginosa* did not become resistant to sub-inhibitory concentrations of RP444 after 30 rounds of selection, whereas gentamicin did, as evidenced by growing at 4096 times the initial MIC after 30 days (Figure 15.6). Moreover, with respect to Gram-positive resistance, *S. aureus* also did not become resistant to RP444, as underscored by growing at 256 times the initial MIC after 30 days. Gentamicin-resistant *P. aeruginosa* from passage 30, with a MIC of 512 µg/mL, was found to be susceptible to RP444 at 8 µg/mL, and similarly clindamycin-resistant *S. aureus* from passage 30, was susceptible to a MIC RP444 of 8 µg/mL. Therefore, there is no cross resistance to dAMPs with the gentamicin-resistant *P. aeruginosa* or clindamycin-resistant *S. aureus*.

dAMPs Reduce Inflammation and Bacterial Burden in Murine Keratitis

RP444 demonstrated statistically significant *in vivo* efficacy in reducing corneal bacterial burden in a murine *P. aeruginosa* model of bacterial keratitis (Clemens et al., 2017). The number of viable bacteria

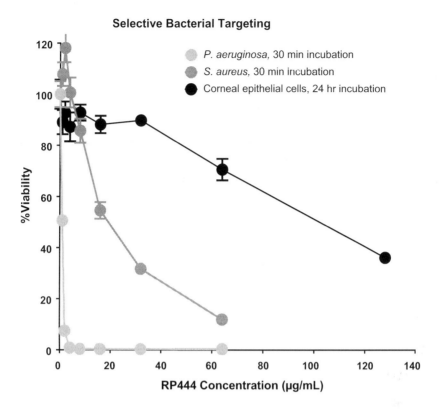

FIGURE 15.5 RP444 rapidly eradicate *P. aeruginosa* and *S. aureus* within 30 minutes with limited cytotoxicity to mammalian cells following 24-hour incubation. Bacteria cell viability was performed using bioluminescent strains of *P. aeruginosa* 19660 and *S. aureus* 49525 using an IVIS Lumina imaging system. The cytotoxicity to human corneal epithelial cells (hTCEpi) was determined using an MTT assay. Data represent the mean of three measurements.

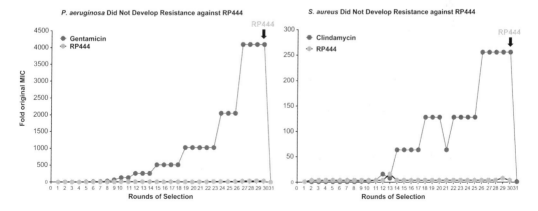

FIGURE 15.6 *P. aeruginosa* and *S. aureus* do not develop resistance against dAMPs. After 30 days of passaging *P. aeruginosa* ATCC 27853 with gentamycin and *S. aureus* ATCC 29213 with clindamycin, resistance was observed at 4096 and 256 times the MIC, respectively. RP444 was able to treat both resistant strains with both experiments yielding the original MIC values, therefore no cross-resistance occurs. Assay was performed in duplicate.

recovered from the 0.0064 and 0.064% dAMP-treated groups on day 3 ($p < 0.001$ and $p < 0.001$) and day 5 post-infection ($p < 0.01$ and $p < 0.001$; Figure 15.7, left), was significantly lower than from the PBS-treated control. There was no difference in the number of viable bacteria between the control and 0.0002% dAMP-treated groups at all time points.

RP444 demonstrated significant *in vivo* efficacy in reducing inflammatory cell infiltration in the *P. aeruginosa* model of bacterial keratitis (Figure 15.7, right). At day 1 post-infection there was no significant difference between the vehicle and RP444-treated corneas at any of the three concentrations. At day 3 post-infection, MPO activity in corneas treated with 0.00002, 0.0064 and 0.064% RP444 was dose-dependently lower than treatment with the PBS vehicle ($p < 0.05$, $p < 0.001$ and $p < 0.001$ for 2, 64 and 640 µg/ml RP444, respectively). At day 5 post-infection, MPO activity was lower in corneas treated with RP444, reaching statistical significance in the 0.0064 and 0.064% RP444-treated groups ($p < 0.001$).

Formulation Considerations

The importance of the right formulation and delivery method in topical pharmaceuticals is critical. The optimal formulation for topical delivery of dAMP is determined by the properties of the dAMP, the selected excipients, type of formulations and the target sites. Given the hydrophilic properties of the dAMPs and the direct cytolytic action of the dAMPs with minimal eukaryotic toxicity, topical gel formulations are commonly used for dAMP administration when targeted for skin indications.

In an evaluation of the preferred formulation of an AMP derived from LL-37, termed P60.4Ac, for topical administration, Haisma et al. (2016) evaluated three formulations: a water-in-oil cream with lanolin (Softisan 649), an oil-in-water cream with polyethylene glycol hexadecyl ether (Cetomacrogol), and a hydroxypropyl methylcellulose (hypromellose) 4000 gel. The formulations underwent a battery of tests including cytotoxicity using human epidermal models, antimicrobial activity against mupirocin-resistant and -sensitive MRSA strains, and 3-month peptide stability assessment. Formulations containing up to 2% P60.4Ac did not damage the keratinocytes in the epidermal model, as assessed by histologic evaluation and metabolic activity. The antibacterial activity of P60.4Ac was affected by the different formulations; the LC_{99} values for P60.4Ac, the lowest concentration of the peptide required to kill 99% of the bacteria, were similar for phosphate-buffered saline and the hypromellose gel; however the LC_{99} values for P60.4Ac in the Cetomacrogol and Softisan creams were substantially higher, indicating that the antibacterial activity of P60.4Ac was significantly reduced in the Cetomacrogol and Softisan creams but not in the hypromellose gel.

P60.4Ac, formulated at 0.5%, remained stable at both the evaluated conditions of 4°C and ambient temperature (15 to 25°C) for up to 3 months in the hypromellose gel and the Softisan cream (Haisma

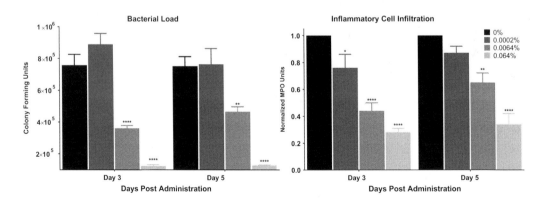

FIGURE 15.7 RP444 topical treatment reduces bacterial load and inflammatory cell infiltration in a murine *P. aeruginosa* keratitis model. (Left) Viable bacterial counts in infected corneas treated with RP444 at 0.0064 and 0.064% were significantly lower than in PBS-treated animals at days 3 and 5 post-infection. (Right) A dose-dependent decrease in inflammatory cell infiltration, as measured by myeloperoxidase (MPO) activity, was observed in RP444-treated animals at days 3 and 5 post-infection. Statistical significance was *$p < 0.5$, **$p < 0.01$ and ***$p < 0.001$.

et al., 2016). However, the Softisan cream stored at 4°C produced a separation of the oil and water phases at the end of the third month. The P60.4Ac peptide content in Cetomacrogol cream stored at ambient temperature decreased to 70% after 1 month with a further reduction to 47% following 3 months. Upon refrigeration, the Cetomacrogol cream formulation was stable during the first month, however, the peptide content was reduced to 66% following 3 months. Given the totality of the antimicrobial, cytotoxicity and stability data, hypromellose gel was concluded to be the preferred formulation of the synthetic peptide P60.4Ac for topical application of MRSA-infected epidermal wounds.

The antimicrobial peptides NP101 and NP108 have modest antimicrobial activity against *Escherichia coli*, *S. aureus* and *P. aeruginosa* with MIC values of 0.31 mg/ml for NP101 and 0.25–0.5 mg/ml for NP108. The peptides were impregnated into freeze-dried wafers prepared from a natural polymer for potential topical utility and displayed potent antibacterial activity when assessed by standard diffusion assay. The targeted approach for direct topical delivery of AMPs through their incorporation within freeze-dried wafer formulations represents an alternate rational approach for treating skin infection (O'Driscoll et al., 2013).

DRGN-1, a cationic AMP–derived peptide from the Komodo dragon, is under evaluation for use as a topical wound treatment against both *P. aeruginosa* and *S. aureus* infections (Chung et al., 2017). DRGN-1 was formulated at 0.1% in 1% hypromellose in 10 mM phosphate buffer and applied every 48 hours followed by covering with a Tegaderm-type semi-occlusive dressing (3M, St. Paul, Minnesota) for 6 days from the first treatment. Treatment with DRGN1 promoted healing, compared to placebo control, in a polymicrobial (*P. aeruginosa* and *S. aureus*) biofilm-infected murine wound model (Chung et al., 2017).

To aid in development of an effective topical microbicide against *Trichomonas vaginalis*, a cecropin peptide, D2A21, was formulated in a hydrophilic gel containing 0.5% or 2% D2A21. A 0.5% or 2% w/v D2A21 peptide solution was prepared in normal saline and 3.25% w/v hydroxyethylcellulose polymer (Lushbaugh et al., 2000). The gel was stored at 2–8°C until used. Estrogenized mice pre-treated with *L. rhamnosus* were pre-medicated with intravaginal placebo gel, 0.5% or 2% D2A21 gel, prior to *T. vaginalis* challenge. It was found that 2% D2A21 gel was more efficacious (10% infected) than placebo gel (53% infected) in preventing vaginal *T. vaginalis* infections in mice. The 0.5% D2A21 gel was not significantly different from placebo in preventing *T. vaginalis* infection.

Conclusion

Designed antimicrobial peptides are synthetic analogues of naturally occurring antimicrobial peptides that provide the first line of defense against invading pathogens. Endogenous AMPs have inherent microbiocidal activity due to their cationic amphipathic structure but are now also recognized as critical immune regulators that guard against infections and support healing, while suppressing inflammation. The increasing emergence of antibiotic resistance highlights the need for innovative alternatives that provide rapid and complete microbicidal activity with minimal toxicity, and exhibit limited selection for microbial resistance. Here, topically applied designed antimicrobial peptides have been rationally designed from endogenous antimicrobial peptides, and evaluation has shown them to exhibit rapid bactericidal activity, broad-spectrum effectiveness against isolates and biofilm, selective targeting of bacterial cell walls, and reduced likelihood of developing bacterial resistance. Further preclinical evaluation and formulation development will enable these to be translated clinically as potential topical therapeutics for the treatment of infectious and inflammatory skin diseases.

REFERENCES

Arrowood, M.J., Jaynes, J.M., Healey, M.C. 1991. In vitro activities of lytic peptides against the sporozoites of *Cryptosporidium parvum*. *Antimicrob Agents Chemother.* 35(2):224–7.

Ballweber, L.M., Jaynes, J.E., Stamm, W.E., Lampe, M.F. 2002. In vitro microbicidal activities of cecropin peptides D2A21 and D4E1 and gel formulations containing 0.1 to 2% D2A21 against *Chlamydia trachomatis*. *Antimicrob Agents Chemother.* 46:34–41.

Barr, S.C., Rose, D., Jaynes, J.M. 1995. Activity of lytic peptides against intracellular *Trypanosoma cruzi* amastigotes in vitro and parasitemias in mice. *J Parasitol.* 81(6):974–8.

Bhat, S., Milner, S. 2007. Antimicrobial peptides in burns and wounds. *Curr Protein Pept Sci.* 8(5):506–20.

Blondelle, S.E., Pérez-Payá, E., Houghten, R.A. 1996. Synthetic combinatorial libraries: Novel discovery strategy for identification of antimicrobial agents. *Antimicrob Agents Chemother.* 40(5):1067–71.

Chung, E.M., Dean, S.N., Propst, C.N., Bishop, B.M., van Hoek, M.L. 2017. Komodo dragon-inspired synthetic peptide DRGN-1 promotes wound-healing of a mixed-biofilm infected wound. *NPJ Biofilms Microbiomes.* 3:9. doi:10.1038/s41522-017-0017-2.

Clemens, L.E., Jaynes, J., Lim, E., Kolar, S.S., Reins, R.Y., Baidouri, H., Hanlon, S. et al. 2017. Designed host defense peptides for the treatment of bacterial keratitis. *Invest Ophthalmol Vis Sci.* 58(14):6273–81.

Duplantier, A.J., van Hoek, M.L. 2013. The human cathelicidin antimicrobial peptide LL-37 as a potential treatment for polymicrobial infected wounds. *Front Immunol.* 4:143.

Dürr, U.H., Sudheendra, U.S., Ramamoorthy, A. 2006. LL-37, the only human member of the cathelicidin family of antimicrobial peptides. *Biochim Biophys Acta.* 1758(9):1408–25.

Elder, M.J., Stapleton, F., Evans, E., Dart, J.K. 1995. Biofilm-related infections in ophthalmology. *Eye (Lond).* 9(Pt 1):102–9.

Ganz, T. 2003. Defensins: Antimicrobial peptides of innate immunity. *Nat Rev Immunol.* 3(9):710–20.

Gordon, Y.J., Romanowski, E.G., McDermott, A.M. 2005. A review of antimicrobial peptides and their therapeutic potential as anti-infective drugs. *Curr Eye Res.* 30(7):505–15.

Gunshefski, L., Mannis, M.J., Cullor, J.S., Schwab, I.R., Jaynes, J.M., Smith, W.L., Mabry, E. et al. 1994. In vitro antimicrobial activity of Shiva-11 against ocular pathogens. *Cornea.* 13(3):237–42.

Haisma, E.M., Göblyös, A., Ravensbergen, B., Adriaans, A.E., Cordfunke, R.A., Schrumpf, J., Limpens, R.W. et al. 2016. Antimicrobial peptide P60.4Ac-containing creams and gel for eradication of methicillin-resistant *Staphylococcus aureus* from cultured skin and airway epithelial surfaces. *Antimicrob Agents Chemother.* 60(7):4063–72.

Hancock, R.E., Sahl, H.G. 2006. Antimicrobial and host-defense peptides as new anti-infective therapeutic strategies. *Nat Biotechnol.* 24(12):1551–7.

Huang, L.C., Redfern, R.L., Narayanan, S., Reins, R.Y., McDermott, A.M. 2007. In vitro activity of human beta-defensin 2 against *Pseudomonas aeruginosa* in the presence of tear fluid. *Antimicrob Agents Chemother.* 51(11):3853–60.

Jaynes, J.M., Burton, C.A., Barr, S.B., Jeffers, G.W., Julian, G.R., White, K.L., Enright, F.M. et al. 1988. In vitro cytocidal effect of novel lytic peptides on *Plasmodium falciparum* and *Trypanosoma cruzi.* *FASEB J.* 2(13):2878–83.

Jenssen, H., Hamill, P., Hancock, R.E. 2006. Peptide antimicrobial agents. *Clin Microbiol Rev.* 19(3):491–511.

Kagan, B.L., Ganz, T., Lehrer, R.I. 1994. Defensins: A family of antimicrobial and cytotoxic peptides. *Toxicology.* 87(1–3):131–49.

Lushbaugh, W.B., Blossom, A.C., Shah, P.H., Banga, A.K., Jaynes, J.M., Cleary, J.D., Finley, R.W. 2000. Use of intravaginal microbicides to prevent acquisition of *Trichomonas vaginalis* infection in Lactobacillus-pretreated, estrogenized young mice. *Am J Trop Med Hyg.* 63(5–6):284–9.

Ma, J., Kennedy-Stoskopf, S., Jaynes, J.M., Thurmond, L.M., Tompkins, W.A. 2002. Inhibitory activity of synthetic peptide antibiotics on feline immunodeficiency virus infectivity in vitro. *J Virol.* 76(19):9952–61.

Meyer-Hoffert, U., Zimmermann, A., Czapp, M., Bartels, J., Koblyakova, Y., Glaser, R., Schröder, J.-M. et al. 2011. Flagellin delivery by *Pseudomonas aeruginosa* rhamnolipids induces the antimicrobial protein psoriasin in human skin. *PLoS One.* 6(1):e16433. doi:10.1371/journal.pone.0016433.

O'Driscoll, N., Labovitiadi, O., Cushnie, T., Matthews, K., Mercer, D., Lamb, A. 2013. Production and evaluation of an antimicrobial peptide-containing wafer formulation for topical application. *Curr Microbiol.* 66(3):271–8.

Schagen, S.K. 2017. Topical peptide treatments with effective anti-aging results. *Cosmetics.* 4:16.

Schittek, B., Paulmann, M., Senyürek, I., Steffen, H. 2008. The role of antimicrobial peptides in human skin and in skin infectious diseases. *Infect Disord Drug Targets.* 8(3):135–43.

Vazirani, J., Wurity, S., Ali, M.H. 2015. Multidrug-resistant *Pseudomonas aeruginosa* keratitis: Risk factors, clinical characteristics, and outcomes. *Ophthalmology.* 122:2110–4.

Visser, M., Stephan, D., Jaynes, J.M., Burger, J.T. 2012. A transient expression assay for the in planta efficacy screening of an antimicrobial peptide against grapevine bacterial pathogens. *Lett Appl Microbiol.* 54(6):543–51.

Yeaman, M.R., Yount, N.Y. 2003. Mechanisms of antimicrobial peptide action and resistance. *Pharmacol Rev.* 55(1):27–55.

Zhang, L., Falla, T.J. 2009. Cosmeceuticals and peptides. *Clin Dermatol.* 27(5):485–94.

Zasloff, M. 1987. Magainins, a class of antimicrobial peptides from Xenopus skin: Isolation, characterization of two active forms, and partial cDNA sequence of a precursor. *Proc Natl Acad Sci USA.* 84(15):5449–53.

Zasloff, M. 2002. Antimicrobial peptides of multicellular organisms. *Nature.* 415(6870):389–95.

16

Understanding Fragrance: From Chemistry to Emotion

Adelino Kaoru Nakano

CONTENTS

Introduction

More than 5 million years ago, the end of the Ice Age revealed a whole new world of abundant forests and vegetation to which our ancestors had to adapt. Over the course of human evolution the brain has almost tripled in size and man learned how to manufacture and use tools, how to make fire and better use the surrounding resources for daily living. Our food sources have changed from those obtained by the hunter–gatherer to the far more complex sources obtained from farming and food processing. We now live in built environments and hierarchical societies with highly structured lives involving education leading into working lives followed by retirement. Whilst there have been vast changes throughout our evolution, the importance of our senses to our daily living has remained. Whilst once we used sight, hearing, taste, touch and smell (olfaction) primarily to hunt and protect ourselves, these senses are now used in a myriad of different circumstances and purposes within our modern lives. As the sensory experience has expanded exponentially, so has our brain expanded to process and translate the information into our decision making.

There are many examples of the fundamental importance of our sense of smell to our everyday lives. These include the attraction between men and women, and the bonding between parents and babies. Amongst mammals, human beings have the most fragile development path, requiring years of total care before they are sufficiently well developed to survive independently. This requires a very significant 'commitment' from the parents to provide care over this prolonged period, and there is evidence that smell plays a role in driving this commitment. An interesting research study demonstrated this role of smell. Babies of only a few weeks of age and children between 2 and 4 years old were each bathed with a neutral odourless soap, then wore a white t-shirt for 2 days. The shirts were taken to a test lab where men and women, both parents and not, smelled the worn shirts and an unworn control shirt and then registered their preferences. Interestingly women didn't notice much difference among the shirts, but there was a positive trend towards the blank shirt. However, men strongly preferred the babies' shirts. The investigators suggested that the babies' smell may be important in protecting the baby by forming a bond with the males in the household and thus reducing any tendency towards aggressiveness.

A similar study was performed to assess the importance of smell to attraction between men and women. Again t-shirts were used to absorb smells that could then be assessed by volunteers of the opposite sex. Men wore a white shirt for a number of days including sleeping overnight. They were instructed to refrain from wearing perfume and cologne, smoking and eating strong-smelling foods (spices, condiments, onion, etc.). The shirts are then put into ziplock bags and given to different women to smell. Results showed that women preferred the shirts of those men with whom they were more genetically compatible. When the study was repeated with women wearing shirts, men most commonly chose the shirts that had been worn by women who were in the peak of the ovulation cycle. These findings demonstrate the importance of smell and the role of pheromones in human evolution.

Throughout human history, mankind has been using this information unconsciously to justify somehow the use of perfumes and why we want to 'drive' our smell. The reasons why we wear perfume vary according to demographic, psychological and chronological parameters. The use by men was socially rejected by most of the cultures and this mindset had changed by the 20th century in which the primarily thinking was the women attraction. On the other hand, reasons for women can vary – they can choose one perfume believing that can represent a state of humour or even influence their current mood state. Some other reasons are situation related or expected aim (like romantic encounter) and to have more self-confidence.

Olfaction is one of the most intriguing senses, as a clear understanding of how smell is perceived by our receptors and interpreted by our brain remains elusive. Olfactory receptor genes form the largest known multigene family in the human genome and are distributed on all but a few chromosomes. The expression of these genes can vary greatly amongst individuals. The complexity of the olfactory system accounts for the ability to perceive the wide range of odours in the environment and the different olfactory experience of individuals. Thus the cosmetic industry has developed fragrances that cater to the range of olfactory perceptions and preferences.

Odorants can enter the body via two ways: by air inspiration through the nostrils or by the retro-nasal route through diffusion in the throat or airways (Figure 16.1). The first route is most important for direct smell and fragrance perception, whereas the second route is more important for tasting food. Once the odorant molecules are in the nasal cavity, they reach the olfactory epithelium located in its roof, just below and between the eyes. The epithelium, which has a surface area of around 10 cm^2 and is just a few micrometers thick, contains specific receptor cells projected like hairs (known as cilia). The molecules diffuse across the mucus on the epithelium surface interacting with odour-binding proteins that can remove excess odorant molecules or transport them to olfactory receptors. These receptors promote a biochemical cascade that leads to a signal in the orbitofrontal cortex of the brain, where the consciousness of the odour happens (odour perception). In addition, signals from the olfactory bulb to the piriform cortex and amygdala are related to emotions and memory associated with the odour. When a smell occurs, these regions together with the thalamus compare the inputs of different senses and provide information of the immediate environment from an olfactory viewpoint. The brain also processes the information from the receptors in the context of the odour memory databank to help recognize the smell.

In the past it was believed that there was a specific receptor for each type of activation. However, now it's known that the brain receives a pattern of activation of different receptors and tries to look it up in its memory databank to make the odour recognition. Even what we consider to be a very simple smell of a defined single molecule can activate different olfactory receptors and then create a signal to the brain that will be totally distinct to another molecule.

The genome contains codification to create more than 1000 different types of olfactive receptors. However, human genes create only around 400 and their expression can be very different among individuals. It is very likely that individuals can have different smell activation and consequently quite different olfactory perception of our world.

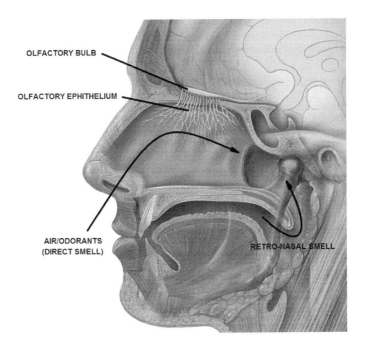

FIGURE 16.1 A simple schematic of how the olfactory system functions. (Adapted from Lynch, P.L., Jaffe, C.C., *Yale University Center for Advanced Instructional Media Medical Illustrations*, Creative Commons Attribution 2.5 License 2006; 1987–2000 [Available from: http://patricklynch.net].)

There are three main theories as to the mechanism by which a molecule can activate a receptor or group of receptors:

- *Steric theory*: Each molecule has a specific size and shape, and thus can be considered to be the key to activate a specific receptor (like a key in a lock). This model is mostly used to explain different enzymatic reactions and processes but fails to clarify why very similar-shaped molecules have very distinctive odours.
- *Vibrational theory*: Each molecule has a characteristic vibration pattern on the infrared spectrum, which can activate receptors and promote the perception of different odours. This model fails to explain the interactions of small molecules with strong characteristic odours such as menthol and carvone, or two enantiomers with very distinctive odour profiles but similar infrared spectra.
- *Vibrational-induced electron tunneling spectroscopy theory*: This is based on simple vibrational theory but with different explanation basis to support it. The odour molecule and the receptor interact with each other. The vibrational energy of the donor should be compatible with the energy gap of the receptor so that electrons can travel via 'inelastic electron tunneling', triggering signal transduction. Some studies support this theory, but it has not been conclusively proven.

What Is a Perfume?

Perfume by definition is a mixture of different ingredients comprised of aromatic compounds like essential oils, resins, waxes or synthetic aromatic ingredient; solvents to promote better dissolution of some solid materials; and additives for stability improvement like ultraviolet (UV) filters, antioxidants, emulsifiers, and so on.

The word 'perfume' comes from the Latin *per fummum* meaning 'through smoke'. This term originated by the Mesopotamians around 4000 years ago where incenses was used as the first air care.

The Egyptians used perfumed water made from flowers, herbs and spices for both ceremonial and personal purposes. Egyptian Pharaohs and their tombs were adorned with fragrances to enhance their journey to and keep their skin silky smooth in the afterlife. Persian physician and alchemist Ali al Husayn ibn Sina (Avicenna, 980–1037) is credited with the introduction of distillation of fragrant oils from flowers. The modern perfume emerged in Europe around the 14th century, when Queen Elisabeth of Hungary wore scented oil made by diluting essential oils in alcohol. This forerunner to eau de cologne was known as 'Hungary Water'. Catherine de Medici (queen of France from 1547 to 1559) brought her perfumer, Rene de Florentin, from Italy when she married King Henri II in 1533, and is credited with initiating the establishment of France as the world leader in modern perfumery. It is rumored that she wore perfumed gloves to hide the smell of the poisons she frequently administered to those who displeased her within her court. With the Renaissance, perfume became linked to fashion and prestige. The role of the 'perfumer' who creates the accords and the scents from a combination of scientific and artistic skills became highly valued. The perfumer's ability to translate vague and emotional descriptors into complex mixtures of ingredients which give a particular olfactive perception is integral to the modern cosmetic industry. Creation of a scent that can represent 'a modern woman who is self-confident and successful but also has a romantic and fragile personal side' or 'a moment in a rainy afternoon in the forest landscapes of the countryside' is very challenging. Yet success can generate global icons such as Chanel No. 5, which was launched in 1924 and remains an iconic perfume almost 100 years later. Roy Genders' Perfume through the Ages provides a fascinating insight into the history of fragrance.

Fragrances are described according to three parameters: concentration, odour families and notes.

- *Concentration* – Refers to the concentration of perfume oil in the alcohol dilution. These levels are not strictly defined in regulatory aspects and the concentration ranges are updated according to modern developments.
    - Parfum: >20%
    - Eau de parfum (EdP): 10–20%
    - Eau de toilette (EdT): 5–15%
    - Eau de cologne (EdC): 3–8%
    - Splash: 1–3%
- *Odour families*: Refers to the classification and description of a single fragrance raw material, accord or perfume into defined groups (or so-called families). This classification is vague because one floral material can contain traces of other odours like green or wood.

Families based on the main ingredient:

- *Citric*: Highly volatile compounds derived from citrus fruits in general and provide freshness and brightness to the fragrance.
- *Green*: Green, resembling wet green leaves, wet earth, plant stalks, violet leaves.
- *Fruity*: Typical fruit notes like apple, pear, banana, melon, peach and so on. Depending on its intensity can be on top or heart notes.
- *Floral*: Smells like a certain flower and can come with the name of the plant, for example, rose, orange blossom.
- *Woody*: Gives structure and 'body' to the fragrance and has typical smell from different woods, such as patchouli, cedar, sandalwood.

Families based on the accord structure/notes:

- *Floriental*: Combination of notes composed of a main floral aspect (such as jasmine, gardenia, freesia, orange blossom) with sweet, amber and oriental background. Considered to be the modern sensual and seductive but lighter than the orientals.

- *Fougère (French for 'fern')*: The name is derived from the fragrance Fougère Royale introduced by the Paris perfume house Houbigant in 1882. Although originally introduced as a perfume for women, it is now mostly used in fragrances for men. This note is described as lavender top and base of oakmoss, coumarin, woods.
- *Oriental (drawn from ingredients used in the early fragrances of India/Arabia)*: Combinations of sweet balsam, resin, amber and spices. Generally used to bring warmth and sensuality. Typical ingredients are vanilla, amber, balsam, coumarin, sandalwood.
- *Chypre (French for 'Cyprus', pronounced 'sheep-ra')*: François Coty introduced the fragrance Osmothèque in 1917 which presented the contrast between citrus top and oakmoss/woody base. The modern 'chypre' family is compounded (but not limited) by the following ingredients: rose, bergamot, jasmine, patchouli, leather.

Notes: This description comes from music (like musical notes) and the notes develop or reveal with time.

- *Top notes (or head notes)*: Immediately perceived scents composed by typical small organic molecules with high volatility (up to 15 min).
- *Middle notes (or heart notes)*: Scents that are noticed just after the volatilization of the top notes (from 15 min up to 6 hr).
- *Base notes (or bottom notes)*: Scents that comprise the 'body' of the fragrance and are noticed after the evaporation of middle notes. Typical ingredients have low volatility and can last several hours or even days on skin.

The total fragrance impression is described as a scent pyramid (Figure 16.2). The top notes are at the top, because they are highly volatile compounds perceived for a short time, then come in order, the middle and base notes.

Fragrance raw materials can be divided into either natural or synthetic ingredients.

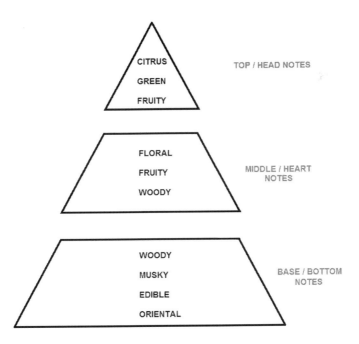

FIGURE 16.2 The relationship of typical ingredients used as part of each note of the scent pyramid.

Natural Ingredients

Living organisms produce different chemicals during their metabolism and many of the odorous compounds produced have a role in chemical signaling for communication and defense. Examples are flower perfumes that attract insects to promote pollination and pheromones for animal attraction. Plants are by far the most used natural sources for natural fragrance ingredients. Many different parts of the plant are used depending on the nature of each plant and the particular scent to be extracted. Some examples:

- *Flowers*: Rose, jasmine, lily, osmanthus, geranium, orange blossom, vanilla blossom, ylang-ylang blossom, mimosa and others. Generally orchids are not used since only a few species can have a nice perceptible scent and also their growth is very limited. In some cases, the flower buds can be used such as clove.
- *Fruits*: Typical citric fruits (lemon, orange, tangerine, lime, grapefruit, etc.), apple, strawberry, cherry, blackberry, and so forth.
- *Leaves*: Patchouli, rosemary, citric leaves, lavender leaves.
- *Bark/wood*: Cinnamon, sassafras, sandalwood, rosewood, agarwood, cedar, pine.
- *Roots or rhizomes*: Ginger and vetiver.
- *Seeds*: Mostly spices used in daily cuisine, such as tonka bean, coriander, cardamom, pepper, cocoa, caraway.
- *Scented resins*: From some plants and trees, for example, pine, myrrh, amber and fir.

Extraction Processes

Depending on the extraction process used to isolate the aromatic compounds from plants, different products are produced: concretes, absolutes, pomades, expression and essential oils.

- *Concrete*: This process is normally applied for flowers, herbs and leaves. The aromatic compound is extracted by use of a hydrocarbon-based solvent such as pentane, hexane and pentane. The final product is a waxy, dark material normally with a high content of high molecular weight materials free from the original solvent (but some of its residue can remain). The yield of this process is around 0.2 to 0.4%.
- *Absolute*: This process involves extraction from flowers, herbs and leaves by ethanol using a concrete as the starting material. Absolute ethanol is added to a concrete in a tank at a defined temperature. After a certain time the contents are filtered to remove the insoluble components and the alcohol is distilled off the low molecular weight materials dissolved to generate the absolute.
- *Pomade*: This is a process, called enfleurage, normally used to extract aromatic compounds from flowers. Purified fat is spread over flower petals/whole flowers and left to let the essential oils leave the petals and infuse the fat with their flower essence. The flower petals are then removed by hand and the process repeated with fresh petals/flowers until the fat is saturated with fragrance to form an enfleurage pomade. This material is then used in a cold alcoholic extraction process to form an absolute and the leftover fat can be used for fragrant soap.
- *Expression*: This is the simplest cold press process where the substrate is squeezed and compressed, and the oils are collected. This process is normally used for citric fruits in which there is a high content of citric oils in the peel.
- *Essential oils*: The aromatic compounds are isolated using the distillation process and all parts of the plant can be used. Normally used for extraction of low molecular weight, highly volatile compounds are extracted via this process. Care must be taken, as some components can denature or be degraded due to temperature conditions.

Animal sources that can be used traditionally come from fecal or gland essences. In the past, products were sourced from the urine and feces of horses, cows and other animals. The ethics of sourcing products from animals has been widely discussed worldwide, and many of these ingredients have or are being replaced by synthetic alternatives like musks.

- *Civet*: Collected from the odorous sacs of civets (a small, mostly nocturnal mammal native to tropical Asia and Africa).
- *Musk*: From the musk gland, which lies in a sac between the genitals and the umbilicus in the male musk deer (genus *Moschus*). The secretions are most likely used to attract mates. Musk was sold as an aphrodisiac by 7th century Arab and Byzantine perfume makers, and is still associated with sensuality. Nowadays, a synthetic substitute is normally.
- *Honeycomb*: Extracted honey or beeswax.
- *Castoreum*: Oily secretion from the castor sacs of the mature North American beaver *Castor canadensis* and European beaver *Castor fiber*. Similar to musk and associated with sensuality.
- *Lichens*: Oakmoss (*Evernia prunastri*) is a species of lichen, a fungus that grows on oak trees in the mountainous temperate forests of the Northern Hemisphere. It is commonly used to contribute a delicate forest-like, rich and earthy fragrance, especially in chypre and Fougère types of perfumes. Treemoss (*Evernia Furfuracea)* is a species of lichen that grows on pines and firs, and also provides an earthy aroma.
- *Ambergris*: From the digestive system of the sperm whale. It has a sweet earthy scent that was used by perfumers as a fixative to prolong the scent, but has been replaced by synthetics.

Synthetic Ingredients

Generating ingredients via synthetic pathways offers several potential advantages: lower cost, highly reproducible process, safety/regulatory compliance, higher purity, avoidance of using endangered species or unethical processes/conditions, high production yields compared to natural harvesting sources, and the possibility of creating new olfactory identities.

The extraction processes used for natural ingredients involve different solvents and require specific intermediate steps to remove the solvent in the final product (concretes, absolutes). If a small fraction of solvent remains, this can cause regulatory restrictions or health issues.

The distillation processes for essential oils production are costly (temperature, vacuum) and the final yield is normally only a few percent of the starting material. Thus several tons of plant materials are required to produce a few kilos of aromatic ingredient. This has an environmental impact since a lot of land is used to plant the interested species and a lot of labour is also required for harvesting. Some essential oils that are highly used by the fragrance industry (e.g. from patchouli and vetiver) are produced only in certain parts of the world limiting their accessibility and consequently their availability. Their production is highly sensitive and dependent on natural conditions in which the quality specification and price can vary according to each harvesting. Cultivation in more accessible areas may not be successful, as the variation in soil and climatic conditions can result in a final ingredient with a totally different smell compared to the original one, even when the same isolation process and plant species are used. This complicates the strategy for sourcing commodity materials that are in high demand by the industry.

Citral is an interesting example of what is called a 'nature-like' ingredient, that is, when a synthetic reaction is used to create an ingredient normally obtained via isolation from natural sources. BASF developed the chemical pathway (summarised in Figure 16.3) to synthesize citral (responsible for the natural aroma of citrus fruits and lemongrass), and through minor modifications of the chemical structure can also produce synthetic linalool (smells of lavender) and geraniol (smells of roses). Using this process BASF can produce 40,000 metric tons of citral annually, whilst cultivation from lemongrass would require 40,000 hectares of land.

Other fragrance ingredients that are now routinely produced synthetically are lilial (lilly aldehyde), linalool, geraniol, musk derivatives (musk ketone, musk xylene, moskene, etc.), civetone (a macrocyclic ketone and main scent of civet) and muscone (15-membered ring ketone that provides the odour of musk). The goal is to develop a synthetic pathway with the minimum number of synthetic steps, a high process yield and minimum residue molecules. This will also simplify the purification process. Compared to the

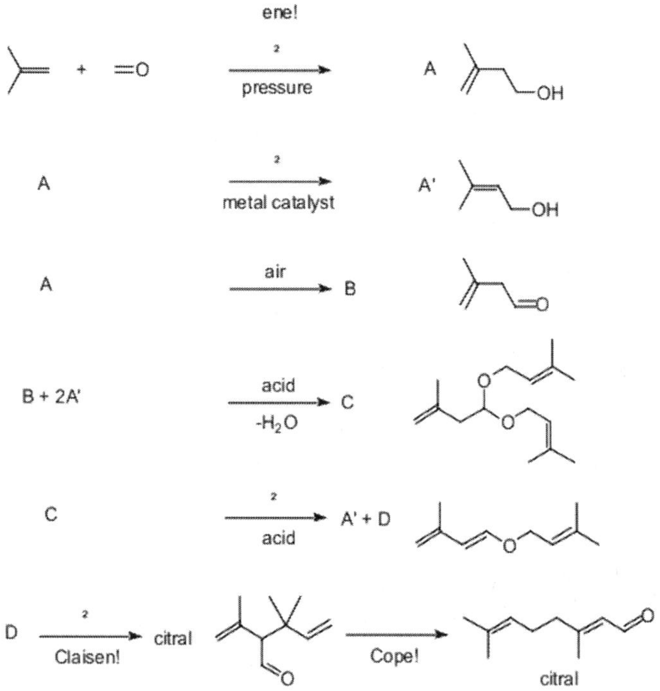

FIGURE 16.3   Chemical pathway to synthesize citral.

isolation of a single molecule from a natural source, there is considerable skill required to develop the best chemical reaction pathway with the lowest number of steps and high process yield to form a few different residue molecules. In addition the purification process should not be time-consuming or costly, but it is essential that impurities are removed, as even small amounts of these will incur regulatory restrictions or have a health/environmental impact. Whilst there are many benefits in terms of control of supply of the material, the final cost of the molecule should not be significantly more expensive compared to natural sources.

Whilst the synthetic molecules can offer many advantages, the olfactory properties may not be identical to the natural ingredient. This can arise because naturals have complex compositions, as the natural material is made of a number of different chemical structures. In contrast, the goal of the synthetic pathway is to generate a single molecule or at most a mixture of hardly similar molecules such as enantiomers. This can result in a more 'hollow', 'whiter', 'shallow' kind of smell (using perfumery wording) compared to the scent generated by naturals. It is very difficult to compare and match single molecules to complex odours. Table 16.1 summarizes the differences (advantages and disadvantages) of synthetic and natural ingredients.

TABLE 16.1

Summary of Differences between Natural and Synthetic Ingredients

	Natural	**Synthetic**
Advantages	Marketing claims: natural plant source	Marketing claims: avoid animal source
	Renewable source (if properly managed)	Simple composition
	Complex odours	Reproducible quality
		Easier management of stability issues
		Higher purity
Disadvantages	Regulatory aspects	Less complex odours
	Compliance/irritancy effects	

Formulation of Fragrances in Cosmetic and Personal Care Products

As fragrances are complex mixtures composed mainly of different aromatic ingredients (oils, waxes, resins, essential oils and so on), stabilizers and solvents, the analysis and study of individual component's properties can help inform the characteristics of the final fragrance. To assist in the appropriate formulation it is essential to have an understanding of the physicochemical properties of each single material and/or the simple mixtures of a few ingredients (known as 'accords'). Knowledge of these properties will allow design of the formulation to ensure the fragrance will be easily incorporated and stable in the final base. The main physicochemical properties used to assess materials are briefly described.

> *Boiling point*: Determined at air pressure, ground level. This provides an indication of evaporation of the fragrance material.
>
> *Water solubility*: Water solubility in milligrams per litre (mg/l) at room temperature.
>
> *clogP*: Calculated value of the distribution of a material between octanol (non-polar) and water (polar). High clogP indicates a lipophilic material that will be most soluble in oils; low clogP indicates hydrophilic material that will be most soluble in aqueous vehicle components.
>
> *Diffusivity*: Determined by a panelist who assesses how much they can smell at a certain distance away from the material. This is an interesting parameter to evaluate how fast and how far a material can expand through the ambient air.
>
> *Tenacity/persistence*: The capacity of an ingredient to remain detectable over time. It is related to the volatility and is therefore dependent on parameters such as the molecular weight (higher weight confers lower volatility) and polarity (intermolecular interactions that bind the material to the base tend to be more prevalent in polar materials).
>
> *Impact*: Relates the intensity of smell compared to the concentration of the material (potency). Normally assessed by a panelist who rates the intensity of the odour to a range of concentrations of a material.
>
> *Threshold*: Minimum concentration with a perceivable odour. This is an important parameter, as it defines the threshold concentration necessary to be used in a fragrance formulation.

Typical Formulations Incorporating Fragrance

The simplest formulation using fragrances is an alcoholic dilution to prepare a perfume. This consists of ethanol (around 80% concentration), water and the fragrance oil. As the final packaging is usually transparent glass, stabilizers are added to protect against interaction with sunlight and prevent any visual change (colour, precipitation, cloudiness, etc.) that will result in consumer rejection of the product. Typical stabilizers are UV filters (benzophenone-3, homosalate and others), chelating agents to reduce availability of heavy metals (ethylenediaminetetraacetic acid [EDTA]) and antioxidant agents (butylated hydroxytoluene [BHT], citric acid). Many perfumes contain colourants or dyes that can add further complexity as some colourants, such as red and blue dyes, have limited photostability requiring a better stabilization system. Perfumes are best prepared by first adding the fragrance oil to alcohol, followed by the addition of the water and other ingredients. The final solution is then chilled to around 5°C and filtered to remove any solid residues or solid material that was not properly solubilized, and also to avoid future precipitation.

Emulsions are systems formed by three main ingredients: oil phase, water phase and a surfactant/emulsifier to stabilize the emulsion. The type of emulsion formed will depend on the ratio of the oil and water, and the properties of the emulsifier: an oil-in-water (O/W) emulsion consists of droplets of oil dispersed in a continuous phase of water, and vice versa for a water-in-oil (W/O) emulsion. Most of the creams or lotions in the personal care category are O/W emulsions, thus the fragrance oil is in the dispersed phase of the emulsion.

The odour profile of a cream product can be altered if there are any changes that lead to chemical or physical changes in the cream formulation. The stability of the cream will depend in the first case on the quality of the ingredients used. Normally the fatty material trends to rancidity and this can cause a bad oxidized odour smell typically noticeable in high temperatures. This can be minimized or avoided having the same stabilizers mentioned before for alcoholic dilutions. The second factor influencing the stability will be the addition of active ingredients responsible for any action on the skin (like alpha hydroxy acids, aluminium salts, perms, tints) or in the cleaning process at home (like silion, alkaline ingredients to remove fat, calcium salts, enzymes), usually most reactive type of material which might contribute to the fragrance instability.

Stability of Fragrance in Personal Care Products

The stability of fragrance materials is highly dependent on the formulation environment, and in particular, the pH of the formulation. Personal care products exhibit a range of pH environments with most skin care products being slightly acidic to mimic the natural skin pH of 5.5 to 6.5, whilst soaps tend to be alkaline due to the nature of the anionic surfactants used in their preparation. Figure 16.4 shows the typical pH of a range of consumer products that incorporate fragrances.

At alkaline pH (such as soap bars), vanillin, eugenol (and derivatives), heliotropin and citral may undergo a colour change due to acetyl formation from the reaction of aldehyde and alcohol groups, but

FIGURE 16.4 Typical pH range of consumer products that incorporate fragrances.

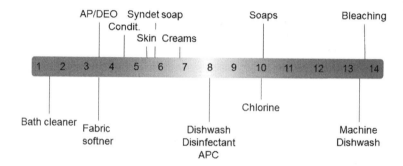

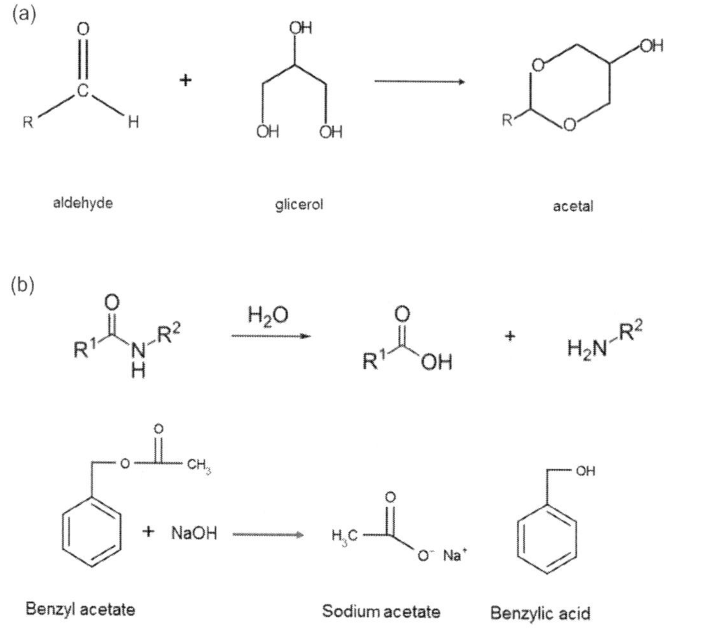

FIGURE 16.5 (a) A colour change due to acetyl formation from the reaction of aldehyde and alcohol groups. (b) Ester hydrolysis can occur in an acidic environment.

mainly to glycerol derivatives (Figure 16.5a). This process can happen mainly on bar soap bases in which the pH is quite high (around 10). Vanillin, eugenol (and derivatives), heliotropin and citral are examples of ingredients that can suffer colour change in alkaline base, whilst ester hydrolysis can occur in an acidic environment (Figure 16.5b). Some personal care products (depilatories, bleaches, perms, tints) have extreme pH variations and a highly oxidizing environment. Few ingredients are stable in these situations since the aldehydes are often attacked in highly alkaline formulae. Phenylethyl alcohol is one of the few fragrance ingredients that is reasonably stable in different bases, including extreme environments. Highly oxidizing formulae like bleaches and cleaners associated with high pH normally restrict the available ingredients to be used.

A common reaction between fragrance ingredients is the Schiff base formation whereby an amine group (either aromatic or aliphatic) reacts with the carbonyl group present on an aldehyde via nucleophilic addition (Figure 16.6). Vanillin is very sensitive to form this reaction, which results in a brown-yellowish colour.

Table 16.2 shows examples of typical stability issues that affect fragrances in personal care or home care formulations.

Tensoactive-based formulations (such as shampoos and liquid soaps) are composed of a primary surfactant, secondary surfactant (or co-surfactant), viscosity builder and additives. The surfactant molecules associate to form micelles (as described in Chapter 9). The fragrance oil ingredients will be retained in the hydrophobic inner part of the surfactant micelles. This can result in changes to the formulation properties such as a change in viscosity and possible increase in the cloud point for clear formulations. Predicting and/or identifying stability issues are very important, as they will require reformulation.

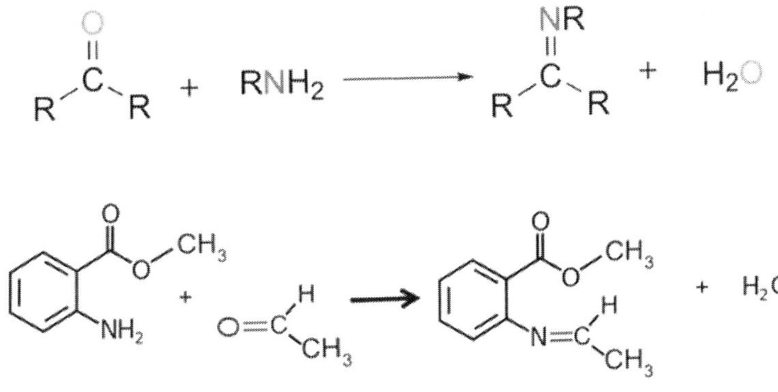

Methyl anthranilate

FIGURE 16.6 Schiff base formation whereby an amine group (either aromatic or aliphatic) reacts with the carbonyl group present on an aldehyde via nucleophilic addition.

TABLE 16.2

Summary of Main Reactions of Fragrance Ingredients Depending on the Product

Product	Active Ingredient	Possible Issue
Antiperspirant/deodorant	Aluminium chloride/zirconium chloride derivatives	Acid catalyzed reactions
Permanent wave	Thiolgycolic acid	Nucleophilic attack, sulphur malodours
Talcum	Talc	Loss of perfume performance
Bar soap	Soap	Alkaline pH
Laundry detergent	Soap, enzymes(?)	Alkaline pH, degradation of some fragrance ingredients
Fabric softener	Quats	Acid pH
Laundry bleach	Soap, bleach, enzymes(?), oxygen(?)	Alkaline pH, oxidation environment
All-purpose cleaners	Bleach, abrasives	Oxidation, cloud point
Bath cleaners	Acid or base, bleach	Oxidations

Stability Testing

There is no standard or regulation that stipulates protocols for stability testing of fragrances so the procedure is normally developed based on the formulation itself. Similar test conditions are used for personal care, fine fragrance or home care products. These generally include a range of temperatures, temperature cycles and UV light exposure:

Temperature:

5°C (control)

20–25°C (ambient)

40–45°C (high temperature – 2 to 3 months)

50°C (high temperature – 3 to 4 weeks)

Temperature cycle: freeze/thaw; cold/warm
Light exposition:

- Artificial UV light
- Natural sunlight
- Showcase light

High temperature (50°C) testing provides an aggressive stability test on a reference sample as an initial indication of stability of the formulation in extreme conditions. Stability of the sample in these conditions is a very positive indication. The stability study should include both test samples and a 'placebo' formulation (formula without the fragrance) to determine the impact of the fragrance in the formula. The product should be tested in neutral packaging (glass) and in the final packaging material to determine packaging/product interaction. The parameters assessed are generally physical in nature: colour change, turbidity, film formation or any other change to the appearance; odour profile (characteristics) and intensity. A minimal change in fragrance intensity and/or characteristics after 4–8 weeks can be expected, but a major change is an indication of an unstable sample and the need for reformulation. Stability considerations and the formulation design are always a compromise between the formulation cost, aesthetics and product life cycle.

New Technologies in Fragrances

There is a strong drive throughout all industries to produce products in more sustainable ways with minimal impact on the environment. In the fragrance and fragrance-related industries this includes sustainable harvesting of raw materials, optimization of their manufacturing processes, reduction of gas emissions and developing synthetic pathways that use minimal energy and solvents. Some of the initiatives that have been implemented to produce natural fragrance ingredients in more sustainable ways include:

- Close collaboration with local producers to optimize their production via education, investments in infrastructure and social programs, fair trade, and ethical production policies.
- Improved extraction processes with better yields and lower residue generation.
- Reuse of byproducts from extraction processes via physical, chemical or biotechnological processes converting disposable starting material to a high-value ingredient for the cosmetic or other industries.
- Reduce/eliminate the production of ingredients from animal sources or endangered flora.

In addition the efficiency of producing synthetic ingredients has been improved by:

- Optimization of the synthetic pathway to increase production yield and reduce byproduct residues.

- Creation of high-impact materials requiring much lower concentration in the final product.
- Change the production route from extraction to synthesis when there is a gain in sustainability.
- Use of biotechnology within the production process to convert raw material into a desired molecule via controlled metabolism by selected microorganisms (bacteria, algae, fungus, etc.).

The move towards sustainable production of cosmetics and fragrances is being driven by both consumers and the industry. Whilst there has been a strong trend towards natural products for many years, consumers need to be assured that these are obtained from renewable sources with minimal adverse environmental impact. Consumer advocate and environmental advocate groups have been very vocal in exposing poor practices and can severely damage brand image.

Delivery Systems for Fragrances

Encapsulation systems are the most common fragrance delivery systems. They protect the fragrance, enhancing its stability by reducing susceptibility to oxidation, hydrolysis, chemical interaction and evaporation. In addition, encapsulation can control release of the fragrance to create a new olfactory experience for consumers. There are three main ways to encapsulate fragrances:

- *Spray-drying*: A pre-mix of the fragrance, additives, base (most commonly used material is a modified starch) and solvent (usually water) is injected as liquid under pressure into a heated nozzle which evaporates the solvent to generate a powder as the final product. This encapsulated material can be incorporated in any anhydrous system such as detergent powders, talcum, diapers, feminine hygiene and some antiperspirant/deodorant products. When the starch comes in contact with water it dissolves and releases the fragrance. This normally serves as proof that the application product is working (like diapers) or can be a protection signal (like antiperspirant/deodorant). This technology is not a traditional encapsulation technology *per se* since the fragrance is diluted and dispersed in a powder environment.
- *Polymerization*: The fragrance is embedded in a polymeric capsule composed of water and an insoluble material such as gum, gelatin, urea formaldehyde, alginate and melamine. The fragrance is dispersed in a solution containing all ingredients and the polymerization process is initiated by addition of a hardener to the mixture. The capsule shell becomes rigid when dried, therefore these systems can be applied in any product which will be used to deposit a certain amount on a defined surface: body lotions, antiperspirant/deodorant, conditioners, fabric softeners and so on.
- *Fragrance precursors*: These are 'one step behind' that when activated form the final fragrance ingredient by a chemical transformation. Activation of the precursor can be by exposure to light, water or other triggering pathways as discussed earlier. The advantage is that the chemical material can be directly incorporated in a base making it easier to formulate. However, the regulation requires the molecule to have a smell (precursors normally don't smell), therefore this approach demands a lot of investigation. In addition the requirement for additional extensive safety and toxicity studies led many in the industry to reduce their investment in this focus.

Other types of triggering systems based on pH/electrolyte (change in pH or concentration of electrolytes) can be used to cause fragrance release. However, these technologies tend to be more expensive and are therefore not normally seen in fragrance applications. Many other mechanisms can be used as release triggers such as temperature change, enzymatic environment, light, oxygen or pressure.

The Fragrance Experience

Fragrances are expected to evoke emotions. Research groups in neurophysiology, psychology and related areas investigate how these emotions can be measured and, in association with industry researchers, how these emotions can be activated by certain accords or fragrance hedonics. Biofeedback measurements that are commonly measured include heart rate and rhythm (electrocardiogram [ECG]), blood pressure, prosody (vocal stress) and electrical activity of the brain (electroencephalogram [EEG]). In addition questionnaires are administered to collect subjective assessment of the fragrance experience. It is the combination of the information gathered from these sources that is used to correlate the physical and emotional response to the applied fragrance product.

Some interesting fragrance experiences are being developed to accompany new technologies. For example, in recent years a number of companies have been trying to 'send' a fragrance via the Internet. This system works by having a defined number of accords in small equipment at the destination that is actuated to release the fragrant scent into the air when a message is received. Another interesting system is to create a beyond 3D experience in movie theatres by spraying certain smells in the air during a defined scene in the movie. This brings deeper movie immersion to the audience since an additional human sense is stimulated. Other examples of innovative fragrant experiences are brand olfactive signatures, textiles that release fragrances upon mood variation and small electronic devices (such as an MP3 player) that can release any fragrance upon a selection in a menu on the device.

Final Comments

Olfaction is the only one of the five senses that can influence mood and remind us of memorable events. This is because the olfactory bulb is linked to our limbic system, which is the part of the brain responsible for memory and emotion. Thus a smell can trigger a response almost immediately due to an associated memory.

Fragrance, initially considered as a futile adornment, is now a market segment which generates more than €5 billion worldwide for fine fragrances (and related formats) and is growing around 6% every year. This number does not include the application in personal care, home care products and differentiated applications for entertainment and olfactive marketing industries. The size of the market clearly demonstrates the importance of getting the fragrance right in cosmetic products and supports the attention it is given by the industry. It involves multidisciplinary areas, from a perfumer who mixes and blends ingredients and materials to create fragrances; to lab technicians, chemists, researchers, designers, computer programmers, engineers and several others who contribute new technologies for applying fragrances. It is the coming together of art and science that is constantly creating new fragrance experiences for the consumer.

BIBLIOGRAPHY

Anonis, D., Perfumes and creams. *DCI*, 107(3), 44–48, 164–166, 1970.

Aubert, A., Social and psychological influence of odors: A bio-cultural perspective. In: *Biology of Odors: Sources, Olfaction and Responses*, L.E. Weiss, J.M. Atwood (Eds.), Nova Science Publishers, New York, 2011.

Baser, K., Buchbauer, G., *Handbook of Essential Oils: Science, Technology and Applications*, CRC Press, Boca Raton, FL, 2010.

Behan, J., Perring, K., Perfume interactions with sodium dodecyl sulphate solutions. *Intl. J. Cos. Sci.*, 9, 261–268, 1987.

Blakeway, J., Bourdon, P., Seu, M., Studies in perfume solubilization. *Int. J. Cos. Sci.*, 1, 1–15, 1979.

Davies, E., The sweet scent of success. *Chem. World*, Feb, 40–44, 2009.

Ericksen, M., *Healing with Aromatherapy*, McGraw-Hill Professional, New York, 2000.

Firestein, S., How the olfactory system makes sense of scents. *Nature*, 413, 211–218, 2001.

Fischer, E., Fieber, W., Navarro, C., Sommer, H., Benczedi, D., Velazco, M.I., et al., Partitioning and localization of fragrances in surfactant mixed micelles. *J. Surfact. Deterg.*, 12, 73–84, 2009.

Genders, R., *Perfume through the Ages*, Putnam, New York, 1972.

Gottfried, J. (Ed.), *Neurobiology of Sensation and Reward*, CRC Press, Boca Raton, FL, 2011.

Handa, S., Khanuja, S., Longo, G., Rakesh, D. (Eds.), *Extraction Technologies for Medicinal and Aromatic Plants*, International Centre for Science and High Technology, Trieste, 2008.

Herman, S., Fragrancing emulsions. *Cosmetics & Toiletries*, 109(8), 71–75, 1994.

Herz, R., Eliassen, J., Beland, S., Souza, T., Neuroimaging evidence for the emotional potency of odor-evoked memory. *Neuropsychologia*, 42, 371–378, 2004.

Kemp, C., *Floating Gold: A Natural (and Unnatural) History of Ambergris*, University of Chicago Press, Chicago, 2012.

Lynch, P.L., Jaffe, C.C., *Yale University Center for Advanced Instructional Media Medical Illustrations*, Creative Commons Attribution 2.5 License 2006; 1987–2000 [Available from: http://patricklynch.net].

Mookherjee, B., Patel, S.M., Trenkle, R.W., Wilson, R.A., Fragrance emission from the skin. *Cosmetics & Toiletries*, 113(7), 53–60, 1998.

Nakano, A., Carrasco, C., Bellon, P., Olfactory stimulation and psychophysiological measurement of hexenyl methylbutanoate by EMG technique and other physiological parameters, 27th IFSCC Conference, Brazil, 2013.

Nakano, A., Carrasco, C., Bellon P., Human calming method: Emotional assessment of perfumes, 27th IFSCC Conference, Brazil, 2013.

Rouquier, S., Taviaux, S., Trask, B.J., Brand-Arpon, V., van den Engh, G., Demaille, J., et al., Distribution of olfactory receptor genes in the human genome. *Nat. Genet.*, 18(3), 243–250, 1998.

Schleidt, M., Hold, B., Attila, G., A cross-cultural study on the attitude towards personal odors. *J. Chem. Ecol.*, 7, 19–31, 1981.

Segot, E., Jacquin, J., Abriat, A., Fanchon, C., Aubert, A., How women under everyday stress perceive the benefits of fragranced cosmetic product use, 25th IFSCC Congress, Barcelona, Spain, 2008.

Sell, C., *Understanding Fragrance Chemistry*, Allured, Carol Stream, IL, 2008.

Soussignan, R., Schaal, B., Marlier, L., Jiang, T., Facial and autonomic responses to biological and artificial olfactory stimuli in human neonates: Re-examining early hedonic discrimination of odors. *Physiol. Behav.*, 62, 745–758, 1997.

Vernon, A., James, D., Olfaction: Anatomy, physiology and behavior. *Env. Health Persp.*, 44, 15–21, 1982.

Winston, R., *Human Instinct*, Bantam Books, London, 2002.

17

Sunscreens

Zoe Webster

CONTENTS

Introduction

Sunscreens are broadly defined as materials that prevent ultraviolet (UV) light from reaching the skin. The mechanism of action varies from materials that reflect the UV photons through to highly complex organic molecules that work by absorbing the energy of the photons and converting it into less harmful radiation, usually infrared.

Sunscreens are not applied directly to the skin and in fact most regions of the world are governed by regulations which limit percentage inclusion levels of sunscreens in products. In the European Union (EU) products with sun protection claims are regulated as cosmetics and this is also the case in many areas of the world. The United States of America (USA), Canada and Australia are notably different in that products with sun protection claims are regulated as drugs and are therefore more closely controlled and are costly and more complex to develop. The sector is also cluttered with patents in particular covering benefits of a specific blend of sunscreens in combination with particular solvents and emollients.

The sunscreens themselves all have unique characteristics, and understanding them individually and how they behave after blending is key to developing a successful sun protection formulation. A requirement for all sun protection formulations is that the sunscreen system remains solubilized (for organic/ absorbers) and does not crystallize under any condition during the life of the product. For inorganic/ reflector sunscreens the requirement is for the particles to remain dispersed evenly throughout the formulation and not agglomerate as the product ages. Further challenge is provided by the need for sun protection products to deliver a multitude of additional claims and deliver skin care benefits over and above the sun protection factor (SPF), water resistancy and UVA claims.

In addition to the considerable technical challenges and complex regulatory and patent landscape through which the formulator has to navigate the demands of the 21st century consumer that the product is aesthetically pleasing and is a pleasure to use and would prefer to not be aware they are wearing the product at all whilst deriving all of the benefits. No longer will the heavy oily creams of the past sell however well they protected us. All of this ensures that the job of a sun protection formulator remains highly challenging.

History and Evolution of Sun Protection

Sun protection in the form of products in a bottle that are applied to the skin originate from within the 20th century. We also employ other measures to protect ourselves such as wearing clothing and staying in the shade. There is little evidence that ancient man used products to protect the skin but our ancestors certainly knew of the effects of sun exposure. Paintings on a 5th century vase held in the Metropolitan Museum of Arts in New York show shepherds wearing full body covering garments and hats with a large brim. The writings of the ancient Greeks indicate that a tan was unacceptable and that ladies would cover themselves up whilst in the sun to protect themselves[1].

It is unclear whether the drive for protection was driven by fashion or to protect health, but it was most likely a combination of both. Certainly as recently as the mid-1900s the full implications of too much or indeed too little sun exposure were not fully understood.

Sun protection products started to appear in the early part of the 20th century and the brand names they sold under are still familiar to us today. In 1935 Eugène Schueller, founder of L'Oreal, formulated the first radiation filtering product, 'Ambre Solaire Huile'[2]. In 1938 a chemistry student, Franz Greiter, experienced the full effects of sun exposure whilst climbing the Piz Buin Peak in the Alps and was inspired to develop a cream he called Gletscher Crème (Glacier Cream) launched in 1946[3]. Boots the Chemist launched their its own label Soltan Brand in 1939 with a product that protected against sunburn[4].

None of the products at this time quantified the amount of protection provided and it was not until 1956 when Schulze introduced the concept of the sun protection factor (SPF) and measured the effectiveness of sun protection products when applied at an even rate[5].

Products continued to appear during the intervening years and in the post-war era a tan started to become a sign of wealth and became fashionable. Many of the early products had little or no protection against UV and were used purely to promote a tan. A Consumer Reports study from the late 1940s showed that, of 61 suncare products, 5 gave excellent protection, 13 gave good protection, and 31 gave no protection[6]. As recently as the early 1990s, SPF 2 sun oil was still a common sight on shelves and in Mediterranean countries SPF 0 was also available (author's experience).

The advent of the SPF 30 and higher is a relatively recent occurrence. Products started appearing in mainstream brands throughout the 1990s and initially formed a small proportion of sales. SPF 15 was labelled maximum and SPF 6 as high[7], and there was still a prevailing mentality that protection was only required to the extent that the product prevented sunburn but only to the extent that the wearer would still develop a nice tan.

Claims were SPF, which is primarily a measure of how long the wearer is protected before developing sunburn, and water resistance, a property that had been introduced into formulations during the 1970s. The protection was delivered via the use of UVB sunscreens, and a relatively small palate of materials was used at the time.

The dangers of UVA were well documented by this time and UVA sunscreens began to be more widely used. What did not exist was a mechanism for communicating to the consumer the level of UVA protection provided by the product.

Broadly speaking, two differing approaches emerged to address this. A method called PPD (persistent pigment darkening) uses UVA radiation to develop a tan within human skin, and the level of protection is indicated by a protection grade. In the UK, an *in vitro* approach was developed, and in 1992, scientists at Boots, in conjunction with Brian Diffey, launched the star rating system which appears as a symbol on the pack. The system dominates in the UK, and is used extensively under license. In a bid to provide a simple European-wide system of indicating level of UVA protection, COLIPA (The European Cosmetic, Toiletry and Perfumery Association) introduced an in vitro method for testing products. In simple terms the test determines the amount of UVA protection in relation to the UVB protection. Provided the proportion is above 1/3, a symbol may be used on the pack.

From the year 2000 onwards, sun protection products became increasingly complex technically and the scientific community had started to understand the delicate nature of many sunscreen molecules and how easily they could be damaged and rendered ineffective when exposed to UV light. The resilience of a sunscreen molecule became known as photostability. In practice a formulation that was not photostable would degrade on the skin and the level of protection provided would decrease during wear. At the same time the need to provide much higher levels of UVB and UVA protection became mainstream. The use of photostable sunscreens helped the formulator to produce an effective product at SPF 50 and above, which would remain effective on the skin. Photostable claims appeared first on Ambre Solaire and later on other key brands. *In vitro* UVA testing methods were updated to take account of the photostability of the system.

Products that provide protection over an extended period of time and hence do not need to be reapplied as frequently as the standard sun protection product appeared in several brands although they remain niche, and many of the big global brands have not ventured into this area.

Sunscreens

Cosmetic scientists have available a wide range of sunscreen materials for inclusion in formulations which will enable their product to provide the wearer with protection from the damaging effects of UV radiation.

Sunscreens can be categorized in a variety of ways. The most common is to split them into two sections depending on how they work. The first, and largest, group is frequently referred to as chemical or organic filters and their mode of action is absorption (Figure 17.1). They are typically large organic molecules with multiple unsaturated bonds. When subjected to UV radiation the molecule absorbs the photon(s), which increases the energy of the molecule from the ground state to the higher level of energy called an excited state. The molecule will then attempt to return to the lower energy ground state level and, if successful, returns back to the same form it was in prior to irradiation, ejects the extra energy as infrared radiation, and is ready to receive more photons and continue the cycle once more.

Molecules can only increase or decrease in energy according to certain rules and once in an excited state they will always seek to eliminate the extra energy and return to the ground state as soon as possible. There are many mechanisms whereby this can happen and there is potential for a molecule to achieve this by breaking up and forming smaller molecules which are not UV absorbers. This effect is accelerated by continued exposure to UV when the molecule is being bombarded with higher energy levels than it can accept. This is not an ideal feature in a sunscreen material because the primary molecule no longer exists and therefore the efficacy of the sunscreen reduces each time a molecule breaks up and the wearer is no longer fully protected. The resiliency to this effect varies according to the structure of the specific sunscreen and is known as photostability.

The images in the Appendix (Figures 17.4 to 17.27) show the chemical structure and UV profiles for the most commonly used absorber sunscreen materials. The dotted red line on the graph shows the UV profile of the material prior to irradiation with UV. The solid red line shows the profile absorbed after the material has been subjected to an irradiation dose of SPF×MED (1 minimal erythema dose passes through sunscreen onto skin). The larger the gap between the two lines the less photostable than the material. This is only demonstrated here for individual sunscreens and the photostability of many molecules can be improved by combining them with other sunscreens and cosmetic materials.

The second category and the smallest are the inorganic or reflector sunscreens. These function somewhat obviously by reflecting UV light. They are particulate materials and the wavelength of UV light reflected is governed by both the chemistry of the materials and also the size and form of the particles. Two materials dominate this sector: titanium dioxide and zinc oxide.

Titanium dioxide is widely used as a pigment and is bright white because it very effectively reflects the whole of the visible spectrum of radiation. The pigmentary material has a particle size of approximately 0.25 microns and provides virtually no protection against UV light. To produce it in a form that will reflect UV light it has to be micronized to a particle size in the region of 30–40 nm and at this size it does the opposite to its pigmentary cousin and does not reflect visible light and is therefore invisible when applied to the skin.

In practice, many UV grades of titanium dioxide do give some white colour when applied to the skin and this is due to two factors.

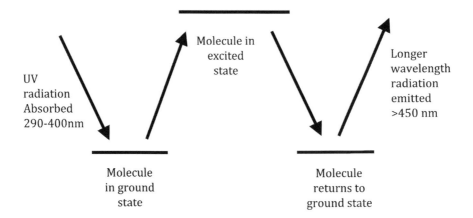

FIGURE 17.1   Organic UV sunscreen, mechanism of absorption.

1. Production of this type of material will always result in a range of primary particle sizes and some will be large enough to reflect visible light which sits next to UVA on the electromagnetic spectrum. The greater the UVA protection afforded by the material, the larger the particle size that is required and the more likely it is to be white on the skin.

2. The primary particles have a tendency to clump together and form agglomerates to reduce any static forces. The agglomerates act as if they were primary particles and according to their size if they are large enough will reflect visible light. Many of the commercially available titanium dioxide materials are sold as dispersions, which have been designed to hold as many of the particles in suspension in their primary form as possible and prevent agglomeration.

The UV-grade titanium dioxide that is commonly available reflects primarily in the UVB region, although each grade will have a slightly different profile and efficiency (Figure 17.2).

Titanium dioxide is usually produced with a coating that provides two benefits:

1. Helps prevent agglomeration.
2. Prevents the raw titanium dioxide from generating free radicals when exposed to UV light.

Zinc oxide is a very useful chemical and is widely used in the electronics industry as well as an antibacterial and deodorizing agent in many cosmetics and pharmaceutical applications. When it is micronized it gains the property of reflecting UV radiation. The particle size required is small enough to ensure that under EU legislation the word 'nano' is included in the International Nomenclature of Cosmetic Ingredients (INCI) name for zinc oxide. Zinc oxide reflects both UVA and UVB radiation in roughly equal amounts and therefore has a unique spectrum within the family of sunscreens discussed here of being a flat straight line with a slight kink at around 380–400 nm (Figure 17.3).

Zinc oxide is a very recent addition to the list of sunscreens permitted for use in the EU and was officially included in Annexe VI to regulation EC No 1223/2009 of the European Parliament and of the Council On Cosmetic Products on the 21 April 2016.

In common with titanium dioxide, zinc oxide is sold in both powder format and as a dispersion.

Each sunscreen material has a unique set of properties and characteristics, and will absorb (or reflect) UV energy across a specific range of wavelengths resulting in a characteristic extinction spectrum for the material. The relative strength of the effect can also be compared by standardizing the percentage

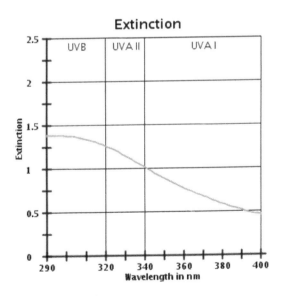

FIGURE 17.2 Titanium dioxide spectrum. (From www.sunscreensimulator.basf.com/Sunscreen_Simulator.)

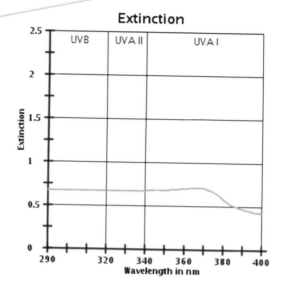

FIGURE 17.3 Zinc oxide spectrum. (From www.sunscreensimulator.basf.com/Sunscreen_Simulator.)

of material in solution. The λ_{max} can also be determined and is quoted in Table 17.1 for commonly used sunscreens.

The solubility of each individual sunscreen in a variety of emollients and solvents varies considerably and part of the art of creating an effective suncare product is to ensure that all sunscreens are solubilized at point of manufacture and stay solubilized throughout the life of the product.

The majority of sunscreens both organic and inorganic are used in the oil phase of a formulation. The efficacy of these varies, although high UVB- and UVA-related claims can be achieved with relative low inclusion levels of materials. Some materials are water soluble and larger quantities are needed in the formulation to deliver a relatively small contribution to the SPF or UVA claims of a product. The efficacy of these materials is susceptible to product pH, which must be carefully controlled and remain stable throughout the life of the product. The water-soluble sunscreens have a key advantage over the oil soluble materials in that they do not negatively affect the texture of a product as many of the oily sunscreen materials can. The efficacy of these materials is closely linked to the pH of the formulation.

The choice of sunscreen system is also influenced by the end user and the type of claims required. Products for sensitive skin or for use with babies and children frequently utilize the inorganic sunscreens, which are perceived as natural and more gentle. They can be used at higher levels with lower risk of causing irritation or provoking an allergic response. The organic/chemical filters do carry a higher risk of causing irritation or, in some cases, provoking allergic reaction, and this is managed via capping of usage levels via legislation and ongoing monitoring of the materials. Many brand owners will put additional restrictions over and above the regulatory minimum to ensure their products are as safe for their intended use as possible. Whichever system is used the safety profile of the final formulation is a function of the other ingredients included, so sunscreen selection is only one factor in the decision-making process.

Organic sunscreens can also be used in these product types, although there is a restricted palette of materials to select from. Most sunscreens absorb or reflect predominantly in either the UVB (290–320 nm) or UVA (320–400 nm) part of the spectrum. There are a few which have activity across the full range of 290–400 nm and these are often referred to as broad-spectrum sunscreens. This is summarised in Table 17.1 for the commonly used sunscreens.

The use of sunscreens within formulations is highly regulated across all world markets and the regulations control which materials can be used and limits on percentage inclusion. Current inclusion levels for the five main areas (EU, United States, Canada, Australia and Japan) are included in Table 17.1. In the United States, Canada and Australia the products are classified as over-the-counter (OTC) drugs and

TABLE 17.1

Sunscreen Agents in Use Worldwide

INCI Name	Soluble in	Form	Chemical Structure	UVA or UVB	Max Wavelength	Max Permitted Levels %					UV Absorbance Profile
						EU	USA	Canada	Australia	Japan	
Bis-Ethylhexyloxyphenol methoxyphenyl triazine	Oil	Pale yellow powder	Figure 17.4	UVA/UVB	341 nm	10	—	—	10	10	Figure 17.5
Butyl methoxydibenzoyl-methane	Oil	Pale yellow powder	Figure 17.6	UVA	355 nm	5	3	3	5	10	Figure 17.7
Diethylamino hydroxy-benzoyl hexyl benzoate	Oil	Yellow crystalline powder	Figure 17.8	UVA	354 nm	10	—	—	10	10	Figure 17.9
4-Methylbenzylidene camphor	Oil	Off-white powder	Figure 17.10	UVB	303 nm	4	—	—	4	—	Figure 17.11
Benzophenone-3	Oil	Pale yellow crystalline powder	Figure 17.12	UVB/UVA	287 nm	10	6	6	10	5	Figure 17.13
Benzophenone-4	Water	Pale yellow crystalline powder	Figure 17.14	UVB/UVA	285 nm	5	10	10	10	10	Figure 17.15
Diethylhexyl butamido triazone	Oil	White powder	Figure 17.16	UVB	311 nm	10	—	—	—	—	Figure 17.17
Ethylhexyl methoxycinnamate	Oil	Pale yellow liquid	Figure 17.18	UVB	310 nm	10	7.5	7.5	10	20	Figure 17.19
Ethylhexyl salicylate	Oil	Colourless liquid	Figure 17.20	UVB	306 nm	5	5	5	5	10	Figure 17.21
Ethylhexyl triazone	Oil	Pale yellow powder	Figure 17.22	UVB	312 nm	5	—	—	—	3	Figure 17.23
Octocrylene	Oil	Clear yellow viscous liquid	Figure 17.24	UVB	302 nm	10	10	10	10	10	Figure 17.25
Phenylbenzimidazole sulfonic acid	Water (salt form)	White powder	Figure 17.26	UVB	307 nm	8	4	4	4	3	Figure 17.27

Sources: BASF Raw Material Data Sheet[8]; DSM Raw Material Data Sheet[9]; Aako Raw Material Data Sheet[10]; Charts and Notes C 2014 DR COSREG Ltd (Summit Events)[11]; www.sunscreensimulator.basf.com/Sunscreen_Simulator.

development of new filters is both time consuming and expensive. Costs are in the multi-million-dollar order of magnitude and a timeline of between 3 and 10 years is typical to enable the completion of all of the necessary tests and to get the product approved by regulators. For this reason, new materials do not regularly enter the market. Only a handful of materials have been approved since 2000 and these have not been included in all global regions at this time.

Regulatory Landscape

Sun protection products are heavily regulated worldwide. In most global regions there are regulations applicable to the sunscreen materials themselves, which is surprisingly inconsistent in some instances. The maximum inclusion levels for commonly used sunscreens in five main regions is included in Table 17.1 and it can be seen that in some regions certain materials are not permitted, whilst in another region the sunscreen may be used at a significant level. There are frequently further restrictions on combinations of sunscreens.

The marketing of sun protection products is also closely regulated. As mentioned earlier, in Europe sun protection products are usually regulated as cosmetics, whereas in the United States and Canada they are governed as controlled drugs. The regulations are dynamic, in particular the positive lists of sunscreens permitted for use. There is an ongoing review of data for materials already approved, and materials are removed or inclusion levels and conditions revised on an ongoing basis. For this reason, it is essential to establish the regions of intended sale for a product at the earliest stage of development possible and to obtain the latest regulatory status before commencing any formulation work. It is also advisable to manage expectations of the marketing and brand management team at the earliest opportunity. A common request from marketing teams part way through a development when the full fiscal potential of a product has been established is to expand the plans to send it to other global markets, which is not always possible.

A brief summary of the regulatory requirements for four main global regions is discussed next. Countries that sit outside of these regions and do not have any local legislation usually affiliate themselves to one of the main regions. This is also a dynamic situation and establishing the current status prior to beginning formulation work is vital.

Europe

Sun protection products are regulated as cosmetics according to Regulation (EC) No 1223/2009[13]. The list of permitted sunscreen materials is under Annex VI. There have been a number of amendments to the original regulation. The most recent is Commission Regulation (EU) 2016/621 on 21 April 2016. It permits the use of zinc oxide (both nano and non-nano).

The regulation replaces Directive 76/768/EC, which was adopted in 1976. The regulation is published on the web site https://cosmeticseurope.eu/, as are any amendments.

The trade association created to provide the links between the regulation and industry changed its name from COLIPA (European Cosmetic, Toiletry and Perfumery Association) to Cosmetics Europe in 2012.

All cosmetic products to be sold in the EU since 11 January 2012 must be registered in the Cosmetic Products Notification Portal (CPNP) before being placed on the market.

Regulation (EC) No 1223/2009 covers many aspects pertaining to a product including the use and display of certain efficacy claims.

The recommendation on the efficacy of sunscreen products and the claims made relating to them was adopted in 2006 and sets out the

- Claims which should not be made in relation to sunscreen products
- Precautions to be observed including application instructions

- Minimum efficacy standard for sunscreen products in order to ensure a high level of protection of public health
- Simple and understandable labelling to assist in choosing the appropriate product

The testing methods used to support the three key claims found on sun protection products do not sit within the regulation and it is not therefore mandatory to use these methods to generate data to support product claims. The test methods for SPF and UVA claims, which are covered in more detail in a later section covering testing, are considered as the reference methods within the EU and are used by all major players in the European Sun Protection market. They are offered as a validated suite of tests by most contract research organisations (CROs). The standardization of methods provides an increased level of consistency across brands.

United States

In the United States, sun protection products are controlled as drugs and governed by the Food and Drug Administration (FDA). The requirements are defined in Federal Regulation 21 CFR Parts 201, 310, and 352 Sunscreen Drug Products for Over-the-Counter Human Use[14].

There are similarities between the US and EU regulations in particular the testing protocols. There is naturally more complexity and controls required when marketing a drug than a cosmetic, and the ongoing support for the product in market is considerable. In common with the EU, there is a list of permitted sunscreens (referred to as active ingredients) which are allowed in addition to permitted combinations of these actives. They must be United States Pharmacopoeia (USP) grade.

The process for approval of new sunscreen actives has been particularly laborious and a gap of 10 years between approval in the EU and other regions is typical. In fact the last new sunscreen approval in the United States was in 1999 and the palette available for the formulator to use in products for U.S. compliance is therefore much smaller and restrictive. In 2014, the Sunscreen Innovation Act was introduced as an amendment to the Federal Food, Drug, and Cosmetic Act to address this issue and provide an expedited process for the review and approval of OTC sunscreens[15]. The new law gave the FDA one year to respond to the existing backlog of sunscreen ingredient approval requests, and then 18 months to reply to any future applications[16].

The testing approaches for SPF and water-resistance claims are included within the regulations. The methods of testing for SPF and water resistance are similar to the EU methods but are mandatory and must be performed by an FDA-approved testing facility.

The standard testing for UVA in the United States prior to 2011 had been the persistent pigment darkening (PPD) method. The protocol was similar in design to SPF testing involving the irradiation of human subjects with the key difference that the UV source was UVA with no UVB component. This generated pigmentation in the skin and is recognized and used in regions outside of the United States. The UVA spectrum applied is not found in nature so there are questions over whether the test is truly representative of the protection claimed, and there have been criticisms made that the method is unethical due to irradiating humans with high doses of UV radiation that are linked with causing cancer. Other regions use *in vitro* techniques, and in 2011 an *in vitro* technique was adopted by the United States. The claim that could be made is 'broad spectrum'.

As previously discussed the development of a U.S. OTC drug product is more complex than a cosmetic, and manufacturing is also tightly controlled by the regulation. Manufacturing facilities must be FDA approved and there is an extensive suite of paperwork required to support the product in market. Considerable expertise is required together with specialist quality management support.

The stability of a US drug product includes the usual parameters that are required for a cosmetic such as emulsion stability, changes in viscosity or pH alongside ongoing analysis of test products to support the presence of each of sunscreen active throughout the real life of the product. The products are released for sale using accelerated data and labelled with an expiry date. Ongoing monitoring is mandatory and any changes to the stability of the product may result in the requirement to shorten the period under which the product is considered safe to use.

Canada

Sun protection products are governed by Health Canada's Health Products and Food Branch and the latest monograph, Sunscreen Monograph version 2.0, was issued on 7 July 2013, replacing the Sunburn Protectants Monograph of 12 October 2006[17].

The monograph describes the requirements necessary to receive market authorization (i.e. a Drug Identification Number [DIN] or a Natural Product Number [NPN]), for topical sunscreen products. The monograph identifies the permitted medicinal and nonmedicinal ingredients, concentrations, indications, directions and conditions of use for these products to be licensed without the submission to Health Canada of additional evidence. It also contains the test methods recommended to be used to comply with the requirements of the monograph.

Food and Drug Regulations are administered by the Therapeutic Products Directorate (TPD) and the Natural Health Products Regulations are administered by the Natural Health Products Directorate (NHPD). This includes requirements related to labelling, manufacturing and product specifications. Additional information on labels, outside of those specified in the monograph, such as additional directions for use and/or cosmetic claims, are acceptable as long as they meet the Guidelines for Cosmetic Advertising and Labelling Claims, or are not false, misleading or counterintuitive to the use of the product.

Many elements of the Sunscreen Monograph mirror the requirements in the U.S. FDA regulations. In practice most formulations developed for the United States are compliant in Canada as well, although they cannot share either primary or secondary packaging due to the need to display the Drug Identification Number on the packaging in Canada.

Australia

Products containing sunscreens in Australia and New Zealand are governed by the Australian Government Department of Health Therapeutic Goods Administration. The regulation is called the Australian regulatory guidelines for sunscreens (ARGS) AS/NZS 2604:2012[18].

Different types of sunscreen products are regulated by two regulatory organisations: the TGA (Therapeutic Goods Administration) and NICNAS (National Industrial Chemicals Notification and Assessment Scheme).

Therapeutic sunscreens are classified as

* Primary sunscreen products with an SPF of 4 or more
* Secondary sunscreen products except those regulated as cosmetics
* Primary or secondary sunscreen products with an SPF of 4 or more that contain an insect repellent

Products regulated as cosmetics are defined in section 5 of the NICNAS Act 1989. The definition covers multiple classes of products that are not discussed here. Of interest for this text are sunscreen-containing products which have an SPF of less than 4 or contain the sunscreens for a purpose which is not to protect the user. The most common example is where sunscreens are included in products such as shampoo or bath gels to protect the colour from fading under UV light.

The preceding is a broad-brush view, and when formulating a product for the Australian and New Zealand markets, the regulations must be consulted carefully to ensure that the product is developed in line with the correct regulation.

Products regulated under the TGA are in effect drugs and the regulation has many similarities with the US FDA regulations. The stability testing required must ensure that the shelf life can be determined with adequate certainty and an expiry or use-by date, and the data generated is used to support the application for registration of the product. The standard is AS/NZS 2604:2012 and the test methods for SPF and UVA (broad spectrum) are ISO 24444 and ISO 24443.

Formulation of Sun Protection Products

Introduction

The sections covered so far have illustrated some of the complexity and challenges faced by the formulator of a sun protection product. So far we have focused on the regulatory controls and mandatory technical requirements. An additional layer of complexity has arrived during the 21st century thanks to the trend (or curse depending on your viewpoint) to offer the consumer every possible permutation and variation of what is essentially the same thing.

A review carried out in 2015 on a major UK sun-care brand revealed that it was offering 17 different variants of an SPF 15 beach product. Every product offered SPF 15, UVA protection indicated via the Boots star rating and water-resistant claims. A few of the variants could be accounted for by pack size, which is a useful differentiator, and it is easy to see how the consumer would want a 400 ml variant and a 50 ml variant for different circumstances. It is also easy to see that some customers may prefer a spray format to a traditional lotion and to identify products that are designed for their children rather than products for themselves. The remaining variants had only subtle differences and could often only be differentiated by a single benefit or claim (personal communication).

It is also no longer acceptable to provide products which just deliver sun protection benefits. They must include multiple other benefits such as moisturisation, antioxidant protection, allow skin to breathe, reduce wrinkles, diminish age spots, and have a light and elegant texture. Consequently, the suncare formulator faces the challenge of combining all of these characteristics into a single stable and efficacious formulation.

As many of the most effective sun filters are oil soluble and have a heavy texture or require high levels of oils to dissolve them and keep them dissolved, high SPF products with an elegant texture are particularly challenging to formulate.

The predominant technology used most in the sector is an oil-in-water emulsion, and most of the major brands have a chassis formulation that is essentially a base formulation on which other products in the range are based. The reasons for this are various.

The main reason is that sun protection products are challenging and expensive to formulate and, unlike a simple skin care formula which could be created using multiple formulation routes, there are not many emulsification systems available to the suncare formulator which provide all of the benefits required. The SPF of a sun protection formulation is as much a function of the base as it is of levels of sunscreen added. An efficient base enables the SPF and other UV claims to be achieved with lower levels of sunscreen materials and gives the following additional benefits:

1. Costs (both formulation and testing) are kept to a minimum.
2. Fewer chemicals applied to the skin so consumer safety is improved.
3. More options for obtaining a light texture.
4. Keeps ratio of oil to water as low as possible and improves stability.
5. Minimises risk by using a starting point with known claims and stability profile negating the need to start from scratch each time.

Tools and Resources

Supplier Guide Formulations

There are a multitude of formulations available online and in raw material supplier's brochures which can provide direction and inspiration for the formulator. This is a great resource and they often include the latest materials and technology. Many of these formulations have been *in vivo* SPF tested which gives a higher level of confidence than *in vitro* data. When using these formulations as a starting point there are however a few things to consider:

1. *In vitro* SPF results are by no means accurate (see claims section) and the only reliable indicator of SPF is if it has been tested *in vivo*.
2. Formulations published by raw material suppliers will typically be formulated using a restricted palette of materials favouring those sold by the company.

Online Sunscreen Simulator

A very useful resource is provided by BASF in the form of its BASF Sunscreen Simulator[12]. This is an online tool which enables combinations of sunscreen materials to be added into a calculator which gives predicted results for SPF, UVA, Boots star rating and other useful information. As with all tools this is an essential part of the formulator's toolbox, but must be treated with caution and respect, and it does not negate the requirement to prepare a formulation on the bench and test it thoroughly.

Things to consider are

1. The indicated SPF is based on the sunscreen combination only and does not take account of the other materials in the formulation and the efficiency of the base. It is therefore useful in a broad sense but does not guarantee that once the sunscreen blend is included in a formulation that it will deliver the required SPF.
2. The particulate materials zinc oxide and titanium dioxide are represented on the calculator by a single spectrum, which is broadly representative of the spectrum of the materials. Different grades do however have slightly different spectrums, so once again the results are useful in a broad sense but may not be reproduced in the formulation.

In Vitro Testing

The *in vitro* tests that are used to support UVA claims can also be used as a tool to help the formulator. They do produce accurate results for UVA/UVB ratios and determination of photostability, which is of course what they are designed for. The main limitation of *in vitro* techniques is in determining an accurate indication of SPF delivered by the formulation. The variation in SPF values from an *in vitro* scan compared with *in vivo* results is so unpredictable it is impossible to generalize with a percentage.

The power in the technique is in using the scan data to compare to data from a formulation with known *in vivo* SPF. Changes to profiles can therefore indicate if the SPF is increased or decreased against the standard and is a useful guide.

Thus formulating from scratch carries more risk and is frequently costly with several iterations of test, reformulate, retest and so on before the product is finalized. Time is also precious with speed to market a prime consideration, so it is therefore favourable to use a known starting point which has already been tested and is why most large companies use the chassis approach. Small changes to an existing formulation frequently result in a different product from a commercial viewpoint, but technically the changes to the chassis are small and the level of confidence that the SPF test will provide the expected result is increased.

Formulation Types

Emulsions

Emulsions are by far the most popular formulation type used for sun protection products. The format is able to support a broad range of textures ranging from thick creams up to almost water-thin sprayable lotions, and allows for a wide range of additional actives to be included in both the oil and water phases. Oil-in-water (O/W) and water-in-oil (W/O) are both widely used although O/W predominates. Each type of emulsion delivers different benefits.

W/O formulations have the built-in benefit of water resistance without the need to add in specific film formers or hydrophobic materials that are required in the O/W systems. W/O is also less susceptible to microbial attack and therefore lower levels of preservative can usually be used. This improves the safety profile of the product and minimizes the risk of interaction of the preservative with the emulsion and other ingredients. As a group W/O emulsions have a tendency to feel heavy and are more occlusive on the skin and do not enable a wide range of elegant textures to be formulated.

O/W formulations, the largest group in the market, allow for more flexibility in terms of texture and are used in products ranging from light moisturizing creams to products with the highest permitted SPF, UVA levels and water resistance. The challenge with this type of emulsion is that at high SPF ranges the proportion of oil phase to water phase can be in the region of 50:50 and the products are therefore difficult to stabilize. Water-resistance claims with this type of system are delivered by including film formers and resins, which are often sticky in nature, and offset the elegant feel from the base formulation. Improvements to texture can be achieved by addition of low levels of silicone-based materials, powders, and careful choice of esters and solvents. Ethanol is also frequently added to improve texture and has the added benefit of providing some preservation, and levels of other preservatives can be reduced or, in some cases, removed. The negative with ethanol is that at higher levels it can be very drying when applied to the skin.

Oils

Suncare oils have a significant place in history when it was fashionable to lie on a beach covered in glossy oil and slowly turn a shade of lobster pink. Typical formulations were SPF 2 and even SPF 0. Today it is still not possible to deliver a high SPF, high UVA protection product using this vehicle, and many brands, in Europe in particular, set a minimum SPF of 15 for their range of products and no longer offer an oil. The low SPF obtained with oils is mainly caused by the inability of the medium to provide an even film on the skin and thus UV actives are not held in place rendering them unable to do their work and absorb the UV light thereby protecting the skin. Inclusion of oil-soluble film formers and some soluble waxes and materials that provide substance and help the oil form a film on the skin provide higher SPF values.

Gels

The gel format does not form a significant proportion of marketed suncare products but can be both oil based and water based. The gelled oil format usually starts out as a sun protection oil which is then thickened with a gelling agent. The addition of the gelling agent renders the oil more substantive and often enables the product to sit more effectively when applied to the skin thereby delivering a higher SPF than the non-gelled oil. Water based gels can only be prepared using a restricted palette of water-soluble filters. The format does not therefore support development of high SPF products or of broad-spectrum UVA claims.

Single-Phase Spray

The single-phase spray format originated from the sun oil and most of the commercially available products of this type are an oil phase with additional ethanol. The ethanol acts as a polar solvent and flashes off on application leaving a good film on the skin. This format can be found for sale in the EU up to SPF 50. The highest claim category of SPF 50+ (EU) is still very challenging with this format.

Sticks and Compacts

Sun sticks have been a core product in most ranges for many years. They are close relatives to lipsticks and can include organic or physical filters, or a combination of both. They are best known for their use on the cricket field where they are responsible for coloured stripes on the faces of the players. They are a format in which it is relatively easy to obtain a high SPF. Wax-based systems can also be found poured into compacts. This format is seen more commonly in Australian brands than in any European ranges.

Aerosols

Two types of aerosol spray products exist:

1. *A true aerosol in which the propellant is part of the formulation.* These are difficult to formulate because the most popular propellants for cosmetic use are non-polar solvents, which have a negative effect on the solubility of the sunscreens in the formulation. The formulator needs to be expert in both formulating sun protection formulations and in aerosol technology.
2. *Bag on valve.* In this format the formula is a standard formulation with no additional solvent to act as propellant. The product is then put in a 'bag' and attached to the valve. The propellant is typically compressed air, which sits inside the can but outside the bag. When the valve is pressed the compressed air expands forcing the product out in a fine spray. Key advantages are that the formulation is easier to develop and often a formulation, which works in a pump spray, will work in an aerosol. A negative is that this format is typically more expensive than the true aerosol route.

In all types of aerosol special consideration has to be given, as with other products, to the particle size of the spray and the risk of inhalation during use.

To produce a sunscreen product in mousse format the aerosol is used in much the same way as mousse for use on hair. For reasons that are unclear, the format has not become popular with the public in the same way that the spray format has been and remains a niche product. Formulating for this format carries all of the complexities of formulating an effective sun protection formula plus the added challenge of ensuring that the foam has the required characteristics of density, softness, spreadability and stability.

Selection of Formulation Type

Selection of the appropriate format for a product is in part a commercial decision dependent on the type of product the marketing team has requested and the target consumer, overlaid with any regulatory and patent constraints and choice of pack type. There is likely to be additional constraints supplied by the brand owner such as policies on animal testing and rejection of certain unwanted ingredients in their products.

Selection of Materials (O/W Emulsions)

Emulsification System

The choice of an emulsifier is dependent on a range of factors. One of the first considerations is whether the system should contain anionic, cationic or non-ionic surfactants. There is no single best choice and the selection will depend on what else is to be included in the formulation and if the components require a particular type of system.

Oil phase concentration is also critical for sun protection formulations. In particular for high SPF products the oil phase is present as a significant proportion of the total formula and an O/W emulsifier has to hold this extremely large amount of material as an internal phase and keep it stable. Many primary emulsifiers work better when combined with a co-emulsifier and this needs to be considered. The pH range required for some sunscreens to function is critical and it is therefore vital that the emulsifier can be effective at the same pH range.

Viscosity build is also a key consideration. If the product is destined to be a thin spray then the emulsifier must be powerful enough to deliver the bonding required to hold the oil and water phases as an emulsion but without delivering high build. Emulsifiers that work best for sun protection are tolerant to electrolytes and the emulsions do not break when applied to the skin. There are some product types, usually sport lotions for men, that do use this technology but they generally does not support the development of high SPF formulations.

Processing factors must also be taken into consideration. How does the emulsifier system behave under the temperatures and shear rates it will experience during manufacture? Ideally, the lab process should mimic as closely as possible the process which will be used to manufacture the product at commercial batch size.

Film Former

The key to an efficient suncare formulation is to ensure that the sun filter molecules are spread on the skin in an even film and remain in place for the duration of the sun exposure. Film formers are therefore a vital ingredient and work alongside the emulsification system. Many film formers are also co-emulsifiers, although they do not need to be. Film formers are used in O/W emulsion systems and single-phase systems to enhance or provide water resistance. The negative of the benefits provided is that these materials can feel very sticky and occlusive on the skin. Inclusion levels are therefore kept to the minimum required to deliver the desired benefit.

Sunscreen System

The choice of sunscreen system is infinitely variable and can seem daunting to a formulator faced with a blank sheet of paper. Variability can be reduced by eliminating certain materials based on:

- Regulatory restrictions.
- Formulation type (water soluble sunscreens will not be an option for an oil based product.
- Cost – Sunscreen materials are relatively expensive and usually form a large proportion of the final cost.
- Patent restrictions – There are a significant number of patents covering blends of sunscreens usually in combination with another material which brings an additional benefit. The patents owned by the large brands and fast moving consumer goods (FMCG) companies are rigorously upheld, and the patent landscape frequently changes. It is advisable to check with a patent expert prior to carrying out any expensive formulation development or testing.
- Level of UVA and UVB protection required.
- Customer-imposed restrictions such as a company policy that avoids certain types of materials or caps inclusion levels.
- Target consumer – Products for babies, children and adults with sensitive skin typically use the sunscreens with the best safety profile.
- Packaging type – Both the format and choice of material must be considered.
- Aesthetic properties – The texture and skin feel of sunscreens varies and due to the high inclusion levels required the choice will significantly influence the aesthetic properties of the final formulation.
- Purchasing restrictions – In some instances the department or company developing the product will not want to source materials from a particular supplier for commercial reasons. Usually sunscreens can be obtained from multiple sources so an alternative can be found. This can be a problem if the sunscreen required is a specific grade and only available from a single supplier.

Once the aforementioned considerations have been reviewed and addressed, and the available palette has been established, use of the tools and resources listed earlier can be used to help build a sunscreen system or indeed several systems to use in the formulation. Usually the blend will include one or more UVB sunscreens in combination with one or more UVA sunscreens.

Oils and Emollients

Oils and emollients are key components in a suncare formulation and provide multiple benefits. They are solvents for the powder and crystalline sun filters. They can also be the main vehicle for carrying the sunscreens. Careful selection enables manipulation of the texture of the product. Use of a dry-feeling ester for example will offset the oily feeling of some of the liquid sun filters. Esters with long alkyl chain length are less irritating than some shorter chain materials and are a better choice for products for sensitive skin and children, which need to be milder.

The polarity of the final system is critical for keeping the sunscreens solubilized. Some oils can cause a shift in the UV absorbance curve of the sunscreen system. This may be positive or negative. Generally non-polar materials will cause a shift to a lower wavelength. This is because the non-polar material stabilizes the ground state of the sunscreen and it is more reluctant to move to the excited state that is required for correct function and protection[19].

Waxes

Waxes may of course be emollients but natural waxes are often used. The inclusion of wax will usually increase the viscosity of the emulsion and improve stability, so they are not used in sprayable formats. Use of waxes may also boost the SPF of a product by contributing a level of structure to an emulsion that remains when the product is applied to the skin. They are not included at high levels because they can produce a draggy or heavy feel on the skin.

Stabilizers/Other Materials

There are a number of other materials which must be included if the formulation is to be successful.

Preservatives – Choices depend on regulations, target consumer and sometimes company policy. A suite of preservatives is usually required because sun products are notoriously difficult to preserve and it is advisable to discuss requirements with an expert in preservation.

Chelating agent – Most commonly EDTA (ethylenediaminetetraacetic acid) or a derivative is used. This is included primarily to mop up any stray heavy metal ions and prevent them from catalyzing oxidation, which may result in the formation of brightly coloured oxidation products.

Antioxidant – Included to protect against oxidation of sunscreen actives and to prevent rancidity of oils and other unsaturated compounds.

Water-soluble emollients – Included to help with skin feel. The two most commonly used are glycerin and 1,3-butylene glycol. Both materials provide moisturisation benefits that are important in suncare to help offset the drying effects of UV exposure. They also help with low temperature stability of the final formulation by acting as an antifreeze in the water phase.

Optional Ingredients

pH adjuster – Required for some sunscreens and film formers. Triethylamine (TEA) and sodium or potassium hydroxide are typical materials. The SPF of the final formulation can be influenced by the choice of pH adjuster.

Fragrance – Most large fragrance houses have expertise in developing fragrance compounds for use in sun protection formulations. It is critical that any fragrance used has been specifically developed for this use to prevent risk of photoinduced reactions. Fragrance can also negatively affect the SPF delivered by a formulation so it is vital to ensure that if a fragrance is to be used it is included at the required level before any *in vivo* testing is carried out.

Ethanol – Typically not included in emulsions above 15%. Benefits are enhanced skin feel and increased stability. The ethanol acts as an additional solvent and can also replace some or all of the preservative system. The negative aspect of ethanol is that it can cause drying out of the skin.

Product Testing

Stability Testing: Products Regulated as Cosmetics

The protocols for testing compatibility of a sun protection formulation product are the same in principle as that used for any cosmetic product. The product is subjected to a set of conditions over a fixed time period and the product must remain fit for purpose. This is not the same as saying that there must be no changes at all to the product or formulation; emulsions are inherently unstable, but the object of the exercise is to ensure that the product remains safe and effective for the consumer to use throughout the period both before and after sale.

Success criteria and the level of change that is acceptable vary, but a small change in colour is often acceptable whereas a weakening of the preservation system with time for a product with a long shelf life would require reformulation.

Packaging Compatibility Testing

Key to success of formulating any suncare product, in fact any consumer product, is to ensure that that the intended packaging is considered from the outset and final packaging is used wherever possible for stability testing and any consumer testing.

The challenge with stabilizing sun protection formulations is due to the complex nature of the material blend and the more aggressive nature of the materials. Incompatibility with packaging materials is common, therefore it is essential that the formulation is tested in final pack or at least packs of the correct material at the earliest opportunity. Some commonly encountered compatibility issues are listed next:

1. Plasticisation of some plastics.
2. Peeling of labels due to migration of sunscreen through plastic.
3. Rusting of metal parts.
4. Degradation of labels and print materials due to contact with the formulation.
5. Coloured spots in products caused by impurities or contact with metals in packaging components.
6. Crystallization of sunscreens onto dip tubes (in spray formats).

Stability Testing

The testing approach used by most companies is essentially a 13-week accelerated test that enables the shelf life of the product to be predicted from the data. In the EU, products which have a shelf life of less than 30 months must carry an expiry date. This causes supply chain and distribution problems if the product does not sell within the expiry period, which frequently results in stock destruction. Most brand owners wish to avoid this and a typical brief will require the product to be stable for 30 months or longer if it can be justified.

Products are released for sale based on the 13-week accelerated test, although testing will continue beyond the 13 weeks to obtain real-time information.

During development it is important to obtain stability information on prototype formulations and rapid testing is carried out in glass containers. This predictive testing is a subset of the full protocol and is typically carried out in glass jars. Batches of formulation used for predictive testing are usually generated on a small scale in the laboratory. It is helpful if the manufacturing process mimics as closely as possible the intended process in the factory. The following test conditions are the recommended minimum that should be considered for predictive testing (Table 17.2).

The aforementioned criteria are very stringent, and if the success criteria are met for all of the tests, then it is a good indicator that the formulation will be stable throughout the extended conditions of a full test.

TABLE 17.2

Formulation Stability Criteria

Test	Indicator	Details	Required Result
Fridge 4°C	Baseline for comparison	2 weeks	No changes
Centrifuge (emulsion only)	Emulsion and suspension stability	Centrifuge at 3000 rpm for 10 minutes; initial and after 2 weeks stored at 40°C	No separation
Microscope	Emulsion structure and stability; compare to initial	Instant results	No crystallization, no change to emulsion structure
50°C	Severe test for emulsion and fragrance stability	7 days	No changes
Freeze–thaw cycle	Potential instability from separation and formation of crystals in emulsions and aqueous/alcoholic formulations Potential crystal formation in alcoholic or oil systems	Cycles of 12 hours at −10°C to 40°C Test for 5 cycles	No changes

Stability Protocol

For the formal stability test it is important that the product is manufactured at a representative batch size. If small-scale laboratory production does not represent the product that will be produced during production, then there will be little value in the results. Tests must also be carried out in final packaging and usually in glass as well. This can be helpful to identify changes that are caused by the packaging rather than changes that are inherent in the formulation and occur irrespective of packaging type.

Sufficient packs must be tested at the start of the trial and it is preferable to have more than one at each location. This helps eliminate what can only be described as the random unexpected result. It is rare but occasionally a product inexplicably develops a problem that is unexpected and does not fit the profile of the product. This can be due to anomalies or contamination in an individual bottle or even the top not being correctly applied. By looking at several packs this can be eliminated as a cause.

The following parameters are monitored throughout the storage of the product:

Appearance

Colour

Odour

pH (if applicable)

Viscosity (if applicable)

Weight loss

Preservative efficacy

Colour stability

UV profile

Changes to packaging

Microscopy (if required)

Other parameters which may be evaluated are:

Evacuation of product

Nitrosamine test (if applicable)

TABLE 17.3

Stability Testing Schedule

	4°C	23°C–25°C	30°C	40°C	50°C	–10°C–40°C	40°C/80% humidity	Light Testing*
Initial	X	X						
Day 1	X	X				X		X
Day 3	X	X				X		X
Week 1	X	X	X	X	X	X	X	X
Week 2	X	X	X	X	X	X	X	X
Week 4	X	X	X	X			X	X
Week 8	X	X	X	X			X	X
Week 13	X	X	X	X			X	X
Month 6	X	X	X					
Year 1		X						

Note: P.

*Light source to be appropriate to conditions under which the product will be subjected.

The exact testing protocol and conditions vary between companies and is not regulated. The tests should however include a range of temperatures and samples should be inspected at realistic time periods. The criteria shown in Table 17.3 are representative of the test protocols used by cosmetic manufacturers.

Pass–Fail Criteria

In a perfect world the product would remain completely stable throughout the test with no changes to any parameter. This of course rarely occurs and changes will be observed. The extent to which the changes are relevant depends on the product, and the type and extent of the change. Classification of change is also subjective and can be difficult. What may be termed a minor change in colour to one evaluator may seem more significant to another evaluator. Odours are particularly subjective. For that reason it is recommended that there be a grading system for changes and that the personnel carrying out the test are trained to ensure consistency.

The extent to which a change impacts on final stability depends also on where the product is to be sold. If the destination is Siberia where temperatures are frequently below zero then changes at 40°C will have less impact. Overall it is preferable for the product to be stable over the full range of temperatures and conditions.

Specific Testing Criteria and Guidance

Temperature Cycling

The temperature cycling test is particularly relevant for sun protection products that are likely to be subjected to large changes in temperature by nature of how they are used. Products are frequently taken onto a beach during the day get heated by the sun and are then taken indoors or left in vehicles and cool down overnight only to be subjected to the same cycle the following day. Subjecting products to a cycle of temperatures typically between –10°C and 40°C with one full cycle per day is a severe test that stresses the formulation. It is more appropriate for emulsions but can be a useful indicator for single-phase oils and alcoholic systems which may crystallize out of solution at the lower temperature. If a sun protection formula cannot survive one week under this condition it does not mean it will be unstable in the market but is a good indicator that the product is weak on stability and could manifest as a problem later.

UV Profile Testing

Testing to ensure that the UV profile of the product remains stable throughout the stability test is usually carried out using the *in vitro* UVA protocol and comparing the shape of the curve to the curve obtained when the sample was fresh. Any changes may indicate degradation of one or more of the sunscreens. This is usually accompanied by crystals appearing in the product. The levels of sunscreens can be determined by analysis, but this is not usually required for a cosmetic.

Microscopy

Another key stability issue is that many sunscreens are powders in their raw form, which require solubilizing before they can work effectively. If the oil phase is not correctly formulated, then the materials can crystallize out if conditions permit. This is observed more commonly at low temperatures and will occur if the solubility of the sunscreen in the oil phase is low. Some liquid sunscreens are good solvents for powder sunscreens as are many oils with high polarity. Choice of oil phase is therefore as much about polarity of the system as it is of ensuring that the volume of oil is sufficient to keep the powders in solution.

Crystals can be observed under the microscope long before they can be felt or seen on the skin. Any form of crystallization is undesirable for a number of reasons:

1. When the crystals reach a certain size they can be felt upon application as grit. Consumers often think they have got sand in the product and complain to the retailer.
2. In sprays and oils this can lead to blocking of the spray pump.
3. Consumer safety – If there are concentrated pockets of neat sunscreen material this may cause an adverse reaction when applied to the skin. More critically, any crystals signify a loss in performance of either the UVA or UVB element and the product will not provide the claimed level of protection.

In a clear spray format or alcoholic liquid where low temperature has caused crystallization there is a chance that the crystals will return to solution if warmed. This often happens during use.

In an emulsion, once a sunscreen has started to crystallize it rarely stabilizes or reverses, and the crystals once seeded will grow. The seeds may also act as nucleation sites for any other filters present. The only solution is reformulation.

Microscopy is also used to examine the structure of an emulsion system and monitor any changes, for example in droplet size distribution. Most laboratory-based microscopes are linked via a camera to a computer enabling photographs to be taken at the selected time points and storage conditions and which form part of the stability data pack.

Stability: OTC Drug

When developing a sun protection product for the United States or Canada additional work is required to ensure that the product is effective and stable, and to enable it to be registered for sale. The cosmetic characteristics are investigated as per the stability protocol for a cosmetic. The additional testing required is regarding the chemical stability of actives. These actives are those defined by the regulation, i.e. the sunscreen actives and possibly the preservatives.

Efficacy Testing

SPF and Water Resistance

Although test methods have only been standardized relatively recently, the use of SPF values as a means to indicate the level of protection afforded by a product against the burning effect of the sun has been recognizable for decades. This is a relatively simple concept for the consumers to understand because they can see and feel burning and therefore relate the effect of the product to a physiological effect.

In December 2010, the Standard EN ISO 24444:2010 'Cosmetics – Sun protection test methods – In vivo determination of the sun protection factor (SPF)' was published by the European Standardisation Organisation (CEN). This standard replaced an earlier method, the International Sun Protection Factor Test Method (2006), and includes the testing of water resistance.

The methods used for *in vivo* determination across the main regulatory regions of the world are similar. The US FDA method is in fact very similar to the EU protocol. However, testing must be carried out by FDA-approved contractors to comply with the requirements.

The standard is FDA 2011 21 CFR Parts 201 and 310 – Labelling and Effective Testing; Sunscreen Drug Products for Over-the-Counter Human Use – Food and Drug Administration (U.S.), and includes two elements:

- Static determination.
- Water resistance 40 minutes/water resistance 80 minutes (water resistant testing includes static determination).

In Australia and New Zealand the standard is AS/NZS 2604:2012 – Australian/New Zealand Standard, and includes:

- Static determination.
- Water resistance 40 minutes, water resistance 80 minutes, 2 hours, 3 hours, 4 hours (water resistant testing includes static determination)

UVA Testing

Damage from UVA is not immediately visible to the consumer and generally falls into the category of causing long-term skin damage. The awareness of this damage is very well established through promotion from sun protection brands, cancer charities and media attention. It is still however very difficult to communicate the level of UVA protection provided by a product to the consumer and until 2010 there was no standardized method in the EU.

In the UK the Boots star rating system[20] is widely used and has been since it was launched in 1992 (www.boots-uk.com/about-boots-uk/company-information/boots-heritage/). It is an *in vitro* test. The level of UVA protection is determined relative to the amount of UVB protection and is represented on the pack with a symbol. The system is used under license by proprietary and private label brands. The symbol is not used widely on products outside of the UK.

In 2010, COLIPA (as it was then called) published an *in vitro* test method which was similar in approach to the Boots star-rating approach. This was then updated and in June 2012 the European Committee for Standardisation (CEN) published the Standard EN ISO 24443:2012 Cosmetics – Sun protection test methods – In vitro determination of sunscreen UVA photoprotection.

A key difference between the two test methods is that whereas the Boots system permits different levels of UVA protection to be represented via the use of different numbers of stars, the EN ISO 24443:2012 method results in a pass or fail. To pass, the product must provide approximately 1/3 UVA protection in relation to the UVB protection. A pass is then represented on the pack a symbol. A copy of the standard can be purchased from the national standardisation organisations in the EU or from the International Organisation for Standardisation (ISO).

This standard test method replaces the earlier reference method (Guidelines – Method for in vitro determination of UVA protection, 2011). The EN ISO test method is now considered as the reference method within the EU. Cosmetics Europe therefore recommends cosmetic manufacturers use this standard to determine the UVA protection factor.

In the United States, an *in vivo* method for determining UVA protection has been in use for many years. It is a method called PPD (persistent pigment darkening) and uses UVA radiation to develop a tan within human skin and the level of protection indicated on the pack with a protection grade of PA followed by plus (+) symbols. The higher the number of plus symbols the greater the level of UVA protection provided. The method is widely used across the globe but has a number of shortcomings. First, the spectrum used does not appear in nature and therefore is not representative. Second is the ethical element. Subjects are exposed to high levels if UVA of wavelengths known to cause DNA damage which is linked to cancer. Many companies have refused to use this test because of this.

In 2011 the United States took an *in vitro* method into law. The permitted claim is similar to the EU approach in that it is a single statement not a range of levels as with the star rating system. UVA protection is stated as broad spectrum on the pack. The regulation is FDA 2011 21 CFR Parts 201 and 310 – Labeling and Effective Testing; Sunscreen Drug Products for Over-the-Counter Human Use – Food and Drug Administration (U.S.) Critical Wavelength – Necessary for Broad Spectrum claims.

The Australian and New Zealand standard is AS/NZS 2604:2012 – Australian/New Zealand Standard; Sunscreen Products – Evaluation and Classification.

Appendix: UV Absorbers

Bis-Ethylhexyloxyphenol methoxyphenyl triazine

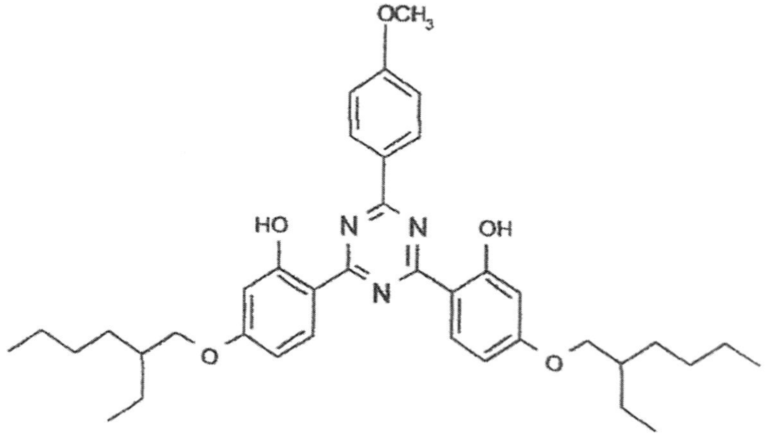

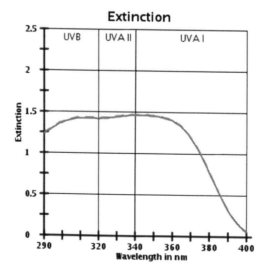

FIGURE 17.4 Chemical structure.

FIGURE 17.5 Spectrum.

Butyl methoxydibenzoylmethane

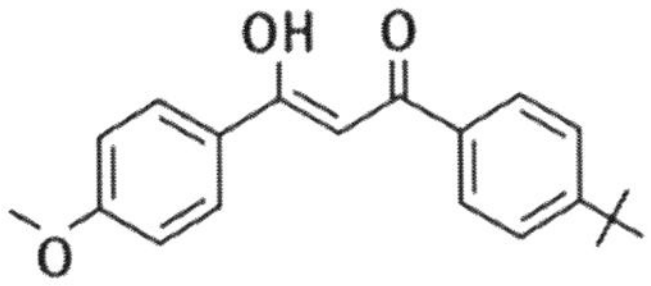

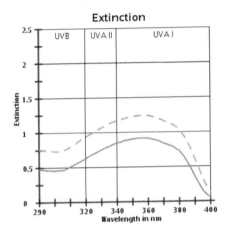

FIGURE 17.6   Chemical structure.

FIGURE 17.7   Spectrum.

Diethylamino hydroxybenzoyl hexyl cenzoate

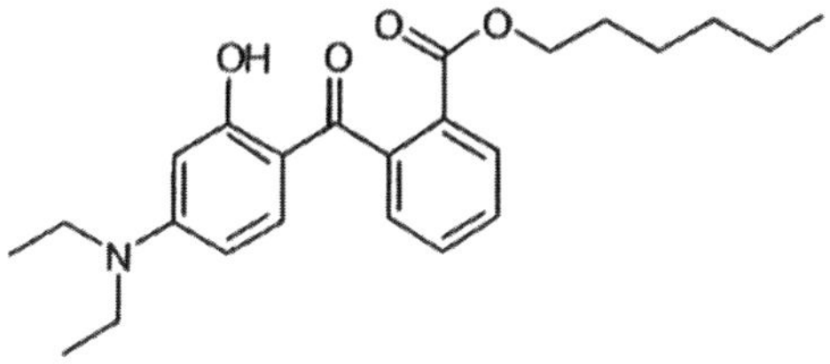

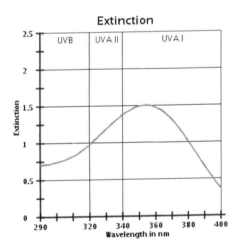

FIGURE 17.8   Chemical structure.

FIGURE 17.9   Spectrum.

4-Methylbenzylidene camphor

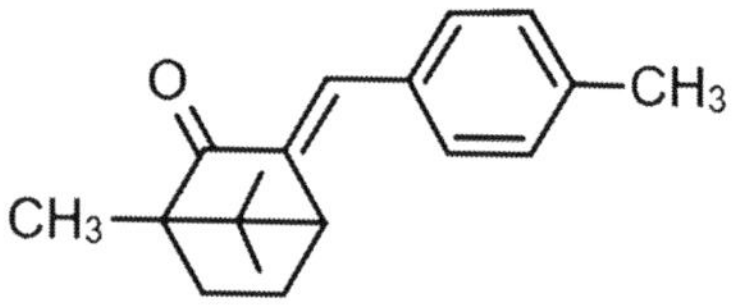

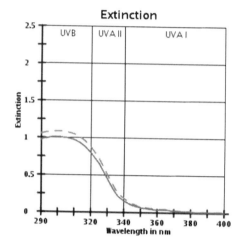

FIGURE 17.10   Chemical structure.

FIGURE 17.11   Spectrum.

Benzophenone-3

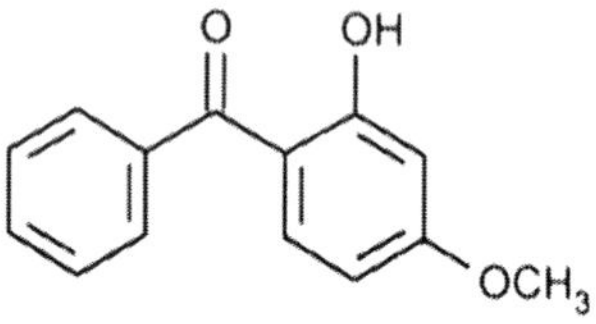

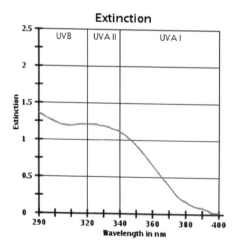

FIGURE 17.12   Chemical structure.

FIGURE 17.13   Spectrum.

Benzophenone-4

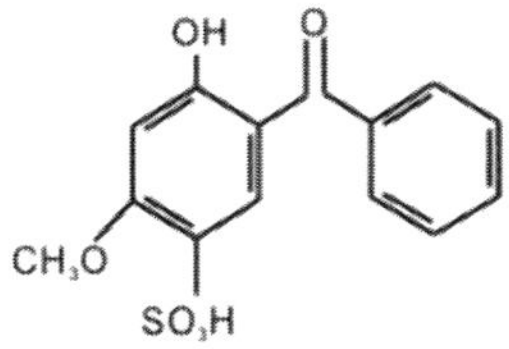

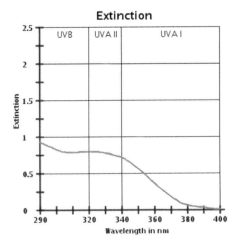

FIGURE 17.14 Chemical structure.

FIGURE 17.15 Spectrum.

Diethylhexyl butamido triazone

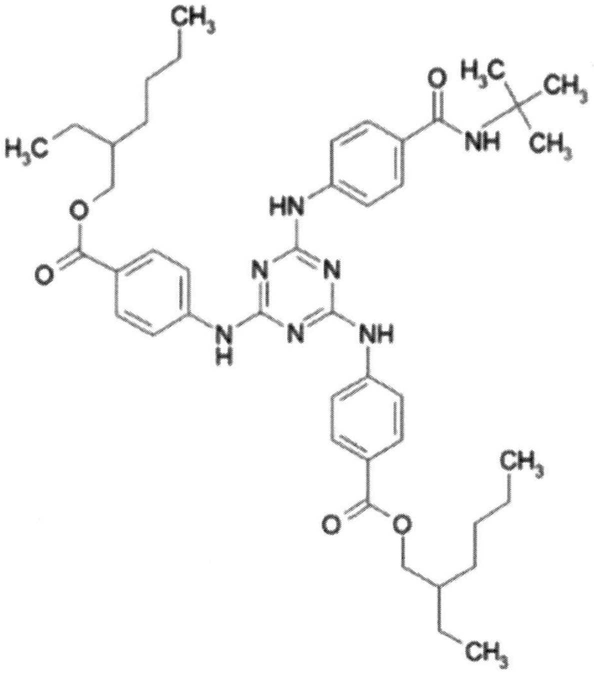

FIGURE 17.16 Chemical structure.

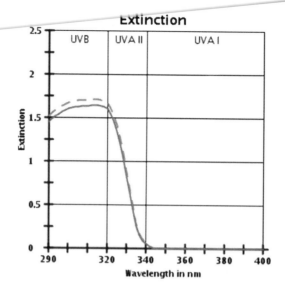

FIGURE 17.17 Spectrum.

Ethylhexyl methoxycinnamate

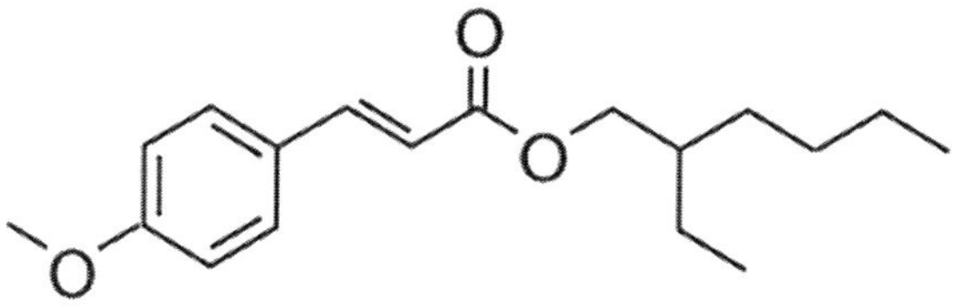

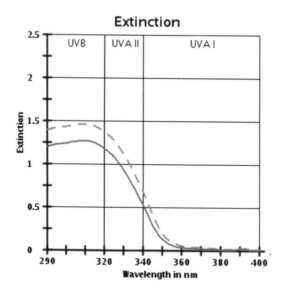

FIGURE 17.18 Chemical structure.

FIGURE 17.19 Spectrum.

Ethylhexyl salicylate

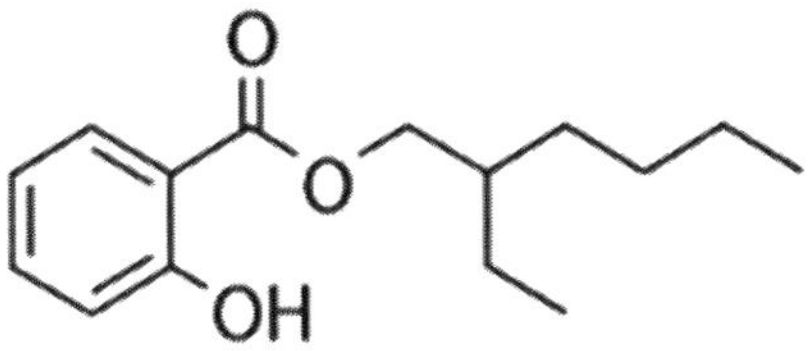

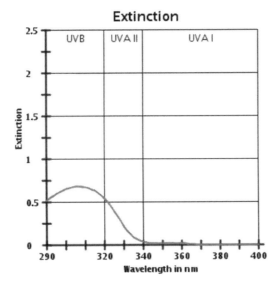

FIGURE 17.20 Chemical structure.

FIGURE 17.21 Spectrum.

Ethylhexyl triazone

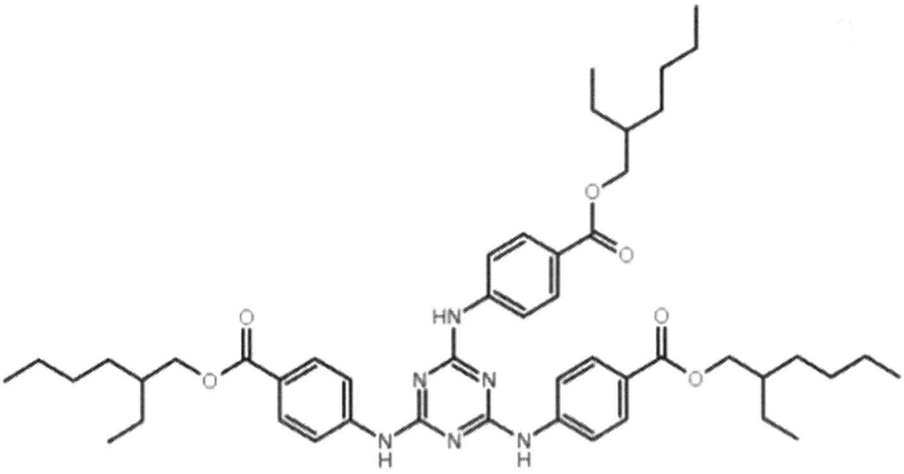

FIGURE 17.22 Chemical structure.

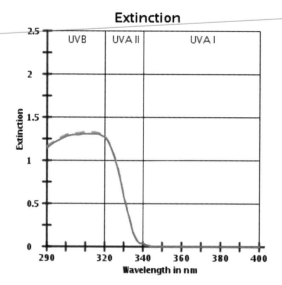

FIGURE 17.23 Spectrum.

Octocrylene

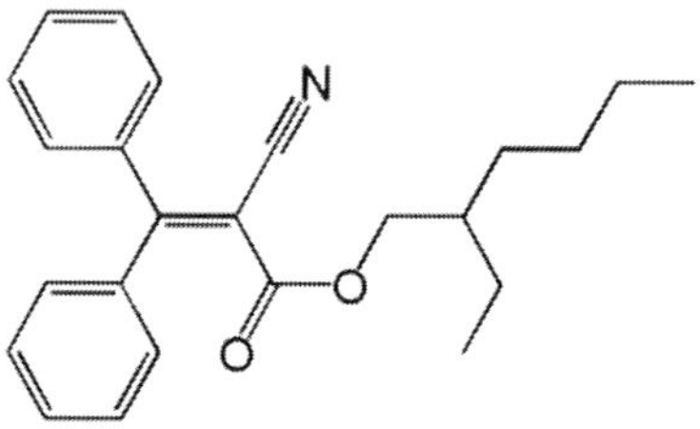

FIGURE 17.24 Chemical structure.

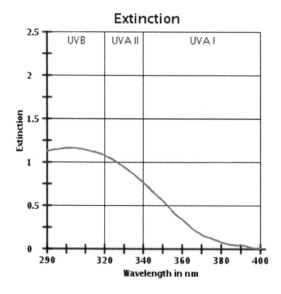

FIGURE 17.25 Spectrum.

Phenylbenzimidazole sulfonic acid

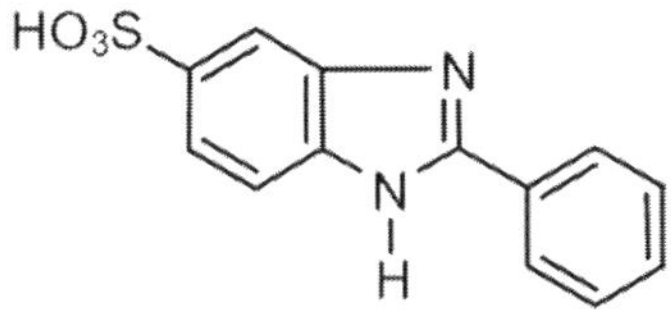

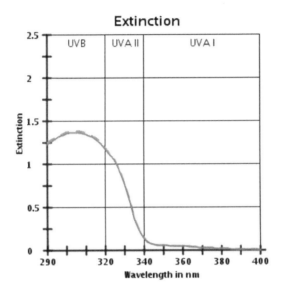

FIGURE 17.26 Chemical structure.

FIGURE 17.27 Spectrum.

NOTES

1. Giacomoni, Paolo U. (2005). Sun Protection: Historical Perspective, in *Sunscreens: Regulations and Commercial Development*, 3rd edition, N. A. Shaath, ed. Boca Raton, FL: Taylor & Francis Group, p. 72.
2. www.garnier.fr/l-histoire-ambre-solaire.
3. www.pizbuin.com/en/our-heritage/.
4. www.walgreensbootsalliance.com/about/history/.
5. Schulze, R. Einige Versuche und Bemerkungen zum Problem der Handel Süblichen Lichtschutzmittel. *Parfüm, Kosmetik* 37 (1956).
6. Grönlund, M. *Cosmetics & Toiletries*, 110, 77 (1995).
7. Walgreens Boots Alliance, archive photographs.
8. BASF Raw Material Data Sheet.
9. DSM Raw Material Data Sheet.
10. Aako Raw Material Data Sheet.
11. Charts and Notes C 2014 DR COSREG Ltd (Summit Events).
12. UV profile generated from the BASF Sunscreen Simulator. www.sunscreensimulator.basf.com/Sunscreen_Simulator.
13. Regulation (EC) No 1223/2009.
14. Federal Regulation 21 CFR Parts 201, 310, and 352 Sunscreen Drug Products for Over-the-Counter Human Use.
15. H.R. 4250 – Summary. 113th United States Congress (2013–2014). Retrieved 30 July 2014.

16. Sifferlin, Alexandra (16 July 2014). We're One Step Closer to Better Sunscreen. *Time Magazine*. Retrieved 30 July 2014.
17. Health Canada Sunscreen Monograph v2.0.
18. AS/NZS 2604:2012.
19. Agrapidis-Paloympis, L., Nash, R.A., Shaath, N. The effects of solvents on the ultraviolet absorbance of sunscreens. *J Soc Cosmetic Chem*, 38, 209–221 (1987).
20. www.boots-uk.com/about-boots-uk/company-information/boots-heritage/.

18

Advanced Formulation Techniques Including Innovative Materials

**Bozena Michniak-Kohn, Tannaz Ramezanli,
Frank Romanski, Cliff Milow and Kishore Shah**

CONTENTS

Introduction

Normal healthy skin is characterized by effective barrier properties, adequate cell differentiation and proliferation, optimal moisturization levels, and slightly acidic surface pH (pH 4.5–6.0). Cosmetic/pharmaceutical actives together with the formulation excipients are designed to either maintain or restore skin health and barrier properties in an effective manner. In order to create even more efficacious products, novel actives as well as formulation ingredients are constantly being introduced to the consumers/patients. In this chapter, readers will be provided with three examples where innovative materials and formulation techniques have been used to improve the delivery of actives. The first case describes the development of novel polymers to form carriers called TyroSpheres that significantly improve stability of actives as well as their solubility and target delivery into skin layers with little or no systemic uptake into the body. The second example presents a novel patented polymer with unique skin bioadhesive properties that forms an 'invisible' patch on the skin with unique delivery properties of actives. The third section discusses novel formulation excipients such as emulsifiers, sunscreens and new emollients as well as poloxamers for topical formulations. It is important to note that, although the work presented here on TyroSpheres discusses pharmaceutical actives, TyroSphere technology has the potential to significantly impact cosmetic science since the carriers are able to target skin layers and prevent actives from being transported in measurable amounts into the lower dermis and also into the blood capillaries minimizing systemic uptake.

Case 1: Tyrosine-Derived Nanoparticles for Topical Delivery

A unique class of polymeric nanospheres, made of a family of biocompatible, amphiphilic tyrosine-derived ABA-triblock copolymers, has been developed at the New Jersey Center for Biomaterials, Rutgers University. The chemical structure of these copolymers is composed of hydrophobic B-block oligomers of desaminotyrosyl-tyrosine ester and diacid – made from naturally occurring metabolites – and hydrophilic PEG A-blocks. Tyrosine-derived polymers, which are biocompatible and not cytotoxic, have been developed and explored for medical applications by Kohn et al. also at the New Jersey Center for Biomaterials (Yu and Kohn, 1997). These PEG-*b*-oligo(DTR-XA)-*b*-PEG triblocks are, in general, amorphous with glass transition temperatures (Tg) in the range of –32°C to –34°C, and a melting transition between 47°C and 54°C. Following the addition of this copolymer to an aqueous environment, it self-assembles into polymeric micelles referred to as TyroSpheres. The hydrophilic PEG segments and the amorphous oligo(DTR-XA) segment creates a flexible structure, ensuring the self-assembly of the polymers into a dynamic and non-frozen structure in aqueous media (Nardin et al., 2004; Sheihet et al., 2005).

Evaluation of this family of polymers has shown that the triblock polymer PEG_{5K}-b-oligo (desaminotyrosyl-tyrosine octyl ester suberate)-b-PEG_{5K} (DTO-SA/5K) is optimal for encapsulation efficiency, has the lowest polydispersity index (0.17), a reproducible hydrodynamic diameter of 60–70 nm, and very low critical aggregation concentration (2.6×10^{-7} g/mL measured by static light scattering), making this polymer the lead candidate evaluated for nanosphere applications. The cytotoxicity of TyroSpheres was examined on a variety of cell lines and these nanospheres were found not toxic at concentrations up to 11.5 mg/mL (Sheihet et al., 2005, 2007).

TyroSpheres are fabricated by the solvent displacement method. In this technique, the polymer is dissolved in a semi-polar water miscible solvent, for example, dimethylformamide. Next, it is added dropwise to the aqueous phase, which results in the formation of nanoparticles. The organic solvent is then removed by ultracentrifugation. Co-dissolving the drug and polymer in the organic phase allows drug loading in the nanospheres. The preparation is then purified by filtration (using 0.22 μm polyvinylidene fluoride filters), ultracentrifugation, re-dispersion in phosphate-buffered saline (pH 7.4) followed by another filtration (Sheihet et al., 2007; Zhang et al., 2014).

There are several advantages with the fabrication process of TyroSpheres: the process is generally simple and relatively easy to scale up for large batch production in industry. Chlorinated solvents and high temperatures are avoided in the preparation of nanoparticles. The PEG segments of the block copolymers (positioned at the periphery of the nanoparticles) are able to stabilize the nanosphere dispersion; therefore, external surfactants/stabilizers are not used in the fabrication of TyroSpheres.

Scientists at the New Jersey Center for Biomaterials have investigated the applicability of TyroSpheres as suitable carriers for several hydrophobic compounds including paclitaxel, Nile red, curcumin, vitamin D3 and cyclosporine A. Incorporation of these compounds into TyroSpheres did not show significant impact on the particle sizes of the carriers. In a previous study computational modelling was employed to study drug–polymer interaction in the TyroSpheres (Costache et al., 2009). The authors demonstrated that the binding affinity of drug to DTO-SA copolymers depends on their hydrophobic compatibility and other physical interactions, such as hydrogen binding and π–π stacking.

Polymeric nanoparticulate–based drug delivery systems have been investigated by several groups for the treatment of dermatological diseases (Zhang et al., 2013). The rationale is that these nanoparticles can protect the drug from degradation, provide sustained/controlled release of their cargo, and improve partitioning and permeation into the skin. To evaluate the potential use of TyroSpheres as a skin delivery system, fluorescent dyes, for example, Nile red (log D: 3.10) and 5-dodecanoylaminofluorescein (DAF, logD: 7.54), were loaded into nanospheres separately, and their percutaneous penetration using human cadaver skin was assessed and compared to controls comprising solutions in propylene glycol. The studies were conducted using vertical Franz diffusion cells. It was found that Nile red and DAF delivery to skin strata was 9.0 and 2.5 times enhanced, respectively, when formulated into TyroSpheres relative to propylene glycol (Sheihet et al., 2008). Moreover, viscous formulations of TyroSpheres – that are more suitable than liquid dispersions for topical applications – were prepared using hydroxypropyl methylcellulose (HPMC) and propylene glycol. In an ex vivo study on human cadaver skin, the viscous formulations of Nile red TyroSpheres resulted in similar dye deposition in the stratum corneum and viable epidermis compared to a TyroSphere liquid dispersion. Remarkably, in an *in vivo* study on pigs, topically applied Nile red–loaded TyroSphere in a viscous formulation resulted in 40% higher dye delivery to skin than that obtained from Nile red–TyroSphere liquid dispersion (applied via Hill Top Chambers® in an occlusive patch test system) (Batheja et al., 2011).

Among the drugs that have been loaded in TyroSpheres, a significant effort has been made to develop and characterize paclitaxel-TyroSphere formulations aimed at treating skin disorders such as psoriasis. Paclitaxel is a mitotic inhibitor drug and is marketed for cancer therapy. Paclitaxel also prevents cellular over-proliferation and therefore can potentially be used to treat psoriasis. However, the poor solubility of paclitaxel and its toxicity limits the medical applications of this drug. TyroSpheres were able to load paclitaxel with up to 8.4% w/w loading efficiency and provided substantial enhancement of its solubility (1160 µg/mL). In a 72-hour *in vitro* drug release study using dialysis cassettes, a sustained release pattern was observed with paclitaxel-TyroSpheres. In 72 h about 44 and 58% of the drug was released from the paclitaxel-TyroSpheres with 5.0 and 8.4 wt% drug loading, respectively, while no burst release was observed. A viscous formulation of paclitaxel-TyroSpheres was prepared by adding 1% HPMC to the formulation, which showed a similar drug release profile to the TyroSphere liquid formulation. Following a skin permeation study using human cadaver skin, paclitaxel–TyroSpheres in both aqueous dispersion and gel-like formulation delivered significant amounts of the drug to the epidermis, and the delivery to the receptor compartment – representing the systemic circulation – was minimal (Kilfoyle et al., 2012).

To further illustrate the efficiency of TyroSpheres as a topical drug delivery system for the treatment of psoriasis, Nile red–loaded TyroSpheres were applied to human cadaver skin, normal biopsy skin and psoriatic biopsy skin to compare the permeability of the dye in the different skin types. Following fluorescent analysis of skin cross sections, the penetration effect of Nile red on psoriatic skin was 5- to 10-fold greater than that in the normal biopsy skin or human cadaver skin samples. Moreover, in fluorescent images of psoriatic skin biopsies the dye was detected in the basal layer of the epidermis (where psoriasis originates) but did not appear to be transported further into the skin (Kilfoyle, 2011).

Cyclosporine A (CSA) was another model drug that was studied for topical delivery using TyroSpheres. CSA is a potent immunosuppressive drug that has been found very efficient in several skin disorders, including psoriasis, atopic dermatitis and alopecia areata. The drug's large molecular size (1202 Da) and low water solubility poses difficulty in the formulation and dermal penetration of this compound. Using TyroSpheres, a solubility of 8.7 mg/ml and loading efficiency of up to 50% was obtained. Stability assessment of CSA-TyroSphere liquid dispersions showed that the TyroSpheres enhanced the stability of CSA in the formulation at 4, 25 and 37°C compared to free CSA (not loaded in TyroSpheres). The TyroSphere formulation was stable during 6 months storage at 4°C. It is noteworthy that many other particulate systems such as PLGA nanoparticle solutions are not stable in the solution for more than a few weeks. Dermal penetration of CSA-TyroSpheres in a gel-like formulation was also evaluated on human cadaver skin. Following 6 h application of the formulation, drug recovery from dermis was 22 µg per gram of skin and no drug was detected in the receptor compartment (analyzed by liquid chromatography-mass spectroscopy) (Goyal et al., 2015).

In conclusion, TyroSpheres provide a nanotechnology platform for loading and delivery of various lipophilic compounds. The TyroSphere topical delivery system can be used to formulate various lipophilic active pharmaceutical/personal care ingredients and possesses potential application in areas such as treatment of psoriasis, melanoma, acne and contact dermatitis, hair loss treatment, and sunscreen applications.

Case 2: PharmaDur® Polymer in Topical Cosmetic Products – Chemistry and Exemplary Formulations

The use of film-forming polymers in topical skin care formulations to provide certain benefits such as delivery of actives, skin tightening, and so forth has been known for a long time. However, there are significant challenges facing their use on skin. Since the skin is a living substrate, that is, it breathes, perspires, stretches and contracts, the formed film must accommodate the various functions of the skin during different body activities and environmental conditions such as temperature and humidity. In order to be compliant to these challenges, the formed polymeric film on skin needs to be bioadherent and have the viscoelastic properties, breathability, and moisture vapour permeability properties that are similar to skin. In the absence of such compliance, the film tends to feel uncomfortable or tight, peel, and flake during use. Retention of the formed polymeric film on skin for the desired duration of time is a commonly encountered problem.

A graft copolymer that is designed to address these and other issues is commercially available under the trade name PharmaDur (www.pharmadur.com; Polytherapeutics, Inc.). The key molecular characteristic of the copolymer is a combination of selected hydrophilic and hydrophobic polymeric moieties (Figure 18.1). PharmaDur copolymer has an INCI name of dimethylacrylamide/acrylic acid/polystyrene ethyl methacrylate copolymer. The copolymer is utilized in the formulation of several commercially available skin care products including prescription dermatological products, face make-up foundations and antiaging wrinkle reducer for eyes. When a dermatological vehicle, for example, a cream, lotion, or gel, formulated with the copolymer is applied to skin, it forms an imperceptible and invisible hydrogel film ('Virtual Patch'). The PharmaDur polymer film is non-tacky, non-greasy, has a very smooth feel, does not exhibit flaking or skin stretchiness, and is breathable. More detailed discussion of PharmaDur is provided by Shah (2006).

To summarise, the key product performance enhancing benefits of formulations of skin care products with PharmaDur can be summarised as follows:

- Forms imperceptible hydrogel film on skin ('Virtual Patch')
- Retains active at the site of application and provides sustained release of active agent
- Provides long-lasting treatment (once/day application)
- Significantly enhances the efficacy of penetration enhancers in skin permeation of actives
- Skin moisturizing
- Prevents transfer of active to fingers, clothing, towels and bed linen
- May decrease skin irritation potential of active

The following exemplary skin care formulations demonstrate the effect of PharmaDur copolymer in such products.

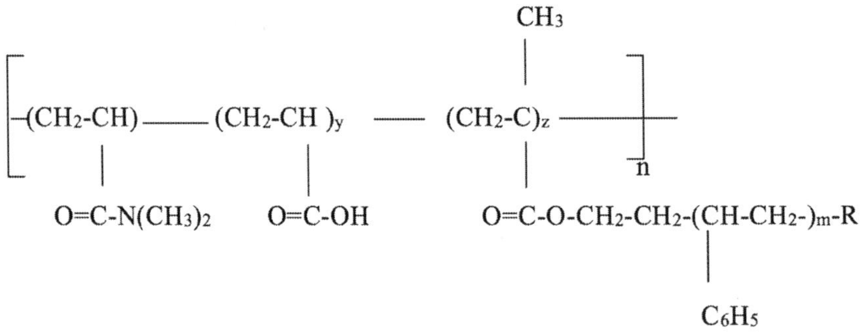

R - alkyl radical

n, m, x, y, & z- positive integers having unspecified different large values

FIGURE 18.1 PharmaDur graft copolymer.

Wrinkle Reduction

Belfer and colleagues (unpublished results) have developed a composition, Botanicin™, for facial wrinkle reduction therapy. Botanicin is a serum comprising a muscle inhibitory vegetal extract (Gatuline Expression; Gattefosse), a dermal stimulating peptide (Matrixyl 3000; Sederma) and the PharmaDur copolymer. Botanicin represents a unique composition for wrinkle reduction therapy, which provides an enhanced and long-lasting efficacy of a synergistic blend of two botanically derived cosmetic ingredients. In addition, this composition elicits immediate user perceived firming and skin tightening effect at the site of its application.

Belfer (2009) has previously reported that when a dermal matrix–stimulating peptide is combined with a myoinhibitory botanical extract such as *Spilanthes acmella oleracea*, the mixture upon application to skin enhances the speed of dermal matrix rejuvenation. Repeated treatment to facial skin of such a composition results in faster reduction of fine lines and wrinkles. The hypothesis leading to the Belfer discovery originates from universal knowledge about the wound healing process within collagenous elements of mammalian tissues, and in particular, external skin. During cutaneous wound healing the collagen fibres within the injured dermis reorganize and repair best when they are not subjected to disruptive movement or under contractile tension. Furthermore, mechanical stimulation of dermal cells causes them to secrete more elastase and collagenase, and this reduces the skin's ability to generate collagen fiber precursors. The objective of combining the myoinhibitory vegetal extract with the peptide is to provide the most favorable environment within which compositions for rejuvenating the dermis can have maximum effect. In order for the healing of expression lines to take place in the fastest manner, the contractibility of muscle fibres must be suppressed quickly, and intensely, over a sustained period of time. Once this condition is met, repair of the dermal matrix can occur with efficient delivery of matrix-stimulating peptides. *Spilanthes acmella oleracea* helps to create such an environment. This discovery has been utilized in some commercially marketed antiaging, wrinkle-reduction cosmetic products.

Belfer and colleagues (unpublished results) have found that unexpected benefits result from combining the peptide and *Spilanthes* mixture with the PharmaDur copolymer delivery system. Results of preliminary consumer evaluation (Table 18.1) of Botanicin formulation with and without soft focus (SF) particles have demonstrated long-lasting reduction in both static (relaxed) and dynamic (expression, e.g. smiling) wrinkles for 6–8 hours. A photographic observation of a subject (Figure 18.2) clearly demonstrates the dynamic wrinkle reduction effect of Botanicin.

Sun Protection

Thau and Shah (unpublished results) have developed two types sun protection formulations utilizing the PharmaDur copolymer in oil-in-water (O/W) and high internal phase water-in-oil emulsion (W/O) bases. The formed polymer film on skin allows uniform distribution of the ultraviolet (UV) filters and photostabilizers on skin for more efficient and effective sun protection of skin.

The O/W formulation (Table 18.2) was prepared using a self-emulsifying system (Crystalcast; MMP, Inc.) (Mercier et al., 2010) consisting of a blend of betasitosterol, sucrose stearate, sucrose distearate, cetyl alcohol and stearyl alcohol.

TABLE 18.1

Results of Botanicin Clinical Trial

Trial 1 using Botanicin Complex: 6 volunteers age 45–68, focused on crow's feet and other zones, using placebo control, questionnaire and photo analysis.	**Trial 2 using Botanicin SF**: 6 volunteers age 58–68, focused on multiple wrinkle zones considered problem areas, no control serum, compared before and after photos after 3 minutes, analyzed by expert evaluator and self-observation.
• 6/6 saw immediate improvement in dynamic wrinkle appearance • 6/6 felt immediate tightness • 6/6 saw improvements lasting at least 6 hours • 4/6 saw long-term improvement at 1 month	• 5/6 saw instant improvement in selected area • 1 volunteer saw no change but evaluator did • 5/6 felt instant tightening or tingling sensation

CONTROL: Smile
w/o Treatment

BOTANICIN™: Smile
15 Minutes after Treatment

FIGURE 18.2 Effect of Botanicin on dynamic wrinkles.

TABLE 18.2

O/W Sunscreen Emulsion with PharmaDur

Aqueous Phase		Oil Phase	
Ingredient	**%**	**Ingredient**	**%**
Deionized water	61.35	Homosalate	10.00
Polyglycerin-3	2.00	Octyl salicylate	3.00
PharmaDur copolymer	1.00	Octyl 4-methoxycinnamate	7.50
Na$_4$ EDTA	0.10	Octocrylene	2.75
DMDM hydantoin	0.30	Butyl methoxydibenzoylmethane	2.00
		Crystal cast	5.00
		Isododecane and isononyl isononanoate	5.00

A control formulation without the PharmaDur copolymer and Crystalcast was also prepared for comparison purposes. Sun protection factors (SPF) of both the control and PharmaDur-containing emulsions were determined by standardized *in vitro* testing on precision cast sample film (1.3 mg/cm^2) on poly(methyl methacrylate) plates (9 samples/formulation). SPF of the control formulation was found to be 5.0 (SD 0), whereas that of the PharmaDur-containing formulation was 17 (SD 1). Thus, a pronounced enhancement in SPF of the test formulation containing PhamaDur was evidenced. The enhancement in SPF can be attributed to uniform film formation.

High internal phase W/O emulsions (HIPEs) are those that contain more than about 70% by volume of the dispersed aqueous phase. The droplets present in HIPE are deformed from the usual spherical shape into polyhedral shapes and are locked in place. Thau (2013) has reviewed the chemistry and preparation of HIPEs. HIPE lotion containing PharmaDur copolymer was prepared utilizing micronized-treated zinc oxide, other sunscreen actives and other conventional ingredients (Table 18.3).

PharmaDur copolymer is known to form polymeric micelles on account of its hydrophilic–hydrophobic structural characteristic and therefore may stabilize the zinc oxide dispersion. One of the purposes of preparing the PharmaDur HIPE lotion, with a high loading of zinc oxide (14.5%), was to evaluate the clarity of the formed film on skin, which relates to the dispersion uniformity and stability. Appearance of this test formulation on skin after drying was compared with that of a leading marketed sunscreen product brand, which had the same actives, including zinc oxide and equal concentrations in the products (Figure 18.3). Identical amounts of the test and the leading brand formulations were applied on equal surface areas on the back of the hands of a subject.

The test formulation exhibited optical clarity of appearance indicating absence of any particulate agglomeration, whereas the leading marketed product having the same concentration (14.5%) of zinc oxide had a discernible bluish haze. The difference in appearance of the two products on skin may be attributed to zinc oxide dispersion stabilization by PharmaDur in the test formulation.

TABLE 18.3

PharmaDur Sunscreen with Zinc Oxide

	Ingredient	%
Phase A	Octyl salicylate	5.0
	Octyl 4-methoxycinnamate	7.5
	$C_{12}C_{15}$ alkyl benzoate	9.5
	Polyglyceryl-4 diisostearate/polyhydroxystearate/sebacate	3.0
Phase B	Deionized water	40.0
	Glycerin	2.0
	Sodium chloride	1.0
Other	Aqueous 10% PharmaDur solution	15.0
	Micronized zinc oxide (Z-Coat HP, BASF)	14.5
	Preservative blend (Germaben-II, ISP)	1.0

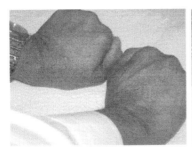 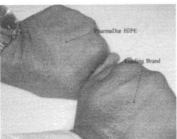

Appearance before application   Appearance after application

FIGURE 18.3 Appearance on skin of sunscreens with high loading of zinc oxide. Comparison of the test formulation containing PharmaDur with that of the leading marketed product.

Anti-Cellulite

Cellulite is an accumulation of fatty tissue irregularly distributed in the thighs and buttocks of women. It is well documented that these female fat cells (adipocytes) are special fat cells that are reserved fat deposits for pregnancy and lactation. This fatty tissue is under hormonal control and is not used for general energy supply, as is the fat on the abdomen and elsewhere. Therefore, diet, exercise and other metabolic activities alone cannot reduce cellulite. It is known that xanthines can reduce fatty tissue in the underlying skin if applied topically. Various products, for example, ROC Cream with caffeine, are commercially available for topical application. Such products are claimed to reduce the appearance of cellulite if used in conjunction with appropriate diet and exercise. The rate of permeation of xanthine through the skin is an important consideration in formulation of such products. For effective reduction of cellulite, xanthine has to be delivered at the site for an extended period of time, and under appropriate metabolic and dietary conditions. Higher xanthine skin permeation rates would be advantageous if deeper fatty tissues in the dermis are to be affected by topical application. Caffeine is the most widely used xanthine in anti-cellulite product formulations. Michniak-Kohn et al. (2005) have reported enhancement in skin permeation of caffeine from its aqueous saturated solutions by incorporating a combination of PharmaDur copolymer and oleic acid as an enhancer. The skin permeation study was carried out with Franz diffusion cells using human cadaver skin as the control membrane. It was found that while 2 and 5% oleic acid increased the rates of caffeine permeation through skin, additional incorporation of 2% PharmaDur resulted in further enhancement in the rate by a factor of about 2 (Figure 18.4).

Such combinations of PharmaDur copolymer and oleic acid can be used to formulate anti-cellulite caffeine formulations that can potentially exhibit significantly enhanced performance effectiveness.

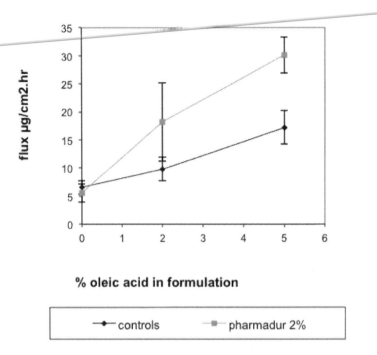

FIGURE 18.4   Effect of PharmaDur and oleic acid on skin permeation rates of caffeine.

Case 3: Innovative Ingredients and Excipients in Topical Formulations and Their Use in Cosmetics, Sunscreens and Topical Pharmaceuticals

Challenges of Launching New Cosmetic Materials

Citing the Personal Care Product Council website, the leading trade association for the industry, there are over 21,000 ingredients available globally to the formulation chemist. A saturated market mandates new chemistry be functionally unique or a significant improvement upon existing chemistry to generate interest with finished goods companies. A strong business case with thorough market research is mandatory, because a significant investment will be made based on the results. Development costs, efficacy testing, legal review, regulatory fees, safety/toxicology, marketing expenditures, manufacturing setup and supply chain planning are just a few of the challenges facing BASF and other suppliers when preparing to launch a new material. Obstacles could arise at any step and derail the launch, sometimes well into the process, resulting in a loss of all the resources, time and money allocated to the project.

External factors can also influence a successful product introduction: the economy, competition, evolving regulatory environment and customers' desire to streamline coded raw materials are a few cases. A current example affecting the industry is an environmental concern about palm or palm kernel oil, the feedstock for many personal care raw materials. In the search for sustainable solutions, the industry moved away from petroleum-based products to vegetable-based products. Palm oil offers ideal stock to fractionate the cuts needed for emollients, emulsifiers and surfactants. Unfortunately, the growing demand for palm tree oil has led to poor harvesting management and deforestation of tropical forest. BASF is committed to sustainability and since 2009 has been a member of the Roundtable on Sustainable Palm Oil (RSPO). By 2014, BASF was purchasing 100% from its members. However, a trend that is slowly creeping in the finished goods sector is to shy away from palm-derived materials, even when harvested responsibly. This is a legitimate concern for suppliers, whose options for sustainable feedstock are already limited, and an excellent example of how an external circumstance influences internal decisions.

High-Pressure, High-Shear Technology

Emulsion chemistry, the stabilization of two immiscible substances with polymers and emulsifiers, is the basis for a wide range of personal care products such as creams, lotions, conditioners, pomades, mascaras and foundations. Typical processing for emulsions involves high temperatures, blade mixing and homogenizers to create opaque emulsions with droplet sizes in the ~50 μm range. BASF offers a unique process called high pressure high shear (HPHS) which utilizes proprietary equipment to emulsify at extremely high pressure, up to 25,000 psi, over patented high shear blades to create nano-emulsions with narrow particle size distribution in the 200–500 nm range. This process allows large quantities, up to 50%, of hydrophobic materials like petrolatum to be stabilized and become infinitely water dispersible. Figure 18.5 shows a traditional process emulsion of 50% petrolatum and BASF's Sansurf® Pet 50, which is both pourable and easily water dispersible. This technology allows BASF to create a line of products that are multi-functional and unique to the personal care market. Ideal applications include wipes, cold process and ease of addition for sensitive formulations.

Emulgade® Sucro Plus

Natural claims and products are no longer a niche market confined to small boutiques and health food stores. Many global companies have added natural lines to their products or minimally natural claims to existing products. National retailers such as Walmart and Target are demanding sustainability trails back to the origin of the feedstock and manufacturing details. Ingredients in personal care are under increased scrutiny via NGOs such as the Environmental Working Group or government watch lists like Proposition 65 in California. One ingredient that is becoming more scrutinized is ethylene oxide, which very generally speaking is used to make oils and waxes more water-soluble. A vast majority of emulsifiers and surfactants in the personal care industry are ethoxylated. BASF recognized the need for natural-based emulsifiers and added to its list of 274 Natural Product Association–certified ingredients with the launch of an exclusive sucrose-based, non-ethoxylated emulsifier: Emulgade Sucro Plus (INCI: sucrose polystearate and cetyl palmitate). This product can serve as a primary emulsifier at higher concentrations or the secondary in lower. It supports the formation of lamellar micelle structure (Figure 18.6), that is liquid crystals, which create a unique sensorial experience and have been suggested to lead to better penetration of active molecules. Emulgade Sucro Plus gives formulators a natural option to create luxurious emulsions with high-end consumer appeal.

Challenges of Launching New Sunscreen Materials

Many of the aforementioned challenges are the same for UV filters. In the United States, UV filters are regulated as drugs and therefore fall under the regulatory jurisdiction of the Food and Drug Administration

FIGURE 18.5 BASF's Sansurf Pet 50 using high pressure high shear (HPHS) technology (left) compared to a traditional emulsion containing 50% petrolatum using traditional processing (right).

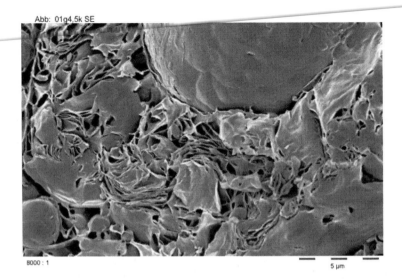

FIGURE 18.6   Emulgade Sucro Plus lamellar structure in basic emulsion under Cryo-SEM.

(FDA). In recent discussions between the industry and the FDA, it appears that the barrier to entry for the U.S. market on new sunscreens will ultimately be as rigorous as a pharmaceutical new drug application (NDA), a barrier very challenging to overcome for suppliers and brand-owners alike.

Tinosorb® M

In North America when cosmetic formulation chemists think of 'physical' UV filters, they think of titanium dioxide and zinc oxide. Inorganic minerals are typically used as pigments but have been micronized to be more transparent upon application to skin. While these filters have advantages over organic filters such as reflecting and absorbing UV and a better safety profile, there are some disadvantages as well. Whitening on the skin (particularly for darker skin tones) and formulation incompatibilities with common ingredients are the two main drawbacks.

Tinosorb M (USAN: bisoctrizole; INCI: methylene bis-benzotriazolyl tetramethylbutylphenol) is an organic, nanosize particle that offers extreme broad-spectrum UV protection (Figure 18.7). It is provided as a 50% aqueous dispersion and is extremely easy to formulate into a product, just add at the end of the formula with minimal mixing or shear. It has none of the whitening or formula incompatibilities associated with the mineral filters. Approved up to 10% use level in all global regions except the United States and Canada, it is currently pending FDA approval via a time and extent application. As of this writing, it is hopeful that Tinosorb M will be approved by the FDA and available to U.S. consumers by 2017.

Challenges of Launching New Pharmaceutical Excipients

The development and launching of new pharmaceutical excipients, commonly referred to as 'novel excipients', is very challenging from both a supplier's perspective, as well as for the pharmaceutical formulator. In addition to the aforementioned development and production costs, toxicology packages, and economic factors outlined for cosmetic ingredients, there are additional regulatory hurdles required. Excipients, either novel or traditional, are not regulated individually by the FDA, but rather as a portion of the overall submission. This causes challenges associated with the inherent risk assessment of the drug submission.

Trade organizations and pharmacopeias manage excipients from a quality perspective individually by monograph compliance. Most notably, the USP–NF for the U.S. market, the European Pharmacopoeia (Ph. Eur.) for the European Market and the JP/JPE for the Japanese market. These

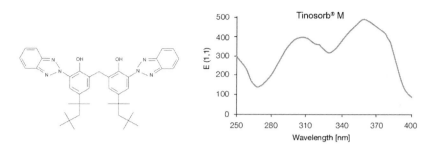

FIGURE 18.7 Methylene bis-benzotriazolyl tetramethylbutylphenol chemical structure (left) and Tinosorb M UV absorbance spectrum (right).

monographs contain all of the required specifications, as well as the validated testing procedures for ensuring the identification and purity of pharmaceutical excipients. In addition, a general requirement of good manufacturing practices (GMPs) is required. This is typically overseen by the IPEC (International Pharmaceutical Excipient Council) Federation, and excipient manufacturing sites are thus certified as IPEC-GMP compliant sites. Ingredients that meet the test requirements of the USP–NF, Ph. Eur. or JPE are typically identified as such on the Certificate of Analysis (CoA); this marking implies GMP manufacturing by the supplier. Monograph listings within the major three compendia require additional time, effort and establishment of testing; this also adds to the quality and regulatory challenges of novel excipients.

From a regulatory perspective, a typical pharmaceutical formulator utilizes the Inactive Ingredient Database, published on the FDA website (http://www.accessdata.fda.gov/scripts/cder/iig/index.Cfm) as a search reference for excipients previously utilized within approved drugs. The maximum concentration of a given excipient in an approved drug for a specific type of formulation is listed within the database. For example, if a formulator was to use cetostearyl alcohol (Kolliwax® CSA 50), searching for this term in the database would yield numerous listings including use in tablets, creams, foams, lotions, ointments and suppositories. Excipients are all listed with their CAS number and a UNII number for reference, and finally, the maximum concentration of use is listed. As of the writing of this chapter, the listing for cetostearyl alcohol in a topical cream is listed as 11.2% w/w. A formulator may certainly utilize 12, 13, 15% or greater of cetostearyl alcohol in a given cream formulation, but it may invite additional questions regarding safety and efficacy of the formulation, and therefore functions as an artificial ceiling for use at most pharmaceutical companies. This is unfortunately in contrast to the original rationale for publishing of this database. Furthermore, the use of 'novel excipients', which are defined as ingredients that have not been previously approved in an NDA or ANDA (abbreviated new drug application), and therefore do not appear in the database, would require further scrutiny form the regulatory authorities for further proof of safety. Again, this is a level of risk most pharmaceutical formulators are unwilling to take. As previously stated, since ingredients are not regulated by the FDA individually but rather as a portion of a filing, a supplier needs to wait for a novel excipient to be utilized and approved within a pharmaceutical company's drug filing before it will become widely accepted by the industry. This renders innovation on the supplier side extremely challenging.

Kollicream® 3C

For the development of a pharmaceutical product, there are three key areas of the utmost importance. First, the drug must be delivered to the target area. This can include dermal delivery, transdermal delivery, follicular delivery and others methods of transport. Second, the ingredients must be inherently mild and not add further irritation to the skin during treatment. Finally, the ingredients should add to the overall patient compliance by being formulated into a product with positive sensory properties. Each of these three core areas are addressed by the utilization of Kollicream 3C as an emollient and penetration enhancer. Kollicream 3C is the trade name for cocoyl caprylate/caprate, an ester blend of lipids with notably excellent properties for dermal and transdermal delivery.

Preliminary studies with one active pharmaceutical ingredient suggest that Kollicream 3C may act as a dermal penetration enhancer, analogous to other well-known penetration enhancers, such as isopropyl myristate (Kollicream IPM). Contrary to the perception of penetration enhancers causing an irritation effect, Kollicream 3C has been clinically proven to be mild. Independent, clinical allergy and irritation testing was performed on 500 patients manifesting signs of chronic dermatitis and allergic reactions; these tests were performed using patch tests for 48–72 hours, and evaluated with a 3-point visual scoring system by a trained dermatologist. Of all 500 patients, zero patients exhibited any irritation during or following exposure.

For sensory evaluation, Kollicream 3C offers the lightness and comparably fast absorption rate of isopropyl myristate, with a slightly slower spreading value (800 mm²/10 min vs. 1200 mm²/10 min). In addition to delivering the drug effectively, it leaves the skin feeling smooth, soft and with little residual oil; a comparison of Kollicream 3C to other common pharmaceutical emollients is shown in Figure 18.8.

Newly listed within the FDA Inactive Ingredient Database, Kollicream 3C is an emollient growing in interest within the pharmaceutical industry. The core benefits of penetration enhancement, clinically proven mildness and a strong sensory profile render it one of the most interesting new additions to the pharmaceutical formulator's toolbox.

Kolliphor® P–Grade Poloxamers (P 188, P407)

Kolliphor P–grade surfactants are the trade name for poloxamer chemistries. Poloxamers are a unique, ethylene oxide/propylene oxide (EO/PO) chemistry combined into an ABA-type copolymer of PEO (polyethylene oxide) and PPO (polypropylene oxide), which produces a unique range of multi-functional non-ionic surfactants. There are currently five compendial poloxamers: 188, 237, 338, 407 and 124. The numbering system is such that the first two numbers represent the average molecular weight of the PPO portion when multiplied by 100, for example, Kolliphor P 407 has an average molecular weight of the PPO portion at 4000 g/mol. The second number is the molar percentage of PEO; as the PEO portion of the molecule is the more hydrophilic of the poloxamer, this gives a rough estimate of the hydrophilicity of the overall molecule. Using our previous example, Kolliphor P 407 exhibits approximately 70% PEO portion and is relatively hydrophilic. The aforementioned poloxamers, Kolliphor P 188, P 237, P 338 and

Emollient	Chemical name	µ mPa s	MW	Spreading mm²/10min	Spreading class	Emollience
Kollicream® IPM	Isopropyl myristate	5-6	270	1200	Fast	Dry, Light
Kollicream® 3C	Coco-caprylate/caprate	11	335	800		
Kollicream® OA	Oleyl alcohol	33	270	700	Medium	Medium
Kollicream® DO	Decyl oleate	15.9	415	700		
Kollicream® OD	Octyl dodecanol	58-64	300	600		
Kollisolv® MCT 70	Medium-chain triglycerides	25-33	500	550		
Kollicream® CP 15	Cetyl palmitate	(s)	480	Solid	Slow (molten)	Rich

FIGURE 18.8 Comparison of common emollients utilized in pharmaceutical formulations, highlighting the low viscosity, high spreading value of Kollicream 3C.

P 407 are solids at room temperature, and commonly supplied as prills; while the Kollisolv® P 124, also being the most hydrophobic, is dispensed at room temperature as a liquid.

Of the poloxamers relevant to the development of semi-solid, topical or transdermal formulations, Kolliphor P 188 and Kolliphor P 407 are very commonly used as solubilizers, gelling agents, plasticizers (for films) and emulsifiers. Kollisolv P 124 is more commonly used in liquid systems, where its unique surfactant and solubilizing capacity is utilized. Due to the thermo-reversible gelling properties of Kolliphor P 188 and Kolliphor P 407, these have the advantage of being the primary gelling agents for a topical semi-solid, which offers an interesting alternative to carbomer chemistry. Specifically, at concentrations above 15% w/w in water, poloxamers will gel at room temperature into a clear-glass, hydrophilic gel with solubilizing properties; solutions can even be tailored to be applicable as a liquid and gel at body temperature. Since poloxamer gels are non-ionic in nature, the use of salt-form active ingredients or pH changes have a significantly less drastic effect than on traditional carbomer-type gels. Figure 18.9 shows an image of a poloxamer gel containing BASF L-Menthol Pharma.

While the functional properties of the Kolliphor P–brand poloxamers are numerous, another core area fundamental to skin-based formulation is the inherent mildness of the surfactants. For example, Kolliphor P 188 is often used in products for the treatment of minor wounds, scrapes and cuts either as occlusive barriers or as a wound-cleaning product. Several published articles allude to several benefits of poloxamers for the protection of cells, but these claims are difficult to establish. Nevertheless, what is clear is the mildness of the product itself. For example, using an industry standard test, the transepithelial permeability assay (TEP), monolayers of Madin-Darby Kidney epithelial cells are grown into monolayers onto a multi-well plate. Concentrations of the excipient are diluted to 1:100 through 1:300 in water and briefly exposed (15 min) to the monolayer of cells, where an irritation response on the cellular level will lead to the disruption of the tight junctions between the cells. The degree of tight junction degradation is linearly correlated with the irritation potential of the test substance and porosity of the cell monolayer. The porosity of the cell monolayer is measured by quantifying the amount of applied fluorescein solution that passes through. After extensive *in vitro* testing, it was found that the Kolliphor P 188 offered exceedingly low irritation compared with other widely accepted topical ingredients. These data are shown Figure 18.10, where Kolliphor SLS (or sodium lauryl sulphate) is used as the positive control, and Kolliphor PS 80 (or polysorbate 80) and Kolliphor PS 60 (or polysorbate 60) are shown as a comparable and mild topical excipient. Kolliphor TPGS (vitamin E polyethylene succinate, USP, tocofersolan) and Soluplus® (polyvinyl caprolactam-polyvinyl acetate-polyethylene glycol graft copolymer) are two additional amphiphilic excipients developed by BASF for solubilization. Clearly the mildness of Kolliphor P 188 is evident, as the mildness is shown across all concentrations as seen in Figure 18.10.

In summary, the utilization of Kolliphor P–grade surfactants is in its relative infancy, and novel uses of these interesting and unique block copolymers are being discovered each day. With a multitude of applications for topical, semi-solid, liquid and transdermal applications for skin, the list will continue to grow.

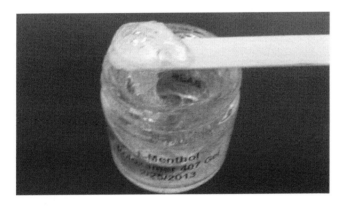

FIGURE 18.9 Image of a Kolliphor P407 (poloxamer 407)–based gel containing L-menthol.

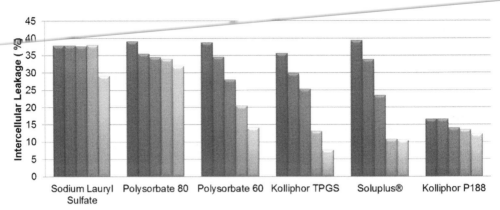

FIGURE 18.10 Irritation potential of selected surfactants at 1:100, 1:150, 1:300, 1:350 and 1:300 w/w concentrations as measured by percent of intercellular leakage from the industry standard transepithelial permeability assay.

REFERENCES

Batheja, P.; Sheihet, L.; Kohn, J.; Singer, A. J.; Michniak-Kohn, B., Topical drug delivery by a polymeric nanosphere gel: Formulation optimization and in vitro and in vivo skin distribution studies. *J Control Release* **2011**, *149* (2), 159–67.

Belfer, W. A., U.S. Patent 7,566,464 (July 28, **2009**).

Costache, A. D.; Sheihet, L.; Zaveri, K.; Knight, D. D.; Kohn, J., Polymer-drug interactions in tyrosine-derived tri-block copolymer nanospheres: A computational modeling approach. *Molecular Pharm* **2009**, *6* (5), 1620–7.

Goyal, R.; Macri, L.; Kohn, J., Formulation strategy for the delivery of cyclosporine A: Comparison of two polymeric nanospheres. *Sci Rep-UK* **2015**, *5*, 13065.

Kilfoyle, B. E., Tyrosine-derived Nanoparticles for the Topical Treatment of Psoriasis. PhD thesis, Rutgers University, New Brunswick, NJ, **2011**.

Kilfoyle, B. E.; Sheihet, L.; Zhang, Z.; Laohoo, M.; Kohn, J.; Michniak-Kohn, B. B., Development of pacli-taxel-TyroSpheres for topical skin treatment. *J Control Rel* **2012**, *163* (1), 18–24.

Mercier, M. F.; Thau, P.; Chase, J., U.S. Patent 7, 754,775 (July 13, **2010**).

Michniak, B.; Thakur, R.; Shah, K. R., Evaluation of Novel Polymer for Transdermal Permeation of Caffeine in Presence of Different Concentrations of Chemical Enhancer, Poster #323, Proceedings of Controlled Release Society 32nd Annual Meeting, Miami Beach, FL, June **2005**.

Nardin, C.; Bolikal, D.; Kohn, J., Nontoxic block copolymer nanospheres: Design and characterization. *Langmuir* **2004**, *20* (26), 11721–5.

Shah, K.R., PharmaDur® bioadhesive delivery system, in John J. Wille, ed., *Skin Delivery Systems: Transdermals, Dermatologicals, and Cosmetic Actives*, Blackwell, Ames, IA, **2006**, pp. 211–22.

Sheihet, L.; Dubin, R. A.; Devore, D.; Kohn, J., Hydrophobic drug delivery by self-assembling triblock copo-lymer-derived nanospheres. *Biomacromolecules* **2005**, *6* (5), 2726–31.

Sheihet, L.; Piotrowska, K.; Dubin, R. A.; Kohn, J.; Devore, D., Effect of tyrosine-derived triblock copolymer compositions on nanosphere self-assembly and drug delivery. *Biomacromolecules* **2007**, *8* (3), 998–1003.

Sheihet, L.; Chandra, P.; Batheja, P.; Devore, D.; Kohn, J.; Michniak, B., Tyrosine-derived nanospheres for enhanced topical skin penetration. *Int J Pharm* **2008**, *350* (1–2), 312–9.

Thau, P., High Internal Phase Water in Oil Emulsions, in Nava Dayan, ed., *Apply Topically: A Practical Guide to Formulating Topical Applications*, Allured Books, Carol Stream, IL, **2013**, pp. 155–6.

Yu, C.; Kohn, J., Copolymers of tyrosine-based polycarbonate and poly(alkylene oxide). U.S. Patent 5,658,995, **1997**.

Zhang, Z.; Tsai, P. C.; Ramezanli, T.; Michniak-Kohn, B. B., Polymeric nanoparticles-based topical delivery systems for the treatment of dermatological diseases. *Wires Nanomed Nanobi* **2013**, *5* (3), 205–18.

Zhang, Z.; Ramezanli, T.; Tsai, P.-C., Drug delivery systems based on tyrosine-derived nanospheres (TyroSpheres™), in Jose L. Arias, ed., *Nanotechnology and Drug Delivery*, Volume 1: *Nanoplatforms in Drug Delivery*, CRC Press, Boca Raton, FL, **2014**, pp. 210–32.

19

Advanced Nanoformulation Technologies in Cosmetic Science

Reinhard H.H. Neubert

CONTENTS

Introduction

Nanosized colloidal systems (NSCSs) have structures below a particle or droplet size of 500 nm. In previous years, the focus for the application of nanocarriers was primarily placed on the parenteral and oral application. However, NSCSs applied to the skin are now the centre of attention and are expected to be increasingly applied because the skin offers a lot of advantages for the administration of such systems.

For the cosmetic use of NSCSs, one has to differentiate between the desired effects: the effect on the skin or the effect within the different layers of the skin.

The stratum corneum (SC), the main barrier of the skin, has to be overcome if the cosmetic active should penetrate into one of the skin layers. The SC is one of the tightest barriers of the human body. Therefore, it is the primary goal of new NSCSs in cosmetics to overcome this protective and effective barrier. For that purpose in cosmetics new NSCSs such as *nanoparticular systems (nanoparticles and nanocrystals), nanoemulsions, microemulsions,* and special *vesicular systems* (e.g. *unilamellar liposomes*) are being developed and investigated. NSCSs are particularly valuable for the application of extremely hydrophilic cosmetic actives such as peptides and for extremely lipophilic cosmetic actives such as ceramides (see also Neubert (2011)).

This chapter evaluates the preparation, characterisation and potential of NSCSs for cosmetic applications.

Technology and Equipment

Special technical procedures and equipment are necessary for the preparation and production of *nano-emulsions* (NEs), *nanoparticles* (NPs) and *nanocrystals* (NCs). In contrast, the preparation as well as the production of liquid colloidal nanosized formulations, defined as *microemulsions* (MEs) in the literature, is very simple and no energy-consuming procedure is necessary.

NEs are produced using high-energy methods such as ultrasonication and homogenization with high-pressure homogenizers or microfluidizers. In the last years, there have been attempts to produce NEs using low-energy procedures such as the phase inversion temperature and the solvent evaporation method (Klang et al., 2015).

NP-based delivery systems for cosmetic actives can be prepared as described in the literature using emulsion polymerization, interfacial polymerization, and denaturation or desolvation of the used polymeric materials (Kreuter, 1991), nanoprecipitation (Palacio et al., 2016), and combination methods such as high-pressure emulsification–solvent evaporation (Jaiswal et al., 2004).

Lipid-based nanoparticular delivery systems for cosmetic actives are prepared as described in the literature (Müller et al., 2007). The process involves preparation of pre-emulsion by melting of the lipid phase and dissolving the cosmetic active. The pre-emulsion is dispersed by high-speed stirring in hot aqueous stabilizer phase and homogenized by high pressure. After cooling NP crystals are obtained (Müller et al., 2007).

NCs are produced using two approaches: (1) bottom-up and (2) top-down technologies (Müller et al., 2011). The basic principle of the bottom-up technology is the precipitation process. It has to be controlled to produce NCs in nanosized dimension and to obtain either crystalline or amorphous materials. To date, there are some interesting new bottom-up technologies such as high-gravity controlled precipitation, sonocrystallization, confined impinging liquid jet precipitation and multi-inlet vortex mixing. The top-down technologies cover milling procedures such as pearl milling, high-pressure homogenization and combination technologies (Müller et al., 2011).

Nanoformulations in Cosmetic Science

Nanoemulsions

Recently, the dermal use of nanoemulsions (NEs) was comprehensively reviewed (Klang et al., 2015). NEs were first known as *submicron emulsions*. In comparison to microemulsions, NEs are not thermodynamically stable. They are just kinetically stable, and their appearance is transparent to translucent (Anton and Vandamme, 2011). NEs have to be stabilized using suitable surfactants, and high energy is necessary for their production (see Table 19.1). It is stated that is used for NEs in the range <100 nm (Klang et al., 2012). However, NEs used in cosmetics are mainly oil-in-water (O/W) emulsions and have droplets sizes <500 nm. Therefore, lipophilic cosmetic actives can be incorporated into the inner phase of the NEs.

Dynamic light scattering, and microscopic techniques such as cryo-transmission electron microscopy (Cryo-TEM) and atomic force microscopy (AFM) are used to characterize NPs for droplet sizes as well as polydispersity (Klang et al., 2012).

The classical lecithin-based NEs are produced using high-energy emulsification. However, there are attempts in the literature to use low-energy methods for the production of NEs. The phase inversion method and the solvent evaporation method are used for this purpose (Tadros et al., 2004; Sole et al., 2010; Fernandez et al., 2004).

Stabilization of NEs is a key issue in the production of NEs. Monitoring of the appearance, of the pH and of the zeta potential can be used to optimize the stability of NEs (Klang et al., 2011a). In the past, NPs were stabilized using lecithin as surfactant (Klang et al., 2015; Klang and Valenta, 2011). Nowadays, a lot of mild and skin-friendly surfactants such as alkyl glucosides, sucrose fatty acid esters and polyglycerol fatty acid derivatives are available as alternatives.

Another important issue in the potential of NEs concerns penetration enhancement of cosmetic actives. The best way is to incorporate penetration enhancers such as propylene glycol or oleic acid into the NEs (Anton and Vandamme, 2011). Co-surfactants such as alkyl glucosides (Klang et al., 2015; Hoeller et al., 2009) and cyclodextrins (Klang et al., 2011a) were also used in order to improve penetration rates of actives

TABLE 19.1

Different Properties of Microemulsions, Nanoemulsions and Standard Macroemulsions

Property	Microemulsion	Macroemulsion	Nanoemulsion
Appearance	Transparent	Milky	Transparent to translucent
Thermodynamic	Stable	Unstable (kinetically stabilized)	Unstable (kinetically stabilized)
Formation	Spontaneous	Energy needed	High energy needed
Interfacial tension	Tending towards 0 mN m^{-1}	~50 mN m^{-1}	>50 mN m^{-1}
Flow behaviour	Newtonian flow behaviour (viscosity: 80–200 mPa sec)	Non-Newtonian thixotropic flow behaviour (high viscosity)	Near-Newtonian flow behaviour
Microstructure	Dynamic (fluctuating structures)	Single structures (up to coalescence)	Single structures
Optically isotropic	Yes	No	No
Droplet size of the internal and colloidal phase	10 nm–50 nm	>500 nm	< 500 nm < 100 nm*

* Data from Mason TG et al., 2006, *J Phys Condens Matter* 18: R635–R666.

into the skin from NEs. In addition penetration enhancement may be due to an occlusion of the skin surface caused by the lipophilic components of the NEs (Zhou et al., 2010). Furthermore, an interaction of the surfactants of the NE with the SC lipids could cause penetration enhancement of actives incorporated into NEs.

No relationship has been clearly identified between the droplet size and the penetration enhancement caused by the NEs. It was found that the bioavailability of tocopherol was 2.5-fold higher when incorporated into a NE in comparison with a microsized emulsion (Kotyla et al., 2008). Skin hydration as well as the skin penetration of a model active (Nile red) was enhanced when nanosized droplets were used in an O/W cream compared to a standard O/W cream (Zhou et al., 2010). Puglia et al. reported that the skin penetration of glycyrrhetic acid was enhanced if it was incorporated into a NE compared to O/W emulsion (Puglia et al., 2010). Thermodynamic activity of the incorporated active can be changed by incorporation into NEs (Cevc and Vierl, 2010).

In the past, NEs were developed for the following cosmetic actives: glycyrrhetic acid (Puglia et al., 2010), lutein (Mitri et al., 2011) and terpenes (Mou et al., 2008). Stable NEs with a droplet size of 136 nm were prepared for cosmetic use based on swiftlet nest material containing a glycoprotein as the main component (Taib et al., 2015).

Microemulsions

The term 'microemulsion' (ME) is traditionally used in the literature. However, MEs are colloidal formulations having specific properties. Because of these properties MEs have many advantages as colloidal vehicle systems for cosmetic application in comparison to the standard macroemulsions and NEs (see Table 19.1). These include the thermodynamic stability, the simplicity of manufacture, excellent solubilization properties for both hydrophilic and lipophilic cosmetic actives, and excellent penetration properties (Heuschkel et al., 2008). Application of MEs as vehicle systems for cosmetic actives is an innovative way to enhance the availability of substances in the respective target compartments of the skin. Despite the advantages of MEs, their use in cosmetics is not widespread. One of the disadvantages of MEs in the past has been the use of relatively high concentrations of emulsifiers. To ensure high tolerability, the concentration of surfactants should be below 20%, and only skin-friendly emulsifiers should be used.

The excellent penetration-enhancing properties of MEs for cosmetic agents are widely accepted, although the specific mode of action as colloidal carriers is not yet fully understood. It is assumed that various factors interact depending on the composition and the resulting nanostructure of the MEs. The penetration-enhancing properties of MEs cannot be attributed to the interactions of the surfactants with the SC.

Crucial for the penetration-enhancing effect of MEs is the fact that continuous and spontaneous fluctuations of the domains within the MEs allow a high mobility of the active agent, which improves the diffusion process (Heuschkel et al., 2008), MEs are very effective carrier systems for extremely both lipophilic and hydrophilic cosmetic actives. Mild and highly skin compatible surfactant systems are now commercially available. Research activities have shown that individual colloidal systems have to be developed for each cosmetic active. However, this fact can be realized because a high number of innovative combinations of highly skin compatible surfactant systems can be used for that purpose.

Microemulsions for Highly Lipophilic Cosmetic Actives

MEs were developed for extremely lipophilic cosmetic actives which were in the past not able to penetrate into the skin or whose penetration into the skin had to be improved significantly by incorporation into MEs. These actives were incorporated into the colloidal lipophilic phase of the O/W MEs.

Fatty acids, in particular *linoleic acid*, which is an essential component for the organization and stabilization of the skin barrier, are used in skin care products. The development of well-tolerated MEs containing linoleic acid as the active ingredient is described in the literature (Sahle et al., 2012). A comprehensive physiochemical characterisation of a novel ME system was performed using different techniques. The potential of the developed ME system compared to a cream as a suitable carrier for the dermal delivery of linoleic acid was determined. Penetration studies showed higher linoleic acid concentrations after administration of the MEs in all skin layers independent of the time of incubation. Up to 23% of the applied dose reached the skin from the MEs, whereas at most 8% of the active ingredient could be detected after applying the cream. Particularly, the percentage of the linoleic acids penetrated from the MEs in the SC and the viable epidermis differed significantly ($p < 0.01$) when compared to a cream. Furthermore, linoleic acids accumulated in the epidermis at longer incubation times (see Figure 19.1). Using the MEs, the penetration of linoleic acids was significantly enhanced ($p < 0.01$). Hence, the MEs might be an innovative vehicle for the delivery of linoleic acids to the epidermis making use of their barrier-improving and possible anti-inflammatory effects (Goebel et al., 2010).

Ceramides (CERs) are integral parts of the SC lipid lamellae that play a major role in the barrier properties of the epidermis. Studies have shown that several skin diseases, such as psoriasis and atopic dermatitis (AD), are associated with depletion of CERs in the SC lipid matrix indicating that these lipids may help to restore the barrier function in aged and affected skin. Among the different CER classes, the short-chain CERs [AP] and [NP], by virtue of their head group structure, are substantially important to form super-stable lipid lamellae. However, their dermal application is limited due to their extreme lipophilicity and their poor penetration into the SC from conventional dosage forms. Thus, stable MEs for CER [AP] and [NP] were developed using polyglyceryl fatty acids (Sahle et al., 2012) and lecithin (Sahle et al., 2013), respectively, and Miglyol® 812 and water 1,2-pentanediol mixture as amphiphilic, oily and hydrophilic components.

Microemulsions for Highly Hydrophilic Cosmetic Actives

The development of colloidal carrier systems for extremely hydrophilic cosmetic actives such as *peptides* is a special challenge because these components are not able to diffuse into or to penetrate through the SC.

In the literature, the improvement of the skin penetration of the hydrophilic *dipeptide N-acetyl-L-carnosine* is described. MEs were developed for this dipeptide. N-acetyl-L-carnosine was incorporated into the aqueous colloidal phase of the MEs. The penetration of the peptide in *ex vivo* human skin was investigated for MEs developed. Approximately fivefold and higher dipeptide concentrations in the SC and the viable skin layers were detected at all experimental periods when the MEs with the dipeptide were used (Goebel et al., 2012). MEs containing several peptides are now in the late development stage. Furthermore, MEs with a hydrophilic model active were stabilized using carrageenan (Valenta and Schultz, 2004).

Microemulsions for Other Cosmetic Actives

MEs based on a vegetable protein surfactant were developed as potential dermal delivery systems for the cosmetic active *dihydroavenanthramide* (DHAvD). They are composed of water; isopropylphthalimide; isobutylphthalimide; sodium cocoyl hydrolyzed wheat protein (Gluadin WK®); poloxamer 407; and

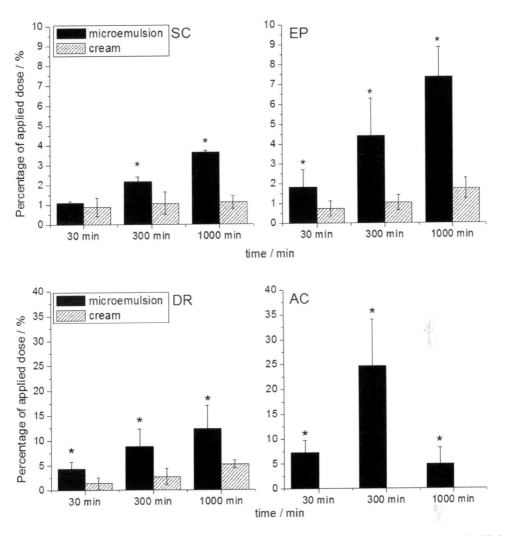

FIGURE 19.1 Comparison of relative amounts of linoleic acids in the stratum corneum (SC), the viable epidermis (EP) in the dermis (DR) and the acceptor compartment (AC) following application of a microemulsion and the cream at different incubation times. (Data given as mean ± standard deviation, $n = 9$; $*p < 0.01$ versus cream.) (From Goebel ASB et al., 2010, *Eur J Pharm Sci* 75: 162–172.)

propylene glycol (PrG), butylene glycol or pentylene glycol (PeG) as co-solubilizers. Three pseudoternary phase diagrams were constructed and the influence of the 1,2-alkanediol on ME formation was investigated. With increasing chain length of the glycol, the single phase and optical isotropic phase area increased. PeG-containing systems were characterized in detail by electrical conductivity and differential scanning calorimetry. The methods indicated the presence of water continuous ME. Two selected formulations of similar composition but with a different glycol compound, namely PrG-MEs and PeG-MEs, underwent a detailed physicochemical characterisation by TEM, conductivity, viscosity, refractive index and temperature stability. They were identified as low viscous, isotropic oil-in-water MEs with Newtonian flow. The subsequent penetration studies with DHAvD were performed in a finite dose mode by means of Franz diffusion cells equipped with human breast skin. Both formulations showed sufficient penetration into viable skin layers and particularly high permeation rates. Compared to a previously investigated glycol-containing cream, the MEs revealed a smaller fraction of the model drug within viable epidermis and dermis, but a strongly increased amount in deeper skin layers (Heuschkel et al., 2009).

Nanoparticles and Nanocrystals

Nanoparticles

The use of nanoparticles (NPs) in cosmetics appears to be an interesting alternative to the application of other nanosized colloidal systems.

In principle, two directions were identified concerning the use of NPs in cosmetics:

1. Application of NPs in cosmetic formulations
2. Follicular application of NPs

The advantages of the dermal application of NPs are still a matter of debate. The cosmetic application of NPs seems to be preferable. In the literature, there are reports that NPs having diameters between 6 and 36 nm may penetrate via the lipophilic penetration pathway and through aqueous pores, respectively (Hua, 2015). However, these results need to be confirmed.

The development of solid lipid NPs (SLNs) and nanostructured lipid carriers (NLCs) for cosmetic application in combination with occlusion is an interesting research topic (Müller et al., 2007; Muller et al., 2002). SLNs as well as NLCs can be used to chemically stabilize cosmetic actives in the formulation (Müller et al., 2007). However, it is necessary to validate the results concerning occlusion, skin hydration and penetration of cosmetic actives from these systems in comparison to relevant standard creams and other colloidal systems such as MEs and nanosized liposomes.

Furthermore, pectin-based NPs were developed and loaded with retinyl palmitate (RP) (Ro et al., 2015). These NPs had a diameter of 530 nm. RP was deposited throughout the skin layers and these NPs loaded with RP are expected to have advantageous antiwrinkle effects.

New and promising results were obtained by studying the penetration of NPs into the hair follicles, where a high increase of penetration depth was observed. Within the hair follicles the NPs can also be used as a depot for the active agents (Lademann et al., 2007). However, massage after application of the NPs is necessary. Based on the surface structure of the SC of human hair follicles, it was assumed that the movement of the hairs caused by massage pumps the NPs deeper into the hair follicles (Knorr et al., 2009; Lademann et al., 2008; Lademann et al., 2006).

Titanium Dioxide and Zinc Oxide

Many cosmetic sunscreen formulations contain NPs of titanium dioxide (TiO_2) and zinc oxide (ZnO) as physical light filters. The diameters of these nanosized materials range from 30 to 150 nm. TiO_2 and ZnO NPs are colourless, water insoluble, and reflect and scatter, respectively, UV radiation more efficiently than microparticles. A matter of debate in the literature is whether the metal oxide–based NPs show toxic effects to human skin and penetrate through the human SC. This issue has been intensively reviewed in the literature (Noynek et al., 2007; Schilling et al., 2010; Nohynek and Dufour, 2012; Gulson et al., 2015). In these reviews it was stated that water-insoluble NPs such as TiO_2 and ZnO NPs do not penetrate into or through the human skin. Both metal oxides have low systemic toxicity and they are well tolerated on the skin. The *in vitro* genotoxic and photogenotoxic profiles of TiO_2 and ZnO NPs show no consequences to human health (Schilling et al., 2010).

A research strategy was presented in order to evaluate the potential and the limitations of dermal use NPs in sunscreens (Gulson et al., 2015). In the literature, there is one report that small amounts of zinc from ZnO NPs (diameter >100 nm) in sunscreen appeared in the blood and urine of humans (Turci et al., 2013). However, the overwhelming amount of the applied [68]Zn was not absorbed and the amount of the tracer detected in the blood was just 1/1000th of the total Zn in this compartment after a 5-day application. Furthermore, a comparative *in vitro* study was published. It was shown that the spreadability was better and the *in vitro* sun protection factor was higher when TiO_2 and ZnO NPs were present in the cosmetic formulation in comparison to sunscreen cream containing conventional ZnO particles (Gulson et al., 2010; Singh and Nanda, 2014).

Nanocrystals

Nanocrystals (NCs) also have some advantages for application in cosmetics. In the literature, NCs having diameters <100 nm are defined as smartCrystals® (Shegokar and Müller, 2010). In general, NCs can be used for poorly water-soluble drugs because NCs show the following properties: (1) increased saturation solubility, (2) increased dissolution velocity, and (3) increased adhesiveness to surfaces and biomembranes (Müller et al., 2011). NCs have to be stabilized using suitable surfactants in order to prevent aggregation in the formulation.

The key question concerning the use of NCs in cosmetics is whether these advantages can be helpful for poorly water-soluble cosmetic actives too, because the solubility of poorly water soluble cosmetic actives in cosmetic formulations can be easily controlled using the following approaches:

1. Use of different cosmetically relevant solvents as well as co-solvents
2. Incorporation into liquid colloidal formulations such as MEs

However, NCs can be used in cosmetics in order to chemically stabilize cosmetic actives in cosmetic formulations.

In the literature, there is one report that dermal smartCrystals were used for quercetin (Hatahet et al., 2016). The smartCrystals were produced using a combination of bead milling and subsequent high-pressure homogenization at 300 bar. The NCs were physically stable in a gel formulation over 3 months and showed no cellular toxicity. Quercetin had sufficient antioxidant properties when formulated as NCs.

An interesting dermal application of a nanosuspension (NS) of resveratrol was published with particle diameters from 150 nm to 220 nm (Kobierski et al., 2009). This NS was prepared using high-pressure homogenization and stabilized using Tween, poloxamer and plantacare. The particles of the NS were crystalline.

Vesicular Vehicle Systems in Cosmetics

In the past, colloidal lipid aggregates such as *liposomes* (LIPs) were developed as vesicular carrier systems in cosmetics, too.

LIPs consist of lipids, typically cholesterol and phospholipids, but also other amphiphilic components are possible. Depending on the preparation method used, different LIPs result. Multivesicular vesicles (MVVs) contain several vesicles embedded in one surrounding spherical lipid bilayer, and multilamellar vesicles (MLVs) comprise several concentric bilayers. Large unilamellar vesicles (LUVs) and small unilamellar vesicles (SUVs) are made up of only one lipid bilayer. However, just the *SUVs represent nanosized vesicles*. The objective of the cosmetic application of LIPs appears to be an optimal localized effect. It is possible to influence the physicochemical properties of the LIPs to receive a specialized system in terms of penetration of the cosmetic active and protection of the active agent.

Classical LIPs (MLVs) are already used in cosmetics. However, there are several new vesicle types, depending on the additives used for the vesicle preparation: *transfersomes, flexosomes, ethosomes, niosomes, vesosomes, invasomes, cerasomes* and *polymerosomes*.

Several results were published regarding the potential of nanosized liposomal carrier systems for cosmetic application, particularly elastic ultra-deformable vesicles (transfersomes) and ethosomes (Hua, 2015; Cevc and Gebauer, 2003; Cevc, 2004).

In principle, there are two ways to influence drug penetration across the SC by ultra-deformable vesicles. Discussions are still occurring about whether the ultra-deformable vesicles are penetrating across the SC as intact aggregates or there is an incorporation of the liposomal lipids into the SC lipids (El Maghraby et al., 2004; El Maghraby et al., 2006). Therefore, research work is necessary to study the mechanisms of the penetration of LIPs as well as transfersomes into and through the SC until new effective liposomal formulations for cosmetic use can be developed, and the potential of these systems has to be validated. Furthermore, stability of the LIPs has to be taken into account. Rovisomes were shown to have a high stability and carry the cosmetic actives into deeper skin layers (Blume, 2000). Modifications of SUVs were carried out in order to control the penetration of cosmetic actives from this type of LIP

(Blume et al., 2003). It could be shown that actives penetrate from negatively charged liquid-state SUVs into deeper skin layers in contrast to positively charged ones. It was also demonstrated that SUVs which were stabilized by PEG coating did not penetrate into human abdominal skin. Furthermore, it could be shown that the penetration of actives can be controlled by the use of co-solvents such as pentylene and propylene glycol (Blume et al., 2003). New sunflower lipids (Hydro-Tops) were used to prepare nanovesicles having better penetration profiles than conventional ultraflexible LIPs based on non-GMO–derived soybean lipids (Blume, 2012).

Nanosized liposomes with a diameter of about 70 nm were already prepared for coenzyme Q10 (Xia et al., 2006). The penetration of semisolid fluorinated-DPPC liposomes (diameter = 70–350 nm) loaded with Na fluorescein was measured using mass spectrometry (Mahrhauser et al., 2015). After 24 h just a very low amount (0.0023% of the applied dose) of these liposomes permeated across the human skin.

Recently, it was stated in a review that that large vesicles (diameters >600 nm) do not carry cosmetic actives into deeper skin layers. They tend to remain in or on the SC. In contrast, nanosized vesicles (diameters <300 nm) appear to be able to carry cosmetic actives to some extent into deeper layers of the skin (Hua, 2015).

Conclusions

During the last years many new insights into the nanostructure of the SC, in particular of its lipid matrix, prompted the development of new delivery systems for cosmetic actives such as NSCSs. The ultimate goal is to understand the interactions between the lipid bilayers of the SC and the different cosmetic NSCSs, thereby creating the most efficient and mild NSCS and causing the least damage in respect of the SC barrier.

Taken together, the presented NSCSs have different potentials concerning the application of cosmetic actives. MEs appear to be most promising for the delivery of cosmetic actives. Most challenging is the use of NPs concerning follicular application of cosmetic actives as well as concerning deposition of these systems in the hair follicles.

Nevertheless, it is necessary (1) to fully understand the transport mechanism of each NSCS category, (2) to study the nanotoxicology of these systems and (3) to simplify the production procedures of NPs and NCs.

REFERENCES

Anton N, Vandamme TF (2011) Nano-emulsions and micro-emulsions: Clarifications of the critical differences. *Pharm Res* 28: 978–985.

Blume G (2000) Liposomes = Liposomes? *SÖFW-J* 126: 14–17.

Blume G (2012) Analysing a new generation of liposomes. *Pers Care* February 2012: 37–40.

Blume G, Sacher M, Teichmüller D, Schäfer U (2003) The role of liposomes and their future perspective. *SÖFW-J* 129: 10–14.

Cevc G (2004) Lipid vesicles and other colloids as drug carriers on the skin. *Adv Drug Deliv Rev* 56: 675–711.

Cevc G, Gebauer D (2003) Hydration driven transport of deformable lipid vesicles through fine pores and the skin barrier. *Biophys J* 84: 1010–1024.

Cevc G, Vierl U (2010) Nanotechnology and the transdermal route: A state of the art review and critical appraisal. *J Control Rel* 141: 277–299.

Fernandez P, André V, Rieger J, Kuehnle A (2004) Nano-emulsion formation by emulsion phase inversion. *Colloids Surf A Physicochem Eng Asp* 251: 53–58.

Goebel ASB, Knie U, Abels C, Wohlrab J, Neubert RHH (2010) A novel microemulsion system for improved and specific skin penetration of linoleic acids. *Eur J Pharm Sci* 75: 162–172.

Goebel ASB, Schmaus G, Neubert RHH, Wohlrab J (2012) Dermal peptide delivery: Carnosine. *Skin Pharm Physiol* 25: 281–287.

Gulson B, McCall M, Korsch M, Gomez L, Casey P, Oytam Y, Taylor A et al. (2010) Small amounts of zinc oxide particles in sunscreens applied outdoors are absorbed through human skin. *Toxicol Sci* 118: 140–149.

Gulson B, McCall MJ, Bowmann DM, Pinheiro T (2015) A review of critical factors for assessing the dermal absorption of metal oxide nanoparticles from sunscreens applied to humans, and a research strategy to address current deficiencies. *Arch Toxicol* 89: 1909–1930.

Hatahet T, Morille M, Hommoss A, Dorandeu C, Müller RH, Begu S (2016) Dermal quercetin smartCrystals®: Formulation development, antioxidant activity and cellular safety. *Eur J Pharm Biopharm* 102: 51–63.

Heuschkel S, Goebel A, Neubert RHH (2008) Microemulsions – Modern colloidal carrier for dermal and transdermal drug delivery. *J Pharm Sci* 97: 603–631.

Heuschkel S, Wohlrab J, Neubert RHH (2009) Dermal and transdermal targeting of dihydroaven-anthramide D using enhancer molecules and novel microemulsions. *Eur J Pharm Biopharm* 72: 552–560.

Hoeller S, Sperger A, Valenta C (2009) Lecithin based nanoemulsions: A comparative study of the influence of non-ionic surfactants and the cationic phytosphingosine on physicochemical behaviour and skin permeation. *Int J Pharm* 370(1–2): 181–186.

Hua S (2015) Lipid based delivery systems for skin delivery of drugs and bioactives. *Front Pharmacol* 6: 1–5.

Jaiswal J, Kupta SK, Kreuter J (2004) Preparation of biodegradable cyclosporine nanoparticles by high-pressure emulsification-solvent evaporation process. *J Control Rel* 96: 169–178.

Klang V, Matsko N, Raupach K, El-Hagin N, Valenta C (2011a) Development of sucrose stearate-based nanoemulsions and optimisation through gamma-cyclodextrin. *Eur J Pharm Biopharm* 79: 58–67.

Klang V, Matsko NB, Valenta C, Hofer F (2012) Electron microscopy of nanoemulsions: An essential tool for characterisation and stability assessment. *Micron* 43: 85–103.

Klang V, Schwarz JC, Valenta C (2015) Nanoemulsions in dermal delivery. In *Percutaneous Penetration Enhancers: Chemical Methods in Penetration Enhancement.* Maibach H, Dragicevic-Curic N (Eds.), Springer Verlag, Berlin and Heidelberg, pp. 255–266.

Klang V, Valenta C (2011) Lecithin-based nanoemulsions. *J Drug Del Sci Tech* 21(1): 55–76.

Knorr F, Lademann J, Patzelt A, Sterry W, Blume-Peytavi U, Vogt A (2009) Follicular transport route – Research progress and future perspectives. *Eur J Pharm Biopharm* 71: 173–180.

Kobierski S, Ofori-Kwakye K, Müller RH, Keck CM (2009) Resveratrol nanosuspensions for dermal application – Production, characterization, and physical stability. *Pharmazie* 64: 741–747.

Kotyla T, Kuo F, Moolchandani V, Wilson T, Nicolosi R (2008) Increased bioavailability of a transdermal application of a nano-sized emulsion preparation. *Int J Pharm* 347(1–2): 144–148.

Kreuter J (1991) Nanoparticle-base drug delivery systems. *J Control Rel* 16: 169–176.

Lademann J, Knorr F, Richter H, Blume-Peytavi U, Vogt A, Antoniou C, Sterry W et al. (2008) Hair follicles – An efficient storage and penetration pathway for topically applied substances. *Skin Pharmacol Physiol* 21: 150–155.

Lademann J, Richter H, Schaefer UF, Blume-Peytavi U, Teichmann A, Otberg N, Sterry W (2006) Hair follicles – A long-term reservoir for drug delivery. *Skin Pharmacol Physiol* 19: 232–236.

Lademann J, Richter H, Teichmann A, Otberg N, Blume-Peytavi U, Luengo J, Weiss B et al. (2007) Nanoparticles – An efficient carrier for drug delivery into the hair follicles. *Eur J Pharm Biopharm* 66: 159–164.

El Maghraby GMM, Williams AC, Barry BW (2004) Interactions of surfactants (edge activators) and skin penetration enhancers with liposomes. *Int J Pharm* 276: 143–161.

El Maghraby GMM, Williams AC, Barry BM (2006) Can drug bearing liposomes penetrate intact skin? *J Pharm Pharmacol* 58: 415–429.

Mahrhauser DS, Reznicek G, Kotisch H, Brandstetter M, Nagelreiter C, Kwizda K, Valenta C (2015) Semi-solid fluorinated-DPPC liposomes: Morphological, rheological and thermic properties as well as examination of the influence of a model drug on their skin permeation. *Int J Pharm* 486: 350–355.

Mason TG, Wilking JN, Meleson K, Chang CB, Graves SM (2006) Nanoemulsions: Formation, structure, and physical properties. *J Phys Condens Matter* 18: R635–R666.

Mitri K, Shegokar R, Gohla S, Anselmi C, Mueller RH (2011) Lipid nanocarriers for dermal delivery of lutein: Preparation, characterization, stability and performance. *Int J Pharm* 414: 267–275.

Mou D, Chen H, Du D, Mao C, Wan J, Xu H, Yang X (2008) Hydrogel-thickened nanoemulsion system for topical delivery of lipophilic drugs. *Int J Pharm* 353: 270–276.

Müller RH, Gohla S, Keck M (2011) State of the art of nanocrystals – Special features, production, nanotoxicology aspects and intracellular delivery. *Eur J Pharm Biopharm* 78: 1–9.

Müller RH, Petersen RD, Hommoss A, Pardeike J (2007) Nanostructures lipid carriers (NLC) in cosmetic dermal products. *Adv Drug Deliv Rev* 59: 522–530.

Muller RH, Radtke M, Wissing SA (2002) Solid lipid nanoparticles (SLN) and nanostructured lipid carriers (NLC) in cosmetic and dermatological preparations. *Adv Drug Deliv Rev* 54 Suppl 1: S131–S155.

Neubert RHH (2011) Potentials of new nanocarriers for dermal transdermal drug delivery. *Eur J Pharm Biopharm* 77: 1–2.

Nohynek GJ, Dufour EK (2012) Nano-sized cosmetic formulations or solid nanoparticles in sunscreens: A risk to human health? *Arch Toxicol* 86: 1063–1075.

Noynek GJ, Lademann J, Ribaud C, Roberts MS (2007) Grey goo on the skin? Nanotechnology, cosmetic and sunscreen safety. *Crit Rev Toxicol* 37: 251–277.

Palacio J, Agudelo NA, Lopez BL (2016) PLA/Pluronic® nanoparticles as potential delivery systems: Preparation, colloidal and chemical stability, and loading capacity. *J Appl Poly Sci* 133: 43828.

Puglia C, Rizza L, Drechsler M, Bonina F (2010) Nanoemulsions as vehicles for topical administration of glycyrrhetic acid: Characterization and in vitro and in vivo evaluation. *Drug Deliv* 17: 123–129.

Ro J, Kim Y, Kim H, Park K, Lee KE, Khadka P, Yun G et al. (2015) Pectin micro- and nano-capsules of retinyl palmitate as cosmeceutical carriers for stabilized skin transport. *Korean J Physiol Pharmacol* 19: 59–64.

Sahle FF, Metz H, Wohlrab J, Neubert RHH (2013) Lecithin-based microemulsions for targeted delivery of ceramide AP into the stratum corneum: Formulation, characterizations, and in vitro release and penetration studies. *Pharm Res* 30: 538–551.

Sahle FF, Wohlrab J, Neubert RHH (2012) Polyglycerol fatty acid ester surfactant based microemulsions for targeted delivery of ceramide AP into the stratum corneum: Formulation, characterisation, in vitro release and penetration investigation. *J Pharm Biopharm* 30: 139–150.

Schilling K, Bradford B, Caselle D, Dufour D, Nash JF, Pape W, Schulte S et al. (2010) Human safety of "nano" titanium dioxide and zinc oxide. *Photochem Photobiol Sci* 9: 495–509.

Shegokar R, Müller RH (2010) Nanocrystals: Industrially feasible multifunctional technology for poorly soluble actives. *Int J Pharm* 399: 129–139.

Singh P, Nanda A (2014) Enhanced sun protection of nano-sized metal oxide particles over conventional metal oxide particles: An in vitro comparative study. *Int J Cosmet Sci* 36: 271–283.

Sole I, Pey CM, Maestro A, Gonzalez C, Porras M, Solans C, Gutierrez JM (2010) Nano-emulsions prepared by the phase inversion composition method: Preparation variables and scale up. *J Colloid Interface Sci* 344(2): 417–423.

Tadros T, Izquierdo P, Esquena J, Solans C (2004) Formation and stability of nano-emulsions. *Adv Colloid Interface Sci* 108–109: 303–318.

Taib SHM, AbdGani SS, Ab Rahman MZ, Ismail A, Shamsudin R (2015) Formulation and process optimizations of nano-cosmeceuticals containing purified swiftlet nest. *RSC Adv* 5: 42322–42328.

Turci F, Peira E, Corazzari I, Fenoglio I, Trotta M, Fubini B (2013) Crystalline phase modulates the potency of nanometric TiO_2 to adhere to and perturb the stratum corneum of porcine skin under indoor light. *Chem Res Toxicol* 26: 1579–1590.

Valenta C, Schultz K (2004) Influence of carrageenan on the rheology and skin permeation of microemulsion formulations. *J Control Rel* 95: 257–265.

Xia S, Xu S, Zhang X (2006) Optimization in the preparation of coenzyme Q10 nanoliposomes. *J Agric Food Chem* 54: 6358–6366.

Zhou H, Yue Y, Liu G, Li Y, Zhang J, Gong Q, Yan Z et al. (2010) Preparation and characterization of a lecithin nanoemulsion as a topical delivery system. *Nanoscale Res Lett* 5: 224–230.

20

Nanocarrier-Based Formulations: Production and Cosmeceutic Applications

D. Knoth, R. Eckert, V. Farida, P. Stahr, S. Hartmann, F. Stumpf, O. Pelikh and C.M. Keck

CONTENTS

Introduction

The word 'nano' is derived from the Greek word *nannos* and from the Latin word *nanus*. The English translation of *nannos* and *nanus* is 'dwarf'. Hence, nanoparticles are tiny particles. In a more scientific explanation 'nano-' is known as a unit prefix and corresponds to a billionth of a whole. This means 1 nanometer (1 nm) corresponds to 0.000000001 (10^{-9}) meters. In fact, nanoparticles are extremely small particles and due to this they possess different properties when compared to larger bulk material of the same material.

Properties of Nanoparticles

Due to the small size, nanoparticles cannot be seen by the naked eye and even normal light microscopy is not able to detect such small sizes. Only modern analytical techniques, for example, electron microscopy, enable the visualization and characterisation of nanoparticles. As these techniques were only invented in the last century, it is not astonishing that research in the field of nanotechnology is relatively novel. Nevertheless, the unique properties of nanomaterial enable the production of products with advanced performance in almost all fields of daily life (Heiligtag and Niederberger, 2013). In addition to nanoparticles that are intentionally produced, there are also natural nanoparticles, that is, particles that are produced by nature. Examples for natural particles include water droplets in the air (i.e. fog) or volcano ash (Griffin et al., 2017; Strambeanu et al., 2015). Hence, nanoparticles have always existed around humankind (Hough et al., 2011).

The use of nanoparticles in cosmetics started before 'nanotechnology' was even invented. For example, carbon black is a colourant that is used in mascara and other decorative products such as eyeliners, eye pencils and eye shadows. It possesses particle sizes in the nanometer range and has been used in cosmetic products for more than 80 years (Katz et al., 2015; Nafisi and Maibach, 2017; Mihranyan et al., 2012). The word 'nanotechnology' was conceived by Japan's Norio Taniguchi in 1974 (Taniguchi, 1974), and both science and technology in this field have developed tremendously since then (Florence, 2007; Forrest and Kwon, 2008; Park, 2007).

Cosmetic use of nanoparticles has also developed and in principle two main types of nanoparticles in cosmetics can be distinguished. The first type of nanoparticles represents particles in the nanometer range that are composed of 100% material. Examples are carbon black, titanium dioxide, metal (i.e. gold, silver or platinum) or fullerenes (Smijs and Pavel, 2011; Contado, 2015; Elder et al., 2005). These particles don't carry any additional actives. The use of carbon black was explained earlier. Titanium dioxide is a pigment and ultraviolet (UV)-light absorber and is used as UV protectant in sunscreens (Smijs and Pavel, 2011; Rechmann, 1974; Howard, 1974; Smijs and Pavel, 2011; Veronovski et al., 2014). Nanoparticles being composed of gold and platinum are used for their antioxidative properties (Watanabe et al., 2009; Kajita et al., 2007; Bladen et al., 2012; Fathi-Azarbayjani et al., 2010; Yoshihisa et al., 2010; Zhou et al., 2013), and fullerenes are used as radical scavengers (Zeynalov et al., 2009; Lens, 2009, 2011; Partha and Conyers, 2009; Takada et al., 2006). Silver nanoparticles are well known for their oligodynamic effect and are mainly used as antibacterial agents (Franci et al., 2015; Rai et al., 2012; Clement and Jarrett, 1994; Ansari et al., 2015). The second type of nanoparticles used in cosmetics is the so-called nanocarriers. These types of particles are used to transport cosmetic and/or cosmeceutical actives into deeper skin layers. The most prominent nanocarriers are liposomes, nanoemulsions, lipid nanoparticles and nanocrystals. Their properties, production and cosmeceutical applications are explained in more detail in this chapter.

Overview of Cosmeceutical Nanocarriers

Liposomes

Liposomes are spherical vesicles with sizes in the range between about 60 and 300 nm (Lasic, 1998). They are composed of phospholipids which form at least one phospholipid bilayer (Figure 20.1). There are different types of liposomes such as multilamellar vesicles (MLV), small unilamellar vesicles (SUV), large unilamellar vesicles (LUV), multivesicular liposomes and the cochleate vesicles (Figure 20.1) (New, 1995; Schubert, 2010; Akbarzadeh et al., 2013). Independent of the type of vesicle, all liposomes contain hydrophilic cores in which hydrophilic actives can be encapsulated. In addition, hydrophobic actives can be incorporated in the bilayer (Figure 20.1). Hence, liposomes are suitable carriers for both hydrophilic and lipophilic actives. Liposomes were first described in the late 1960s by Bangham et al. (Bangham et al., 1974; Gregoriadis, 1976) and since then much research has been done in this field (Diederichs and Müller, 1994; Fahr and Liu, 2007; Fahr et al., 2006; Rahimpour and Hamishehkar, 2012; Touitou et al., 1994; Verma et al., 2003).

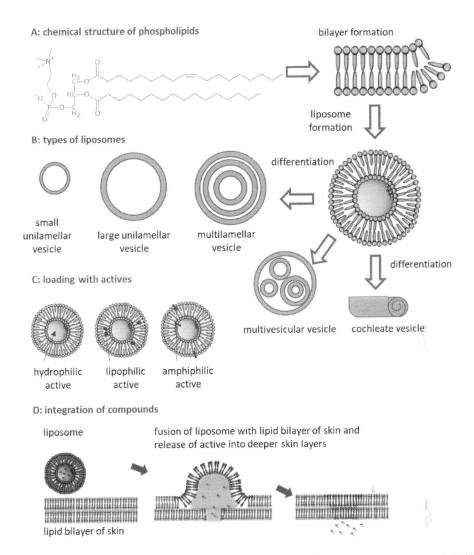

FIGURE 20.1 Scheme of liposomes which are composed of phospholipids. A: chemical structure of phospholipids, B: phospholipids form a lipid bilayer, which forms (depending on the method used), multilamellar vesicles (MLV), small unilamellar vesicles (SUV), large unilamellar vesicles (LUV), multivesicular liposomes and cochleate vesicles. C: Hydrophilic actives can be incorporated into the hydrophilic core, whereas lipophilic or amphiphilic actives can be incorporated into the lipid bilayer. D: Upon dermal application liposomes made from phospholipids fuse with the stratum corneum lipids, the actives are released and penetrate into deeper skin layers.

In cosmetics, in most cases, liposomes are used to improve the dermal penetration of actives. This is best achieved by selecting phospholipids which contain unsaturated fatty acids. Unsaturated fatty acids possess lower melting points, which results in the formation of more flexible liposomal structures (Leekumjorn et al., 2009; Muranushi et al., 1981; van Hoogevest et al., 2013). The chemical composition of phospholipids and especially the fatty acid composition depend on the source of the phospholipid (Figure 20.1). Phospholipids derived from animal sources, for example, egg lecithin, contain high amounts of saturated fatty acids, leading to liposomes with a more rigid structure, whereas plants and plant phospholipids are rich in unsaturated fatty acids and are therefore used in most cosmetic liposomal formulations today (van Hoogevest et al., 2013). In general – as a rule of thumb – the more flexible the liposomal structure, the faster and deeper will be the penetration of actives (van Hoogevest et al., 2013; Badran et al., 2012; Hussain et al., 2017; Li et al., 2015; Raza et al., 2011). The disadvantage of the highly unsaturated phospholipids is their sensitivity against oxidation. This means effective antioxidants (e.g. tocopherols) must always be added to such formulations.

Production

Liposomes can be produced in different ways. Depending on the method used, different types of liposomes will be obtained (Maherani et al., 2011; Patil and Jadhav, 2014). In general, to ensure the formation of liposomes, di-acylphospholipids should be used (van Hoogevest et al., 2013; Huang and Li, 1996). Multilamellar vesicles are formed spontaneously upon hydration of the phospholipids in water, whereas unilamellar vesicles (SUV and LUV) are obtained if ethanolic solutions are slowly added to an aqueous phase (Akbarzadeh et al., 2013). Another method to obtain SUV with defined size and narrow size distribution is the extrusion of MLV through membranes of defined size (Hope et al., 1993; Olson et al., 1979) or the use of homogenizers, for example, microfluidizers or high pressure homogenizers (Gregoriadis, 2007). In cosmetics, most liposomes are produced by using the latter techniques.

Cosmetic formulations containing liposomes are obtained by adding the liposome concentrate to the vehicle by gentle stirring. Any heating should be avoided, as liposomes are heat sensitive. The use of vehicles which contain high amounts of oil should also be avoided, because liposomes can fuse with oil droplets and thus disappear during storage (Hernándes-Caselles et al., 1990).

Dermal Penetration: Mechanism of Action

Upon dermal application it is believed that the flexible liposomes fuse with the stratum corneum lipids. During fusion with the stratum corneum lipids, the cargo of the liposomes (i.e. the active) is released and enabled to penetrate into deeper skin layers (Figure 20.1). The fusion of the phospholipids dilutes the stratum corneum lipids and increases their fluidity. The increase in fluidity depends on the amount of phospholipids, that is, higher concentrations of liposomes lead to better penetration, because the depth of penetration, speed of penetration and the total amount of penetrated active is increased with increasing amounts of phospholipids (van Hoogevest et al., 2013; El Maghraby et al., 2006; Gillet et al., 2011; Sakdiset et al., 2018).

Cosmeceutic Application

Due to their penetration-enhancing properties for both hydrophilic and lipophilic actives, liposomes are a widely applied formulation principle in both cosmetic and cosmeceutical products (van Hoogevest et al., 2013; Betz et al., 2005; Edgar et al., 1991; Lautenschläger, 1990; Weiner et al., 1994). In addition to the penetration-improving properties, liposomes possess other skin carrying properties. Phospholipids are physiological compounds of membranes in all living cells. Due to this they are non-toxic and are extremely well tolerated. Some phospholipids (e.g. soy bean phospholipids) contain unsaturated, essential fatty acids (e.g. linolenic acid and linoleic acid) and thus promote a further skin carrying effect by supplying essential compounds to the skin (van Hoogevest et al., 2013). As a lack of linolenic and linoleic acid is associated with an impaired skin barrier (Elias et al., 1980; Kendall et al., 2017; McCusker and Grant-Kels, 2010) and diseases such as acne, atopic dermatitis or psoriasis (van Hoogevest et al., 2013; Antonio et al., 2014; Briganti and Picardo, 2003; Grattan et al., 1990; Horrobin, 1989; Hua, 2015), phospholipid-containing formulations can also be used to improve such skin conditions. Formulations containing high concentrations of phospholipids (>25%) increase skin hydration and reduce skin roughness, thus providing a significant skin carrying and 'anti-wrinkle' effect (van Hoogevest et al., 2013; Sinambela et al., 2013).

Nanoemulsions

Nanoemulsions were first invented in the 1950s for pharmaceutical use, for example, parenteral nutrition (Hörmann and Zimmer, 2016). Later the beneficial properties of nanoemulsions were also discovered for many other applications routes, such as oral, ocular or dermal applications (Jumaa and Müller, 1998; Friedman et al., 1995; Tamilvanan, 2004; Rai et al., 2018; Lundberg, 1997; Gupta et al., 2016). Nanoemulsions are also known as fine disperse emulsions, mini-emulsions or submicron emulsions. They are emulsions with sizes below 1 μm, typically in the range between 50 and 500 nm (Izquierdo et al., 2002). In cosmetics, typically oil-in-water (o/w) nanoemulsions are used (Yukuyama et al., 2016; Sonneville-Aubrun et al., 2004). Due to the small size, properties of the emulsions are changed, leading to many advantages when compared to larger sized macroemulsions. In contrast to macroemulsions,

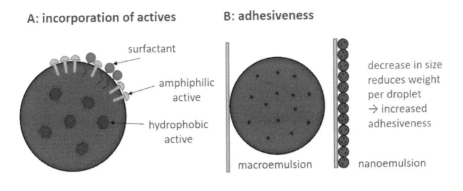

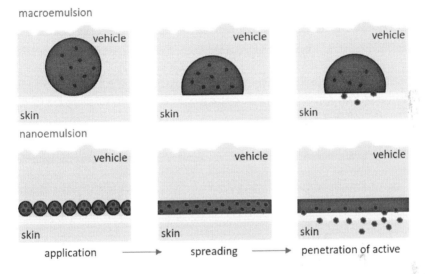

FIGURE 20.2 Scheme of nanoemulsions. (A) Hydrophilic actives are incorporated into the oil droplets and amphiphilic actives can be incorporated into the surfactant layer. (B) The adhesiveness of oil increases with decreasing size of the emulsion droplets, because more droplets (more attaching points) are created which need to carry less weight. (C) Smaller droplets increase the penetration of actives, because the smaller droplets create a lager contact area on the skin from which the active can penetrate.

nanoemulsions possess a much higher physical stability. The reason for this is the Brownian motion of the molecules in the dispersing phase, which bombard the oil droplets being dispersed in this phase, thus leading to a movement of the oil droplets. Due to their lower weight, smaller oil droplets are more affeced by this 'energy-kick' than larger droplets. Hence, the mobility of the oil droplets increases with decreasing size of the droplets. This leads to an increased motion of the small sized particles, which reduces flotation (Stokes' law) and coalescence of the oil droplets. This means the droplets are always homogeneously distributed within the formulation, which enables accurate dosing – not only during the production of nanoemulsion-based formulations but also for dermal application. Nanoemulsions possess a low viscosity and can appear milky (if droplet size is > 100 nm) or even translucent (if droplet size is below 100 nm) (Shah et al., 2010; Solans et al., 2005; Jafari and McClements, 2018). Nanoemulsions must be stabilized by the use of surfactants (Tadros et al., 2004). For skin care products, the use of skin-friendly stabilizers, such as non-ionic surfactants and/or phospholipids, is advisable. When compared to classical macroemulsions or microemulsions, the amount of surfactant needed for the stabilization of nanoemulsions is relatively low (McClements, 2012; Isailović et al., 2016; Müller et al., 2004; Shakeel et al., 2007). Hence, nanoemulsions are especially skin-friendly when compared to other emulsion systems. Nanoemulsions can be used as nanocarriers for improved penetration of hydrophobic and amphiphilic actives, where hydrophobic actives can be dissolved in the oil phase and amphiphilic actives can be incorporated into the surfactant layer (Isailović et al., 2016; Müller et al., 2004) (Figure 20.2).

Production

There are different methods to produce nanoemulsions. In the early years nanoemulsions were produced by high-shear forces, including the use of rotor-stator high-speed stirrers, high-pressure homogenizers, microfluidizers or even combinations of these (Tadros et al., 2004; Müller et al., 2012; Mahdi Jafari et al., 2006; Scholz and Keck, 2015). However, the trend has changed and low-energy processes are more favoured (Yukuyama et al., 2016). Low-energy processes mainly include catastrophic phase inversion techniques. Phase inversion is a well-known phenomenon that occurs mostly when non-ionic stabilizers are used and describes the reversion of oil-in-water emulsions to water-in-oil systems and vice versa (Bauer et al., 1997). Phase inversion processes lead to the formation of very finely dispersed droplets in a continuous phase without the application of high-shear forces. It is therefore not only suitable for low-cost production of nanoemulsions but also useful in the formulation of shear-sensitive actives. Nanoemulsions are produced as concentrates. The oil phase is typically 10% (w/w). If higher amounts are required, emulsions with up to 40% oil can be produced (Müller et al., 2012). Nanoemulsions which are formulated correctly can remain physically stable for more than 10 years (Müller et al., 2012; Keck et al., 2008; Müller and Heinemann, 1993). They are stable against heat and thus – if required – even heat sterilization is possible. However, the liquid character of the oil in which the actives are typically incorporated is not able to protect the incorporated actives from oxygen or light, therefore the chemical stability of chemically labile actives that were incorporated into nanoemulsions is typically less when compared to macroemulsions with larger droplet size. The reason for this is the larger surface to volume ratio (Gupta et al., 2016) (Figure 20.2).

Cosmetic formulations containing nanoemulsions are obtained by adding the nanoemulsion concentrate to the vehicle by gentle stirring. Nanoemulsions can be added to hot or cold formulations, making the production of final products highly flexible. The use of vehicles which contain high amounts of oil should also be avoided, because the nanodroplets can fuse with oil droplets and thus disappear during storage (Yukuyama et al., 2016).

Dermal Penetration: Mechanism of Action

Upon dermal application nanoemulsions are believed to form a thin film on the skin (Figure 20.2). The film is slightly occlusive, thus leading to an increase in skin hydration and to an increase in dermal penetration of the actives (Baspinar et al., 2010). A further promoter of improved penetration of actives is the increased contact area when compared to larger sized droplets (Figure 20.2) (McClements, 2012). The effect can be explained by Fick's law (Formula 20.1), where dQ/dt is the amount of active penetrating out of the formulation, D the diffusion coefficient in the stratum corneum, V_k the distribution coefficient between stratum corneum and vehicle, A the area of skin on which the formulation is applied, d the thickness of stratum corneum and c_v the concentration of dissolved active in the vehicle.

$$\frac{dQ}{dt} = \frac{D \cdot V_k \cdot A}{d} \cdot c_v$$

Formula 20.1 Fick's Law

For optimal penetration a sufficient amount of oil droplets is required. Too low concentrations of oil will not form a film on the skin and thus the area of skin on which the oil with active is in contact with the skin is reduced (Figure 20.2). In practice this can be achieved by calculation of the contact area of a single droplet of a given size and by knowing the surface area of the skin that should be covered by the droplets. The volume of oil droplets and nanoemulsion required from this film can then be calculated. For example, only 10 µl of a nanoemulsion being composed of 10% lipid phase with droplet sizes of about 200 nm can form a monolayer droplet film with a size of about 50 cm^2, which is about the surface area of a face and neck. However, if this nanoemulsion is used as intermediate product in a concentration of only 1% in a final cosmetic product, about 1 ml of the final product needs to be applied on the face and neck to yield the film. Depending on the product this quantity might be feasible. However, in most cases smaller quantities are applied and thus higher concentrations of nanoemulsion need to be incorporated into the final cosmetic product. As a rule of thumb, about 2–5% of nanosuspension (containing 10% lipid phase) should be used for the formulation of nanoemulsion-based cosmetic products to ensure optimal performance. To further promote effective penetration of the active from the oil droplets, the concentration gradient should be high. This means the concentration of dissolved active in the oil droplets of the nanoemulsions should be as high as possible (Pawar and Babu, 2014).

Cosmeceutic Application

In most cases cosmeceutic application of nanoemulsions aims at improved delivery of lipophilic actives, which are dissolved in the oil phase of the emulsion. Examples are antioxidants, such as tocopherols, beta carotene or coenzyme Q10 (Qian et al., 2012; Morais Diane and Burgess, 2014). Another trend is the emulsification of natural oils, such as sacha inchi oil (Golz-Berner and Zastrow, 2011; Tunkam and Satirapipathkul, 2016), *Moringa oleifera* seed oil (Chanchal and Swarnlata, 2008), olive oil

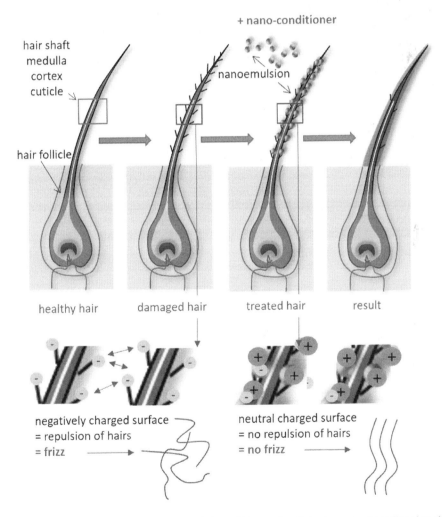

FIGURE 20.3 Principle of use of nanoemulsions as hair conditioners. Small sized nanodroplets adhere into the opened structures of the broken cuticle, form a film and 'repair' the surface of the hair. If positively charged nanoemulsions are used, they compensate the negative charge of the damaged hair, reduce repulsion and frizz.

(Mehmood et al., 2017) and argan oil (Guillaume and Charrouf, 2011; Lococo et al., 2012; Yaghmur et al., 1999) and many others (Rai et al., 2018; Teo et al., 2010; Lohani et al., 2014), which already possess antioxidative or other skin carrying properties. Nanoemulsions can also be used for hair care products (Hu et al., 2012). Due to their small droplet size, nanoemulsions are great conditioners. The surface of damaged hair is hydrophilic, because the lipophilic, outer layer of the hair shaft (the cuticle) is impaired and the inner cortex, which is hydrophilic, is exposed to the environment and not protected. The aim of hair conditioners is the restoration of the lipid layer. Small sized nanoemulsions possess an increased adhesiveness on surfaces, due to the increased surface to volume ratio, hence the weight which needs to be carried per attaching point is less when compared to larger sized droplets (Figure 20.2). In addition, the small size is ideal to intercalate into the opened structures of the broken cuticle (Figure 20.3). Thus, the cuticle defects can be filled up, leading to an efficient restoration of the lipophilic hair structure. The hair appears shiny and healthy. To promote an 'extra-long lasting' hair conditioning effect, the adhesiveness of the nanodroplets can be further increased by using positively charged surfactants for the stabilization of the nanoemulsions. As the hair is negatively charged, especially high adhesiveness can be achieved in this way. The addition of positively charged droplets reduces the all over negative charge of the hair, which is a main cause for frizz (Figure 20.3). In fact, positively charged nanoemulsions are ideal hair conditioners. Nanoemulsions for hair conditioning can be produced by using silicon oil and/or natural oils, such as coconut oil (Sonneville-Aubrun et al., 2004; Glenn et al., 2015; Desenne and Chesneau, 2011), but are especially feasible if oils with less spreadability are used. Many consumers try to avoid silicone-based formulations, which forces manufacturers to use other oils which lack the pleasant feel and spreadability of silicone oils. The use of nanoemulsions (and other nanocarriers) might compensate for the less pleasant feeling of silicone-free formulations, as a film formation is achieved by the fast spreading of the nanosized droplets.

Nanoemulsions possess a highly pleasant aesthetic character and a beautiful skin feeling upon dermal application (Sangwan et al., 2014). The film formation – depending on the oil used for the formulation of the nanoemulsion – can also be used to improve the skin barrier function (Rachmawati et al., 2015; Rähse and Dicoi, 2009; Yilmaz and Borchert, 2006; Trotta et al., 1997; Radomska and Dobrucki, 2000; Gasco et al., 1991). Formulations containing nanoemulsions can be creams or gels and, because nanoemulsions do not cream, they can also be used as sprays (Daniels, 2009; Carlotti et al., 1993). Possible

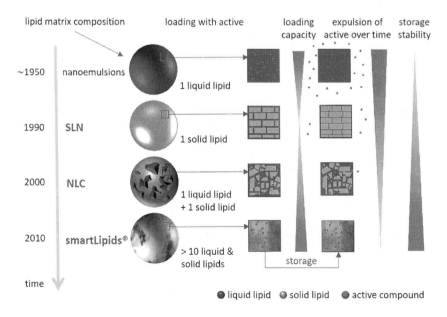

FIGURE 20.4   Scheme of lipid nanocarriers which are derived from nanoemulsions by exchanging the liquid lipid into a solid lipid (SLN). Further improvements are mixtures of liquid and solid lipids (NLC and smartLipids®, which increase drug loading capacity and reduce drug expulsion during storage.

applications, especially for sprays, are skin care products, hair care products and sun-protection sprays (Sonneville-Aubrun et al., 2004; Iskandar et al., 2016; Puglia et al., 2014; Hougaz, 2009).

Lipid Nanoparticles

Lipid nanoparticles were invented in the beginning of the 1990s (Müller et al., 1995, 2000; Priano et al., 2007; Jenning and Gohla, 2001; Gasco, 1993; Lucks and Müller, 1999). They can be best described by being oil-in-water nanoemulsions, where the lipid of the inner phase is exchanged by a solid lipid. Hence, instead of oil droplets, the inner phase is composed of solid lipid particles (Figure 20.4). The advantage over nanoemulsions is therefore the possibility to encapsulate lipophilic actives into a solid matrix, which increases the chemical stability of sensitive actives. For cosmetic use this is especially beneficial for antioxidants, such as tocopherol or retinol (Jenning and Gohla, 2001; Dingler et al., 1999). A further benefit of the lipid nanoparticles is the controlled release of the active from the lipid matrix, which can be modified by modifying the composition of the lipid matrix. Prior to explaining the mechanisms for controlled and tailor-made release profiles, it is important to give a short introduction about the different types of lipid matrices.

In the beginning of the 1990s the so-called solid lipid nanoparticles were introduced. They consisted of only one solid lipid, in which the lipophilic actives were incorporated (Figure 20.4) and were named solid lipid nanoparticles (SLN) (Lucks and Müller, 1999). Solid lipids possess some challenging properties and the most prominent property is that they can change some of their physico-chemical properties during storage. The phenomenon is known as polymorphism and means that the structure of the crystalline lattice undergoes modification over time. The process is thermodynamically driven, hence changes are associated with a loss in energy and an increase in crystallinity (Bunjes et al., 1996). This results in a more dense packaging of the lipid molecules and thus in a reduction of the space between the lipid molecules. As a consequence, upon changes in lipid modification, actives that were located in these interfaces can be expelled from the lipid lattices and will diffuse in the outer aqueous phase, where they typically precipitate (Figure 20.4).

As expulsion of the active should be prevented, lipid nanoparticles which do not undergo modification over time were invented. The so-called second generation of lipid nanoparticles are the nanostructured lipid carriers (NLC) (Müller et al., 2014). These particles are composed not only of one solid lipid, but of a mixture of one solid and one liquid lipid (Figure 20.4). This leads to the formation of small oil regions within the solid lipid matrix, the nanostructures, which prevent – or at least slow down – the modification of the solid lipid over time. This results in a higher storage stability, hence less active is expelled over time. Moreover, due to the addition of oil to the solid lipid, 'more space' is created between the solid lipid molecules, which results in a much higher loading capacity for actives (Figure 20.4) (Müller et al., 2007, 2014; Pardeike et al., 2009). The principle was further exploited by inventing the so-called smartLipids. smartLipids are not only composed of one solid and one liquid lipid but of a mixture of many different liquid and solid lipids, up to 15 lipids (Ruick, 2016). Such mixtures create chaotic structures and re-crystallization or changes in modification over time of a single lipid within the mixture can be greatly prevented (Figure 20.4). Drug loading capacity and storage stability, compared to NLC, are further enhanced (Keck et al., 2015; Ding et al., 2017).

Modifications in the lipid matrix composition of lipid nanoparticles can also be used to create tailor-made release profiles for the active. Fast release is achieved if the active is located in the outer shell of the particles or on the surface, whereas a slow, retarded release is achieved if the active is tightly incorporated into the core of the lipid particles. By incorporating the actives homogenously into the lipid matrix, or by using lipid matrices where the active dissolves only in the oil but not in the solid lipid and by varying the ratio of solid:liquid lipid, almost any release profile can be created (Figure 20.5). For lipophilic actives which do not dissolve in the lipids phase, for example, titanium dioxide nanoparticles, release and/or contact of the active with the skin can even be prevented. The latter is beneficial for sun blockers, which can cause skin irritant effects when in direct contact with skin. In case lipid nanoparticles are used for improved dermal delivery of lipophilic actives, recent studies suggest to use oil-enriched lipid nanoparticles (Keck et al., 2014). These particles contain high amounts of oils, which form an outer liquid shell around a solid core, in which the active is dissolved (Figure 20.5). This leads to a high

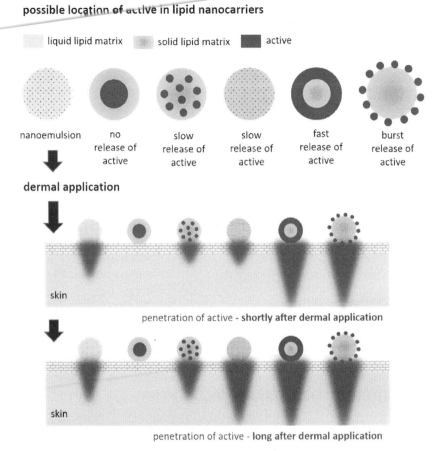

FIGURE 20.5   Different types of lipid nanocarriers and modifications of the lipid matrices allow for a different localization of the active within the nanocarriers (upper). The localization of the active influences the release profile of the active from the carrier and thus the penetration of the active in the skin (lower).

concentration gradient of the active between formulation and skin and thus – based on Fick's law – to an improved dermal penetration of the active.

Production

Lipid nanoparticles can be produced by different techniques. The first method used was the hot microemulsion method. For this a formulation forming a microemulsion at elevated temperatures (above the melting point of the highest melting lipid within the formulation) must be developed. Lipid nanoparticles are obtained by dilution of the hot microemulsion with water, leading to a finely dispersed nanoemulsion. Cooling the hot nanoemulsion leads to the solidification of the oil droplets, resulting in the formation of solid lipid nanoparticles (Cavalli et al., 2002, 2003). This method has been widely used to prepare small sized lipid nanoparticles for pharmaceutical use. However, in cosmetics it has not been used so far, because the formulation of microemulsions requires the use of high amounts of surfactants and very often special types of surfactant that are not skin-friendly (Müller, 1997; Müller et al., 2002).

Another method to produce lipid nanoparticles is the use of high-pressure homogenization. This method is historically used to homogenize milk (Müller and Keck, 2012) and thus it is a well-established method, with many advantages (Müller and Keck, 2004, 2008). The production is fast (typically 3 cycles at 500 bar), highly reproducible, low cost and even sterile production is feasible (Mukherjee et al., 2009). The process can be easily scaled up to larger batch sizes and can be run under GMP conditions. Many machines with different batch sizes are available for both continuous and discontinuous production methods (Shegokar et al., 2011; Jenning et al., 2002). By using high-pressure homogenization instead of hot-microemulsion techniques, the formulations can be produced with less surfactant (Pardeike et al., 2009; Müller et al., 2002), which is especially feasible for skin care products. Based on these advantages, today high-pressure homogenization is the most frequently used technique for the production of lipid nanoparticles for cosmetic and cosmeceutical applications (Müller and Keck, 2012; Shegokar et al., 2011; Schwarz et al., 1994; Al Shaal et al., 2014).

Dermal Penetration: Mechanism of Action

Similar to nanoemulsions, lipid nanoparticles form a thin film on the skin upon dermal application. The so-called invisible patch increases the skin hydration due to occlusive effects which are much more pronounced than is the case for nanoemulsions. This effect was nicely demonstrated by Pardeike et al. for Q10-loaded NLC. NLC increased the skin hydration as effective as liquid paraffin, which is the gold standard to achieve skin occlusion. In contrast, the increase in skin hydration was almost negligible when a Q10-loaded nanoemulsion was used with the same lipid content and similar particle size (Pardeike and Müller, 2007).

In principle the mechanism of action of dermal penetration from lipid nanoparticles is similar to the principle of nanoemulsions, but the penetration is higher due to the higher occlusiveness of the nanolipid carriers. Recent studies suggest that – based on Fick's law – the concentration gradient of dissolved active between formulation and skin should be high to obtain good dermal penetration of the active. This is best achieved if the active is located in the outer shell of the lipid particles or even only associated to the particle (Figure 20.5). Opposite effects, that is, retarded release or even no penetration, can be achieved if the active is tightly incorporated into the core of the particles (Figure 20.5).

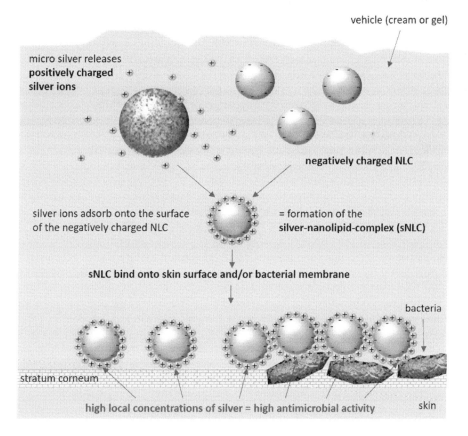

FIGURE 20.6 Scheme of silver-nanolipid-complex (sNLC). Upper: Microsilver in the formulation releases positively charged silover ions, which adsorb onto the surface of the negatively charged lipid nanocarriers, forming the sNLC. Lower: Upon dermal penetration the sNLC adsorb onto the surface of the skin or bacteria, which leads to a high local concentration of silver and a high anti-microbial activity.

Cosmeceutic Application

Lipid nanoparticles are used in many cosmetic and cosmeceutic products (Pardeike et al., 2009; Keck and Müller, 2010; Souto and Müller, 2008). They are used for improved delivery of lipophilic actives, for skin care and as sun protection. Other applications are the encapsulation of perfumes for retarded and long-lasting action of perfumes or the incorporation of insect repellents (Frederiksen et al., 2003). Lipid nanocarriers possess many skin carrying properties. Skin occlusive properties and the increase in skin hydration are associated with a decrease in wrinkles. Thus, lipid nanoparticles can be used efficiently in anti-aging skin care products for both wrinkle reduction by physical means and improved penetration for anti-aging actives (Schwarz et al., 2013). Besides occlusion and the increase in skin hydration they strengthen or even repair the natural lipid film of the skin and thus can also be used to treat symptoms of impaired skin (Gupta et al., 2013). For this it is advisable to use natural lipids, for example, olive oil or hemp seed oil, which contain high amounts of omega 3 fatty acids, which are essential for an intact skin barrier.

A very efficient cosmeceutical application is the use of the so-called silver-nanolipid complex (sNLC), which is a combination of lipid nanocarriers and microsilver (Keck and Schwabe, 2009; Keck et al., 2014). Upon addition of elemental silver, the silver releases positively charged silver ions, which adsorb onto the surface of the negatively charged lipid nanocarriers, thus forming the sNLC (Figure 20.6). Upon contact with skin the sNLC forms the film on the skin as described earlier, which in this case leads to a high local concentration of silver ions on the skin (Figure 20.6). As silver ions possess a high antibacterial activity (oligodynamic effect), bacteria are efficiently killed. Atopic dermatitis is often associated with the occupation of bacteria (e.g. *Staphylococcus aureus* or *Pseudomonas aeruginosa*), which promote the symptoms of atopic dermatitis (e.g. scratching, itching and inflammation) (Keck et al., 2014). As atopic dermatitis cannot be cured to date, these symptoms are typically treated with pharmaceutical drug products containing glucocorticoids (Maia et al., 2000). Recently, a formulation containing sNLC was shown to be as effective as a classical treatment with glucocorticoids and is therefore a nice example how innovative nanocarriers can be used for the formulation of modern, effective cosmeceuticals (Keck et al., 2014; Palmer and DeLouise, 2016).

Nanolipid carriers scatter UV light very efficiently, which means they possess UV-protecting properties, which can further be improved by the incorporation of UV blockers (Müller et al., 2004; Nikolić et al., 2011; Hommoss et al., 2007). A new trend is the combination of antioxidants and UV-protecting actives, because there is scientific evidence that not only UV light, but also visible and especially near-infrared light promote skin aging and cancer due to the formation of free radicals in the skin (Zastrow et al., 2004, 2009a,b,c). Light with longer wavelength, that is, light being different from UV light, cannot be blocked and will always penetrate the skin and create free radicals. Therefore, until now the only possibility to prevent oxidative damage due to light is the application of antioxidants. Recent studies prove that the use of ultra-small lipid nanoparticles, loaded with antioxidants, represent one of the most efficient methods to date to prevent and reduce the amount of free radicals in the skin generated by UV and infrared light. The improved activity is due to a superposition of the light-scattering effect of the lipid particles and the improved penetration of the antioxidants into deeper skin layers from the lipid nanoparticles (Lohan et al., 2015). Future developments aim to further improve the light-protecting effects of the lipid carriers by combining UV blockers and antioxidants in one carrier. As UV blockers should not penetrate the skin, they will be encapsulated into the core of the particles and antioxidants, which should penetrate the skin and reach deeper skin layers, and will be located in the outer shell of the lipid particles. With this it is believed to enable the formulation of highly effective sun care products. Similar to nanoemulsions, lipid nanoparticles also do not cream. This means lipid nanoparticles can be formulated in creams, gels or sprays. The advantage over nanoemulsions is the longer retention time on the skin, that is, formulations containing lipid nanocarriers are 'waterproof' and cannot be easily washed off. The effect was nicely demonstrated by Pyo et al. (2007). In conclusion, lipid nanocarriers are a highly effective formulation principle for the delivery of lipophilic actives to the skin. They can be easily produced, are skin-friendly, and possess many skin carrying properties and an excellent skin feeling upon dermal application. Formulations possess a pleasant feeling with a smooth and soft texture. Formulations containing lipid carriers should contain at least 0.5–1% lipid nanoparticles to ensure the formation of the film on the skin and to unlock the superior potential of these carriers.

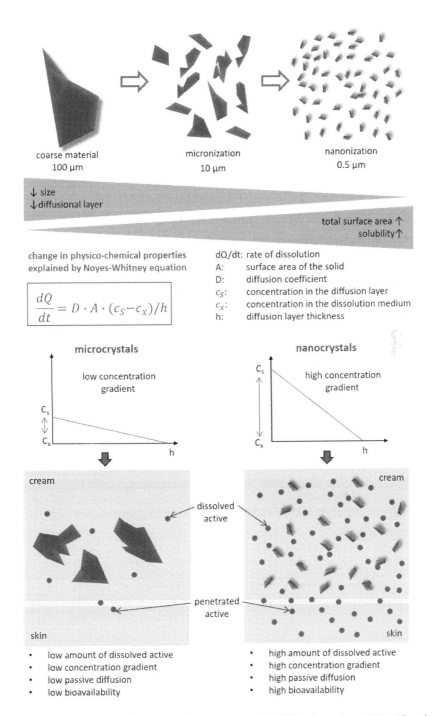

FIGURE 20.7 Scheme of nanocrystals. Nanocrystals are composed of 100% active and are obtained from larger sized bulk material. Due to the small size solubility cs, concentration gradient and dissolution rate dQ/dt are increased, which results in an improved passive diffusion and improved penetration of poorly soluble actives upon dermal penetration.

Nanocrystals

Nanocrystals for pharmaceutical use were invented in the 1990s to overcome poor bioavailability of poorly soluble actives (List and Sucker, 1988; Moschwitzer, 2006; Müller et al., 1996; Liversidge et al., 1991). Nanocrystals are composed of 100% actives and are typically sized in the range between 200 and 800 nm

(Müller and Keck, 2008). Due to their small size, nanocrystals possess different properties when compared to larger sized material. This includes an increased total surface area and an increased curvature of the particle surface. The increased curvature leads to an increased dissolution pressure, that is, the tendency of a molecule to dissolve from a solid surface, which results in an increased kinetic solubility (Figure 20.7). The increased surface area improves the dissolution velocity, which can be explained by the Noyes-Whitney equation (Figure 20.7). The increased amount of dissolved active leads to a high concentration gradient, and thus passive diffusion through membranes is improved with the use of nanocrystals (Fig. 20.7) (Keck and Müller, 2006). The principle is very simple but highly efficient, and has been used for many years in the pharmaceutical field for the formulation of poorly soluble actives (Keck and Müller, 2006). About 10 years ago the principle was first applied for improved dermal penetration of poorly soluble actives and has since been used in various cosmetic and cosmeceutical applications (Müller et al., 2011; Junghanns et al., 2008).

Production

Nanocrystals can be produced by different techniques involving both bottom-up and top-down techniques. Bottom-up techniques, that is, the precipitation of dissolved actives, need to use organic solvents which must be removed upon the precipitation step. As this is costly and not environmentally friendly, this method did not gain much acceptance for the production of nanocrystals in the cosmetic field. Top-down production techniques include bead milling (BM) and high-pressure homogenization (HPH). In the pharmaceutical field BM is the most frequently applied technique. However, in cosmetics the combination of

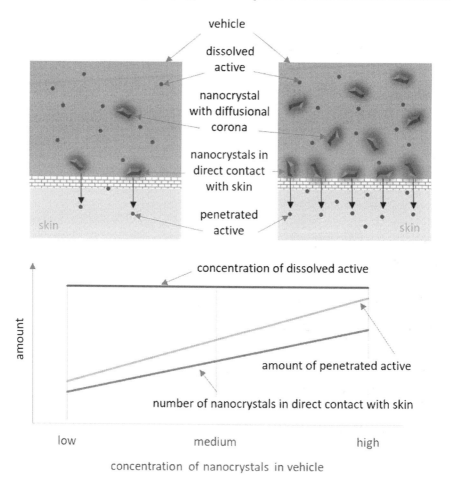

FIGURE 20.8 Influence of the number of nanocrystals that get in contact with the skin on dermal penetration of poorly soluble actives.

BM and HPH with reduced pressures is the most frequently applied technique (Junghanns and Müller, 2008; Shegokar and Müller, 2010; Keck et al., 2013). The so-called smartCrystal® technology uses the advantages of both technologies, while avoiding, or at least reducing, their drawbacks. The combination reduces the long milling times of the BM process and avoids an increase in temperature (which leads to agglomeration of the nanocrystals) during HPH, due to the reduced pressures applied. Particles produced via this technique, depending on the material used, possess sizes typically in the range between 200 and 600 nm and narrow size distributions.

All techniques are wet milling techniques. This means nanocrystals are dispersed in water and hence, upon nanonization of the bulk material, nanosized suspensions, that is, nanosuspensions, are obtained. The nanocrystals need to be stabilized with the use of surfactants. For dermal application, of course, skin-friendly stabilizers, for example, non-ionic surfactants or phospholipids, should be used. If nanosuspensions are formulated correctly they do not produce any sediment and the nanocrystals remain homogeneously distributed within the aqueous phase. The physical stability is typically at least 24 months.

Dermal formulations containing nanocrystals can be obtained by adding the nanosuspensions as the intermediate product to the final dermal formulation, that is, the vehicle. Any heat during the production must be avoided, because heat temporally increases the solubility of the nanocrystals, which is reduced upon cooling. This creates an oversaturated and thermodynamically highly instable system. Hence, during storage the dissolved active will precipitate by forming large crystals with poor solubility, leading to a loss of the special properties of the nanocrystals (Keck, 2006). Nanocrystals can be incorporated into creams and gels and can also be formulated as sprays.

Dermal Penetration: Mechanism of Action

As mentioned earlier, improved dermal penetration of poorly soluble actives from nanocrystals is mainly due to the increased concentration gradient, which is achieved by the increase in kinetic solubility. As the increase in solubility is size dependent, it is advisable to use small sized nanocrystals, that is, sizes below 400 nm (Pelikh et al., 2018). Recent studies proved the influence of the number of nanocrystals that get in contact with the skin upon dermal application on the penetration efficacy, and revealed that the

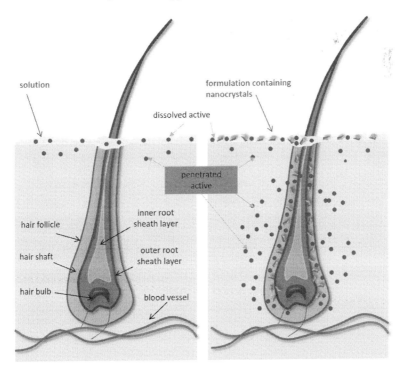

FIGURE 20.9 Principle of improved delivery of actives via hair follicle targeting with nanocrystals.

penetration efficacy is increased with increasing numbers of nanocrystals which get in contact with the skin (Vidlářová et al., 2016). The reason for this is that – similar to the sNLC – high local concentrations of dissolved active are created, which again increase the concentration gradient and with this the passive diffusion of the active into the skin (Figure 20.8).

A very important aspect for effective dermal penetration is the choice of vehicle used for the formulation (Brinkmeier, 2006; Gloor et al., 2000; Langner et al., 2011). Unfortunately until now, there is not much scientific knowledge about the influence of the vehicle on the penetration efficacy of actives from the different nanocarriers. A recent study investigated the influence of the type of vehicle on the penetration efficacy of poorly soluble actives from nanocrystals. Results revealed that hydrogels are not suitable vehicles, whereas oleogels are more suitable. The best penetration was achieved for a cream containing both hydrophilic and lipophilic excipients (Vidlářová et al., 2016).

Cosmeceutic Application

Nanocrystals are the formulation principle of choice for the delivery of poorly soluble actives. In contrast to the other nanocarriers described in this chapter, they possess no additional skin carrying properties. This is because they are composed of 100% actives that is aimed to be delivered into the skin upon dissolution, whereas the other carriers are composed of matrix material, which is mainly responsible for these additional skin carrying properties. However, the benefit of the nanocrystals, which are composed of 100% actives, is that only very small quantities are needed to obtain highly effective formulations. Depending on the material used, nanocrystal amounts of about 0.02–0.2% in the final cosmetic or cosmeceutical product are already sufficient to formulate highly effective dermal products (Keck et al., 2013).

Another advantage of nanocrystals is their possible use to deliver actives via the hair follicle. Studies by Vogt et al. demonstrated that particles, but not liquids, can enter the hair follicle (Vogt et al., 2005; Teichmann et al., 2005, 2006; Patzelt et al., 2011; Patzelt and Lademann, 2013). If the particle is carrying an active, this active can be released in the hair follicle and penetrate the skin by bypassing the stratum corneum (Patzelt and Lademann, 2013; Mak et al., 2012; Lademann et al., 2008) (Figure 20.9). If particles entered the hair follicle, they remain and form a depot, and hence can be used for a long lasting but slow release of actives. Lademann et al. (2011) also showed that larger sized nanoparticles, that is, sizes of about 600 nm, are favoured for hair follicle targeting. As nanocrystals can be produced by different techniques and by varying the production parameters, the production of large sized nanocrystals is easily possible, and the improved uptake of larger sized nanocrystals into the hair follicles has already be demonstrated (Pelikh et al., 2018).

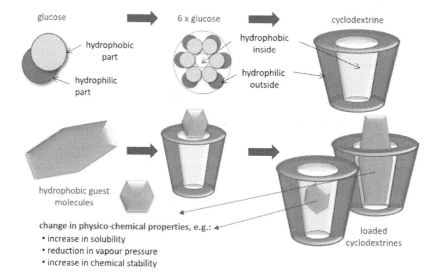

FIGURE 20.10 Principle of cyclodextrins for improved formulation and delivery of poorly soluble actives.

Miscellaneous Nanocarriers

The chapter cannot give a comprehensive overview of all nanocarriers being available in the cosmetic field. Therefore, a selection was chosen and the most prominent carriers were described in detail. Other carriers, also used for cosmetic and cosmeceutical applications, include micelles (Wichit et al., 2012), polymeric nanoparticles (Koosha and Müller, 1988; Lu and Ding, 2008; Somasundaran and Purohit, 2011), dendrimers (Abbasi et al., 2014; Allard and Forestier, 2002; Bahary and Hogan, 1997; Franzke et al., 2000), and new delivery systems that are based on nanostructures, such as smartPearls® (Jin et al., 2015) and smartFilms® (Lemke, 2017; Lemke, et al., 2017).

Last but not least a very special excipient should be mentioned. Per definition it is not a nanocarrier but more a structure that can be used to create 'nanocarrier-like' properties. The so-called cyclodextrins are non-toxic, cyclic polysaccharides and built from six to eight D-glucose units and possess sizes <2 nm. Cyclodextrins have been known for more than 100 years but gained popularity for industrial applications only during the last two decades. The reason was a lack of large-scale production until then (Buschmann and Schollmeyer, 2002). Special properties are derived due to their special structure, which is a rigid cavity formed as an inner hydrophobic core by the glucose molecules (Figure 20.10). In these cavities, which possess sizes below 1 nm, small hydrophobic guest molecules can be enclosed (Figure 20.10). The complex formation between cyclodextrin and guest molecule can modify the physico-chemical properties of the enclosed guest molecules. The most prominent changes are an increase in solubility and a reduction in vapour pressure, as well as an increase in chemical stability of chemically sensitive compounds (Loftsson and Duchêne, 2007; Brewster and Loftsson, 2007; Davis and Brewster, 2004).

Due to their beneficial properties, cyclodextrins are widely used in many different applications, including various cosmetic products. Prominent examples are the encapsulation of tea tree oil to protect it from light and oxidation, or the encapsulation of volatile compounds (perfumes) to prolong the smell by reducing the vapour pressure of the compounds. A further beneficial possibility is to transform liquids into powders, which eases handling and dosing. Also the solubility of poorly soluble actives can be increased by incorporation of such actives into cyclodextrins. For example, menthol was incorporated into cyclodextrins. By doing this the use of ethanol as a solvent for menthol can be avoided, enabling the formulation of ethanol-free mouthwash products (Buschmann and Schollmeyer, 2002). Other examples are antioxidants (e.g. retinol or tocopherol) where encapsulation increases the solubility of lipophilic actives, and promotes a slow and controlled release of the actives, thus enabling a long-lasting effect of the formulation. The most common use of cyclodextrins is their application to remove undesired odours and many products for this are available to refresh the air. However, this effect can also be used in cosmetics, for example, in deodorants to complex perspiration malodours (Buschmann and Schollmeyer, 2002). In fact, cyclodextrins are a highly effective and useful cosmetic excipient to overcome various challenges in the formulation of cosmetic and cosmeceutic products.

Characterisation of Nanocarriers

Most of the special properties of nanocarriers are due to their small size. Therefore, size characterisation of the carriers is highly important. The small sized particles cannot be seen by eye, thus special characterisation methods are required for particle examination. Electron microscopy, atomic force microscopy, and especially dynamic light scattering (DLS) are used to determine the mean particle size (Müller and Schuhmann, 1996). Besides the small mean size, it is often (more) important to know if larger sized particles are also within the formulation. This is important because changes in size are most likely to be associated with a loss of the special 'nano-properties'. Large particles can occur due to physical instability (i.e. Ostwald ripening), during storage or agglomeration of the nanoparticles due to incompatibilities (Keck and Müller, 2006). Large particles can be detected by using light microscopy. The method is simple but effective and should always be used when characterizing nanoparticles, because it enables a fast determination if a 'nano-formulation' is still in nano size or not. More sophisticated methods include laser light diffraction and other techniques, which are highly recommended if particle size analysis of nanoparticles is performed frequently, for example, during the development and/or formulation of nanocarriers (Keck and Müller, 2008; Kübart and Keck, 2013; Keck, 2010). Another useful tool to predict

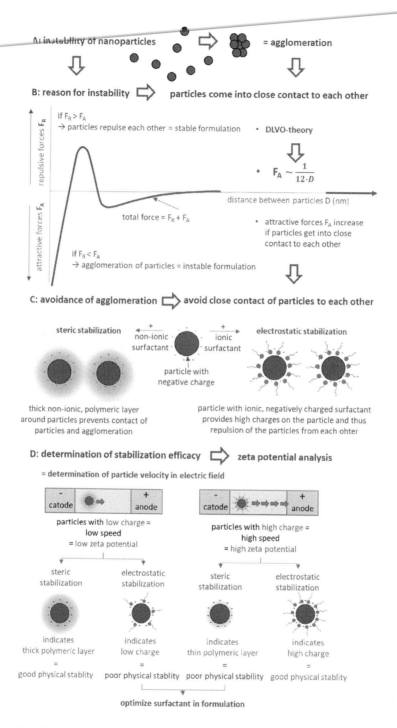

FIGURE 20.11 Principle of zeta potential and zeta potential analysis as tool to predict the physical stability of nanocarriers (explanations cf. text).

the physical stability of nanocarriers is the determination of the zeta potential, which is a measure of the charge of the particles (Clogston and Patri, 2011). If non-ionic surfactants are used for the stabilization of the nanocarriers, steric stabilization is obtained with a thick surfactant layer, thus the zeta potential should be low and almost zero (Figure 20.11). In contrast, ionic surfactants provide stabilization

TABLE 20.1

Summary of Properties of Nanocarriers

Carrier	↑ Skin Barrier Function	↑ Skin Hydration	↑ Lipid Content	Incorporation and ↑ Penetration of Actives		Loading Capacity	Stability
Liposomes	++	+	+/−	Hydrophilic	+++	+	+/−
				Lipophilic	+++		
				Amphiphilic	+++		
Nanoemulsions	+	+/−	+	Hydrophilic	−	+	+/−
				Lipophilic	+		
				Amphiphilic	+		
Nanolipid carrier	+++	++	++	Hydrophilic	−	+++	+
				Lipophilic	+++		
				Amphiphilic	+++		
Nanocrystals	−	−	−	Poorly soluble	++++	100%	+
				Hydrophilic	−		

via electrostatic stabilization. Therefore, if ionic stabilizers are used, the zeta potential should be high (Figure 20.11). Recent literature might be used to exploit this important topic in more detail (Müller and Schuhmann, 1996; Keck, 2010; Müller et al., 1996).

Regulatory Aspects

With the increase in products containing nanoparticles, nanotechnology came into focus of the regulatory authorities. Since then nanotoxicity is an emerging field that aims at evaluating the potential risks that might be associated with nanoparticles used in consumer care products. Until now there are no general, international regulations for nanoparticles that are used in cosmetic products. The U.S. Food and Drug Administration regulates nanoproducts on a case-by-case approach and most countries follow this strategy until now. A different approach has been taken in the European Union since 2011, when the new cosmetic regulation (European Parliament, 2009) came into force. In this regulation nanomaterials are defined as being 'insoluble or biopersistent and intentionally manufactured material with one or more external dimensions, or an internal structure, on the scale from 1 to 100 nm' (European Parliament, 2009). All nanocarriers described in this chapter are soluble (or disintegrate upon contact with skin), not biopersistent and possess sizes well above 100 nm. This means they do not fulfil this definition and thus are not regarded as nanomaterial as defined by the European Committee (Sainz et al., 2015; Keck and Müller, 2013). It is expected that 'nano regulations' for cosmetic products will soon be modified in most parts of the world and might even be harmonized with the European regulations. However, the nanocarriers described in this chapter were proven to be safe and typically possess sizes well above 100 nm. Many studies have proven that nanoparticles with sizes >40 nm cannot penetrate through an intact skin barrier (Rancan et al., 2012; Vogt et al., 2006), making these nanocarriers safe but highly effective for the development of modern, innovative cosmetic and cosmeceutical products.

Conclusions

Nanocarriers such as liposomes, nanoemulsions, lipid nanoparticles and nanocrystals are the most frequently applied 'nano-formulation principles' in modern, cosmetic formulations. Despite their small size, which improves the adhesiveness and thus prolongs the retention time on the skin upon dermal application, which leads to a more pronounced uptake of actives in the skin, all nanocarriers possess different properties. Each nanocarrier system has advantages and disadvantages, hence, like in daily life, nothing is perfect. The challenge when using nanocarriers for dermal application is to know all the different properties of the individual carrier systems. With this knowledge it is possible to select the optimal carrier system for the intended dermal product. Table 20.1 summarizes the properties of the different nanocarriers and visualizes for which purpose they are ideal to use.

Nanocarriers are ideal for improved performance of existing products and superior for the development of more effective cosmetic products. The main focus is the improved delivery of actives into deeper skin layers and improved skin care. Besides this, nanocarriers enable the formulation of highly aesthetic dermal products with high consumer acceptance and good acceptance by regulatory authorities. Due to the positive and unique features of nanocarriers, it is believed that more nanocarrier-based products will enter the cosmetic market in the future.

REFERENCES

Abbasi, E., Aval, S. F., Akbarzadeh, A., et al. (2014) Dendrimers: Synthesis, applications, and properties, *Nanoscale Res Lett*, 9(1): p. 247.

Akbarzadeh, A., Rezaei-Sadabady, R., Davaran, S., et al. (2013) Liposome: Classification, preparation, and applications, *Nanoscale Res Lett*, 8(1): p. 102.

Allard, D., Forestier, S. (2002) Self-tanning cosmetic compositions, US Patent 6399048B1.

Al Shaal, L., Mishra, P. R., Müller, R. H., et al. (2014) Nanosuspensions of hesperetin: Preparation and characterization, *Die Pharmazie*, 69(3): p. 173–182.

Ansari, M. A., Khan, H. M., Khan, A. A., et al. (2015) Silver nanoparticles: 21st century powerful antimicrobial nano-weapon, in *NanoBiomedicine* (Singh, B., Kanwar, J. R., Katare, O. P., Eds.), Studium Press, Houston, TX: p. 53–106.

Antonio, J. R., Antônio, C. R., Cardeal, I. L. S., et al. (2014) Nanotechnology in dermatology, *An Bras Dermatol*, 89(1): p. 126–136.

Badran, M., Shalaby, K., Al-Omrani, A. (2012) Influence of the flexible liposomes on the skin deposition of a hydrophilic model drug, carboxyfluorescein: Dependency on their composition, *Sci World J*, 2012: p. 134876.

Bahary, W. S., Hogan, M. P. (1997) Cleansing compositions with dendrimers as mildness agents, US Patent 5658574A.

Bangham, A. D., Hill, M. W., Miller, N. G. A. (1974) Preparation and use of liposomes as models of biological membranes, in *Methods in Membrane Biology* (Korn, E. D., Ed.), Springer, Boston: p. 1–68.

Baspinar, Y., Keck, C. M., Borchert, H.-H. (2010) Development of a positively charged prednicarbate nanoemulsion, *Int J Pharm*, 383(1–2): p. 201–208.

Bauer, K. H., Frömming, K.-H., Führer, C. (1997) *Pharmazeutische Technologie*, Govi-Verl., Stuttgart, Germany.

Betz, G., Aeppli, A., Menshutina, N., et al. (2005) In vivo comparison of various liposome formulations for cosmetic application, *Int J Pharm*, 296(1–2): p. 44–54.

Bladen, J. C., Norris, J. H., Malhotra, R. (2012) Cosmetic comparison of gold weight and platinum chain insertion in primary upper eyelid loading for lagophthalmos, *Ophthal Plast Reconstr Surg*, 28(3): p. 171–175.

Brewster, M. E., Loftsson, T. (2007) Cyclodextrins as pharmaceutical solubilizers, *Adv Drug Deliv Rev*, 59(7): p. 645–666.

Briganti, S., Picardo, M. (2003) Antioxidant activity, lipid peroxidation and skin diseases: What's new, *J Eur Acad Dermatol Venereol*, 17(6): p. 663–669.

Brinkmeier, T. (2006) *Springer Kompendium Dermatologie*. Springer, New York.

Bunjes, H., Westesen, K., Koch, M. H. J. (1996) Crystallization tendency and polymorphic transitions in triglyceride nanoparticles, *Int J Pharm*, 129: p. 159–173.

Buschmann, H.-J., Schollmeyer, E. (2002) Applications of cyclodextrins in cosmetic products: A review, *J Cosmet Sci*, 53(3): p. 185–191.

Carlotti, M. E., Pattarino, F., Gasco, M. R., et al. (1993) Optimization of parameters in the emulsification process by two different methods, *Int J Cosmet Sci*, 15(6): p. 245–259.

Cavalli, R., Bargoni, A., Podio, V., et al. (2003) Duodenal administration of solid lipid nanoparticles loaded with different percentages of tobramycin, *J Pharm Sci*, 92(5): p. 1085–1094.

Cavalli, R., Gasco, M. R., Chetoni, P., et al. (2002) Solid lipid nanoparticles (SLN) as ocular delivery system for tobramycin, *Int J Pharm*, 238(1–2): p. 241–245.

Chanchal, D., Swarnlata, S. (2008) Novel approaches in herbal cosmetics, *J Cosmet Dermatol*, 7(2): p. 89–95.

Clement, J. L., Jarrett, P. S. (1994) Antibacterial silver, *Met Based Drugs*, 1(5–6): p. 467–482.

Clogston, J. D., Patri, A. K. (2011) Zeta potential measurement, *Methods Mol Biol*, 697: p. 63–70.

Contado, C. (2015) Nanomaterials in consumer products: A challenging analytical problem, *Front Chem*, 3: p. 48.

Daniels, R. (2009) Dermopharmacy: The right galenics for skin disease, *Pharmazeutische Zeitung*, 154.

Davis, M. E., Brewster, M. E. (2004) Cyclodextrin-based pharmaceutics: Past, present and future, *Nat Rev Drug Discov*, 3(12): p. 1023–1035.

Desenne, P., Chesneau, L. (2011) Cosmetic composition in the form of a nanoemulsion containing a volatile linear alkane, US Patent 2011/0150813 A1.

Diederichs, J. E., Müller, R. H. (1994) *Pharm Ind*, 56: p. 267–275.

Ding, Y., Pyo, S. M., Müller, R. H. (2017) smartLipids® as third solid lipid nanoparticle generation – Stabilization of retinol for dermal application, *Pharmazie*, 72(12): p. 728–735.

Dingler, A., Blum, R. P., Niehus, H., et al. (1999) Solid lipid nanoparticles (SLN/Lipopearls) – A pharmaceutical and cosmetic carrier for the application of vitamin E in dermal products, *J Microencapsul*, 16(6): p. 751–767.

Edgar, M., Michel, C., Purmann, T. (1991) Stable small particle liposome preparations, their production and use in topical cosmetic, and pharmaceutical compositions, US patent 5498420A.

Elder, A., Gelein, R., Finkelstein, J. N., et al. (2005) Effects of subchronically inhaled carbon black in three species. I. Retention kinetics, lung inflammation, and histopathology, *Toxicol Sci*, 88(2): p. 614–629.

Elias, P. M., Brown, B. E., Ziboh, V. A. (1980) The permeability barrier in essential fatty acid deficiency: Evidence for a direct role for linoleic acid in barrier function, *J Invest Dermatol*, 74(4): p. 230–233.

European Parliament. (2009) Regulation (EC) no 1223/2009 of the European Parliament and of the council of 30 November 2009 on cosmetic products, Official Journal of the European Union.

Fahr, A., Liu, X. (2007) Drug delivery strategies for poorly water-soluble drugs, *Expert Opin Drug Deliv*, 4(4): p. 403–416.

Fahr, A., van Hoogevest, P., Kuntsche, J., et al. (2006) Lipophilic drug transfer between liposomal and biological membranes: What does it mean for parenteral and oral drug delivery? *J Liposome Res*, 16(3): p. 281–301.

Fathi-Azarbayjani, A., Qun, L., Chan, Y. W., et al. (2010) Novel vitamin and gold-loaded nanofiber facial mask for topical delivery, *AAPS PharmSciTech*, 11(3): p. 1164–1170.

Florence, A. T. (2007) Pharmaceutical nanotechnology. More than size. Ten topics for research, *Int J Pharm*, 339(1–2): p. 1–2.

Forrest, M. L., Kwon, G. S. (2008) Clinical developments in drug delivery nanotechnology, *Adv Drug Deliv Rev*, 60(8): p. 861–862.

Franci, G., Falanga, A., Galdiero, S., et al. (2015) Silver nanoparticles as potential antibacterial agents, *Molecules*, 20(5): p. 8856–8874.

Franzke, M., Steinbrecht, K., Clausen, T., et al. (2000) Cosmetic compositions for hair treatment containing dendrimers or dendrimer conjugates, US Patent 6068835A.

Frederiksen, H. K., Kristensen, H. G., Pedersen, M. (2003) Solid lipid microparticle formulations of the pyrethroid gamma-cyhalothrin – Incompatibility of the lipid and the pyrethroid and biological properties of the formulations, *J Control Release*, 86: p. 243–252.

Friedman, D. I., Schwarz, J. S., Weisspapir, M. (1995) Submicron emulsion vehicle for enhanced transdermal delivery of steroidal and nonsteroidal antiinflammatory drugs, *J Pharm Sci*, 84(3): p. 324–329.

Gasco, M. R. (1993) Method for producing solid lipid microspheres having a narrow size distribution, US Patent 5,250,236.

Gasco, M. R., Gallarate, M., Pattarino, F. (1991) In vitro permeation of azelaic acid from viscosized microemulsions, *Int J Pharm*, 69(3): p. 193–196.

Gillet, A., Lecomte, F., Hubert, P., et al. (2011) Skin penetration behaviour of liposomes as a function of their composition, *Eur J Pharm Biopharm*, 79(1): p. 43–53.

Glenn JR., R. W., Kaufman, K. M., Hosseinpour, D. (2015) Method of treating hair with a concentrated conditioner, US Patent 2015/0359727 A1.

Gloor, M., Fluhr, J., Thoma, K. (2000) *Dermatologische Externatherapie*. Springer, Berlin.

Golz-Berner, K., Zastrow, L. (2011) Cosmetic skin care complex with anti-aging effect, US Patent 7,968,129 B2.

Grattan, C., Burton, J. L., Manku, M., et al. (1990) Essential-fatty-acid metabolites in plasma phospholipids in patients with ichthyosis vulgaris, acne vulgaris and psoriasis, *Clin Exp Dermatol*, 15(3): p. 174–176.

Gregoriadis, G. (1976) The carrier potential of liposomes in biology and medicine (second of two parts), *N Engl J Med*, 295(14): p. 765–770.

Gregoriadis, G. (Ed.) (2007) *Liposome Technology*, Vol. 1: *Liposome Preparation and Related Techniques*, Informa Healthcare, New York.

Griffin, S., Masood, M. I., Nasim, M. J., et al. (2017) Natural nanoparticles: A particular matter inspired by nature, *Antioxidants (Basel)*, 7(1): p. 3.

Guillaume, D., Charrouf, Z. (2011) Argan oil and other argan products. Use in dermocosmetology, *Eur J Lipid Sci Technol*, 113(4): p. 403–408.

Gupta, A., Eral, H. B., Hatton, T. A., et al. (2016) Nanoemulsions: Formation, properties and applications, *Soft Matter*, 12(11): p. 2826–2841.

Gupta, S., Bansal, R., Gupta, S., et al. (2013) Nanocarriers and nanoparticles for skin care and dermatological treatments, *Indian Dermatol Online J*, 4(4): p. 267–272.

Heiligtag, F. J., Niederberger, M. (2013) The fascinating world of nanoparticle research, *Mater Today*, 16(7–8): p. 262–271.

Hernándes-Caselles, T., Villavaín, J., Gómez-Fernádez, J. C. (1990) Stability of liposomes on long term storage, *J Pharm Pharmacol*, 42(6): p. 397–400.

Hommoss, A., Peter, M., Müller, R. H. (2007) Sun protection factor (SPF) increase using nanostructured lipid carriers (NLC). International Symposium on Controlled Release of Bioactive Materials, 34. Long Beach, CA.

Hope, M. J., Nayar, R., Mayer, L. D., et al. (1993) Reduction of liposome size and preparation of unilamellar vesicles by extrusion techniques, in *Liposome Technology* (Gregoriadis, G., Ed.), CRC Press, Boca Raton, FL: p. 123–139.

Hörmann, K., Zimmer, A. (2016) Drug delivery and drug targeting with parenteral lipid nanoemulsions – A review, *J Control Release*, 223: p. 85–98.

Horrobin, D. F. (1989) Essential fatty acids in clinical dermatology, *J Am Acad Dermatol*, 20(6): p. 1045–1053.

Hougaz, L. (2009) Sunscreen aerosol spray, US Patent 20090061001A1.

Hough, R. M., Noble, R. R. P., Reich, M. (2011) Natural gold nanoparticles, *Ore Geol Rev*, 42(1): p. 55–61.

Howard, G. M. (1974) Sunburn preparations, in *Perfumes, Cosmetics and Soaps: Modern Cosmetics* (Howard, G. M., Ed.), Springer, Boston: p. 402–424.

Hu, Z., Liao, M., Chen, Y., et al. (2012) A novel preparation method for silicone oil nanoemulsions and its application for coating hair with silicone, *Int J Nanomed*, 7: p. 5719–5724.

Hua, S. (2015) Lipid-based nano-delivery systems for skin delivery of drugs and bioactives, *Front Pharmacol*, 6: p. 219.

Huang, C.-H., Li, S. (1996) Computational molecular models of lipid bilayers containing mixed-chain saturated and monounsaturated acyl chain, in *Handbook of Nonmedical Applications of Liposomes* (Lasic, D. D., Ed.), CRC Press, Boca Raton, FL: p. 173–193.

Hussain, A., Singh, S., Sharma, D., et al. (2017) Elastic liposomes as novel carriers: Recent advances in drug delivery, *Int J Nanomed*, 12: p. 5087–5108.

Isailović, T., Đorđević, S., Marković, B., et al. (2016) Biocompatible nanoemulsions for improved aceclofenac skin delivery: Formulation approach using combined mixture-process experimental design, *J Pharm Sci*, 105(1): p. 308–323.

Iskandar, B., Karsono, Silalahi, J. (2016) Preparation of spray nanoemulsion and cream containing vitamin E as anti-aging product tested in vitro and in vivo method, *Int J PharmTech Res*, 9(6): p. 307–315.

Izquierdo, P., Esquena, J., Tadros, T. F., et al. (2002) Formation and stability of nano-emulsions prepared using the phase inversion temperature method, *Langmuir*, 18(1): p. 26–30.

Jafari, S. M., McClements, D. J. (2018) *Nanoemulsions: Formulation, Applications, and Characterization*, Academic Press, London, UK.

Jenning, V., Gohla, S. H. (2001) Encapsulation of retinoids in solid lipid nanoparticles (SLN), *J Microencapsul*, 2(18): p. 149–158.

Jenning, V., Lippacher, A., Gohla, S. H. (2002) Medium scale production of solid lipid nanoparticles (SLN) by high pressure homogenization, *J Microencapsul*, 19(1): p. 1–10.

Jin, N., Keck, C. M., Staufenbiel, S., et al. (2015) smartPearls – Novel dermal delivery system for amorphous cosmetic and pharma actives. Annual Meeting of the Society for Dermopharmacy. Berlin, Germany, 16–18 March: p. 4.

Jumaa, M., Müller, B. W. (1998) The effect of oil components and homogenization conditions on the physico-chemical properties and stability of parenteral fat emulsions, *Int J Pharm*, 163(1–2): p. 81–89.

Junghanns, J.-U. A. H., Müller, R. H. (2008) Nanocrystal technology, drug delivery and clinical applications, *Int J Nanomed*, 3(3): p. 295–309.

Kajita, M., Hikosaka, K., Iitsuka, M., et al. (2007) Platinum nanoparticle is a useful scavenger of superoxide anion and hydrogen peroxide, *Free Radical Res*, 41(6): p. 615–626.

Katz, L. M., Dewan, K., Bronaugh, R. L. (2015) Nanotechnology in cosmetics, *Food Chem Toxicol*, 85: p. 127–137.

Keck, C. M. (2006) Cyclosporine nanosuspensions: Optimised size characterisation & oral formulations, PhD thesis, Freie Universität Berlin.

Keck, C. M. (2010) Particle size analysis of nanocrystals: Improved analysis method, *Int J Pharm*, 390(1): p. 3–12.

Keck, C. M., Anantaworasakul, P., Patel, M., et al. (2014) A new concept for the treatment of atopic dermatitis: Silver-nanolipid complex (sNLC), *Int J Pharm*, 462(1–2): p. 44–51.

Keck, C. M., Baisaeng, N., Durand, P., et al. (2014) Oil-enriched, ultra-small nanostructured lipid carriers (usNLC): A novel delivery system based on flip-flop structure, *Int J Pharm*, 477(1–2): p. 227–235.

Keck, C. M., Chen, R., Müller, R. H. (2013) SmartCrystals for consumer care & cosmetics: Enhanced dermal delivery of poorly soluble plant actives, *H&PC Today*, 8(5): p. 18–24.

Keck, C. M., Koppe, F., Tiede, H., et al. (2008) Parenteral lipofundin nanoemulsions: 20 years long-term stability. 7th European Workshop on Particulate Systems. Berlin, Germany: p. 51.

Keck, C. M., Lohan, S. B., Bauersachs, S., et al. (2015) smartLipids® Q10: Highly effective protection from light as innovative anti-aging formulation for the skin. Menopause, Andropause, Anti-Aging-Congress, Vienna, Austria, 10–12 December.

Keck, C. M., Müller, R. H. (2006) Drug nanocrystals of poorly soluble drugs produced by high pressure homogenisation, *Eur J Pharm Biopharm*, 62(1): p. 3–16.

Keck, C. M., Müller, R. H. (2008) Size analysis of submicron particles by laser diffractometry – 90% of the published measurements are false, *Int J Pharm*, 355(1–2): p. 150–163.

Keck, C. M., Müller, R. H. (2010) Silber-Lipid-Zwerge - Ein neues Therapiekonzept gegen Neurodermitis, *Quantos* 1: p. 6–11.

Keck, C. M., Müller, R. H. (2013) Nanotoxicological classification system (NCS) – A guide for the risk-benefit assessment of nanoparticulate drug delivery systems, *Eur J Pharm Biopharm*, 84(3): p. 445–448.

Keck, C. M., Schwabe, K. (2009) Silver-nanolipid complex for application to atopic dermatitis skin: Rheological characterization, in vivo efficiency and theory of action, *J Biomed Nanotechnol*, 5(4): p. 428–436.

Kendall, A. C., Kiezel-Tsugunova, M., Brownbridge, L. C., et al. (2017) Lipid functions in skin: Differential effects of n-3 polyunsaturated fatty acids on cutaneous ceramides, in a human skin organ culture model, *Biochim Biophys Acta*, 1859(9 Pt B): p. 1679–1689.

Koosha, F., Müller, R. H. (1988) Production of polymeric drug carriers for drug targeting, *Acta Pharm Technol*, 34: p. 24.

Kübart, S. A., Keck, C. M. (2013) Laser diffractometry of nanoparticles: Frequent pitfalls & overlooked opportunities, *J Pharm Technol Drug Res*, 2(1): p. 17.

Lademann, J., Knorr, F., Richter, H., et al. (2008) Hair follicles – An efficient storage and penetration pathway for topically applied substances. Summary of recent results obtained at the Center of Experimental and Applied Cutaneous Physiology, Charité -Universitätsmedizin Berlin, Germany, *Skin Pharmacol Physiol*, 21(3): p. 150–155.

Lademann, J., Richter, H., Schanzer, S., et al. (2011) Penetration and storage of particles in human skin: Perspectives and safety aspects, *Eur J Pharm Biopharm*, 77(3): p. 465–468.

Langner, A., Borchert, H.-H., Mehnert, W., et al. (2011) *Biopharmazie*. WVG, Stuttgart.

Lasic, D. (1998) Novel applications of liposomes, *Trends Biotechnol*, 16(7): p. 307–321.

Lautenschläger, H. (1990) Liposomes in dermatological preparations, *Cosmet Toiletries*, 105: p. 89–96.

Leekumjorn, S., Cho, H. J., Wu, Y., et al. (2009) The role of fatty acid unsaturation in minimizing biophysical changes on the structure and local effects of bilayer membranes, *Biochim Biophys Acta*, 1788(7): p. 1508–1516.

Lemke, S. (2017) Cellulosebasierte Filme (smartFilms®) als alternative orale oder perorale Applikationsform; Herstellung und Prüfung, PhD Thesis, Freie Universität Berlin.

Lemke, S., Strätling, E.-J., Welzel, H.-P., et al. (2017) Cellulosefaserbasierte Trägermatrices (smartFilms) zur Applikation von Inhaltsstoffen sowie deren Herstellung, DE Patent Application 102016000541A1.

Lens, M. (2009) Use of fullerenes in cosmetics, *Recent Pat Biotechnol*, 3(2): p. 118–123.

Lens, M. (2011) Recent progresses in application of fullerenes in cosmetics, *Recent Pat Biotechnol*, 5(2): p. 67–73.

Li, J., Wang, X., Zhang, T., et al. (2015) A review on phospholipids and their main applications in drug delivery systems, *Asian J Pharm*, 10(2): p. 81–98.

List, M. A., Sucker, H. (1988) Pharmaceutical colloidal hydrosols for injection, GB Patent 2200048 A.

Liversidge, G. G., Cundy, K. C., Bishop, J. F., et al. (1991) Surface modified drug nanoparticles, US Patent 5145684A.

Lococo, D., Mora-Huertas, C. E., Fessi, H., et al. (2012) Argan oil nanoemulsions as new hydrophobic drug-loaded delivery system for transdermal application, *J Biomed Nanotechnol*, 8(5): p. 843–848.

Loftsson, T., Duchêne, D. (2007) Cyclodextrins and their pharmaceutical applications, *Int J Pharm*, 329(1–2): p. 1–11.

Lohan, S. B., Bauersachs, S., Ahlberg, S., et al. (2015) Ultra-small lipid nanoparticles promote the penetration of coenzyme Q10 in skin cells and counteract oxidative stress, *Eur J Pharm Biopharm*, 89: p. 201–207.

Lohani, A., Verma, A., Joshi, H., et al. (2014) Nanotechnology-based cosmeceuticals, *ISRN Dermatol*, 2014: p. 843687.

Lu, P., Ding, B. (2008) Applications of electrospun fibers, *Nanotechnology*, 2(3): p. 169–182.

Lucks, S., Müller, R. H. (1999) Arzneistoffträger aus festen Lipidteilchen (feste Lipidnanosphären (SLN)), EP Grant 0605497B2.

Lundberg, B. B. (1997) A submicron lipid emulsion coated with amphipathic polyethylene glycol for parenteral administration of paclitaxel (Taxol), *J Pharm Pharmacol*, 49(1): p. 16–21.

El Maghraby, G. M. M., Williams, A. C., Barry, B. W. (2006) Can drug-bearing liposomes penetrate intact skin? *J Pharm Pharmacol*, 58(4): p. 415–429.

Mahdi Jafari, S., He, Y., Bhandari, B. (2006) Nano-emulsion production by sonication and microfluidization—A comparison, *Int J Food Sci Nutr*, 9(3): p. 475–485.

Maherani, B., Arab-Tehrany, E., R. Mozafari, M., et al. (2011) Liposomes. A review of manufacturing techniques and targeting strategies, *Curr Nanosci*, 7(3): p. 436–452.

Maia, C. S., Mehnert, W., Schäfer-Korting, M. (2000) Solid lipid nanoparticles as drug carriers for topical glucocorticoids, *Int J Pharm*, 196(2): p. 165–167.

Mak, W. C., Patzelt, A., Richter, H., et al. (2012) Triggering of drug release of particles in hair follicles, *J Control Release*, 160(3): p. 509–514.

McClements, D. J. (2012) Nanoemulsions versus microemulsions: Terminology, differences, and similarities, *Soft Matter*, 8(6): p. 1719–1729.

McCusker, M. M., Grant-Kels, J. M. (2010) Healing fats of the skin: The structural and immunologic roles of the omega-6 and omega-3 fatty acids, *Clin Dermatol*, 28(4): p. 440–451.

Mehmood, T., Ahmad, A., Ahmed, A., et al. (2017) Optimization of olive oil based O/W nanoemulsions prepared through ultrasonic homogenization: A response surface methodology approach, *Food Chem*, 229: p. 790–796.

Mihranyan, A., Ferraz, N., Strømme, M. (2012) Current status and future prospects of nanotechnology in cosmetics, *Prog Mater Sci*, 57(5): p. 875–910.

Morais Diane, J. M., Burgess, J. (2014) Vitamin E nanoemulsions characterization and analysis, *Int J Pharm*, 465(1–2): p. 455–463.

Moschwitzer, J. (2006) A process for the production of ultrafine submicron suspensions, DE Patent 102005011786A1.

Mukherjee, S., Ray, S., Thakur, R. S. (2009) Solid lipid nanoparticles: A modern formulation approach in drug delivery system, *Indian J Pharm Sci*, 71(4): p. 349–358.

Müller, B. W. (1997) Mikroemulsionen als neue Wirkstoff-Trägersysteme. *Pharmazeutische Technologie: Moderne Arzneiformen* (Müller, R. H., Hildebrand, G. E., Bauer, K. H., Eds.), WVG, Stuttgart, Germany.

Müller, R. H., Becker, R., Kruss, B., et al. (1996) Pharmaceutical nanosuspensions for drug administration than systems with increased saturation solubility and dissolution rate, DE Patent 4440337A1.

Müller, R. H., Gohla, S., Keck, C. M. (2011) State of the art of nanocrystals – Special features, production, nanotoxicology aspects and intracellular delivery, *Eur J Pharm Biopharm*, 78(1): p. 1–9.

Müller, R. H., Harden, D., Keck, C. M. (2012) Development of industrially feasible concentrated 30% and 40% nanoemulsions for intravenous drug delivery, *Drug Dev Ind Pharm*, 38(4): p. 420–430.

Müller, R. H., Heinemann, S. (1993) Fat emulsions for parenteral nutrition II: Characterisation and physical long-term stability of Lipofundin MCT LCT, *Clin Nutr*, 12(5): p. 298–309.

Müller, R. H., Jenning, V., Mäder, K., et al. (2014) Lipidpartikel auf der Basis von Mischungen von flüssigen und festen Lipiden und Verfahren zu ihrer Herstellung, EP Grant 1176949B1.

Müller, R. H., Keck, C. M. (2004) Challenges and solutions for the delivery of biotech drugs – A review of drug nanocrystal technology and lipid nanoparticles, *J Biotechnol*, 113(1–3): p. 151–170.

Müller, R. H., Keck, C. M. (2008) Second generation of drug nanocrystals for delivery of poorly soluble drugs: SmartCrystal® technology, *Eur J Pharm Sci*, 34(1): p. 20–21.

Müller, R. H., Keck, C. M. (2012) Twenty years of drug nanocrystals: Where are we, and where do we go? *Eur J Pharm Biopharm*, 80(1): p. 1–3.

Müller, R. H., Mäder, K., Gohla, S. (2000) Solid lipid nanoparticles (SLN) for controlled drug delivery – A review of the state of the art, *Eur J Pharm Biopharm*, 50(1): p. 161–177.

Müller, R. H., Mehnert, W., Lucks, J. S., et al. (1995) Solid lipid nanoparticles (SLN)—An alternative colloidal carrier system for controlled drug delivery, *Eur J Pharm Biopharm*, (41): p. 62–69.

Müller, R. H., Nitzsche, R., Paulke, B.-R. (1996) Zetapotential und Partikelladung in der Laborpraxis. Paperback APV, 37. WVG, Stuttgart.

Müller, R. H., Petersen, R. D., Hommoss, A., et al. (2007) Nanostructured lipid carriers (NLC) in cosmetic dermal products, *Adv Drug Deliv Rev*, 59(6): p. 522–530.

Müller, R. H., Radtke, M., Wissing, S. A. (2002) Solid lipid nanoparticles (SLN) and nanostructured lipid carriers (NLC) in cosmetic and dermatological preparations, *Adv Drug Deliv Rev*, 54: p. S131–S155.

Müller, R. H., Schmidt, S., Buttle, I., et al. (2004) SolEmuls® – Novel technology for the formulation of i.v. emulsions with poorly soluble drugs, *Int J Pharm*, 269(2): p. 293–302.

Müller, R. H., Schuhmann, R. (1996) Teilchengrößenmessung in der Laborpraxis. Paperback APV, 38. WVG, Stuttgart.

Müller, R. H., Wissing, S., Mäder, K. (2004) UV radiation reflecting or absorbing agents, protecting against harmful UV radiation and reinforcing the natural skin barrier, US Patent 6,814,959 B1.

Muranushi, N., Takagi, N., Muranishi, S., et al. (1981) Effect of fatty acids and monoglycerides on permeability of lipid bilayer, *Chem Phys Lipids*, 28(3): p. 269–279.

Nafisi, S., Maibach, H. I. (2017) Nanotechnology in cosmetics. *Cosmetic Science and Technology* (Sakamoto, K., Lochhead, R. Y., Maibach, H. I., et al., Eds.), Elsevier, Amsterdam: p. 337–369.

New, R. R. C. (1995) Influence of liposome characteristics on their properties and fate, in *Liposomes as Tools in Basic Research and Industry* (Philippot, J. R., Schuber, F., Eds.), CRC Press, Boca Raton, FL: p. 18–21.

Nikolić, S., Keck, C. M., Anselmi, C., et al. (2011) Skin photoprotection improvement: Synergistic interaction between lipid nanoparticles and organic UV filters, *Int J Pharm*, 414(1–2): p. 276–284.

Olson, F., Hunt, C. A., Szoka, F. C., et al. (1979) Preparation of liposomes of defined size distribution by extrusion through polycarbonate membranes, *Biochim Biophys Acta*, 557(1): p. 9–23.

Palmer, B. C., DeLouise, L. A. (2016) Nanoparticle enabled transdermal drug delivery systems for enhanced dose control and tissue targeting, *Molecules*, 21(12).

Pardeike, J., Hommoss, A., Müller, R. H. (2009) Lipid nanoparticles (SLN, NLC) in cosmetic and pharmaceutical dermal products, *Int J Pharm*, 366(1–2): p. 170–184.

Pardeike, J., Müller, R. H. (2007) Coenzyme Q10-loaded NLCs: Preparation, occlusive properties and penetration enhancement, *Pharm Tech Eur*, 19(7): p. 46–49.

Park, K. (2007) Nanotechnology: What it can do for drug delivery, *J Control Release*, 120(1–2): p. 1–3.

Partha, R., Conyers, J. L. (2009) Biomedical applications of functionalized fullerene-based nanomaterials, *Int J Nanomed*, 4: p. 261–275.

Patil, Y. P., Jadhav, S. (2014) Novel methods for liposome preparation, *Chem Phys Lipids*, 177: p. 8–18.

Patzelt, A., Lademann, J. (2013) Drug delivery to hair follicles, *Expert Opin Drug Deliv*, 10(6): p. 787–797.

Patzelt, A., Richter, H., Knorr, F., et al. (2011) Selective follicular targeting by modification of the particle sizes, *J Control Release*, 150(1): p. 45–48.

Pawar, K. R., Babu, R. J. (2014) Lipid materials for topical and transdermal delivery of nanoemulsions, *Crit Rev Ther Drug Carrier Syst*, 31: p. 429–458.

Pelikh, O., Stahr, P., Eckert, R., et al. (2018) Dermal drug delivery with nanocrystals – Size matters. 11th World Meeting on Pharmaceutics, Biopharmaceutics and Pharmaceutical Technology. Granada, Spain, 19–22 March.

Pelikh, O., Stahr, P., Gerst, M., et al. (2018) Nanocrystals for improved dermal drug delivery, *Eur J Pharm Biopharm*, 128: p. 170–178.

Priano, L., Esposti, D., Esposti, R., et al. (2007) Solid lipid nanoparticles incorporating melatonin as new model for sustained oral and transdermal delivery systems, *J Nanosci Nanotech*, 7(10): p. 3596–3601.

Puglia, C., Damiani, E., Offerta, A., et al. (2014) Evaluation of nanostructured lipid carriers (NLC) and nano-emulsions as carriers for UV-filters: Characterization, in vitro penetration and photostability studies, *Eur J Pharm Sci*, 51: p. 211–217.

Pyo, S. M., Meinke, M. C., Keck, C. M., et al. (2007) Dermal delivery systems – Nanoemulsion vs. nanostruc-tured lipid carriers (NLC). International Symposium on Controlled Release of Bioactive Materials, 34. Long Beach, CA: p. 24–26.

Qian, C., Decker, E. A., Xiao, H., et al. (2012) Inhibition of β-carotene degradation in oil-in-water nanoemul-sions: Influence of oil-soluble and water-soluble antioxidants, *Food Chem*, 135(3): p. 1036–1043.

Rachmawati, H., Budiputra, D. K., Mauludin, R. (2015) Curcumin nanoemulsion for transdermal application: Formulation and evaluation, *Drug Dev Ind Pharm*, 41(4): p. 560–566.

Radomska, A., Dobrucki, R. (2000) The use of some ingredients for microemulsion preparation containing retinol and its esters, *Int J Pharm*, 196(2): p. 131–134.

Rahimpour, Y., Hamishehkar, H. (2012) Liposomes in cosmeceutics, *Expert Opin Drug Deliv*, 9(4): p. 443–455.

Rähse, W., Dicoi, O. (2009) Produktdesign disperser Stoffe. Emulsionen für die kosmetische Industrie, *Chemie Ingenieur Technik*, 81(9): p. 1369–1383.

Rai, M. K., Deshmukh, S. D., Ingle, A. P., et al. (2012) Silver nanoparticles. The powerful nanoweapon against multidrug-resistant bacteria, *J Appl Microbiol*, 112(5): p. 841–852.

Rai, V. K., Mishra, N., Yadav, K. S., et al. (2018) Nanoemulsion as pharmaceutical carrier for dermal and trans-dermal drug delivery: Formulation development, stability issues, basic considerations and applications, *J Control Release*, 270: p. 203–225.

Rancan, F., Gao, Q., Graf, C., et al. (2012) Skin penetration and cellular uptake of amorphous silica nanopar-ticles with variable size, surface functionalization, and colloidal stability, *ACS Nano*, 6(8): p. 6829–6842.

Raza, K., Singh, B., Mahajan, A., et al. (2011) Design and evaluation of flexible membrane vesicles (FMVs) for enhanced topical delivery of capsaicin, *J Drug Target*, 19(4): p. 293–302.

Rechmann, H. (1974) Titanium and its importance today, *Pigm Resin Technol*, 3(9): p. 12–18.

Ruick, R. (2016) *SmartLipids - die neue Generation der Lipidnanopartikel nach SLN und NLC*, Freie Universität Berlin, Berlin, Germany.

Sainz, V., Conniot, J., Matos, A. I., et al. (2015) Regulatory aspects on nanomedicines, *Biochem Biophys Res Commun*, 468(3): p. 504–510.

Sakdiset, P., Okada, A., Todo, H., et al. (2018) Selection of phospholipids to design liposome preparations with high skin penetration-enhancing effects, *J Drug Deliv Sci Technol*, 44: p. 58–64.

Sangwan, Y., Hooda, T., Kumar, H. (2014) Nanoemulsions: A pharmaceutical review, *Int J Pharm Prof Res* 5: p. 1031–1038.

Scholz, P., Keck, C. M. (2015) Nanoemulsions produced by rotor-stator high speed stirring, *Int J Pharm*, 482(1–2): p. 110–117.

Schubert, R. (2010) Liposomen, in *Innovative Arzneiformen* (Mäder, K., Weidenauer, U., Allhenn, D., Eds.), WVG, Stuttgart: p. 162.

Schwarz, C., Mehnert, W., Lucks, J. S., et al. (1994) Solid lipid nanoparticles (SLN) for controlled drug deliv-ery. I. Production, characterization and sterilization, *J Control Release*, 30(1): p. 83–96.

Schwarz, J. C., Baisaeng, N., Hoppel, M., et al. (2013) Ultra-small NLC for improved dermal delivery of coe-nyzme Q10, *Int J Pharm*, 447(1–2): p. 213–217.

Shah, P., Bhalodia, D., Shelat, P. (2010) Nanoemulsion: A pharmaceutical review, *Syst Rev Pharm*, 1(1): p. 24–32.

Shakeel, F., Baboota, S., Ahuja, A., et al. (2007) Nanoemulsions as vehicles for transdermal delivery of aceclof-enac, *AAPS PharmSciTech*, 8(4): p. E104.

Shegokar, R., Müller, R. H. (2010) Nanocrystals: Industrially feasible multifunctional formulation technology for poorly soluble actives, *Int J Pharm*, 399(1–2): p. 129–139.

Shegokar, R., Singh, K. K., Müller, R. H. (2011) Production & stability of stavudine solid lipid nanoparticles – From lab to industrial scale, *Int J Pharm*, 416(2): p. 461–470.

Sinambela, P., Egorov, E., Löffler, B. M., et al. (2013) Human in vivo study: Dermal application of rutin smartCrystals® & peptide-loaded liposomes to decrease skin roughness, 40th Annual Meeting of the Controlled Release Society (CRS), Honolulu, Hawaii, 21–24 July.

Smijs, T. G., Pavel, S. (2011) Titanium dioxide and zinc oxide nanoparticles in sunscreens: Focus on their safety and effectiveness, *Nanotechnol Sci Appl*, 4: p. 95–112.

Solans, C., Izquierdo, P., Nolla, J., et al. (2005) Nano-emulsions, *Curr Opin Colloid Interface Sci*, 10(3–4): p. 102–110.

Somasundaran, P., Purohit, P. (2011) Polymer/surfactant interactions and nanostructures: Current development for cleansing, release, and deposition of actives, *J Cosmet Sci*, 62(2): p. 251–258.

Sonneville-Aubrun, O., Simonnet, J.-T., L'Alloret, F. (2004) Nanoemulsions: A new vehicle for skincare products, *Adv Colloid Interface Sci*, 108–109: p. 145–149.

Souto, E. B., Müller, R. H. (2008) Cosmetic features and applications of lipid nanoparticles (SLN, NLC), *Int J Cosmet Sci*, 30(3): p. 157–165.

Strambeanu, N., Demetrovici, L., Dragos, D. (2015) Natural sources of nanoparticles, in *Nanoparticles' Promises and Risks* (Lungu, M., Neculae, A., Bunoiu, M., et al., Eds.), Springer, Cham: p. 9–19.

Tadros, T., Izquierdo, P., Esquena, J., et al. (2004) Formation and stability of nano-emulsions, *Adv Colloid Interface Sci*, 108–109: p. 303–318.

Takada, H., Mimura, H., Xiao, L., et al. (2006) Innovative anti-oxidant: Fullerene (INCI #: 7587) is as "radical sponge" on the skin. Its high level of safety, stability and potential as premier anti-aging and whitening cosmetic ingredient, *Fullerenes Nanotubes Carbon Nanostruct*, 14(2–3): p. 335–341.

Tamilvanan, S. (2004) Oil-in-water lipid emulsions: Implications for parenteral and ocular delivering systems, *Prog Lipid Res*, 43(6): p. 489–533.

Taniguchi, N. (1974) On the basic concept of 'nano-technology'. Proceedings of the International Conference on Production Engineering, Japan Society of Precision Engineering. Tokyo, Japan: p. 18–23.

Teichmann, A., Jacobi, U., Ossadnik, M., et al. (2005) Differential stripping: Determination of the amount of topically applied substances penetrated into the hair follicles, *J Invest Dermatol*, 125(2): p. 264–269.

Teichmann, A., Otberg, N., Jacobi, U., et al. (2006) Follicular penetration: Development of a method to block the follicles selectively against the penetration of topically applied substances, *Skin Pharmacol Physiol*, 19(4): p. 216–223.

Teo, B. S. X., Basri, M., Zakaria, M. R. S., et al. (2010) A potential tocopherol acetate loaded palm oil esters-in-water nanoemulsions for nanocosmeceuticals, *J Nanobiotechnol*, 8: p. 4.

Touitou, E., Junginger, H. E., Weiner, N. D., et al. (1994) Liposomes as carriers for topical and transdermal delivery, *J Pharm Sci*, 83(9): p. 1189–1203.

Trotta, M., Morel, S., Gasco, M. R. (1997) Effect of oil phase composition on the skin permeation of felodipine from o/w microemulsions, *Pharmazie*, 52(1): p. 50–53.

Tunkam, P., Satirapipathkul, C. (2016) Preparation of nanoemulsion from sacha inchi oil/water by emulsion phase inversion methods, *Key Eng Mater*, 675–676: p. 57–60.

van Hoogevest, P., Prusseit, B., Wajda, R. (2013) Phospholipids: Natural functional ingredients and actives for cosmetic products, *SÖFW*, 139, 8-2013: p. 9–14.

Verma, D. D., Verma, S., Blume, G., et al. (2003) Particle size of liposomes influences dermal delivery of substances into skin, *Int J Pharm*, 258(1–2): p. 141–151.

Veronovski, N., Lešnik, M., Lubej, A., et al. (2014) Surface treated titanium dioxide nanoparticles as inorganic UV filters in sunscreen products, *Acta Chim Slov*, 61(3): p. 595–600.

Vidlářová, L., Romero, G. B., Hanuš, J., et al. (2016) Nanocrystals for dermal penetration enhancement – Effect of concentration and underlying mechanisms using curcumin as model, *Eur J Pharm Biopharm*, 104: p. 216–225.

Vogt, A., Combadiere, B., Hadam, S., et al. (2006) 40 nm, but not 750 or 1,500 nm, nanoparticles enter epidermal CD1a+ cells after transcutaneous application on human skin, *J Invest Dermatol*, 126(6): p. 1316–1322.

Vogt, A., Mandt, N., Lademann, J., et al. (2005) Follicular targeting – A promising tool in selective dermato-therapy, *J Invest Dermatol Symp Proc*, 10(3): p. 252–255.

Watanabe, A., Kajita, M., Kim, J., et al. (2009) In vitro free radical scavenging activity of platinum nanoparticles, *Nanotechnology*, 20(45): p. 455105.

Weiner, N., Lieb, L., Niemiec, S., et al. (1994) Liposomes: A novel topical delivery system for pharmaceutical and cosmetic applications, *J Drug Target*, 2(5): p. 405–410.

Wichit, A., Tangsumranjit, A., Pitaksuteepong, T., et al. (2012) Polymeric micelles of PEG-PE as carriers of all-trans retinoic acid for stability improvement, *AAPS PharmSciTech*, 13(1): p. 336–343.

Yaghmur, A., Aserin, A., Mizrahi, Y., et al. (1999) Argan oil-in-water emulsions. Preparation and stabilization, *J Am Oil Chem Soc*, 76(1): p. 15–18.

Yilmaz, E., Borchert, H.-H. (2006) Effect of lipid-containing, positively charged nanoemulsions on skin hydration, elasticity and erythema – An in vivo study, *Int J Pharm*, 307(2): p. 232–238.

Yoshihisa, Y., Honda, A., Zhao, Q.-L., et al. (2010) Protective effects of platinum nanoparticles against UV-light-induced epidermal inflammation, *Exp Dermatol*, 19(11): p. 1000–1006.

Yukuyama, M. N., Ghisleni, D. D. M., Pinto, T. J. A., et al. (2016) Nanoemulsion: Process selection and application in cosmetics – A review, *Int J Cosmet Sci*, 38(1): p. 13–24.

Zastrow, L., Ferrero, L., Herrling, T., et al. (2004) Integrated sun protection factor: A new sun protection factor based on free radicals generated by UV irradiation, *Skin Pharmacol Physiol*, 17(5): p. 219–231.

Zastrow, L., Groth, N., Klein, F., et al. (2009a) The missing link – Light-induced (280-1,600 nm) free radical formation in human skin, *Skin Pharmacol Physiol*, 22(1): p. 31–44.

Zastrow, L., Groth, N., Klein, F., et al. (2009b) UV, sichtbares Licht, Infrarot. Welche Wellenlängen produzieren oxidativen Stress in menschlicher Haut? *Hautarzt*, 60(4): p. 310–317.

Zastrow, L., Groth, N., Klein, F., et al. (2009c) Detection and identification of free radicals generated by UV and visible light in ex vivo human skin, *Int J Cosmet Sci*, 31(5): p. 402.

Zeynalov, E. B., Allen, N. S., Salmanova, N. I. (2009) Radical scavenging efficiency of different fullerenes C60–C70 and fullerene soot, *Polym Degrad Stab*, 94(8): p. 1183–1189.

Zhou, Y.-T., He, W., Wamer, W. G., et al. (2013) Enzyme-mimetic effects of gold and platinum nanorods on the antioxidant activity of ascorbic acid, *Nanoscale*, 5(4): p. 1583–1591.

21

From Formulation Design to Production: The Scale-Up Process

Margaret Smith

CONTENTS

Introduction

Scaling up has been defined as 'to increase the size, amount, or importance of something, usually an organization or process'. If only scaling up was a straightforward linear process. Unfortunately it isn't; well certainly not in my experience. There are many pitfalls for the unwary and narrow thinkers. Thinking broadly about scaling up will pay dividends for manufacturer and brand owner alike. In this chapter I can only comment on the world according to Marg, and my world consists of incrementally creating a cosmetic manufacturing facility that produces what our current range of customers want.

Scaling up, to me, takes in the entire world of the production of a cosmetic product from concept to formulation to final manufacture. I am leaving the other aspects of 'scaling up' like selling and distribution to the brand owner, although we do assist in some aspects and checklists. A concept, in a product designer or marketer's mind, has a long journey before being on the shelf of someone's bathroom. Scaling up is simply 'commercialisation', that is, making an idea into a reality, so the product can safely and proudly be sitting in a bathroom, being used happily by a customer who wishes to re-purchase.

To some, scaling up may mean just manufacturing large quantities from a kitchen-size production or a lab sample, yet to me scaling up covers a huge number of functions that need at least some consideration. I have found that if one tries at least to understand that there is a very long list of points to tick off then there is much less of a chance of anything going wrong with the scaling-up process.

Way before Donald Rumsfeld (a former United States Secretary of Defense) infamously declared about unknown unknowns or such, I studied aeronautical engineering and architecture. These fields applied this same concept of analysis (the Johari window) and used it to enhance the safety of the products they developed. Indeed, I learned safety as a second skin. Aeronautical engineers and architects take safety to very high standards. That is my background and as such I never presume what anyone knows or more important what they do not know. The unknown unknowns are usually vastly greater than the known knowns. This is very true indeed of all customers to come to grips with. And, if the manufacturer can have some control over at least part of the development process, that can reduce some of the pressure on commercialisation of a product. A little bit of caution and careful planning may save a crash and burn!

If there is one rule a brand owner customer needs to learn and follow, both in the early days and beyond, it is that a successful product needs time to be developed. A brand owner needs to be patient and considerate. Considerate in the sense that they think of every step, talk and work with their manufacturer to plan and work out what steps are absolutely necessary.

We, as manufacturers, are often (too often) asked if we can copy (insert product name) only to receive an order almost immediately we can give a price. We warn of the pitfalls to the customer without proper process of stability in the pack, and so forth, and yet we are more often than not dismissed. At this point we are very clear on the need for a disclaimer being provided and signed. If the customer has not done their due diligence on the product, then there will be problems somewhere along the line. I am sure of it.

The art of commercialisation is a very broad topic and far too diverse for any one manufacturer to fully flesh out. I shall make an attempt to cover as many aspects and scenarios as I can. None will be highly detailed, simply because every aspect of commercialisation is incredibly individual and my method will differ from others. However, as an overview for anyone new in the industry, this chapter may answer some of the questions that are puzzling you or, better still, may introduce you to a world that you did not know existed.

My Experience

My company began in the 1980s developing and manufacturing syndet bars from scratch. It was a bit of an adventure for me. My late husband had started a manufacturing business making and retailing bulk 'take home in your own container' home and personal care products. That business turned into 100 franchise retail stores until he became a bit sick of it and decided to semi-retire. That was when we began a small company making syndet bars. As a crazy addition, we also took on an old abandoned soap plant in Papua New Guinea. In doing so we were both embarking on projects that neither of us had done before.

Soap making on an industrial scale in a relatively third world country (especially near Rabaul, destroyed by volcanic eruptions early on in the venture) was not something that many people had any knowledge of in Australia, and certainly no one had any idea of in Papua New Guinea. Yet the whole idea of getting it back up and running was very appealing. A year of some very hard graft cleaning all the equipment and lines of solid soap and rust became my basic training for understanding how the scattered bits of machinery fitted together and then how they all worked (sometimes and hopefully).[*]

Running both companies lasted until the early 2000s, when I left Papua New Guinea to concentrate on the Australian factory. The syndet bar production was building numbers strongly even against the tide of washes in blasted plastic bottles (this was the main reason for a solid bar venture … the green revolution that never came), and now loyal customers were asking for more types of products to be made, such as creams, shampoos and other hair products. It seemed that my previous naivety had not taught me any lessons. But as is said 'no pain no gain'. This is where the unknown unknowns became a challenge, and some of what I write here are the steps I learnt from experience, some hard and bitter and some absolutely wonderful.

Slowly over the last 20 years we have gone through all sorts of equipment, from the gas burner under a 44-gallon bin to steam and chilled water jacketed tanks. From hand cartoning to operating three auto-cartoners. Yet it matters not how much new automated equipment one amasses – the next customer will have found a container and formula that requires a multitude of low-tech labour to get it right. If the future product numbers are there, we will invent the next piece of equipment. So far, we have found and used the best equipment that our budget will allow. And of course, everything has been rebuilt or modified to cater to as many different types of products as possible.

My experience in the production of skin care and cosmetics has always been as a contract manufacturer working for a brand owner. Basically, we end up doing the work a brand owner/manufacturer chooses not to do. All brand owners (customers) are different and take on various parts of the project to

---

[*] My husband never saw the first bar of soap, dying unexpectedly early in the adventure.

commercialise a product. To begin the scaling up or commercialisation of an idea or a product, one needs to first examine the two important players in the process, namely the manufacturer, who will be doing most of the work, and the customer or brand owner. It takes time, goodwill and excellent communication to develop a great partnership between the two types of organisation and their individual strategic and, sometimes, quirky needs.

A brand owner who manufactures and packages their own products in-house can have the simplest setup, or indeed the most exclusive and complex. This brand owner can choose a finite range of formulations and packaging types that help limit the amount of equipment needed to blend and fill. Think of a simple brand just selling shampoos and body washes. So simple? They may have a few tanks that need little or no heating, and a simple source of clean water. They may have a very basic filling line, maybe even without labelling as preprinted bottles are easy to obtain. And, most likely, no unit cartoning.

Ah that's the life. The brand owner can have a just-in-time system making enough for what they have as orders, with little warehousing. Their facility can be small in footprint. Even a shed on the property. I have seen these brands and their life is sweet, that is, until they become popular and the demand for additional products from retailers must be met.

The small brand with just a wash and then a shampoo then needs a conditioner, and the flow on from that additional product is a compounding setup that needs heat. Their filling line remains simple, but if the brand grows and gains more outlets, then the filling line needs expanding and more warehousing space is required.

All is still fine, and the good news comes in that their brand is now so popular that it has become noticed worldwide, say in Europe and then in Asia. So much excitement and possibility. But just-in-time for manufacturing is no longer workable. The brand needs stocks that can be drawn off at any time, and sometimes flown in a hurry to the other side of the world. There are masses of new regulations to follow and fulfil, and relationships to create and maintain. All these new opportunities and countries need approvals and Product Information Packages (PIPs). It is discovered, sometimes too late, that the new country bans some of the ingredients in the current product, and it will need to be reformulated and repackaged, and at some point, it sometimes gets too much for the lone brand owner.

With each new product and new market, the facilities and the staff need upgrading and adding to. And sometimes so too does the brand and then the brand owner. Often the brand owner, like a snake in spring, needs to shed their old skin and move or build a new facility. Alternatively they may business plan their expanding operations and just warehouse, and start to talk to contract manufacturers and formulators and consultants to outsource the entire development and manufacture of the product. The brand then concentrates firmly on their style and marketing to make it bigger than Ben Hur to cover all the additional costs. This is much more the normal route for growing brands. All part of 'scaling up'.

Thirty years ago in my experience pretty much all the brands that were on shelf had their own manufacturing factory. These days it is the norm to outsource all the parts of the development and manufacture of cosmetic products from factories anywhere in the world that is cheap and convenient or has the right stuff. From the mid-eighties and onwards the containerisation of trade and goods has meant that the ends of the earth no longer exist. Everywhere was in the neighbourhood, no more than now. Today we can buy almost anything we like, from anywhere we like, and have it delivered in days. We have no idea where it came from in the world and what its journey has been, and to many they just do not care. Anywhere with a decent port and willing workforce (and low taxes) is open for business. Even Australia. However, in Australia's case at the time, most factories and international brands, which depended on local manufacture and distribution, went off its shores years ago, not the other way around. However, things on that front have changed radically in the last few years.

When a brand owner makes the decision to outsource, their own business must change as well. They are no longer totally in control. They become someone's customer. And it means that the brand owner needs to create a very well-versed department to assess and properly partner with, possibly, many different contract manufacturers or even types of manufacturers. The brand owner's new team must still understand manufacturing, not just quality assurance and marketing, to make the best of these new relationships. Once a customer, the brand owner needs to choose the manufacturing type that suits their organisation and product.

Over the years I have been active in outsourced manufacturing, the types and definitions have changed, and sometimes the acronyms my organisation uses are described as being a foreign language. Generally, at the time of writing, these types of manufacturers are the usual:

1. A *toll manufacturer* provides the plant, machinery and labour force to manufacture. The customer generally must provide the final formula, all the packaging and the raw materials. The customer holds ownership of all the intellectual assets (i.e. intellectual property) such as formulas and patents and designs, and assumes all selling risks.

2. A simple *contract manufacturer* still owns the plant and machinery and provides the labour to operate the machinery – similar to a toll manufacturer. The customer can also access the manufacturer's resources to source and supply the materials, sometimes including the packaging to manufacture the products. Again, the customer ultimately assumes the risk of selling the goods.

3. A *CMO* (*contract manufacturing organization*) or a *full-fledge manufacturer* is the most involved of the three organization types. This type of company is responsible for providing the space, machinery and team to manufacture the products as well as all materials and goods necessary for production. *Wikipedia* makes the definition very clear: 'Services offered by CMOs include, but are not limited to: pre-formulation, formulation development, stability studies, method development, formal stability, scale-up, registration batches and commercial production. CMOs are contract manufacturers, but they can also be more than that because of the development aspect.' The customer is very well serviced by this sort of manufacturer but still ultimately assumes the risk of selling the goods.

There is a further differentiation with CMOs with regard to intellectual property (IP; the formula and method in this case), and this is very important to a brand, as the IP is everything when it comes to selling the brand on to a new owner. A new owner may wish to move their manufacturing elsewhere. CMOs (as a distinct subset) are those manufacturers who can assist in commercialisation or scaling up for those entities that do not have clear formulas and packaging in place. There are two more acronyms to consider:

1. *OEM*, or *original equipment manufacturer*, is a term more commonly associated with IT companies. It means that this type of manufacturer has a library of current product type/formulas that a brand owner can choose from and re-label as their own. It is also known as a 'white label' manufacturer. The brand owner will never own the intellectual property, as it will be shared amongst many customers. The brand owner still takes the responsibility of sales.

2. An *ODM*, or *original design manufacturer*, can tailor-make a product for the brand owner. The brand owner has the opportunity to also own the IP and work with the CMO to assist them with much more resources than they have themselves. In return the manufacturer may want a minimum number of units or time before the IP is totally handed over to the brand owner or could negotiate a royalty. The same responsibilities lie on the brand owner in sales.

The brand owner is now a customer of a manufacturer of one or another type; however, I shall continue the journey of scaling up or commercialisation along the route of the ODM, which what is I know and is who I am.

A CMO with formulation capabilities really has two major goals; one is to actually make money out of a partnership. The other can be personal; mine is a professional and formulating challenge. With that the customer has to be worthwhile in respect to not just the ability to pay, but to some charm both in ideas and personality. A new or even current customer does not need to live in the same country as the CMO and, like brand owners, CMOs have the advantage of being available anywhere to serve any customer in the world. Even 10 years ago I did not think we would have over 20% of customers who reside permanently in another country without having a branch of their operations in my country. This just shows how flexible and open one must be to constant change. And the world has changed a great deal in the last decade.

Contract manufacturers need to learn to never presume what a customer knows or, more important, what they do not know. The unknown unknowns are usually vastly greater than the known knowns. The CMO needs to be able to juggle between being helpful and being nosey when interviews are made by potential customers. It is not a one-way exchange. The CMO needs to know they can assist and do the work for a customer who needs to be very clear in communication, and who has expectations or knowledge that allow a productive dialogue, and who will meet the CMO's primary expectation of being paid on time.

A successful CMO with design and formulation capability must also try to be a few concepts ahead of the product designers. At least be able to recognise and understand what the current trends are and how to achieve them. The CMO needs to have their raw material availability worked out so the new wave of material types and regulations does not become an all-engulfing tsunami.

Experiences of the food industry are always a good go-to for information. There are far more experienced food industry consultants and workers than cosmetics, and it is a very comparable industry. It is driven by fashion and is strictly regulated. Food also has a finite shelf life, which is a good lesson to learn and apply to cosmetics. Especially as these days the trend towards natural and organic cosmetics is gaining strength, and the similarities between cosmetics and skin care seems to be getting closer.

For customers and all senior cosmetic staff, reading all the trade magazines and trying to get to a few cosmetic trade shows is just the start. Being Australia based usually means travel to faraway places. After a few trade shows it can become tedious walking amongst the same old stuff. One needs to keep ones eyes open for maybe a couple of little sparkling gems here and there – to see the possibilities that new ingredients may offer. Or it could be a way of finding a better supplier or material, or something profoundly illuminating and unexpected. I find shopping and being a girl tremendously advantageous in this industry.

Brand owners who manufacture their own products, whilst their output is small or their range narrow, can concentrate on doing a few things simply and very well. They need only move towards outsourcing to contractors if something they make proves to be beyond their current capabilities or takes up too much space.

Many CMOs tend to build their operation to become quite large businesses. Their size allows them to cater to different market segments so they can keep most operations in-house – in effect often becoming something of a one-stop shop for the brand owner. Some manufacturers specialise and focus on making similar product types, like bars or simple washes or creams. Others cater to segments such as 'certified organic' or 'natural formulations'. Some target a handful of brand supersizes (those brands with very large volumes and low prices) in the market.

Our facility at Syndet Works caters to a diverse range of customer types and run sizes. We target the prestige end of the market, so run sizes vary greatly but tend to remain below 50,000 units per run. Size does matter and it matters most when scaling up. There is a sweet spot when one hits the right number of units to make it all profitable for all partners. It varies for every customer and product. It seems crazy at times with the wide scope of customers, I am the first to admit; yet, overall, after careful growth, we still have great customers that we have grown with, some since we first began manufacturing and they had their first idea.

To start the complete process of scaling up or commercialising a product within the manufacturing environment, here is a checklist. I shall describe a few parts of this in more detail. In practice we have what we call a 'Good Manufacturing Contract or Process', where a very long table of responsibilities are drawn up, with who needs to do what at each step. When the team of customer and manufacturer go through each step and decide on their jobs, it is very clarifying. The Appendix contains a simple version.

Very Basic Steps in Commercialisation Responsibilities

Customer Responsibilities (May Be Able to Get Assistance from Manufacturer, Consultant or Authorities)

- Supplying the manufacturer with a detailed Product Development Brief including target cost, ingredients claims sought, approximate selling price and intended markets (may affect formulation)
- Marketing opportunities and claims being made (need to ensure brief to manufacturer covers this for formulation reasons and to meet regulations)
- Marketing and distribution channels

- Detailed stability and human safety tests (if required by customer)
- Shipping trials (test shipping of finished goods to see that they arrive in suitable condition)
- Packaging trials (obtaining examples of intended packaging for trialling on manufacturer's equipment to see that it's OK as well as ensuring that packaging is suitable – for example, can a tube hold the required quantity of product, is carton suitable size for tube and so on)
- Regulatory compliance and patent checks (making claims that are illegal or unsubstantiated, confirming no patent or trademark infringements in product or packaging artwork)
- Permits and registration applications
- Insurance – product liability and transportation insurance at the least
- The formula (if the customer is supplying)
- Tweaks to the formula to meet brief (providing feedback on samples to enable manufacturer/formulator to be guided on changes sought)
- Concurrently the package needs its suitability assessed for both the formula and the distribution channels and the filling lines available
- Providing objective feedback in preference to subjective
- Packaging for production supplied in a timely fashion

Manufacturer Responsibilities

- The cost of the formula and filling able to meet customer's target cost
- Materials, availability and susceptibility to price changes
- Product evaluation and feedback forms (especially if samples are being supplied for assessment – helps to establish a framework for objective rather than subjective reviewing)
- Obtaining sign-off from customer of sample approval that will become the 'master sample'
- Ensuring that all communications are properly documented to avoid miscommunications
- Retention of all manufacturing documents including variations
- Pilot lab to manufacture batches
- Packaging for production scale-up tested for compliance with quality standards (e.g. cleanliness) and consistency with earlier test packaging
- Key people present when scale-up batches are being made to ensure that things go correctly and minor formulation changes and tweaks documented
- Comparison of manufactured batches (pilot or scaled up) to reference samples to ensure product within tolerances
- Revisions (to specifications and batch documents) if required
- Final batch
- Filling to standards required
- Shipping to brand owner (unless supplied on an ex-works basis)

Back to the Customer

- Brand owner shipping to distributor
- Distributor shipping to final user

Example 1: Kitchen Bench to Professionally Manufactured

Every week we get queries from home creators of what we call 'kitchen bench' cosmetics. It is not meant to be derogatory, just factual, and leads to understanding where this type of customer is coming from. We may need to hand-hold through some of the processes. These callers are asking for quotes for quantities larger than they can make at home. Or they may have hit either a regulatory roadblock or one of supply. Possibly their partner is sick of the mess in the house ... or all of the above.

Sometimes they are already in a distribution chain or link, and are selling either in a bricks-and-mortar shop or online. Some have not begun and dream to be a brand owner of a cosmetics range. They have a passion. They often have done an online course in cosmetics and are raring to go. Usually the creator of a formula and brand has still a long way to go before getting to the point of scaling up.

It takes a bit of interviewing, at first on the phone, to get some important information from them, but slowly, with time, the enquirer starts to understand that getting an idea or bottle of skin care product to a market is much more than simply asking for a quote for 500–5000 units.

Many times, they have a successful formula or recipe that works very well in small batches (about a kilo) but, unfortunately, has never been trialled at larger scale. Once tried they find the formula and method will not work without modification in batches higher than 20 kilos. Then they may find it cannot be dispensed through commercial equipment. For some CMOs small manufacturing numbers are possible, yet the high labour rates of even the smallest manufacturers labour will mean quite a high per unit cost. This type of scale up needs to be very carefully handled with perhaps reformulation to suit the contractors' equipment or filling into new packaging. At the same time the brand and the customer does need to try to build up numbers (scale up themselves) one way or another to allow economies of scale to manufacture at the right price – that sweet spot I mentioned earlier.

Elements that can create a real problem when scaling up in the production sense have been with availability or consistency of unusual materials used by the home creator (e.g. waste coffee grinds from cafes being different in strength each time and thus batches needing bespoke attention), or with the very nature of the materials and their compatibility with process equipment (e.g. large size salt crystals in a scrub that cannot be processed without breakage in commercial equipment and need to be mixed and filled by hand) or specially created herbal mixes. This type of scale up may not be achieved easily or economically until a reformulation is done or specialised equipment sought.

If a product is developed in-house by a contract manufacturer, the formulation is likely to be the easiest of all for that manufacturer to scale up for compounding, filling and packing, simply because the formulator should be very well versed in the types of equipment they use in full production and can replicate these in the lab. Or at least know the limitations and make some adjustments from lab to tank. The one thing I consistently see as a fault of formulators, both amateur and professional, is the understanding of rheology of the product and the capabilities of the filling machine that will be employed in the job. Drippy or stringy formulations will be extremely problematic for the process engineer in the factory. The experienced in-house formulator will steer far away from formulas that exhibit a strange or amateurish rheology, and try to form a product that is simple to both compound and dispense.

Example 2: The Moveable Product

A brand owner may transfer production from one manufacturer to another, sometimes regularly, and this occurs when a customer or manufacturer parts ways for one of many reasons. Mainly it is because the quantities become larger and the former manufacturer cannot achieve the quantities within the given timelines. Possibly it is because the customer has changed packaging type and they seek a manufacturer who can fulfil the new line. Often they are looking for a better price.

It has to be said that the brand owner really needs to truly understand manufacturing and their specific product to appreciate if they have a good price at the expense of good service and reliability. The new manufacturer needs to be wary as well because customers who start singing to this tune can be very unreliable indeed, and all new customers are lots of work for no reward, and a CMO needs as much stability and reliability as possible.

We had a customer move from their previous manufacturer 'because they could never produce their product on time'. They attempted (unsuccessfully) to insert a clause in our agreement in which we gave a 15% discount every time we did not meet their deadlines. All that sounded sort of fair, until one began working for them and every order was made 5 months in advance, which still sounds fair, yet when it came time for us to receive the packaging so we could produce, the packaging was 'still in transit', their ordered numbers changed from say 20,000 to 5,000, so we were left with materials and labour, or the order was repealed. Their ability to reply to an email or phone call was pretty non-existent. We found that the real reason they left their former manufacturer was that this customer was truly one of the most appalling customers and payers that I have had the sad experience of being party to.

Another example is the customer who 'thinks' they own the formula because they have a product and a piece of paper with ingredients and percentages against them. Until a customer-supplied formula and method of manufacture has been validated properly to confirm it makes what the customer expects, it is nothing worth getting excited about. If the supplied formula produces a suitable stable product, then proceeding to manufacture may involve production of a small pilot batch on new equipment before starting normal sized batches.

Anomalies, or outright fake formulas, occur when the customer has not paid for the formula or the relations between customer and manufacturer have broken down, and the formula is not an actual true formula and method that makes the physical product. In my experience this happens more often than you can imagine. Matching an existing product to a questionable formulation can put the new manufacturer and the customer back to square one. There can be a lot of additional angst if the formulator needs to match to sample, which doesn't correspond to the International Nomenclature of Cosmetic Ingredients (INCI) on the ingredient label or the PIP or similar registration documents. In situations like this one needs to very clearly discuss whether the INCI and ingredient list and registrations change, and all the associated costs taken into account and even maybe claims, or does the end user next buy a product that is fundamentally different to what they have bought for years?

A manufacturer may be able to assist and yet the customer may continue to be lax and frugal and may really well be part of the overall fraud. The manufacturer may get a feeling for the bona fides of each customer and, really, I think it is best to run a mile in the opposite direction when confronted by these scenarios. Too many times have I tried to assist, only to end up in tears.

So far we have the right kind of manufacturer, the best customer and now the formulation is a concept or even in a pot on the bench. The journey has just begun along the list of dot points.

Simple Practicalities in Scaling-Up Production

Stability from Laboratory Sample or Pilot Batch

- Measure Twice, Cut Once! It's an Old Saying But True

I prefer to do a small hot and freeze stability trial of a first-up sample to determine what needs adjusting before handing over to a customer as a sample, and way before producing a small pilot-scale batch, using a smaller version of the batching room equipment, on which to do all stability trials. If the customer is 'in a hurry' and must have it the same week or two weeks after meeting, there will certainly be tears at some point. The CMO cannot do all the tests with their meagre resources, and the brand owner has to be on board and agree to do the above 'measure twice' and pay for it. Better for the customer to pay for a small batch and test it to its limits than to have all the production fail. But then, as I have said, it needs the brand owner to be on board.

- Does My Bum Look Big in This? The Suitability of Final Primary Packaging

Packs should be part of the briefing process, no question, unless the formulator has an idea of what will work and suggests only a single type of pack. Then the pack needs a formula to suit. Someone in the chain needs to take a back seat. However, brand owners and designers must work with their manufacturer as well. It needs to be a mutual discussion about how to get the best out of a formula and within budget. A CMO may not be able to fill bottles or a tube of unusual dimension and shape. A CMO will always be able to advise the best solution or be prepared to upgrade their equipment if a case is made. Then the brand owner can choose where to go to get what they need. CMO manufacturers also know the best packaging suppliers; they use various types every hour of the day. Manufacturers know the practicalities and what is practical ends up looking fantastic.

- *"There is nothing stable in the world; uproar's your only music" (John Keats 1818).* Testing and stability – the truth

Stability tests in temperatures and humidity and light conditions are what we all know and understand. Yet they are just the beginning of a real stability trial. Preservative efficacy is essential, and use in the real world by real people using the product instinctively is even more important. Use tests, for example, when one tells the customer the product is a cleanser and watches how they actually use it, will inform you on how to write your instructions and actually see if the pack that is preferred by marketing actually works. I have seen beloved packs and pumps fail miserably. Development of a use and torture protocol for each type of product is really needed to anticipate all the slings and arrows that may come the way of the product. We are finding more and more that a filled product in all the correct packaging needs shipping trials of various types to be part of the overall stability trials. I have seen more packaging/product failures during this process than any other.

A product that can survive winter or withstand an extremely hot shipping container may be a candidate. Couple those conditions with a bit of rough handling (to account for in-transit abuse) and, if the product (and packaging) survive, you may have a winner or at least one to be an online product contender. Airfreighted products often sit in the unpressurised hull of a plane at −30°C. Under these conditions, it is common for products to blow up and leak under the decreased air pressure. At the other end of the journey it's possible for products to get thrown around again all over miles of conveyor belts, only to be eaten by the neighbour's dog at the base of the letter box. When it comes to stability tests I believe in both the conventional and the slightly off centre.

I sometimes carry a little handbag (not my proper handbag as I know what horrors are likely to befall it) in my car with things like hand creams, skin fresheners, lipstick and mascara, just to see how they fare … not well most of the time. Cars are stressful places for cosmetics and can get even hotter than shipping containers. And the direct piercing sunlight can do some indecent damage to emulsions and waxes. My car mechanic thinks parts of my leather seats and steering wheel are the softest and best leather that he has seen anywhere. The car is pretty old and it gets my spilt emulsions and oils all over the seats. My hand cream is handy so I can top up during the day, and this has turned out to be a way that I know my basic hand cream formulas are doing their job! I do not have a dog so the dog test is one I cannot complete.

We have seen packaging arrive in a seriously heat affected state, packaging that simply can't be used. It's one reason we ask for the packaging to arrive well ahead of the intended fill date. The condition that some packaging arrives in really demonstrates the temperature of the container as it crosses the hot parts of the world or sits on the tarmac or hardstand. I routinely advise that if a brand is sending a full container to a distributor and wants the best delivery, then use a refrigerated (reefer) container. If the container is not temperature controlled, include a maximum/minimum thermometer within it to see just what trials their products go through.

For a simple shipping trial before one decides on a final package, at the very minimum, we advise a dozen filled units in the proposed packaging or a selection of packaging to be posted by air and rail and truck to travel from us to somewhere in the boondocks and back. Fascinating results really informs us all about what the next step in choosing the right package is. Again, a small digital temperature recorder can be very helpful.

Just a note here about stability and heat and cold. I cannot impress upon our customers more forcefully than to be aware of their transport and distributors conditions before they proceed with any new products or markets. Brands and marketing managers need to understand completely how important a stability test is and how it works and what is shelf life.

First, shelf life is not what the retailer wants (they always want a ridiculous number of years). Shelf life is a true estimate of the product life that is derived from the results of a rigorous stability test/trial in the package. Part of that trial should attempt to emulate how the product will be treated during its life by a normal (or abnormal) user.

When a product is given a controlled stability test, say for a tropical climate, it is tested in a high humidity chamber at 40–50°C. If it gets through about 3–4 months without falling apart then all is good … or is it? We have tested shipping containers at greater than 60°C during the day for many months in a car park. A month or even less of that type of treatment and there is none of the anticipated shelf life left. Whatever was planned and tested for has gone. That is if the packaging has not disintegrated before the contents.

Let's be optimistic and assume everything has passed and is looking good along the journey to check the stability of the product. We know the package is right and everything survives the most dreadful of postal services. We are firmly in the goldilocks zone of 'just right'. The manufacturer now proceeds from laboratory scale to production scale. Some of the aspects, just some, that the manufacturer needs to contend with are detailed.

- Which Tanks and Mixing Machines?

The setup of a compounding area will differ immensely from one facility to another. The CMO will have their preferred way of doing everything and there is no right way. The setup and the equipment will suit the manufacturer, and the types of products and packages that their history of customers, and their budget, has determined. It is usually a slow and steady accumulation of machinery and expertise. Yet there are the lucky few that begin with considerable funding and can start fresh. Even so, I think familiarity is a friend to the manufacturer and their staff, and equipment that is friendly, simple and very accustomed to, is likely to be chosen.

If a small-scale trial batch rather than laboratory samples has not been done, best that it is done now. I have never found a sure method that fits all washes, emulsions and pastes that determines the processes and quantities from the lab to the tank. However, after a few years one gets the hang of what works in their plant. If a CMO has sufficient funds, they may be able to outfit their lab with an entire small-scale plant. I cannot think of anything more wonderful, however, I have never had the benefit of this luxury and I am old school so I enjoy the internal and gut calculations needed to work out the final quantities.

Do not be ashamed of starting with 'slowly, slowly' in the addition of gums and emulsifiers for the first-ever batch. There are little tricks of the trade; for example, if you are using a beaker on a hot plate, the need will be for a greater quantity of gum mixing for a much longer period of time than a steam jacketed tank with both contra-rotating scrapers and propellers and inline emulsification. Just try a few experiments, and if you are a CMO, stretch yourself to have tanks of various sizes that have similar properties, then small batches can be simple. We have a range of tanks from 70 to 5000 litres all with similar formats. This makes us pretty sure that if we make something in a little tank, then only time needs to be factored in for a big tank to be OK with larger quantities.

If you are new to the game or do not have technical people with old instincts (note I did not say old staff!), then the way you formulate and the choice of materials is going to be your best friend. Rather than sticking with familiar-style emulsions, with waxes in the oil phase and gums in the water phase, consider other formulating options. For example, there are now several cold process emulsifiers that can make one's life very simple indeed. Or even consider formulating with the last step being the simple addition of a polymer at the end of a batch to reach the viscosity required. It can make life very sweet, but needs planning and thinking about right at the very beginning of a project.

- Will It Go Through the Pumps and Similar Considerations

This is reasonably straightforward. One needs pumps and inline emulsifiers for all the transferring of products from the mixing tank to the filling machine to the package. The type of pump needs consideration regardless of the product. A wash does not want to become full of air and foam up. Neither does a light lotion. This makes it impossible to get the right amount of product into the package. A paste may need to be warmed and put through heated piping and pump heads.

Large granules in the formula may require the product to be hand filled rather than pumped. Be careful of various types of scrubs going through pumps, as they can destroy everything. We have been asked to put sand through in a scrub … a great idea for the ocean and just disaster for the valve. Types of valves, grommets and seals need super selection here and again; prevention at the formulation stage is better than trying to change seals because the oils in the emulsion are corroding them mid-fill.

- Reference Samples and the Necessity to Have Pre-Production Sample Sign-Offs

The only sample to match to is the customer-approved sample and, no matter what, the final production product must look and feel like that sample. Once the first couple of productions are done and complete, and everyone is happy, then a master is kept. We match to that master like it was gold. It is what the final product specification is formed around. It will be a true picture of what a batch will be every time. We keep all customer-approved samples and masters in cool fridges. Not cold, just cool. And they may need changing over periodically.

Sometimes, and if possible always, before filling it is best to have the brand owner, especially if one is newish to the business, come in and test the product and sign off an approval. This can be bothersome and difficult, especially if the customer is some distance away from the manufacturer, or the product needs time to build or develop, so a sign-off immediately is unwise. However, it can make a customer, especially new ones, very involved and relaxed. A bit like 'happy wife, happy life'; I must try to think of a rhyming catch line for 'happy customer'.

- Accuracy of Measuring Equipment

Again, measure twice and cry less; slowly, very slowly, add things and make triply sure that all the measuring equipment is checked and calibrated and checked and calibrated again. We had a big set of scales go almost imperceptibly out of correctness over 2 months before it became obvious … oh, mercy me.

- Proper Documentation of Processes

No CMO or brand owner can now avoid the documentation that is required for the manufacture of even a simple lotion. Even though it begins with groans and moans from everyone in the plant who has never picked up a pen, I have to say that proper and even OCD documentation can be the greatest gift.

When I got my first stability oven duo (hot, cold) I went for it big time and, I admit, I still do to a smaller extent. Absolutely everything we made over this period has been tested in both hot, cold and freezer for anything between 3 and 12 months just because I needed to understand what would happen. And I have found the results both illuminating and curious over the years. The knowledge of all the materials and formulas evaluated over those years could fill many volumes (and I do not intend to bore anyone with that), but it has revealed that, however much I was assured by suppliers, some things just did not work the way they promised. I have to say that yes, I can be a bit focussed, but it can also add interest to a day when next sitting with the supplier with the 'super bulletproof' material that I just shot holes through over the last few months. Which leads me onto the last gasp of what the manufacturer needs to tie up …

- Materials: Pitfalls to be Aware Of

It looks like it's the same, but it's from a different manufacturer. The INCI is the same; the specifications are slightly different, but they have not tested the same things anyway so it's hard to compare. They look similar, and so on.

The only way to handle this is to try a lab sample. Make it with the 'old' stuff then make it again with the 'new' stuff. Then compare and choose. It is the only way.

- Seasonal Variations

The bane of the natural and organic formula and product is one season it's pale and the next it is lurid. And all the supplier can say is 'It's natural!' and that's all we can say. If you choose natural, then the variations of one batch and season to another must be worn with pride. I remember the words of Jim Thompson (businessman and fabric fashion guru of the 1950s and 1960s) on each bolt of Thai silk, which he popularised. He made sure that it was clear to the purchaser that they were buying a natural, hand-spun fabric and that all the weaving imperfections just highlighted its beauty. The Japanese also have a word for it in *wabi*. Just fresh simple and natural. It must be marketed and understood by the brand owner. We have had brands wanting vegan, organic, natural. They really wanted white, unchanging and basically synthetic. They were highly disturbed that plant extracts had colour and that vegetable oils smelt like, well, vegetable oil. What they wanted was a place that represented this style, but who they were really was not on that planet.

Lesson: be who you are. Stick to your own values. Be proud of what you are.

• New Stock versus Old

Not worth considering it. Old stock, whilst fabulous in good red wine and Chinese porcelain and of course people and pussycats, is literally on the nose with materials for cosmetics. We are asked by customers to insure at 3 years on products we make. Yet suppliers of most raw materials provide shelf lives of 1 to 2 years. Go figure.

• Storage of Materials Before Use and After Being Opened

One day I want a cold storage room for all the materials that come through my door, but not until I can get solar panels with a government grant to pay the huge costs for power. Just keeping materials and products cool is the best way. It does not need to be cold, just 12–15°C is ideal; stability in temperature is all one needs.

These are just a few of the details in the woozy world of commercialising or scaling up your idea into a real – on the shelf and being used by actual paying humans – skin care product. If there is one thing I can impress upon any customer and manufacturer is work together. Together you are strong. And then plan, plan, plan. See the GMP agreement contract in the Appendix at the end of this chapter.

My world was shaken only a little when a customer who had spent a lot of money on planning and their consultants presented us with their very attractive, long and detailed Gantt chart. Planned in detail. It spanned a year of their process, this was their map of getting from here to there, so they thought. Not once was the formulation development ever mentioned. Worse still there was no time allocated to even manufacturing the product, just a few months in ordering packaging, and then with a whoosh of Mr. Gantt's magic chart, the product appeared in the retailer's warehouse!

No matter what side you think you are on, whether it is design or formulation or just filling, the whole process of scaling up needs lots of cooperation and discussion and planning to get the best out of any product. If you feel like you are going around in circles, trust me you are and do not feel ashamed. Scaling up is not linear, it just isn't.

And try to have fun.

Appendix: Typical GMP Agreement Contract

1. Subject of GMP Agreement
    a. Scope
        This Agreement relates to the Contract Manufacturing, Contract Manufacturing and Packaging or Analytical Testing services or provision of Service(s) as specified in the List of Products in Attachment 1 for The Client (Client) at the site nominated in section 1.2 of this Agreement.
    b. Contract Acceptor
        Contract Acceptor's site for Manufacturing/Packaging/Testing Operations:
        Company Site Name and Full address: _____
        Manufacturing License Reference Number (if applicable): _____

2. Specification of GMP Responsibilities
    The roles and responsibilities of all parties have been specified in this GMP Agreement in accordance with:
    • The Current Cosmetics Standard ISO 22716 GMP for Cosmetics
    • The requirements documented in the manufacturing/testing specifications
    • and in relevant reference

The roles and responsibilities listed above relate to the Contract Manufacture and the Clients. Any Commercial Agreements must be covered in a separate agreement.

Further details and responsibilities in respect to this Agreement are detailed in the Responsibility Matrix found in Sections 5–17.

The Parties to this Contract must carry out each of the responsibilities (as specified in Sections 5–17) under the GMP Agreement appropriately and in accordance with the requirements listed above.

The Parties to this Contract shall not appoint or commit any related to the manufacturing operations of the supplied Product(s) (as listed in Attachment 1) to a third party, without prior evaluation and written consent of The Client.

3. Definitions

'Business Day' – Shall mean any day except weekends or other national public holiday in Australia.

'Certificate of Analysis' – Shall mean written certification by Manufacturer that the Product meets all the applicable Specifications in accordance with the Product Specifications in Australia where the Product is marketed and presentation of results for all finished products/relevant in-process tests.

'Certificate of Conformance' – Shall mean written certification by Manufacturer that the Product meets all of the applicable specifications and that the batch has been manufactured under conditions complying with all relevant requirements in the country of manufacture and this Agreement.

'Good Manufacturing Practices (GMP)' – Shall mean the norms or other relevant principles governing the manufacture of the Product, ISO 22716 or any other standard agreed by the Parties in the GMP Agreement, as the same may be amended or supplemented from time to time, and the Client's procedures, that may be communicated to Manufacturer.

'Contract Acceptor' – Shall mean Manufacturer Pty Ltd (Manufacturer).

'Client' – Shall mean The Client who orders logistically the Product and therefore manages Manufacturer in charge of sub-contracted Manufacturing Operations.

'Critical Deviation' – A deviation that has compromised the identity, strength, quality, purity, safety or effectiveness of a Product. A deviation where test results are outside the approved Product Specification (i.e. a confirmed Out of Specification Result).

'Independent Analyst Expert' – Shall mean an independent analyst or qualified laboratory agreed between the Parties.

'Major Deviation' – A deviation that has the potential to impact the identity, strength, quality, purity, safety or effectiveness of the Product and/or can represent a compliance risk. These deviations affect only batches that have not yet been released to the market.

'Manufacturing Operations' – Shall mean individually or collectively the *production, packaging, labelling, control and release* of materials and product from Manufacturer and the Client materials all in accordance with the agreed Specifications.

'Materials' – In the context of this GMP Agreement, shall mean raw materials including active ingredients, excipients, primary and secondary packaging materials, leaflets, all packaging components necessary for the packing of bulk product.

'Minor Deviation' – A deviation that does not impact product identity, strength, quality, purity, safety or effectiveness or represent a compliance risk. However, the deviation represents minor departures from SOPs and batch documents. The deviation can be corrected prior to, or by, the next control point in the process.

'Product' – In the context of this GMP Agreement, shall mean Finished Product for human use.

'Regulatory Authority' – Shall mean any Regulatory Authority depending where the Product is marketed and/or manufactured.

'Release for Supply' – Is the formal verification that the finished product meets the approved, product specification and internal standards prior to release on the market.

'Release for Distribution' – Is an additional verification beyond regulatory requirements that the product quality is acceptable and meets the approved product specification as per the GMP Agreement.

'Sanitisation' – Reduce the level of microbiological load on a surface.

'Specifications' – Shall mean the mutually agreed and written manufacturing instructions, processes and test specifications for the manufacturing and testing of the Product.

'Manufacturing Agreement' (MA) – Shall mean the overall governing and binding agreement and covers all aspects of the relationships between the Client and Manufacturer, entered into by and between the Parties in the Manufacturing Agreement of which the GMP Agreement here under is an Annex.

4. Supply and Manufacture

The principles of Good Manufacturing Practices (GMP) ISO 22716 shall govern the Manufacturing Operations of the Product as well as the Product Specifications and formulary.

The Parties represent that they possess the authorisation from the appropriate Regulatory Authorities to manufacture the Product and shall maintain such authorisation at all times while this GMP Agreement is in effect.

5. Premises

Manufacturer shall perform Manufacturing Operations for the Product at its site(s) as set forth in Section 1 of this Agreement.

The premises and equipment used to manufacture the Product must be in compliance with the current regulatory requirements of the Territory where the Product is supplied.

Manufacturer shall not sub-contract any related to the Manufacturing Operations of the supplied Product without the prior written approval of the Client.

6. Storage and Transport

Manufacturer shall be responsible for the storage and transport of the Product according to the product specifications until the Product is picked up by the Client as provided in the Supply Agreement.

7. Regulatory Inspections

Manufacturer shall immediately (within 5 business days) inform the Client of any deficiencies from regulatory inspections which may involve the Clients Product or related processes.

The Client shall promptly notify Manufacturer, if during any regulatory inspection of the Client, there are negative comments made relating to Manufacturer's performance of its obligations under this GMP Agreement.

In either of the above cases, a summary of deficiencies relating to the Client's Product or related processes shall be provided to the other Party in writing within 14 (fourteen) Business Days of receipt if it relates in any way to the Product, the facilities used to manufacture the Product, or the Quality Systems of Manufacturer.

The Parties agree to co-operate on any response if appropriate or required.

In the event that Manufacturer uses third parties for the manufacture of the Product (see Section 4.1. Premises), Manufacturer commits itself to permit the regulatory inspection by any relevant Regulatory Authority of the Territory. Manufacturer commits itself to procure the agreement of that third party to permit this regulatory inspection.

8. Confidentiality

All information and documents passing between the Client and Manufacturer should be kept strictly confidential, during the continuance of this Agreement and the Confidentiality Agreement or thereafter.

9. Liability

Unless Manufacturer can show that it has complied with all relevant statutory and regulatory requirements, has not been negligent, and it has not failed to comply with the requirements of the Agreement, if the Client rejects a delivery of any Product for non-compliance with the applicable standards and specifications, Manufacturer shall at its own cost (and without cost to the Client) correct the defect (if possible, and approved by the Client), remake the lot or, credit the Client the value of the material provided by the Client.

10. Conflict Resolution

In the event that a dispute arises between Manufacturer and the Client concerning the acceptability of a batch of Product:

- The first stage requires direct communication between the responsible quality assurance personnel from all Parties to determine the facts of the matter.
- Manufacturer shall produce an investigation report, which shall be reviewed and approved by its quality unit. This report shall contain complete details of the problem together with any discussion on the validity and weight to be applied to any results of the investigation.
- The investigation report shall be reviewed by the quality unit from the Client. The Parties shall agree on the action to take.
- If these actions fail to achieve resolution, an Independent Analyst Expert shall be retained to determine the appropriate action. This Independent Expert shall be selected jointly by Manufacturer and the Client quality units. The cost to engage the Independent Analyst Expert shall be shared jointly, unless the investigation report indicates negligence on the part of Manufacturer.
- The results from this Independent Expert shall be used to determine responsibilities for the failure, but whatever the outcome the Client retains the right to determine Product's release to market status.

11. Duration

This GMP Agreement shall be effective and remain in full force and effect as long as Manufacturer is manufacturing the Product(s) referenced in Attachment 1 on behalf of the Client.

This GMP Agreement must be reviewed every 2 years from the date of approval.

The termination or expiration of this GMP Agreement shall not affect the rights and obligations of the Parties which have accrued before termination or expiration of this GMP Agreement and which expressly or which by their nature are intended to survive termination or expiration of the GMP Agreement

This GMP Agreement supersedes all previous GMP Agreements signed between the Clients and Manufacturer.

Section No.	Section Name	Client	Manufacturer
11.	Quality System/Quality Assurance		
11.1	A Quality System will be established, documented and maintained at any facility where processing of the Product(s) takes place as a means of ensuring that Product(s) conform to agreed specifications and regulatory requirements.		

(Continued)

Section No.	Section Name	Client	Manufacturer
11.2	Written procedures shall be available for the control, assessment and approval of changes to specifications, test methods, raw materials, packaging components, manufacturing process, facilities equipment or other changes. Notification of any change shall be in writing and include definition of the change and defined actions required as part of the change. Approval of the change must be made by both Parties' quality units. Approval of changes will not be unreasonably withheld.		
11.3	Manufacturer shall obtain all licenses, permits, certifications and other government authorisations and must comply with all statutory and regulatory requirements necessary for it to perform the manufacturing details specified by this Agreement. Copies of documents must be provided to the Client on request.		
11.4	Any changes to the licenses, permits, certification and authorisations which impact the requirements specified in this Agreement between both parties must be notified in writing and the associated agreement will have to be amended accordingly.		
12.	Raw Materials		
12.1	Only approved/qualified raw material suppliers are to be used for the purchase of approved Raw Materials, as specified in approved specifications. Manufacturer will take all responsibility for starting materials used for Production, unless it is provided to Manufacturer by the Client. In this case the Client provides an assurance that all starting materials comply with their Specifications.		
13.	Packaging Materials and Printed Artwork		
13.1	Artwork is to be controlled by the Client.		
13.2	Artwork design is supplied by the Client. The Client is responsible to ensure information on the labels is in accordance with any regulatory requirement and company branding. Manufacturer will refuse using packaging components that do not comply with regulatory requirements. Manufacturer will provide the Client with INCI names for all product ingredients and the way they should be listed on a label. The above does not apply to any product supplied by Manufacturer in bulk.		
13.3	Initiation and coordination of new artwork or revisions. Approval of new or revised artwork from regulatory point of view. The client is obliged to inform Manufacturer about any changes to the artwork or product label information.		
13.4	Packaging component specifications will be established and jointly approved, when relevant. (*Packaging Components Specification Preparation Packaging Components Specification Approval Packaging Components Specification Control)		
13.5	All packaging must be approved by QC prior to production. Supply correct quantity of clean packaging fit for use in production.		
14.	Manufacturing		
14.1	Supply of SDS for Finished Product (additional charges apply).		
14.2	A unique batch number for identification of each batch of Product must be assigned. This number shall appear on all documents issued by Manufacturer.		
14.3	The date of manufacture must be defined as the day, month and year of compounding materials for the Product. The expiry date for Product shall be determined by adding this date to the shelf life. The content of the manufacturing and expiry dates shall include month and year. Product shelf life will be agreed between Manufacturer and a customer.		
14.4	All Products are to be manufactured and packed using: Equipment and procedures specified in authorised specifications and procedures.		

(Continued)

Section No.	Section Name	Client	Manufacturer
14.5	Recording of key/critical information related to the manufacture of the Product in real time in batch documents, logs or similar controlled documents.		
14.6	Any deviation from the manufacturing process or procedures must be investigated and documented in accordance with an approved procedure. The approved procedure must define deviations and classify them. The investigation must assess the impact of the deviation on the Product, corrective action required prior to release of the Product and preventative action to avoid reoccurrence where possible.		
14.7	All deviations and investigations from the manufacturing process for the Product must be recorded or referenced in the Batch Records.		
14.8	Major and Critical deviations must be approved by the Client prior to release of the Product from the Manufacturer site. The Client shall provide written notification including a copy of the deviation report for approval or a summary when the report is not in English. Upon the occurrence of Major or Critical deviations, the Client shall have the final responsibility to determine the significance of the impact of such deviation on the Product as well as, accordingly, determine the further disposition of the affected batches of Product. Solely upon the Client's respective written notification, shall Manufacturer dispose of, or otherwise handle such batches as requested.		
14.9	Any changes to product formulations made by the Client or suggested by Manufacturer will be discussed by both parties and will go through Manufacturer's Change Control System and signed by the Client.		
15.	Quality Control		
15.1	Specifications, test methods and other relevant documentation relating to processes undertaken to test and assess the Product(s) must be approved and complied with for all Products(s). Control Approval Issuance Compliance to written requirements		
15.2	Quality control tests must be performed as per approved specifications.		
15.3	Manufacturer must ensure that all raw materials, semi-finished goods and packaging components used are in accordance with approved specifications and procedures.		
15.4	Finished Product testing shall be performed in accordance with the standard product testing established at Manufacturer.		
15.5	The implementation of testing regimes that result in not all specified tests being performed (e.g. skip lot testing or rotational testing initiatives) must be agreed by the Client in writing prior to implementation.		
15.6	In-process testing shall be carried out and documented as per the In Process Control specification details in the approved batch documents.		
15.7	Any deviation from approved test methods, laboratory processes or standards must be investigated and documented in accordance with an approved procedure. The approved procedure must define deviations and classify them. The investigation must assess the impact of the deviation on the validity of the analytical result, corrective action required prior to release of the Product and preventative action to avoid reoccurrence where possible.		
15.8	A written summary of any major or critical deviation or confirmed Out of Specification result must be provided to the Client for review. Major or critical deviations or confirmed Out of Specification results must be approved prior to batch release. The Client shall have the final responsibility to determine the significance of the impact of such deviation on the Product and determine the further disposition of the affected batches of Product. Solely upon the Client's written notification, Manufacturer shall dispose of, or otherwise handle such batches as requested.		

(Continued)

Section No.	Section Name	Client	Manufacturer
15.9	For non-conforming or rejected Product rework/retesting will be only permitted after written approval from the Client. Prior to commencing rework, reprocessing or retesting activities Manufacturer must provide the Client all information specific to the proposed rework, reprocessing or retesting activities and the results of risk assessment for the proposed rework/retest for review and approval.		
15.10	Retention samples from each batch manufactured shall be stored in secure conditions that are consistent with the specified storage conditions of Material or Product.		
15.11	Sufficient number of Finished Product samples to perform two full tests of Product against the specifications must be kept for the longest available shelf life of the Product plus one year.		
16.	Release		
16.1	Release from Manufacturers site: Manufacturer shall ensure their documented release process is capable of consistently ensuring product that meets the product specification, all internal standards and this Agreement prior to it being released from the site.		
16.2	Release from Manufacture of the Product for supply must include verification that the product complies with the specification and ensure all deviations associated with the batch have been approved in accordance with this Agreement.		
16.3	Release of Product from manufacture for supply must be done by an authorised person nominated to perform Batch Release process, is educated, experienced and or trained for the release of finished product as per the approved specification.		
16.4	Client can request for finished product Certificate of Analysis to be supplied at additional cost.		
17.	Documentation		
17.1	Approved manufacturing documents must reference the correct document and revision number used in the manufacture of the Product.		
17.2	Manufacturer shall create and maintain a master batch document listing all of the steps to manufacture the Product.		
17.3	Manufacturer shall assign a unique batch number for numbering each batch of product. This number shall appear on all documents issued.		
17.4	Quality testing records must be part of the Routine Batch Documentation.		
17.5	Documents and completed records must be retained by Manufacturer for a minimum of 1 year after product's expiry date.		
17.6	**In case of closure or sale of Manufacturer facility, provisions shall be made to transfer all documentation to a new site or secured storage area or transferred to the Client.**		
17.7	The Intellectual Property of the Product is owned by Manufacturer unless the Manufacturing agreements denote otherwise. Then these other documents take precedence.		
18.	Product Complaints		
18.1	Management of complaints file for Finished Product.		
18.2	Investigation into root cause of complaints concerning Product(s). Investigation reports are to be submitted to the Client in writing, submitted within ten (10) working days, referencing Client complaint number where relevant.		
18.3	If Manufacturer receives any complaint from a third party, that concerns or could concern the Client Product, Manufacturer shall communicate them without delay to the Client.		
19.	Product Recall		

(Continued)

Section No.	Section Name	Client	Manufacturer
19.1	**Upon discovery that any batch previously approved by Manufacturer fails to conform to specifications, or has in any way been adulterated, misbranded by Manufacturer, Manufacturer shall immediately notify the Client of such failure and of the nature thereof in detail, including, but not limited to supplying the Client with all investigatory reports, data, communications, out of specifications reports and the results of any outside laboratory testing. All such failures must be investigated promptly, in accordance with the Client's request, and shall co-operate with the Client in determining the cause for the failure and the corrective action required. Manufacturer shall implement corrective action to prevent non-conformance reoccurring.**		
19.2	The Client shall immediately notify Manufacturer of any real or potential recall that may be due to manufacturing operations, product stability, material or test performed by Manufacturer.		
19.3	Final authority to initiate a recall.		
19.4	Maintain a Product Recall Standard Operating Procedure and relevant Insurances for a recall.		
19.5	Conduct investigation into potential recalls in timely manner (less than 5 business days) or timeframes specified by Regulatory Authorities in accordance to agreed procedures.		
19.6	Authorised to liaise with Regulatory Authorities in the event of a recall involving the Product(s).		
19.7	Provide access to all records associated with the batches in the case of recall, customer complaints or government inquires.		
19.8	Maintaining recall records.		
20.	Validation		
20.1	All relevant facility and utilities or equipment that may influence product quality will be validated as per Validation Master Plan.		
21.	Regulatory/Compliance		
21.1	Advertising (including label copy) complies with regulations of the country(ies) in which the market pack is offered for sale.		
21.2	Registration of the manufacturing facility with Regulatory Authorities.		

ATTACHMENT 1

Schedule of Product(s)

LIST OF PRODUCT(S):

Acceptance Authorisation

The following representatives from the Client and Manufacturer duly authorise this Agreement

Title	Name	Signature	Date
	Client (The Client)		
Witness			
	Manufacturer PL		
Witness			

22

Topical Products Applied to the Nail

Apoorva Panda, Avadhesh Kushwaha, H.N. Shivakumar and S. Narasimha Murthy

CONTENTS

Introduction

Nail is a protective cutaneous tissue that covers the tips of the fingers and toes. Human nails are often prone to a wide range of disorders. Nail diseases cannot be ignored as a mere cosmetic problem, as they could cause severe dystrophy of the nail plate and lead to serious consequences. Considering the high prevalence of these disorders, nail formulations constitute a sizable fraction of cosmetic products. As a result there is a huge demand for technologies that can be used to treat nail disorders and to deliver drugs into the nail apparatus.

Nail Apparatus: Anatomy

The nail unit is typically made up of the nail plate, nail bed, nail matrix and hyponychium (Figure 22.1) (Wortsman and Jemec, 2013). The nail plate, which represents a modified version of the stratum corneum, is composed of nearly 25 layers of keratinized flattened dead cells that cover the nail matrix and the nail bed. It is

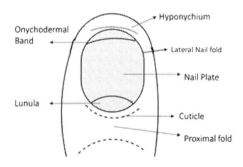

FIGURE 22.1   Structural anatomy of nail apparatus.

a convex, thin yet hard, slightly flexible structure that is curved along the longitudinal as well as the transverse axes (Wortsman and Jemec, 2013). The nail plate is known to grow throughout the life of a person. On an estimated average, the fingernails take about 6 months, while the toe nails take about 12–18 months to grow out completely. However, the shape of the nail plate, its growth and appearance are determined by a number of factors such as gender, age, nutritional status, disease conditions and other environmental factors like temperature. For instance, nail plates are usually longer, broader and thicker in males as compared to females. The nail plate is firmly secured to the nail folds along the lateral and proximal margins. The nail plate is able to impart an appropriate amount of strength to the nail, which is tougher in toe than the fingers (Dawber et al., 1994).

The nail plate basically comprises of three layers, namely the dorsal, intermediate and ventral layers (Kobayashi et al., 1998). The dorsal layer is the topmost layer of the nail plate that acts as a barrier for the entry of microorganism and chemical substances. This layer is known to contain higher amounts of keratin compared to other deeper layers of the nail plate. In contrast, the ventral layer that is attached directly to the nail bed is the most hydrated layer. The majority of the nail plate is constituted by the intermediate layer, as it is almost twice as thick as the dorsal layer. The three layers of the nail plate have different proportions of calcium, phospholipids, disulfides and thiols (Kobayashi et al., 1998). The thicknesses of the dorsal, intermediate and ventral layers of the human nail plate are known to bear a ratio of 3:5:2 (Kobayashi et al., 1999).

Chemical Composition of the Nail Plate

The nail plate chemically comprises of fibrous protein called keratin that provides the necessary mechanical strength to the epithelial cells. Keratins found in humans are either hair keratin or epithelial keratin. Nearly about 80–90% of the keratin present in the nail is the hair keratin, while the rest would be the epithelial keratin. Hair keratin is found to be concentrated in the intermediate layer of the nail plate, while the epithelial keratin is present in the dorsal and ventral layers. Keratin fibres in the nail plate are interconnected through cysteine-rich proteins which are linked via disulphide bridges. This sandwich orientation of keratin fibres imparts the necessary hardness and rigidity to the nail plate (Kobayashi et al., 1999).

Water is a major component of the nail plate that is extremely important to provide the elasticity and flexibility to the nail plate. However, the exact amount of water is still unclear, as the *in vivo* water content of the distal free edges of the nail plate was reported to be 3–30%, whereas the water content in the cut nail clippings was found to be 8–22%. In addition, the nail plate contains amino acids such as glutamic acid (13.6%), serine (11.3%), half cysteine (10.6%), leucine (8.3%), glycine (7.9%), aspartic aid (7%), arginine (6.4%), threonine (6.1%), proline (5.9%), alanine (5.5%), valine (4.2%), tyrosine (3.2%), lysine (3.1%), isoleucine (2.7%), phenylalanine (2.5%), histidine (1%) and methionine (0.7%) (Hossin et al., 2016; Egawa et al., 2006; Stern, 2007). The nail plate is also known to contain small amounts of lipids (0.1–1%) like cholesterol sulphate, ceramides, free sterols, free fatty acids, triglycerides, wax esters and squalene (Helmdach et al., 2000). The lipids that are mainly concentrated in the dorsal and ventral layers of the nail plate are arranged as bilayers with a parallel orientation to the nail surface (Vellar, 1970; Sirota et al., 1988). However, the lipid content in the nail plate varies with gender, and declines in its proportion as one ages. The growing nail plate accumulates other substances such as hormones, isotopes, pollutants and many more all throughout life from the surroundings. Moreover, elemental analysis of the nail plate has revealed the presence of calcium, zinc, sodium, potassium, magnesium, copper, iron and many others (Yazdanparast and Barton, 2006; Murdan et al., 2011).

Nail Bed

The nail bed is a soft, thin and noncornified epithelium that is known to extend from the lunula to hyponychium (Fluhr et al., 2006). The nail bed helps in the growth of the nail plate and acts as a holder for the nail plate. The dorsal surface of the nail bed is comprised of longitudinal ridges that complement similar ridges found on the underside of the nail plate. These ridges are known to ensure the necessary adhesion between the nail plate and nail bed (Jarrett and Spearman 1966). The dermis of the nail bed is very thin and contains a very little amount of fat, sebaceous and follicular appendages (Cecchini et al., 2009; Farren et al., 2004).

Nail Folds

Nail folds are folded skin structures which surround the nail plate. Nail folds usually include the proximal nail folds and lateral nail folds (Wortsman and Jemec, 2013). The lateral and proximal nail folds enclose the lateral and proximal edges of the nail plate, respectively.

> Lateral nail folds are more evident in the toe nails as compared to fingernails. The main function of these lateral nail folds is to ensure firm adherence of the nail plate. A weakened lateral nail fold can often lead to onycholysis (De Berker, 2013).

> The proximal nail fold, also known as the posterior nail fold, is a lip of skin which is formed around the 14 week of embryogenesis. This fold that adheres firmly to the dorsal nail plate acts as a protective layer over the nail matrix. The proximal nail fold along with the cuticle is known to ensure a tight seal against entry of exogenous substances (De Berker, 2013). Any damage to the posterior nail fold can pose the risk of permanent nail scarring or damage.

Eponychium

The eponychium (cuticle) is an epidermal layer that broadens out from the posterior nail fold and adheres strongly to the dorsal layer of the nail plate (Rigopoulos, 2008). The important feature of the cuticle is that it acts as a natural seal against the entry of exogenous substances. However, eponychium loss can occur due to manicures, inflammation or infections, which could be an early indicative sign of chronic paronychia (Rigopoulos, 2008) The loss in eponychium often affects the protective function of the nail folds.

Hyponychium

The hyponychium represents the area of epithelium underlying the distal edge where the nail plate begins to separate from the nail bed (Rigopoulos, 2008) Fungal infections can often start from the hyponychium and progress towards the nail bed and the nail matrix.

Nail Matrix

The nail matrix, often referred to as the nail root, is a half moon–shaped, highly proliferative epidermal tissue located just below the proximal nail fold (De Berker, 2013). Lunula is the distal portion of the matrix that is visible as a white semi-lunar area through the transparent nail plate. The nail matrix serves as an origin to the nail that is prone to damage in cases of surgical and accidental trauma. It is realized that it is hard to repair the nail matrix in case of any such damage (De Berker, 2013).

Onychodermal Band

It is a transverse band between the hyponychium and the nail plate that appears as pink or brown in colour. The colour of the band is influenced by disease or compression that has an effect on the vascular supply. It acts as a physical barrier for the materials penetrating underneath the nail plate (Terry, 1955). Any damage to this band either by loss of cuticle or disease condition can affect the normal functioning of the nail bed.

Nail Fold Vessels

The capillary network of the nail fold is similar to the cutaneous plexus. However, capillary loops of the nail folds grow more parallel as compared to the cutaneous plexus and can be easily seen all throughout the length. The distinctive pattern of the vessels in the nail fold appears to be the distinguishing feature. The capillary networks present in the toe are denser when compared to fingernails but display a reduced flow rate (Richardson and Schwartz, 1984).

Glomus Bodies

Glomus bodies are defined as a cluster or a conglomeration of cavernous blood vessels. These neurovascular bodies are present in the nail bed act as arteriovenous anastomoses (De Berker, 2013). In the nail bed, they are known to connect the arterial side and venous side of the circulations. On an average, there are about 93–501 glomus bodies per cubic centimeter in the nail bed. These bodies are known to regulate capillary circulation in the nail bed.

Diseases Associated with Human Nails

Several diseases that are known to affect the human nails are listed below.

- Onychia is an inflammatory condition in the nail bed and its surrounding tissues that is mainly caused due to infection by microorganisms.
- Onychocryptosis is known as 'ingrown nails'. In this case, the nail is disconnected from the nail bed.
- Onychodystrophy is a condition of distortion of nails as a result of cancer chemotherapy.
- Onychogryphosis is caused by injury to the nail matrix. Common symptoms are thickening and increase in nail curvature.
- Onycholysis is the complete separation of nail plate from the nail bed.
- Onychomadesis is mainly caused by minor injury in the nail bed.
- Onychomycosis is a fungal infection of the toe and fingernails (Kushwaha et al., 2015).
- Onychophosis is a condition of hyperkeratosis in the nail fold.
- Onychoptosis is a periodic shedding of the nail in parts or whole.
- Onychorrhexis indicates brittleness of the nail plate.
- Paronychia is an infection to tissue around the nail caused by bacteria or fungus.
- Koilonychia is known as spoon nail caused by iron deficiency.
- Subungual hematoma is a condition of bleeding below the nail plate.
- Onychomatricoma is a tumour in the nail bed and nail matrix.
- Nail psoriasis is an autoimmune inflammatory disease.

The most common diseases that affect the nail include onychomycosis and nail psoriasis. In this chapter, these two diseases are discussed in detail.

Onychomycosis

Onychomycosis refers to the fungal infection of the nail plate and/or the nail bed. The causative agents include dermatophytes, Candida and nondermatophytic molds (Wortsman and Jemec, 2013). Onychomycosis often causes a gradual destruction of the nail plate. Onychomycosis is classified into four types: (1) distal subungual onychomycosis, (2) white superficial onychomycosis, (3) proximal subungual onychomycosis and (4) Candida onychomycosis (Wortsman and Jemec, 2013). Distal subungual onychomycosis is the most commonly occurring form of the disease that affects the nail plate as well as the nail bed. The fungal infection originates at the hyponychium and the distal or lateral nail bed. The infection may further progress to the nail matrix and eventually to the under surface of the nail plate making it opaque. White superficial onychomycosis refers to the fungal infection of the superficial layers of the nail plate that leaves chalky white patches. Superficial onychomycosis is usually restricted to the toe nails. On the other hand, proximal subungual onychomycosis occurs when fungus infiltrates through the posterior nail fold and spreads to the undersurface of the newly formed nail plate causing a white discolouration in the lanula. Proximal subungual onychomycosis usually affects immunocompromised patients. In contrast, Candida onychomycosis occurs due to infiltration of Candida species through the already infected nails. The most common symptoms of onychomycosis include nail thickening and discolouration. The nails may become thickened, distorted in shape and separate out of the nail bed. In some cases, Candida onychomycosis causes pain or a foul smell.

Nail Psoriasis

Nail psoriasis is known to affect millions of people all around the world (Jefferson and Rich, 2012). The nail matrix, nail bed and nail folds are commonly affected by psoriasis, where the severity ranges from mild to extremely erythrodermic forms. A psoriatic nail displays a broad spectrum of symptoms varying from loosening of the nail plate, common pitting, hemorrhages and discolouration in the nail bed. The major signs of nail psoriasis include yellow-red discolouring, small pits in nails, thickening of skin under the nail, loosening and crumbling of the nail, and eventually nail loss (Jefferson and Rich, 2012). Nail psoriasis if left untreated can lead to functional impairment.

Nail Pitting

Pits found in the nail plate are superficial depressions which may differ in appearance and location. (Wortsman and Jemec, 2013). Nail pitting can be observed in various diseases such as lichen planus, alopecia arreta and chronic eczema. However, nail pitting is more distinctively observed in nail psoriasis.

Discolouration and Onycholysis

Psoriasis of the nail bed produces the characteristic oval, salmon-coloured 'oil drop' discolouration of the nail plate (Wortsman and Jemec, 2013; Egawa et al., 2006) Complete separation of nail plate from the nail bed is the characteristic feature of nail psoriasis.

Subungual Hyperkeratosis

Deposition and agglomeration of cells which do not undergo desquamation lead to subungual hyperkeratosis under the nail plate. The severity of psoriasis decides the extent to which the nail plate is raised from the nail bed. The subungual hyperkeratosis is found in many other nail diseases, but the colour varies among different nail diseases (De Berker et al., 2001). Psoriatic patients with subungual hyperkeratosis usually display a white silvery manifestation that is not commonly observed in other nail diseases.

Splinter Hemorrhages

Splinter hemorrhages occur less predominately in patients with nail psoriasis (Zaias et al., 1996). Clinical signs of splinter hemorrhage include appearance of longitudinal and parallel epidermal ridges, accompanied with dermal grooves on the nail plate (Jefferson and Rich, 2012). However, splinter hemorrhages may also occur with other diseases such as eczema, vasculitis and bacterial endocarditis (Fitzpatrick et al., 1983).

Alternative Abnormalities

A variety of clinical signs can be evident throughout the nail plate depending on the extent of nail matrix being affected by psoriasis. Short-duration lesions will lead to the formation of transverse grooves and ridges on the nail plate known as Beau's lines (Jefferson and Rich, 2012). Alternatively, lesions that last long will often lead to onychorrhexis that is often characterized by the appearance of longitudinal ridges and nail splitting (Doner et al., 2011).

Treatment of Nail Diseases

Therapeutic options available for treating nail diseases broadly include surgical, systemic, topical and biophysical therapies. A schematic of the various therapies available for management of transungual disease is represented in Figure 22.2. Surgical and device-based therapies are employed most often in management of chronic cases of ungual diseases, whereas topical and systemic therapies are used in mild to moderate cases (Doner et al., 2011). Systemic therapy involving oral administration of therapeutic agents is a first-line treatment commonly used when the disease is widespread affecting more than three to four nails of the fingers and toes. In contrast, topical therapy that involves direct application of therapeutic agents to the infected nail is normally employed when few nails are affected. Topical therapy is often recommended with oral therapy to improve the cure rate and enhance the therapeutic outcome. Further, topical therapy for treatment of nail disorders may be either passive or device-based depending on the clinical need. Device-based therapies utilize technology based applications such as iontophoresis and ultrasound to improve the efficacy of the topical therapy. Biophysical techniques that have been used for treatment of nail diseases would include photodynamic and laser therapies (Doner et al., 2011).

Due to various other side effects associated with systemic drug delivery, a change in strategy for the management of nail diseases has been observed in recent years, from the conventional topical therapy to the device-based techniques or applications (Gupta et al., 2013). It is imperative to use device-based therapies to manage diseases of the nail, as they could drastically improve the efficacy of topical therapy.

Conventional Therapies

Nail Avulsion

Nail avulsion involves complete or partial removal of the infected area of the toe or fingernail. Nail avulsion is basically a surgical or chemical procedure of separation of the affected nail plate from surrounding tissue. Avulsion of the nail is a viable option in chronic cases of ungual diseases. The technique is often used when infection spreads to the surrounding tissues making it inevitable for the infected nail bed and plate to be removed (Pandhi and Verma, 2012). Nail avulsion for the treatment of ungual diseases may often involve chemical avulsion or surgical avulsion (Scher, 2007).

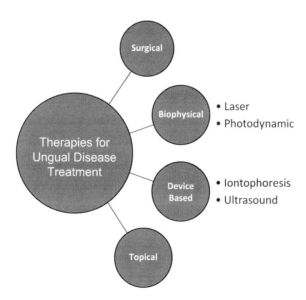

FIGURE 22.2 Schematic of various therapies for ungual disease treatment.

Surgical Avulsion

Depending on the location of the fungal infections in the nail, surgical avulsion may involve distal or a proximal avulsion. The Freer instrument is normally used to perform the surgical avulsion of the nail. Before carrying out the surgical avulsion, the toe or finger is anesthetized using 1% lidocaine. In order to perform the distal avulsion, the instrument is initially passed under the distal free edge of the nail plate and slowly inserted into the nail bed to separate the nail plate from the underlying nail bed. Finally, the instrument is inserted into the proximal nail groove under the posterior nail fold to completely separate the nail plate from the nail matrix. In contrast, proximal nail avulsion is performed in case of distal subungual onychomycosis, where the distal free edge of the nail is free from infection. In this case the procedure is different, as the instrument is inserted in the nail bed until it reaches the distal edge and the nail plate is completely separated from the nail matrix and the nail bed (Albom, 1977). More often, partial avulsion of the nail proves to be beneficial compared to complete avulsion. Complete avulsion sometimes proves to be disadvantageous as it can lead to dislocation and shrinkage of the nail bed. Moreover, complete removal of the nail plate can lead to loss of counter pressure, which can induce expansion of the distal soft tissue and distal edge of the regrowing nail (Kushwaha et al., 2015).

Chemical Avulsion

Chemical avulsion is a more preferred procedure over surgical avulsion, as it is painless. This treatment modality is mostly used for patients suffering from collagen vascular disease, peripheral vascular disease, hemostasis and diabetes mellitus (Albom, 1977). Keratolytic agents such as urea and salicylic acid are often used in this mode of therapy, as they dissolve the bonds found in the nail plate and those between the nail plate and nail bed. Urea (40% w/v) is quite effective in chemical avulsion of dystrophic nail. Chemical avulsion basically employs a two-step treatment in which the first week involves a nail-softening procedure followed by two weeks of treatment so as to completely remove the nail plate. Following the two-step procedure, bifonazole topical cream (1% w/v) is prescribed for the complete eradication of the fungal infection (Albom, 1977). Antifungal drugs and keratolytic agents are used in combination with chemical avulsion to ensure complete eradication of fungal infections. Chemical avulsion therapy has a lower rate of recurrence as compared to surgical avulsion and decreases the risk of hemorrhage. Hence, chemical avulsion is a more preferred method compared to surgical avulsion (Siegle and Swanson, 1982). The only major disadvantage with the chemical avulsion method is that it employs a prolonged treatment duration (Jellinek, 2007; Baran et al., 2008; South and Farber, 1980).

Antifungal Drugs

Antifungal drugs are used via both oral and systemic routes of administration to eradicate fungal infections of the nail. Antifungal drugs administered orally as tablets or capsules constitute the conventional mode of treatment. The drugs used in treating ungual diseases include itraconazole, griseofulvin, ketoconazole, fluconazole and terbinafine (Kushwaha et al., 2015). Most of the antifungals act by inhibiting the synthesis of ergosterol, an important constituent of fungal cell membrane. The disadvantages associated with oral therapy are low bioavailability due to poor drug solubility that could eventually lead to poor drug distribution to the infected nails. Owing to the poor distribution to the infected sites, oral treatment invariably warrants long treatment periods. Moreover, chronic use of oral antifungal drugs could cause serious side effects affecting the liver, cardiovascular system and the gastrointestinal tract (Sheehan et al., 1999; Shivakumar et al., 2012).

Griseofulvin

Griseofulvin is fungistatic in nature and is unable to show its fungicidal activity at higher concentrations. It is occasionally used in the treatment of ungual infections, as it is effective against dermatophytes. The recommended oral dose of griseofulvin is 500 mg per day for 6–9 months for the infection in the fingernail and for a time period of 12–18 months for the infection in the toe nail. The common side effects observed with griseofulvin include urticaria, erythema multiforme, headache, hepatotoxicity and photosensitivity (Kushwaha et al., 2015; Roberts et al., 2003).

Terbinafine

Terbinafine is more effective against dermatophytes than non-dermatophytes. The recommended oral dose of terbinafine is 250 mg per day for 6–8 weeks for fingernail infection and around 12 weeks for toe nail infection. Some of common side effects associated with terbinafine include headache, rashes and problems associated with the gastrointestinal system (Van Duyn Graham and Elewski, 2011). Some rare complications such as Stevens-Johnson syndrome, cholestatic hepatitis and blood dyscrasias are observed with terbinafine.

Itraconazole

Itraconazole is effective against dermatophytes as well as non-dermatophytes. The recommended oral dose for itraconazole is 200 mg per day for 12 weeks in case of toe infection, while the same dose is prescribed for a time period of 6 weeks in case of fingernail infection. The dosage regimen in case of pulsed treatment is 200 mg twice a day for 1 week every month that is continued for a period of 2 months in case of fingernails and 3 months for toe nails. The pulse treatment is known to substantially reduce the systemic exposure of drug and therefore the occurrence of adverse effects. The side effects associated with itraconazole are gastrointestinal disturbances, skin rash and reversible increase in the levels of hepatic enzymes. However, the more serious side effects include idiosyncratic hepatic failure and congestive heart failure (Kushwaha et al., 2015; Shivakumar et al., 2012; Van Duyn Graham and Elewski, 2011).

Fluconazole

Fluconazole is an antifungal agent that is effective against both dermatophytes and non-dermatophytes. The dosage regimen of fluconazole is 450 mg once a week followed by 150 mg a week for a period of 6 months (Kushwaha et al., 2015). The dose is for around 6 to 9 months in case of fingernail infection, while for 9 to 18 months in case of toe nail infection. The clinical cure rates of the nail infection were found to be nearly 90% with fluconazole. The major side effects associated with fluconazole are headache, nausea, pruritus and abnormalities associated with liver enzymes (Kushwaha et al., 2015; Shivakumar et al., 2012; Van Duyn Graham and Elewski, 2011).

Ketoconazole

Ketoconazole is fungistatic and effective against dermatophytes and candida species (Kushwaha et al., 2015). The proposed dosage regimen for this drug is 200 mg a day that needs to be taken with food (Kushwaha et al., 2015). Frequent administration of ketoconazole could lead to an increase in the risk of idiosyncrasy in patients with liver diseases and those consuming alcohol. Other common side effects associated with ketoconazole are fever, pruritus, diarrhea, nausea and vomiting (Kushwaha et al., 2015; Markinson et al., 1997). However, the drug is no longer used in the treatment of onychomycosis due to its fungistatic nature.

Topical Therapy

Topical therapy is said to be more effective compared to oral therapy in the treatment of superficial onychomycosis in which the nail matrix and nail bed are not infected (Shivakumar et al., 2012). However, low permeability of the nail plate is the major obstacle in topical therapy. Generally, molecular size and hydrophilicity of the drug determine the nail plate permeability and therefore the efficacy of topical treatment (Shivakumar et al., 2012). Therefore, topical and oral therapies are used in combination to achieve better efficacy in management of nail infections. Most often nail penetration enhancers are used to improve the effectiveness of topical therapy (Kushwaha et al., 2015). Permeation enhancers can work either by breaking the disulfide bonds or by enhancing the hydration capacity of the nail.

Passive Therapy

Amorolfine

Amorolfine is regarded as a highly efficacious antifungal drug with activity against both dermatophytes and non-dermatophytes fungi (Kushwaha et al., 2015). The drug that belongs to the morpholine class acts by inhibiting the enzymes in the ergosterol biosynthetic pathway. The drug-induced enzyme inhibition hampers the cell multiplication that ultimately causes fungal cell death. Amorolfine nail lacquer (5% w/w) is applied once or twice weekly for a period of 6–12 months in the management of distal and lateral subungual onychomycosis. The product is approved for use in Europe but not in the United States and Canada (Kushwaha et al., 2015). However, the major side effects associated with amorolfine are irritation, burning sensations, pain and redness (Kushwaha et al., 2015; Shivakumar et al., 2012; Hafeez et al., 2014).

Ciclopirox

Ciclopirox is an antifungal with a hydroxyl pyridone ring that is used in the treatment of nail infections. It is an antibacterial that acts by the inhibition of metal-dependence enzymes. Ciclopirox nail lacquer is often prescribed for use in the treatment of mild distal and lateral subungual onychomycosis (Shivakumar et al., 2012; Subissi et al., 2010). The nail lacquer needs to be applied daily for about 24 weeks in the case of nail infection and for 48 weeks in case of toe infections (Kushwaha et al., 2015). Ciclopirox nail lacquers available in the market are either conventional vinyl resin–based or hydroxypropyl chitosan–based (Kushwaha et al., 2015).

Device-Based Therapy for Drug Delivery to Nails

Device based therapy has been found to be effective in enhancing drug delivery through the nail plate. However, these therapies require medical intervention to carry out the treatment procedure as the devices are not supposed to be handled by patients. Device based therapies may include treatment using iontophoresis or ultrasound.

Iontophoresis

Iontophoresis is a physical treatment procedure which facilitates the transport of charged ions through biological membranes on application of an electric current. This method has been especially efficacious for delivery of charged ions through the nail plate. Iontophoresis is applied using a device following application of the gel formulation to the nail plate. Iontophoresis is known to increase the drug delivery by the principle of electrorepulsion. Delivery of terbinafine hydrochloride and ciclopirox olamine through the nail plate was improved using ionotophoresis (Kushwaha et al., 2015; Nair et al., 2009). Iontophoretic efficiency is often dependent on a number of factors such as formulation pH, ionic strength and current density (Kushwaha et al., 2015). As the current density increases, the drug transportation across the nail plate increases proportionally. The maximum current density recommended for transungual application is 0.5 mA/cm^2. Often, nail penetration enhancers have also been used along with iontophoresis to enhance drug delivery through the nail plate (Manda et al., 2012; Repka et al., 2004).

Ultrasound

Ultrasound is another technique that is currently used to improve drug delivery in treatment of ungual diseases. The efficacy of ultrasound is still being explored for treatment of various nail diseases. Ultrasound is known to generate cavitational bubbles that oscillate and implode between the ultrasonic transducers and the nail surface to produce shock waves that mechanically impact the nail plate creating microscopic defects. The efficacy of ultrasound was studied on canine hoof model using blue dye as the marker (Kushwaha et al., 2015). Three energy levels with a power of about 1.5 W/cm were used on the hoof membrane for 120 s (Kushwaha et al., 2015; Roberts et al., 2003). The results indicated a 1.5-fold increase in delivery compared to that achieved by passive treatment. Permeability enhancement was checked after application of low-frequency ultrasound of 20 kHz kept at distance of 13 mm from the bovine hoof membrane (Repka et al., 2004).

Biophysical Therapies

Biophysical techniques are said to demonstrate more efficacy but less adverse effects compared to device-based therapy (Kushwaha et al., 2015; Anderson and Parrish, 1983). Biophysical therapy for the treatment of ungual diseases may include laser therapy and photodynamic therapy.

Laser Therapy

Several laser devices that are approved by the U.S. Food and Drug Administration (FDA) are currently available for treating onychomycosis of the nail. The principle involved in laser therapy is photothermolysis, which refers to selective absorption of short radiation pulse to generate the heat necessary to kill the fungi (Kushwaha et al., 2015). The biggest advantage of laser therapy is that it avoids various systemic side effects associated with conventional drug-based pharmacotherapy. Laser therapy leads to formation of reactive oxygen species, which lead to a transient increase in membrane permeability. The disadvantage associated with laser therapy is that tissue damage can occur if lasers are used at frequencies higher than the threshold levels (Anderson and Parrish, 1983; Gupta and Simpson, 2012). Moreover, frequent usage of laser at infrared wavelength would induce photoaging due to reduced production of type 1 collagen in tissues (Kushwaha et al., 2015).

Commercially available laser systems include:

1. Long-pulse system
2. Short-pulse laser system
3. Q-switched laser system
4. Diode laser system

Long-Pulse System

Fotona Dualis is a long-pulse laser system used in treatment of onychomycosis. In a clinical study, about 162 patients volunteered for testing the efficacy of this system against onychomycosis. The treatment regimen using the laser system was four times in a week for about 12–18 months. Potassium hydroxide microscopy was used to check the fungal infection post treatment. All participants were found to be free of the fungal infection following the treatment regimen (Gupta and Simpson, 2012).

Short-Pulse Laser System

The short-pulse laser system was comprised of a neodymium yttrium aluminium garnet laser system that was operated at 1064, 1320 and 1540 nm (Kushwaha et al., 2015). The highest wavelength approved for the system in the treatment of onychomycosis is 1054 nm. The different models of the laser systems used for the treatment of onychomycosis are Genesis Plus, Pin Pointe Zoom, Joule and Varia (Kushwaha et al., 2015).

Q-Switched Laser Systems

The Q-clear laser is the only approved laser system in this category for treatment of onychomycosis. The duration of pulse with the Q-clear laser is in the range of nanoseconds. Clinical trials were carried out using this laser system involving about 100 patients. The results obtained suggested that after the completion of trials, about 95% of the patients were clear of onychomycosis (Kushwaha et al., 2015).

Diode Laser Systems

Diode laser systems employed for the treatment of fungal infections use semiconductors as the optical gain medium (Kushwaha et al., 2015). The Noveon laser that works on dual wavelength is an example of a diode laser system. Currently, the Noveon laser system is going through a phase II clinical trial in patients. In the study, 26 patients were subjected to dual wavelength treatment at 930 nm for 2 minutes and 4 minutes. The treatment with the system for 180 days indicated that nearly 85% of the patients were clear of the infection (Gupta and Simpson, 2012).

Photodynamic Therapy

The principle of photodynamic therapy involves the interactions between visible spectrum light and photosensitizer agents (Kushwaha et al., 2015). The interaction between photosensitizer and visible spectrum light leads to the production of singlet oxygen (Kushwaha et al., 2015). This singlet oxygen produced has the ability to interact with the cellular components of the fungi eventually leading to fungal cell death (Becker and Bershow, 2013). Methylaminolevulate (16%) and 5-aminolevulinic acid (20%) are two photosensitizing agents used as precursors in the heme biosynthesis pathway that cause accumulation of protoporphyrin IX. The two agents need to be applied 3–5 hours before treatment is initiated and protected from light. The two light options available for the photodynamic therapy are red and blue light (Harris and Pierpoint, 2012). However, in terms of the penetration ability into the nail plate, red light is known to be more effective than blue light. To improve the permeability of the photosensitizing agents, the nail plate has to be pretreated with either 20% or 40% urea cream or ointment (Harris and Pierpoint, 2012).

After photodynamic therapy, few directions need to be followed (Smijs et al., 2004). The area treated should be refrained from sunlight for at least 48 hours. Moreover, heavy physical activities such as hot saunas and exercise should be avoided completely for at least 24 hours after treatment. In addition, skin irritation products and excessive rubbing must be avoided for one week following treatment. Following the photodynamic therapy, patients normally might complain of redness, blisters swelling and bruising (Kushwaha et al., 2015).

Nail Formulations

Topical application of drugs to the affected site constitutes one of the important modes of treatment for the nail diseases. The various topical formulations used in treatment of nail diseases are

- Lacquers
- Gels, ointments, creams
- Lotions
- Patches and films

Nail Lacquers

Nail lacquer for topical application to the nail plate usually contains the active pharmaceutical ingredient (API), polymer responsible for forming the film, plasticizer and solvent. The film that forms after the evaporation of solvent contains a high amount of drug that diffuses across the nail plate. The film thus formed releases the drug in a controlled and sustained manner into the nail apparatus.

Nail lacquers need to be physically and chemically stable so as to enter the contours of the nail plate easily. Once applied, it should dry completely within a few minutes (3–5 min) and adhere firmly to the surface of the nail plate. There are various factors that affect the drug diffusion from the film into the nail bed or nail matrix, including drug solubility in the nail plate, drug solubility in the polymeric film and diffusion coefficient of the drug in the nail plate (Elkeeb et al., 2010). Permeation of the drug from nail lacquers can be often increased by using permeation enhancers. Penlac™ is a lacquer formulation of ciclopirox (8%) that is recommended for treatment of fungal infections of the nail. The nail lacquer solution consists of butyl monoester of poly(methyl vinyl ether/maleic acid) as a water insoluble polymer in a binary solvent system mixture of isopropyl alcohol and ethyl acetate (Shivakumar et al., 2012). EcoNail™ is a nail lacquer containing econazole (5%) that is used to treat fungal infections of the nails. Eudragit RLPO is employed as a film-forming polymer that is dissolved in ethanol as a solvent. Loceryl™ is a film-forming solution containing amorolfine (5%) that is used to treat nail infections. (Shivakumar et al., 2012). The lacquer contains methacrylic acid copolymers as the film former and dibutyl phthalate as the plasticizer in a volatile solvent. Following application, the water-insoluble polymers containing the drug forms a lacquer on the nails on evaporation of the solvents.

Semi-Solids: Gels, Ointments and Creams

Creams, ointments and gels are conventional semi-solid dosage forms used for the treatment of nail diseases.

Gels

Gels retain a lot of water that hydrates the nail plate and therefore enhances drug penetration through the nail plate. Gels are sometimes considered to be better formulations because of their ability to hydrate and swell the nail plate (Crowley et al., 2004). By virtue of their ability to swell and hydrate the nail plate, the gels are known to enable better penetration of the drug. However, a major disadvantage of using gel is that it may get dissipated rather quickly from the surface of the nail. Loprox® Gel is a formulation of ciclopirox that is used in the treatment of fungal infections of the nail (Shivakumar et al., 2012). The gel containing ciclopirox (0.77%), is composed of Carbomer® 980 as the viscosity-imparting agent and sodium hydroxide as the pH modifier in a binary solvent mixture composed of isopropyl alcohol and purified water. The gel is applied twice a day for 4 weeks in case of infections with tinea pedis, tinea cruris or tinea corporis. Gels are ideal formulations when physical techniques such as iontophoresis are employed (Yamamoto et al., 2005).

Ointments

Ointments are available for the treatment of nail diseases like onychomycosis and psoriasis, but their use is limited because their hydrophobic components may fail to hydrate the nail plate and therefore drug permeation is poor. Ointment formulations usually contain some of the common ingredients such as paraffins, fatty acids, oils, higher alcohol and waxes. Some of the active ingredients used in nail ointments include calcipotriol, betamethasone dipropionate and salicylic acid. Ointments are generally used in the treatment of subungual hyperkeratosis. Ointment containing urea (40%) and bifonazole (1%) is recommended for topical treatment of onychomycosis (Shivakumar et al., 2012). The ointment is also used for chemical avulsion of dystrophic nail. Urea being a keratolytic agent is likely to disrupt the disulphide bonds in nail plate and the bonds prevalent between the nail plate and nail bed.

Creams

Creams are emulsion-based dosage forms that are used to treat various nail diseases. They can either be oil droplets dispersed in water (O/W) or water-in-oil type (W/O) (Mogensen et al., 2007). The raw materials used for preparing nail creams include paraffins, higher alcohols, fatty acids, oils and waxes along with a suitable emulsifying agent. Loceryl® cream is an O/W cream that contains amorolfine (0.25%) for the treatment of dermatomycoses and tinea infections (Shivakumar et al., 2012). Loprox® Cream that has the same composition of the Loprox Gel contains ciclopirox olamine (1%). The O/W cream is prescribed for nail infections caused by yeasts, dermatophytes and fungi. The hydrophilic continuous phase of these creams has the ability to hydrate the nail plate and thereby improve the transungual delivery of drugs.

Lotions

Nail lotions are intended for topical application to the nail plate for treatment of various nail disorders. Lotions are concentrated solutions or dispersions of one or more therapeutic actives in an aqueous or a hydroalcoholic vehicle. The nail lotion may not necessarily deposit a film on the nail surface following topical application. One of the preparations available for topical application for the treatment of nail diseases is Loprox® Lotion. The lotion that contains ciclopirox olamine (1%) has the same composition of the Loprox Cream or Loprox Gel (Shivakumar et al., 2012). The lotion is often indicated for treatment of superficial dermatophytes, yeasts and fungal infections of the nail.

Patches and Films

The conventional topical preparations used in the treatment of nail disorders warrant frequent applications, as these dosage forms are likely to be easily dissipated from the nail surface and therefore fail to maintain therapeutic levels of the drug at the infected sites. Transungual films and patches provide better adherence and have shown superior penetration and efficacy compared to the conventional nail formulations. Drug-loaded films used for transungual applications are normally produced using the hot melt extrusion process or film casting technique. A transungual patch (area 2.2 sq.cm) containing 3.63 mg sertarconazole in the DuroTek® adhesive (drug-in-adhesive patch) was designed for treatment of onychomycosis by Trommsdorff Arzneimittel GmbH & Co KG, Germany (Shivakumar et al., 2012). On application for 6 weeks, with the patches being replaced every week, the drug concentration in the nail plate far exceeded minimum inhibitory concentrations for the fungi.

Nail Cosmetics

Most ungual diseases are known to cause damage to the nail plate. The infected nails look ugly, thickened and discoloured, which often pose serious cosmetic and medical problems. Nail infections are known to have a serious emotional and psychological impact on the social life of the patient often affecting the quality of life (QoL). Most topical treatment regimens for ungual diseases exhibit a low efficacy demanding treatment regimen ranging from weeks to months. Cosmetic camouflage with various nail cosmetic products is a viable option for patients to disguise the visual impact.

Whilst nail cosmetics are very frequently used for beautification purposes, their application in the concealment of nail diseases is equally important. Nail cosmetics are also of great value in the management of brittle, soft or split nails (Jefferson and Rich, 2012), thus improving nail health and protecting nails from breakage and subsequent infection and disease. Nail polishes are available commercially for improving the appearance of nails in dyschromias and mild dystrophies. Concealment of nail dystrophy can be achieved by the usage of gel or acrylic nails. Nail prosthetics, which include a fake nail adhered onto a silicone finger glove, are available for use in the case of severe nail dystrophy (Jefferson and Rich, 2012).

Nail cosmetics play an important role in maintenance of brittle, soft and splitting nails. Nails generally have the tendency to either absorb or loose water. Brittle nails are often formed when the water content in the nail plate falls to ~16–18% water content (Jefferson and Rich, 2012). Lacquers and moisturizers keep the nail hydrated by sealing in moisture that would otherwise evaporate. Nail lacquers, gels hardeners, shellacs and elongators can be used to enhance the physical thickness and provide protection for soft or fragile nails (Jefferson and Rich, 2012).

The various nail products that can be used to camouflage ungual diseases include:

- Nail lacquers
- Plastic tips
- Nail wraps and 'no light' gels

- Acrylics
- Sculptured nails
- UV-cured traditional nail gels
- UV-cured shellacs

Nail Lacquers as a Cosmetic Product

Nail lacquers are employed to disguise the manifestation of psoriatic nails such as nail pitting, splinter hemorrhages, onychorrhexis, salmon patches and onycholysis. Nail lacquer can help to maintain the hydration of the nail plate by sealing the moisture that otherwise can evaporate. The major ingredients present in nail lacquers include solvents, resins, thixotropic agents, plasticizers, colourants and colour stabilizers (Bryson and Sirdesai, 2010). The base coat resin provides the necessary adherence onto the nail, whereas the topcoat resin gives the glossy appearance. Two or more varieties of resins having different properties are used depending on the effect required. The pliable resins used in development of nail lacquer may include tosylamide-formaldehyde resin, polyvinyl butyral or polyester resin. Hard glossy resins used to develop lacquers are nitrocellulose, methacrylate polymers or vinyl polymers. Solvents such as ethyl acetate and *n*-butyl acetate are often used to dissolve the resins in the formulations. In order to enhance the flexibility of the resin and impart resistance to chipping, plasticizers such as camphor and sucrose benzoate are incorporated in the formulations. Colourants in the form of oxides, silicates or sulfates are used in the lacquers to conceal the affected nail plate. Thixotropic agents are added to the formulations to typically modulate the flow properties. However, nail lacquers may be associated with a large number of side effects such as yellow staining of the nail, increased friability of the nail plate and brittleness in nails (Rich and Kwak, 2010).

Plastic Tips

Plastic tips are preformed tips made of plastic that are used for elongation of the nails. The plastic tips available are either press-on or pre-glued (Schoon, 2010). Adhesive glue such as cyanoacrylate is applied to the dorsal surface of the nail plate following which the plastic tips are pressed on the glued surface for a few seconds to ensure adequate adhesion. However, the sensitizing potential of cyanoacrylate glues cannot to be ruled out. Although nail tips can cover brittle nails and ensure reinforcement, they are not as flexible as the natural nail plates and are therefore prone to onycholysis on mechanical trauma. Further, the onycholytic nails are prone to more secondary infections with *Candida* and *Pseudomonas* (Rich and Kwak, 2010; Schoon, 2010).

Nail Wraps and 'No Light' Gels

Nail wraps and no light gels are less popular nail enhancement techniques accounting for ~1% of the nail cosmetics worldwide market (Jefferson and Rich, 2012). Nail wraps made of linen, silk or fiberglass are applied onto dehydrated nails using cyanoacrylate glue as an adhesive (Lawry and Rich, 1999). The curing process is hastened using solvents like dimethyl tolylamine or any other tertiary aromatic amines. Once it has dried, extra glue is applied on the surface of the wrap so to provide extra protection. Alternatively, nail wraps may be made of vinyl having an adhesive backing that is activated by a suitable heat source. The 'no light' gels are usually composed of cyanoacrylate monomers that are more often thickened with polymethyl methacrylate (Vickery et al., 2010).

Acrylics

Acrylics are liquid and powder systems that are applied on natural or artificial nail plate (Jefferson and Rich, 2012). Acrylics are nail cosmetics that are durable and chip resistant owing to the unique composition. The surface of the nail plate is dehydrated after which a primer composed of a mixture of hydroxylated monomers or oligomers or carboxylic acids that promote the adhesion is applied. A brush is dipped in the liquid monomer and carefully withdrawn to remove the excess. The liquid monomers are usually comprised of ethyl methacrylate (60–95%) and di or tri functional methacrylate monomers (3–5%) along with dimethyltolylamine (0.75–1.25%) (Jefferson and Rich, 2012). The monomers are used for crosslinking and enhancing durability of the systems, whereas dimethyltolylamine acts as a catalyst. The brush is then drawn through a polymer powder so that small beads are formed at the end of the brush. The polymer powders composed of ethyl methacrylate beads (50–80 μm) are coated with 1–2% of benzoyl peroxide that is used as the polymerization initiator (Jefferson and Rich, 2012). The bead is applied on top of the natural nail plate or plastic nail tips repeatedly. Once all the ingredients are added onto the slurry, polymerization begins immediately and hardens in 2–3 minutes. Nearly 95% of the polymerization is completed within 10 minutes. However, the application of acrylics is time-consuming as it can take about 24–48 hours to cure.

Sculptured Nails

Sculptured nails are also used for elongation of the nail using liquid and powder systems. Mylar©- or Teflon©-coated paper placed underneath the natural nail serves as a template onto which the slurry is applied. If properly applied, sculpted nails fit well and are very difficult to differentiate from natural nails (Jefferson and Rich, 2012). Sculpted nails are custom made and require nearly 2 hours to sculpt 10 fingernails.

UV-Cured Nail Gels

Nail gels for nail enhancement are applied to natural nails or tip overlays or for short extensions. These gels can be rapidly cured, and are durable and chip resistant. UV-cured nail gels photocure within 1–3 minutes upon low intensity light exposure. Nail gels are made of urethane acrylate oligomers and cross-linking monomers (75–95%), and photo initiators (1–4%) with dimethyl tolylamine (0.75–1.25%). Prior to application of the nail gels, a dehydrator is applied to the nail surface followed by a primer gel that is cured under UV light. Following the primer, three coats of gel are applied onto the nail surface with intermittent curing after each application. However, exposure to UV lamp has the potential to cause some serious side effects including skin cancer. In order to avoid the associated risks, dermatologists recommend the use of sunscreen and covering the hands with white cloth while exposed to a UV lamp (Lawry and Rich, 1999). Moreover, the eyes need to be shielded from UV light as the exposure is likely to damage lutein pigment at the back of the eye resulting in macular degeneration.

UV-Cured Shellacs

UV-cured shellacs are nail cosmetics that offer better customer satisfaction, as they are known to dry rapidly. Shellacs are chip resistant and can be worn on natural nails comfortably for about 4–6 weeks after which they need to be soaked in acetone to facilitate removal. Shellacs have the same pigments used in the formulation of nail lacquers (Jefferson and Rich, 2012). Shellacs use a base that contains polymerization photoinitiator along with UV curable methacrylates or acrylate oligomers (Jefferson and Rich, 2012). The application of shellac involves six coating steps that includes two coats of the base polymer followed by two coats of pigmented polymers and finally two clear coats with intermittent UV curing after each application. They use methacrylate or acrylate oligomers and monomers in place of resins and standard solvents. However, great care needs to be taken to prevent sensitization of the skin as patients are likely to develop contact dermatitis to uncured excipients (Lawry and Rich, 1999; Vickery et al., 2010).

Miscellaneous Nail Products

A huge number of nail lacquers available commercially are termed as 'nail treatment products'. These lacquers have additional excipients that provide some therapeutic benefits such as reduced nail breakage, enhanced nail growth and prevention of fungal infection (Jefferson and Rich, 2012). Some nail lacquers termed as nail strengtheners are made of iron or calcium, while nail growth enhancers contain silk protein. These formulations are likely to strengthen the nail plate making them less prone to breakage. The major antifungal active ingredients used in the nail lacquers are undecylenic acid (25%), clotrimazole (1%), tolnaftate (1%) and benzalkonium chloride 0.1% (Jefferson and Rich, 2012).

Nail hardeners are products that help to strengthen fragile or flimsy nails. Nail hardeners comprise of formalin or dimethyl urea or glyoxal that act as a crosslinking agents which reduce flexibility of the nail plate thereby imparting brittleness and yellowing of the nails. Nails are usually treated alternatively with nail hardeners and a non-hardening base coat.

Nail moisturizers are lotions or creams that are used to enhance the hydration of brittle nails. A nail moisturizer typically contains occlusives such as petrolatum or mineral oil or lanolin, humectants like glycerin or propylene glycol, and hydrating agents that would include urea or lactic acid (Jefferson and Rich, 2012).

Even though nail cosmetics are safe to use, they can induce diseases, deformities, or allergic and irritant contact dermatitis. Complete knowledge about techniques and materials used is essential before use by patients. Hence, extensive research needs to be carried out to identify the cosmetic causes of nail diseases.

Future Prospects

Treatment of various ungual diseases depends on a number of factors including location of the infection (toe infection or nail infection) and infectious agents (dermatophytes, non-dermatophytes, fungi and yeast). Different formulations are used to treat nails based on the location and severity of the diseases (Kushwaha et al., 2017). Formulations can be used in combination for treating diseases of the nail, but there is no standard treatment protocol for the treatment of these ungual diseases. Generally, a combination of systemic and topical therapies is used for the treatment of moderate cases of ungual disease, whilst topical therapy alone is restricted to mild cases. On the other hand, topical therapy is limited only to mild cases of ungual diseases. Some of the treatment procedures discussed in this chapter are still in the exploratory stage. Studies recently undertaken have suggested that device-based therapies have an added advantage over traditional therapies. Even though various methods are available for therapeutic treatment of nail diseases, the efficacy and precision of the available methods need to be evaluated.

REFERENCES

Albom, M. J. (1977). Surgical gems: Avulsion of a nail plate. *J Dermatol Surg Oncol.* 3(1): 34–35.

Anderson, R. R., and Parrish, J. A. (1983). Selective photothermolysis: Precise microsurgery by selective absorption of pulsed radiation. *Science.* 220(4596): 524–527.

Baran, R., Hay, R. J., and Garduno, J. I. (2008). Review of antifungal therapy and the severity index for assessing onychomycosis: Part I. *J Dermatol Treat.* 19(2): 72–81.

Becker, C., and Bershow, A. (2013). Lasers and photodynamic therapy in the treatment of onychomycosis: A review of the literature. *Dermatol Online J.* 19(9): 196–201.

De Berker, D. (2013). Nail anatomy. *Clin Dermatol.* 31(5): 509–515.

De Berker, D. A. R., Baran, R., and Dawber, R. P. R. (2001). The nail in dermatological diseases. *Baran and Dawber's Diseases of the Nails and Their Management* (3rd ed.). Baran, R., Dawber, R. P. R., de Berker, D. A. R., Haneke, E., and Tosti, A. (eds.). Malden, MA: Blackwell Science. pp. 172–223.

Bryson, P.H., and Sirdesai, S. J. (2010). Colored nail cosmetics and hardeners. *Cosmetic Dermatology: Products and Procedures* (1st ed.). Draelos, Z. D. (ed.). Hoboken, NJ: Wiley-Blackwell. pp. 206–214.

Cecchini, A., Montella, A., Ena, P., Meloni, G. B., and Mazzarello, V. (2009). Ultrasound anatomy of normal nail unit with 18 MHz linear transducer. *Ital J Anat Embryol.* 114(4): 137–144.

Crowley, M. M., Fredersdorf, A., Schroeder, B., Kucera, S., Prodduturi, S., Repka, M. A., and McGinity, J. W. (2004). The influence of guaifenesin and ketoprofen on the properties of hot-melt extruded polyethylene oxide films. *Eur J Pharm Sci.* 22(5): 409–418.

Dawber, R. P. R., De Berker, D., and Baran, R. (1994). *Science of the Nail Apparatus.* London: Blackwell Scientific. pp. 1–34.

Doner, N., Yasar, S., and Ekmekci, T. R. (2011). Evaluation of obesity-associated dermatoses in obese and overweight individuals. *Turkderm.* 45(3): 146–151.

Egawa, M., Ozaki, Y., and Takahashi, M. (2006). *In vivo* measurement of water content of the fingernail and its seasonal change. *Skin Res Technol.* 12(2): 126–132.

Elkeeb, R., AliKhan, A., Elkeeb, L., Hui, X., and Maibach, H. I. (2010). Transungual drug delivery: Current status. *Int J Pharm.* 384(1): 1–8.

Farren, L., Shayler, S., and Ennos, A. R. (2004). The fracture properties and mechanical design of human fingernails. *J Exp Biol.* 207(5): 735–741.

Fitzpatrick, T. B., Polano, M. K., and Suurmond, D. (1983). *Color Atlas and Synopsis of Clinical Dermatology.* New York: McGraw-Hill. pp. 125–134.

Fluhr, J. W., Feingold, K. R., and Elias, P. M. (2006). Transepidermal water loss reflects permeability barrier status: Validation in human and rodent in vivo and ex vivo models. *Exp Dermatol.* 15(7): 483–492.

Gupta, A., and Simpson, F. (2012). Device-based therapies for onychomycosis treatment. *Skin Ther Lett.* 17(9): 4–9.

Gupta, A. K., Paquet, M., and Simpson, F. C. (2013). Therapies for the treatment of onychomycosis. *Clin Dermatol.* 31(5): 544–554.

Gupta, A. K., and Simpson, F. C. (2012). Medical devices for the treatment of onychomycosis. *Dermatol Ther.* 25(6): 574–581.

Hafeez, F., Hui, X., Selner, M., Rosenthal, B., and Maibach, H. (2014). Ciclopirox delivery into the human nail plate using novel lipid diffusion enhancers. *Drug Dev Ind Pharm.* 40(6): 838–844.

Harris, F., and Pierpoint, L. (2012). Photodynamic therapy based on 5-aminolevulinic acid and its use as an antimicrobial agent. *Med Res Rev.* 32(6): 1292–1327.

Helmdach, M., Thielitz, A., Röpke, E., and Gollnick, H. (2000). Age and sex variation in lipid composition of human fingernail plates. *Skin Pharmacol Appl Skin Physiol.* 13(2): 111–119.

Hossin, B., Rizi, K., and Murdan, S. (2016). Application of Hansen solubility parameters to predict drug–nail interactions, which can assist the design of nail medicines. *Eur J Pharm Biopharm.* 102: 32–40.

Jarrett, A., and Spearman, R. I. (1966). The histochemistry of the human nail. *Arch Dermatol.* 94(5): 652–657.

Jefferson, J., and Rich, P. (2012). Update on nail cosmetics. *Dermatol Ther.* 25(6): 481–490.

Jellinek, N. J. (2007). Nail surgery: Practical tips and treatment options. *Dermatol Ther.* 20(1): 68–74.

Kobayashi, Y., Miyamoto, M., Sugibayashi, K., and Morimoto, Y. (1998). Enhancing effect of N-acetyl-l-cysteine or 2-mercaptoethanol on the in vitro permeation of 5-fluorouracil or tolnaftate through the human nail plate. *Chem Pharm Bull.* 46(11): 1797–1802.

Kobayashi, Y., Miyamoto, M., Suibayashi, K., and Morimoto, Y. (1999). Drug permeation through the three layers of the human nail plate. *J Pharm Pharmacol.* 51(3): 271–278.

Kushwaha, A., Jacob, M., Shivakumar, H. N., Hiremath, S., Aradhya, S., Repka, M. A., and Murthy, S. N. (2015). Trans-ungual delivery of itraconazole hydrochloride by iontophoresis. *Drug Dev Ind Pharm.* 41(7): 1089–1094.

Kushwaha, A., Murthy, R. N., Murthy, S. N., Elkeeb, R., Hui, X., and Maibach, H. I. (2015). Emerging therapies for the treatment of ungual onychomycosis. *Drug Dev Ind Pharm.* 41(10): 1575–1581.

Kushwaha, A. S., Sharma, P., Shivakumar, H. N., Rappleye, C., Zukiwski, A., Proniuk, S., and Murthy, S. N. (2017). Trans-ungual delivery of AR-12, a novel antifungal drug. *AAPS PharmSciTech.* 18(7): 2702–2705.

Lawry, M., and Rich, P. (1999). The nail apparatus: A guide for basic and clinical science. *Curr Probl Dermatol.* 11(5): 161, 163–208.

Manda, P., Sammeta, S. M., Repka, M. A., and Murthy, S. N. (2012). Iontophoresis across the proximal nail fold to target drugs to the nail matrix. *J Pharm Sci.* 101(7): 2392–2397.

Markinson, B. C., Monter, S. I., and Cabrera, G. (1997). Traditional approaches to treatment of onychomycosis. *J Am Podiatric Med Assoc.* 87(12): 551–556.

Mogensen, M., Thomsen, J. B., Skovgaard, L. T., and Jemec, G. B. E. (2007). Nail thickness measurements using optical coherence tomography and 20-MHz ultrasonography. *Br J Dermatol.* 157(5): 894–900.

Murdan, S., Milcovich, G., and Goriparthi, G. S. (2011). An assessment of the human nail plate pH. *Skin Pharmacol Physiol.* 24(4): 175–181.

Nair, A. B., Kim, H. D., Chakraborty, B., Singh, J., Zaman, M., Gupta, A., and Murthy, S. N. (2009). Ungual and trans-ungual iontophoretic delivery of terbinafine for the treatment of onychomycosis. *J Pharm Sci.* 98(11): 4130–4140.

Pandhi, D., and Verma, P. (2012). Nail avulsion: Indications and methods (surgical nail avulsion). *Indian J Dermatol Venereol Leprology.* 78(3): 299–308.

Repka, M. A., Mididoddi, P. K., and Stodghill, S. P. (2004). Influence of human nail etching for the assessment of topical onychomycosis therapies. *Int J Pharm.* 282(1): 95–106.

Rich, P., and Kwak, H. S. R. (2010). Nail physiology and grooming. *Cosmetic Dermatology: Products and Procedures* (1st ed.). Draelos, Z. D. (ed.). Hoboken, NJ: Wiley-Blackwell. pp. 197–205.

Richardson, D., and Schwartz, R. (1984). Comparison of resting capillary flow dynamics in the finger and toe nail folds. *Microcirc Endothelium Lymphatics.* 1(6): 645–656.

Rigopoulos, D. (2008). Acute and chronic paronychia. *Am Fam Physician.* 77(3): 339–348.

Roberts, D. T., Taylor, W. D., and Boyle, J. (2003). Guidelines for treatment of onychomycosis. *Brit J Dermatol.* 148(3): 402–410.

Scher, R. K. (2007). Nail disorders – One of dermatology's last frontiers. *Dermatol Ther.* 20(1): 1–2.

Schoon, D. (2010). Cosmetic prostheses as artificial nail enhancements. *Cosmetic Dermatology: Products and Procedures* (1st ed.). Draelos, Z. D. (ed.). Hoboken, NJ: Wiley-Blackwell. pp. 215–221.

Sheehan, D. J., Hitchcock, C. A., and Sibley, C. M. (1999). Current and emerging azole antifungal agents. *Clin Microbiol Rev.* 12(1): 40–79.

Shivakumar, H. N., Juluri, A., Desai, B. G., and Murthy, S. N. (2012). Ungual and transungual drug delivery. *Drug Dev Ind Pharm.* 38(8): 901–911.

Shivakumar, H. N., Juluri, A., Repka, M. A., and Murthy, S. N. (2012). Approaches to enhance ungual and trans-ungual drug delivery. *Topical Nail Products and Ungual Drug Delivery.* Murthy, S. N., and Maibach, H. I. (eds.). Boca Raton, FL: CRC Press. pp. 87–123.

Siegle, R. J., and Swanson, N. A. (1982). Nail surgery: A review. *J Dermatol Surg Oncol.* 8(8): 721–726.

Sirota, L., Straussberg, R., Fishman, P., Dulitzky, F., and Djaldetti, M. (1988). X-ray microanalysis of the fingernails in term and preterm infants. *Pediatric Dermatol.* 5(3): 184–186.

Smijs, T. G., Haas, R. N., Lugtenburg, J., Liu, Y., Jong, R. L., and Schuitmaker, H. J. (2004). Photodynamic treatment of the dermatophyte trichophyton rubrum and its microconidia with porphyrin photosensitizers. *Photochem Photobiol.* 80(2): 197–202.

South, D. A., and Farber, E. M. (1980). Urea ointment in the nonsurgical avulsion of nail dystrophies-a reappraisal. *Cutis.* 25(6): 609–612.

Stern, D. K. (2007). Water content and other aspects of brittle versus normal fingernails. *J Am Acad Dermatol.* 57(1): 31–36.

Subissi, A., Monti, D., Togni, G., and Mailland, F. (2010). Ciclopirox: Recent nonclinical and clinical data relevant to its use as a topical antimycotic agent. *Drugs.* 70(16): 2133–2152.

Terry, R. (1955). The onychodermal band in health and disease. *The Lancet.* 265(6856): 179–181.

Van Duyn Graham, L., and Elewski, B. E. (2011). Recent updates in oral terbinafine: Its use in onychomycosis and tinea capitis in the US. *Mycoses.* 54(6): 679–685.

Vellar, O. D. (1970). Composition of human nail substance. *Am J Clin Nutr.* 23(10), 1272–1274.

Vickery, S. A., Wyatt, P., and Gilley, J. (2010). Eye cosmetics. *Cosmetic Dermatology: Products and Procedures* (1st ed.). Draelos, Z. D. (ed.). Hoboken, NJ: Wiley-Blackwell. pp. 190–196.

Wortsman, X., and Jemec, G. B. E. (2013). *Dermatologic Ultrasound with Clinical and Histologic Correlations.* New York: Springer. pp. 39–72.

Yamamoto, T., Yokozeki, H., and Nishioka, K. (2005). Clinical analysis of 21 patients with psoriasis arthropathy. *J Dermatol.* 32(2): 84–90.

Yazdanparast, S. A., and Barton, R. C. (2006). Arthroconidia production in trichophyton rubrum and a new ex vivo model of onychomycosis. *J Med Microbiol.* 55(11): 1577–1581.

Zaias, N., Glick, B., and Rebell, G. (1996). Diagnosing and treating onychomycosis. *J Fam Pract.* 42(5): 513–519.

23

Packaging of Cosmetic and Personal Care Products

Antônio Celso da Silva, Celio Takashi Higuchi, Heather A.E. Benson
and Vânia Rodrigues Leite-Silva

CONTENTS

In today's world, and indeed throughout history, people are greatly concerned about looking their best. This underpins both the fashion and beauty industries. Consumers are constantly looking for new cosmetic products that will let them appear younger; a new colour of make-up that can cover blemishes; or even a new fragrance, with the ability to satisfy even the most refined sense of smell. The 'beauty industry' responds by the almost constant launch of new products into the market. It is a highly competitive industry: those who do not constantly innovate will be doomed for failure and most likely end up in bankruptcy. Consequently manufacturers seek to create raw materials for industries that can provide positive and fast results, and are clinically tested.

When a new active ingredient reaches the cosmetic industry laboratory, the concern of the cosmetic formulator is to incorporate it in a safe, efficient vehicle, with a pleasant fragrance and good sensorial properties that will facilitate the sale. Another aspect is the physical and chemical stability, as this formulation will be exposed to various temperatures from its manufacture through transportation and storage at the point of sale, until the arrival in the hands of the ultimate customer.

What about the packaging of the cosmetic product? Normally, it is left to a later stage in product development. However, considering the sale of the product, packaging is one of the most important factors for attracting a customer and, indeed, for many products the packaging is more expensive than the contents.

Historically, packaging was very important because there was a need to protect and transport products. First, natural resources such as leaves, baskets and balloons were used. According to Cavalcanti and Chagas (2006) man began by using leaves, leather, horns and animal bladders, before using ceramics, glass, fabrics, wood and paper until reaching the modern era of cosmetic packaging that involves a wide range of both synthetic and natural materials.

The Egyptians were thought to have first used glass as a packaging material for oils and perfumes around 3000 BC (Moura and Banzato, 2013), and glass continues to be used as a packaging material for many cosmetic products (Figure 23.1).

For many years, little progress was made in the packaging field. The greatest advance occurred around 100 BC when glassmakers mastered glass blow and molding techniques, which provided the production of containers with volumetric capacities, and different formats, for wider production (Evangelista, 2001). By 800 AD the Chinese began to produce paper from flax fibres on a small scale. Production of this material spread to and through Europe, and there are records to indicate that paper began to be produced in England from 1300 AD (Moura and Banzato, 2013). Around 1800 AD in England a technique was developed for preserving food from heat and air-induced degradation. The first packaging to be used with this technique was glass, but this was soon replaced by tin in the form of a can because it was more resistant, especially to mishandling. The beginning of the modern era of packaging had begun (Evangelista, 2001).

Of course packaging must also fulfill other functions including maintaining product security, stability and ease of application. In this chapter, we introduce the important aspects of packaging a cosmetic formulation.

Packaging Functions

Packaging performs a number of important functions:

- *Conditioning* – Primary packaging is in direct contact with the product. Examples of primary packaging: bottles, vials, jars, cases, tubes and trays.
- *Protection* – Secondary packaging does not come into direct contact with the product. The main function of secondary packaging is to protect the primary container, thus avoiding damage during transport or in storage. Examples are cartridges, cribs, partitions and hives.
- *Information* – The primary and/or secondary packaging inform the consumer how to use the product, its components together with warnings and contraindications. In addition to the primary packaging itself, labels, stickers, leaflets and brochures can also be used for consumer information.
- *Sealing* – The primary or secondary packaging must prevent any leakage of the product. This is achieved through the use of lids, stoppers, disks, and so forth.
- *Facilitate use* – The primary or secondary packaging can facilitate the use of the product by the consumer. Examples of this are spray or metering valves, flip-top or disc caps, applicators, and pots and jars that incorporate an ergonomic design.
- *Transport* – A vital function is to protect the product during transportation and storage.

FIGURE 23.1 Examples of glass bottles used as packaging for oils. (From http://jrfmdesign.blogspot.com.br/2010/09/design-de-embalagem.html.)

Cosmetic Packaging: Key Differences with Other Products

The packaging of cosmetic products shares many similar features with packaging for foods, drinks or medicines, as outlined earlier. One area where there is a difference is in the importance of packaging aesthetics. In the case of a pharmaceutical product, the consumer normally discards the secondary packaging, and applies or consumes the drug content as directed within a relatively short period of days or weeks. Pharmaceuticals that are provided on prescription need the secondary packaging mainly to provide function and visual differentiation/ease of identification. In the case of over-the-counter pharmaceuticals, foods and beverages the packaging is optimised to attract customer attention in order to encourage a purchase. Again, once the purchase is made, the product tends to be consumed or used fairly quickly.

Clearly cosmetic products also need attractive packaging as they compete, often with many other similar products, to win a customer's attention. However, it is even more important for cosmetics where the product may be purchased as a gift. An attractively packaged cosmetic product can be on display in the consumer's home for a long period of time.

Thus the cosmetic appearance of the packaging is of paramount importance and there is a pronounced emphasis on the colours, varnish, lacquer, shape and the use of stamping foil gold. These considerations generally result in a higher cost of packaging for cosmetics compared to other products.

FIGURE 23.2 Examples of luxury bottles created as works of art. (From https://adonabela.files.wordpress.com/2013/11/guerlain.jpg.)

The fragrance and cosmetic industries specialize in the art of creating fragrances and beauty products with a brand philosophy that is promoted by the packaging. The richness of detail is synonymous with glamour, through the choice of raw materials of the highest quality. Figure 23.2 shows examples of bottles created as works of art, often by a renowned designer.

Materials Used in Packaging

Packaging can be produced from a variety of materials depending on the desired overall appearance, type of product, stability/compatibility issues related to the formulation and cost constraints. The most common packaging materials used are briefly described (see also Table 23.1).

Plastics

The most commonly used materials for cosmetic packaging are resins, which are classified and identified by numbers (1–7) in a 'triangle' that is usually on the bottom of the pack (Figure 23.3). These symbols provide information that also allows the plastics to be identified for recycling as designated by the triangular arrows.

Polyethylene Terephthalate (PET)

PET is one of the most widely used resins because of its transparency, flexibility and memory (i.e. the ability to return to its original shape following deformation), and good compatibility with most cosmetic ingredients.

TABLE 23.1

Types of Packaging Materials and Their Applications in Cosmetic Products

Type of Material	Examples of Application	Examples of Final Packaging in Cosmetic Products
Plastics	Bottles, jars, lids, self-adhesive labels, heat-shrinkable sleeve, labels, tubes, bungs, cartridges, cases, cribs and flacons	Shampoos, conditioners, children's products, lotions, creams, bath oils, liquid soaps, creams, make-up remover, eyelash mask, lipstick, compact powder, blush and eyeshadow make-up
Glasses	Jars and flacons	Bottles for perfumes, cream jars, bottles of cosmetic treatment lotions
Papers	Cartridges, labels, brochures, boarding boxes and cribs	Secondary packaging for perfumes and treatment creams
Metals	Aluminium trays, covers, rings, cases and covers	Aerosol deodorant (aluminium), lid covers and valves used in liquid soaps, colonies, lotions, coadjuncts in packaging for make-up

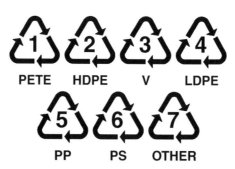

FIGURE 23.3 Symbols used to identify seven commonly used recyclable plastics. (From https://populationeducation.org/6-common-recycling-myths-debunked/.)

FIGURE 23.4 Examples of PET bottles. (From http://www.scamplastic.com.br/embalagens-pet-cosmeticos.)

The good durability of PET is ideal for transparent bottles of shampoos and conditioners that may often slip from the hands and fall to the ground. The bottle will dent, without cracking, unlike other transparent resins.

With the addition of pigments, PET bottles or pots become coloured, transparent or translucent. They can also be completely white by the addition of titanium dioxide or pearlescent by adding mica-based pearl luster pigments such as Iriodin®. The addition of sunscreen to the resin confers protection from light to the inner product without the loss of packaging transparency (Figure 23.4).

High-Density Polyethylene (HDPE)

HDPE is a non-transparent plastic resin that is also widely used in cosmetic packaging of bottles of shampoo, conditioner, lotion and sunscreen. It can be white, coloured or pearlescent by adding pigments and pearl (Figure 23.5).

Polyvinyl Chloride (PVC or V)

PVC offers transparency, but is brittle and the packaging is likely to crack or break if dropped. PVC has little memory and can be coloured with pigments. It is most commonly used in bottles of shampoo, conditioner, lotions, tubes of mascara and lip gloss (Figure 23.6).

Low-Density Polyethylene (LDPE)

LDPE has similar characteristics to HDPE but is more malleable. It is widely used in the packaging of products that require good memory, such as squeeze bottles (Figure 23.7).

FIGURE 23.5 Examples of HDPE bottles. (From https://www.indiamart.com/proddetail/plastic-hdpe-cosmetic-bottle-11857486330.html.)

FIGURE 23.6   Examples of PVC bottles. (From http://www.scamplastic.com.br/fabrica-bisnagas-cosmeticos.)

FIGURE 23.7   Examples of LDPE bottles. (From https://pt.dhgate.com/product/600pcs-soft-style-large-ldpe-eye-dropper/268683529.html.)

Polypropylene (PP)

PP is a translucent plastic resin used in bottles for shampoo, conditioner, jars for creams and products containing a high concentration of vegetable or mineral oil such as bath and sun tanning oils. It has lower porosity than PE, thus reducing the potential for migration of product components through the packaging.

PP is used to produce the majority of lids for containers, which can be flip top, top disc or a blind cover. The discs that are placed under the cover of a cream so that the product has no direct contact with the lid are also generally composed of PP (Figure 23.8).

FIGURE 23.8   Example of PP packaging. (From https://pt.aliexpress.com/item/50pcs-lot-Hot-13-Plastic-Empty-Lipstick-Containers-Packaging-PP-Lipstick-Tube-Lip-Balm-Tube-DIY/32812057893.html.)

FIGURE 23.9 Example of PS packaging. (From https://pt.aliexpress.com/item/10ps-Empty-black-80g-Aluminum-Pot-Jars-Cosmetic-Containers-With-Lids-aluminum-jar-with-window-empty/32814365135.html.)

Polystyrene (PS)

PS is used in packaging for make-up, lipstick, compact powder, blush, eyeshadow and covers (Figure 23.9).

Other Packaging Materials

This category consists of all resins that are not classified in the other categories of plastics and includes ethylene-vinyl acetate (EVA), acrylates, and ionomer resins, such as the Surlyn® brand.

Fabrication of Plastic Containers

There are many manufacturing processes that can be used to form containers, of which the two most important are blowing and injection processes.

The blowing process (Figure 23.10) used to produce bottles and jars consists of placing solid plastic resin into an extruder, heating to the appropriate melting temperature, then extruding in the form of a hose. This hose is held within a two-part mold that has the internal form of the package to be fabricated. Air is then blown via a nozzle into the hose, which expands and takes the form of the mold. After the

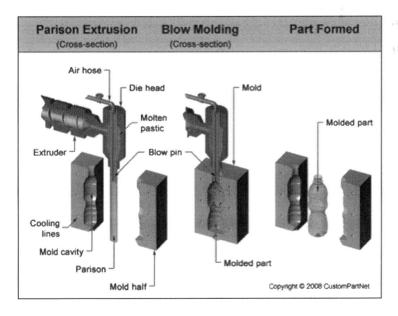

FIGURE 23.10 Extrusion blow molding. (From http://www.custompartnet.com/wu/blow-molding.)

blowing process, a knife cuts the hose outside of the outer upper part of the mold, and the plastic container is then transferred to a tunnel with a low temperature. This solidifies the resin in the shape of the mold. The wall thickness of the container is automatically controlled so that each container is consistent and in accordance with predefined technical specifications. The blowing process is generally used for plastics 1–5.

The injection process also involves placing the solid resin in an extruder and melting at a specified temperature. After fusion, the liquid resin is injected into the interior of a closed mold, which will completely fill the form. This mold tunnel passes through a low temperature at which the resin solidifies. This process is more complex, with controls at various stages, and also involves other equipment that form part of the assembly. The injection process is used for plastic 6, some containers composed of plastic 1 and covers/lids composed of plastic 5.

In both fabrication methods information such as the name of the company, the number of the mold cavity and the resin identification can be embossed onto the packaging. This is achieved by simply embossing the information on the body of the mold. The bottom of the pack is the most common location to place this information.

Glass: Transparent Painted Frosted

Glass is widely used in bottles, pots and jars for fragrances, skin care creams and lotions, foundations, and other cosmetic products. It is highly compatible with cosmetic components, non-porous and elegant. Glass may be coloured by adding pigments, painted, etched, frosted or colour wrapped in plastic film.

Paper

Paper is commonly used for labelling on containers, outer packaging and shipping boxes.

Metals

Metals such as aluminium are most commonly used in aerosol deodorants; covers and valves used in liquid soaps, colognes and lotions; or as adjuncts in packaging for makeup. Occasionally cosmetic products are packaged in metal tubes.

In addition to primary and secondary packaging, there are often other accessories that form part of the whole cosmetic package. These range from embellishments and ribbons on containers to pencil sharpeners, sponges, applicators and brushes. Valves or pumps are often included on metal containers to facilitate their use.

Incompatibility of Packaging Products

Packaging and cosmetic formulations are comprised of multiple chemical components, therefore incompatibilities between these components can occur, and must be predicted or identified and avoided. Specific incompatibilities can result in chemical or physical changes to the cosmetic product. This can affect all aspects from the appearance to the consistency of the product. It is crucial to submit the whole package/product to specific testing to assess both chemical and physical changes resulting from incompatibility between the cosmetic formulation and its packaging. These tests should be initiated when the primary product and packaging are already defined.

An example of compatibility testing is to expose the product to different temperatures, simulating normal exposure during transport, storage and use. Testing therefore comprises a number of different environmental conditions:

1. Room temperature, simulating the product in the consumer's home
2. Controlled temperature (around 45°C), simulating the product being displayed in a store window that is in direct sunlight or being transported in a truck

3. Refrigerated at 10°C, simulating the product in a colder environment
4. Exposure to solar radiation, popularly known as a 'window test', for example, when the consumer leaves the product in the bathroom window

When beginning the testing of a product, a spreadsheet for recording all measurements is prepared. Daily assessments are conducted over a period of 7 days. After this period, weekly assessments are made for a period of 90 days. If there is no interaction between the product and packaging over this period, then there is a high probability that they are compatible.

Experiments performed at different temperatures are usually testing the sensory aspects of the product within the packaging (e.g. colour, odour and appearance, pH and viscosity). The packaging colour, overall appearance, migration, and increase or decrease in size and/or weight, among other variations, must be evaluated. The results recorded at each time point are compared with the initial product (Mutsuga et al., 2005).

In addition to the chemical compatibility between the product and its packaging, the physical characteristics of the product must also be evaluated. For example, depending on the viscosity of the product, the ideal package for a high-viscosity product is a wide-mouth jar, and for a medium or low viscosity a bottleneck with lower inner diameter is preferred.

The physical compatibility test does not require the same monitoring as the chemical compatibility test; however, it is of paramount importance to its completion, so as to prevent any future problems in the use of the product or during the filling process on the production line (Rosa, 2008).

Interaction of Packaging with Equipment in the Production Line

The process of developing a new cosmetic product including its packaging requires a team effort, involving engineering and production personnel.

This will allow the team to highlight and resolve potential problems in the filling process, such as verifying the availability of equipment to package the product in the proposed packaging. Many aspects must be considered such as ensuring that accessories such as a filling nozzle will be compatible with the diameter of a bottle cap closure.

Quality Control of Packaging

In the case of packaging, quality is linked to compliance with the rules, procedures and technical specifications, associated with fitness for purpose and consumer satisfaction. Quality control (or inspection) of the packaging involves the evaluation of each batch by sampling the receipt of products, aiming to detect defects, and can even be done in advance by the provider, prior to the arrival of the consignment to the factory.

Inspections

Inspections are ways by which one evaluates a batch of packaging. For this to be possible, a 'technical design' brief, which is a detailed document containing a list of all the parts of the piece with their measurements and respective tolerances, is used. Every package has its own technical design that must be supplied by the manufacturer of the packaging and must be updated for every minimal change.

Assessments can be made by

- *Variables.* Variables are evaluated defects that may vary, also known as dimensional. Normally used for functional or destructive testing. Examples: weight, volume, height, width, thickness.
- *Attributes.* Visual inspections are made to detect subjective or pattern defects. Typically maximum and minimum allowances to the patterns are provided for the inspection process. Colour, stain, dirt, markings, bubbles and illegibility are some examples (ANVISA, 2008).

Sampling Plan

The sampling plan defines how many units should be analyzed in each batch received. Sampling can never be targeted, that is, samples should not be taken from a single box or package. Packages must be picked at random and be representative of the full batch.

Defects

Defects are the focus of the inspections after the batch has been sampled. Defects in packaging may hinder their use, damage the company's image, create a risk to consumer health, damage important information on the package, or may cause problems in the production line. Defects in a package differ from pre-defined standards and specifications, and are typically classified into three levels, namely (1) critical, (2) major or severe and (3) minimum or small.

Critical are the most serious defects and therefore have lower tolerances. They prevent the use of the packaging, endanger the health of the consumer, prevent legal information, fail to comply with current legislation, and could result in the company being fined and cause problems with law enforcement agencies.

Major or severe defects do not have the same intensity as critical defects but are still considered a serious defect. The packaging can be used but may degrade the image of the company, cause problems in the production line and raise doubts about the information.

Minimum or small defects are only noticeable to the trained eye, but not to the consumer. However, these defects may turn into a serious or critical defect if not corrected in time.

Technical Specification

The technical specification is an official document of the company that defines both visual and dimensional parameters for a package. It should be simple to read and be available to the department conducting the quality control of the packaging. It is prepared by the packaging development department and, in the case of a new packaging and/or a new mold, it should include the participation of the supplier. It is also a technical document to be submitted to qualify a new supplier, whether or not it is for exclusive packaging. It must be periodically updated and the documented copies controlled by the technical department.

A specification document must provide a full description of the package including information such as the class to which the packaging material belongs, and the visual and dimensional parameters with the respective tolerances.

Microbiological Control

Cosmetic formulations, depending on the components used, can have a high likelihood of microbiological contamination. Obviously, for such products there should be greater concern with regard to the control of microbiological contamination in the primary packaging.

In most cases the high temperatures used during the manufacturing process reduce the possibility of initial contamination within the final product. Typically, contamination occurs during storage of the packaging, either by the supplier or the product manufacturer.

Thus, it is important to periodically perform microbiological analysis on stored containers and require the manufacturers of the supplied bottles and jars to place them upside down in clean plastic bags, boxed and with proper identification.

Qualification of Supplier

In the same way as suppliers of raw materials, packaging suppliers must be pre-qualified. The technical department of the company should complete the qualification process. If there is a department of packaging it will be responsible for visiting the supplier, evaluating whether the supplier meets the technical specification of the package to be provided and providing approval (or disapproval) of the supplier. Supplier qualification should be periodically reviewed, in accordance with a time schedule established by the customer and accepted by the supplier.

Importance of Packaging in Development of a New Product

The development of a new product begins with the preparation of a briefing. The more detailed the briefing, the greater the likelihood of success of the new product development process.

The initial concern is the type of material in the primary packaging, as it will have direct contact with the product.

As outlined earlier, compatibility tests simulate situations, such as locations and temperatures to which the product will be exposed. Just as important as these tests is the definition of weights or volumes that will be declared on the packaging. In terms of the declaration of mass (weight), it is necessary to know the density of the product, since there can be variations in the volume during the filling process. Equally important is the expansion chamber (headspace) required to support volume changes in the package due to the variation of temperature in the various environments during the product life.

Logistics and Storage

In a cosmetics factory, the storage space needed for packaging is one of the largest uses of floor area, often comparable with the manufacturing area, a fact that should be taken into consideration when determining the plant size.

To facilitate logistics, the storage area containers should be close to the product packaging area, and the quality control department of packaging should also be in close proximity to the receiving area.

Wherever possible, all packaging should be purchased from the same manufacturer to facilitate a perfect coupling between all components such as lid and jar or lid and pot. Sometimes tenths of a millimeter can make the difference between a good and a bad coupling of two packages, interfering directly with the quality of the final product.

To maintain the quality of the packaging during storage, the stacking information contained on the outside of the boxes must be observed.

Importance of Packaging in Marketing and Selling the Product

In addition to the cosmetic product, you are selling beauty, well-being, self-esteem and seduction. Packaging is the first contact with the consumer and the product must convey these messages. Practicality is also important; the packaging should be ergonomic and dispense the product in an easy manner with the correct amount for the use of the product.

Existing technological resources currently allow technicians to create the most fascinating packaging, letting the consumer be truly enchanted. Since the packaging is such a visual part of the product, it should have a significant role in each stage of the product's development. For this reason, technicians are required to have timely training and current knowledge of the process.

Making the analogy to the human body, the product is 'the heart', but the packaging is 'the skin', so it is critical to create a good first impression.

XI Eco Packing for Cosmetic Products

Over the years, designers have realized the importance of packaging and design, with the brand delivering a strong message to the consumer. Plastic has become more economically viable, enabling innovative ideas. However, concern about the environment is rapidly escalating due to discarded plastic waste. Consequently the use of biodegradable packaging and the concept of recycling are increasingly important and actively promoted by packaging associations around the world (ABRE).

Pinatti (2000) defined 'ecologically correct packaging' as being produced with single, simple materials, easy to acquire and produce, with less energy use, that would enable its reuse or post-use recycling. Ecologically correct packaging should be compatible with the product, meet the minimum needs of society and be sustainable.

Thus, the determination of a precise meaning for 'ecologically correct packaging' depends on the assessment of how much this item 'costs' to the environment from the quantification of what natural resources it consumes; energy; and emissions its production generates for water, air and soil (Garcia, 2005).

In view of this, it is concluded that it is very difficult, if not impossible, to present a single and absolute definition for what would be ecologically correct packaging, since there is no universal ideal packaging. Perhaps the aim should be to use the most environmentally appropriate packaging against a specific set of values (Amaral, 2008).

REFERENCES

ABRE (Brazilian Packaging Association). http://www.abre.org.br/setor/apresentacao-do-setor/a-embalagem/funcoes-das-embalagens/. Accessed on 21 November 2016.

Amaral, L.A. O Ecodesign da embalagem. II Encontro De Sustentabilidade Em Projeto Do Vale Do Itajaî, 2008.

ANVISA. Agência Nacional de Vigilância Sanitária. Guia de controle de qualidade de produtos cosméticos/ Agência Nacional de Vigilância Sanitária. 2a edição, revista – Brasília: Anvisa, 2008.

Cavalcanti, P.; Chagas, C. A história da embalagem no Brasil. São Paulo: Grifo Projetos Históricos e Editoriais, 2006.

Evangelista, J. Tecnologia de Alimentos. Quarta reimpressão da segunda edição. São Paulo: Editora Atheneu, 2001.

Garcia, E. Eficientes, limpas e económicas. http://www.revistapesquisa.fapesp.br/index.php?art=1650&bd=1&pg=2&lg=. Accessed on 25 May 2006.

Moura, R.; Banzato, J.M. Embalagem, unitização e conteinerização. Quarta edição. São Paulo: IMAM, 2013.

Mutsuga, M.; Tojima, T.; Kawamura, Y.; Tanamoto, K. Survey of formaldehyde, acetaldehyde and oligomers in polyethylene terephthalate food-packaging materials. *Food Additives and Contaminants*, 22 (8), p. 783–789, August 2005.

Pinatti, A.E. O desígnio de embalagens de consumo e meio ambiente. Tese de Doutorado FAU/USP. São Paulo, 2000.

Rosa, Faena M. Simulação numérica da migração de elementos metálicos e do monômero ξ-caprolactama de embalagens poliméricas irradiadas para simulantes de alimentos. 2008. 61 f. Dissertação- Instituto de pesquisas energéticas e nucleares IPEN, São Paulo, SP, 2008.

24

Sensory Analysis Applied to Cosmetic Products

Regina Lúcia F. de Noronha, Heather A.E. Benson and Vânia Rodrigues Leite-Silva

CONTENTS

The sensory profile of a cosmetic product, that is, its fragrance, consistency, appearance (colour, transparency), and the feel on the skin and hair (stickiness, ease of spread, foam creaminess, etc.), as well as the benefits it provides (moisturization, gloss, clean feel, film forming, etc.), are essential for its acceptance and repeated use by the consumer.

Sensory analysis is an extremely useful and relevant tool for developing cosmetic products that meet the consumer's expectations and deliver their promised benefits, with adequate sensory characteristics that enhance consumer acceptance.

The methods of sensory analysis can be used for different decision-making processes, from learning and determining the market and the competition sensory profile, to developing and refining prototypes, and assessing formulations potentially more successful among consumers. However, choosing the most appropriate methods to be used and the technical requirements for their execution, and analysis of the data collected, requires a clear understanding of the objectives, possibilities and restrictions of each method.

The objective of this chapter is to inform the reader about the different sensory analysis methods used in the cosmetic industry and to instruct the reader on the premises and technical rigor in handling and analyzing data. In this chapter, we provide examples that will be useful for demonstrating and understanding the technical content and will facilitate the readers use of sensory analysis in developing and evaluating sensory attributes of cosmetic products in which they might be interested.

Introduction

The importance of the sensory characteristics of cosmetic products and their perception by consumers is of paramount importance when considering the formulation of cosmetic products that will be successful in the market. Indeed Wortel and Wiechers (2000) stated that whilst efficacy and safety are important considerations, sensory characteristics represent the most important aspect of a product's sales potential. If the consumer does not like how the cosmetic product feels on the skin, they will not continue to use it and make further purchases of the product, or potentially limit further purchases of the entire brand.

Sensory evaluation studies of cosmetic products have been developed and refined, and are routinely performed by scientists to evaluate cosmetic formulations (Lukic et al., 2012; Ozkan et al., 2012; Gilbert et al., 2012; Iwata and Shimada, 2013). These studies are indispensable for the objective assessment of products and also provide the opportunity to elucidate the most appropriate words and phrases to be used for the advertising of new products. Recently there has been a trend towards complementing sensory studies with the evaluation of the influence of products and fragrances on mood and emotions. This provides a broader knowledge of perceptions and sensations, and thus an overall view of the consumers' assessment of the product, allowing optimization of the experience of use and assessment of the drivers for purchase and subsequent repurchase of products released in the market (Retiveau et al., 2004; Churchill and Behan, 2010; Jaeger et al., 2013).

Schifferstein and Desmet (2010) stated that consumers experience different kinds of emotions in their relationships with cosmetic products. These emotions can be elicited by the products' sensory characteristics, the cosmetic effects of product usage (efficacy), memories, associations and the social implications of product usage. It is indeed a complex relationship, but one that requires some qualitative and quantitative measures in order to facilitate the design and optimization of successful cosmetic products.

This chapter addresses the theoretical concepts and practical examples of sensory analysis of cosmetic products with particular focus on the following key questions:

- How does the skin react to different cosmetic products?
- How do you choose the most appropriate methodology to objectively measure relevant sensory attributes and consumer perception of a cosmetic product?
- How do you measure the pleasure – the hedonic response – caused by the application of a cosmetic product on skin?
- How do you choose the appropriate study to properly substantiate sensory claims?

This chapter aims to guide the reader in selecting the most adequate methodology(s) with regard to the stages and/or objectives of the cosmetic product development project. The methods are illustrated with examples, and further reading is recommended to permit the reader to gain a detailed understanding of the different applications of sensory analysis in the cosmetic field and how to correctly perform the analyses.

Touch and Haptics: How Do We Feel? What Do We Feel? Why Are These Sensations So Important for Cosmetic Formulation and Sensory Marketing?

Although sight and smell contribute, touch is the most important sense in the perception of a cosmetic product. Cutaneous receptors comprise the various types of cutaneous mechanoreceptors that sense touch, texture, pressure and vibration; thermoreceptors that sense temperature; and nociceptors that sense pain. When stimulated, the receptors trigger nerve impulses that travel to the somatosensory cortex in the parietal lobe of the brain, where they are transformed into sensations. Sensitivity to touch varies greatly among different parts of the body, with very sensitive areas such as the fingers and lips, corresponding to a disproportionately large area of the sensory cortex (Frings, 2012). An important aspect of the sense of touch is that receptors are primarily stimulated by change (change in texture, pressure, etc.)

and is sequential in nature. This characteristic can be used strategically for sensory marketing and product development.

Whilst the final site of application may involve any tactile surface on the human body, the first point of contact with the cosmetic product generally begins with the hands, by grabbing, applying and spreading over the area of final application (face, body, hair, feet and/or hands). Thus Peck (2010) indicated the hands as the primary source of input to the perceptual tactile system. Gibson (1966) adopted the term 'haptics' to refer to the functionally discrete system involved in the seeking and extraction of information (perception) by the hand. Therefore, when developing a cosmetic product, the sensory aspects that will be perceived by the hands (haptics) and by the body (touch) in the application area should be taken into consideration.

Raw Materials: Sensations Provided on Skin

The skin sensory characteristics of a cosmetic product are due to the sensory qualities contributed by each ingredient and the physical form in which they are formulated into the final product (Wortel and Wiechers, 2000). The sensory properties and skin performance of emulsions is highly dependent on the emollients in the formulation (Salka, 1997). They determine the consistency and spreadability, which are very important properties to achieve adequate efficacy and consumers' acceptance of the products (Parente et al., 2008). Lipophilic emollients are one of the most commonly used ingredients of cosmetic emulsions because they offer a wide range of sensations and properties when applied on the skin, including waxy, greasy, oily, shiny, sticky and velvety (Kamershwarl and Mistry, 2001).

Parente et al. (2008) used a descriptive sensory analysis to characterize the sensory properties of four commonly used emollients: cyclomethicone (CM), dimethicone (DM), isopropyl myristate (IPM) and octyldodecanol (OD); and four emollients from ratite oil: emu oil (EO), nandu oil (NO), olein (glyceryl trioleate: O15) and estearin (E15). A trained sensory panel composed of 12 assessors evaluated 5 attributes (spreading difficulty during application, gloss and stickiness immediately after application, residue and oiliness immediately after application, and after 5 and 10 min) using a structured 10-point scale (0=nil; 10=high).

Parente et al. (2008) observed that although CM and DM are both silicones, they showed very different sensory profiles (Table 24.1), which suggests they could provide a wide range of characteristics when used in cosmetic emulsions. EO and NO showed similar sensory profiles, reflecting their similar composition. In general, the authors concluded that the solid content of an emollient at skin temperature

TABLE 24.1

Average Scores and ANOVA Results Followed by Fisher's Least Significant Difference Test

Emollient	Difficulty of Spreading	Gloss	Stickiness	Residue			Oiliness		
				t0	t5	t10	t0	t5	t10
ÑO	2.2[c]	9.3[a]	3.5[c]	9.5[a]	8.7[a]	8.2[a]	9.4[a]	9.1[a]	8.3[a]
CM	0.9[de]	3.0[d]	0.6[c]	3.2[e]	1.3[c]	0.3[f]	3.8[c]	1.3[f]	0.5[c]
DM	8.9[a]	8.0[b]	8.6[a]	8.5[bc]	7.7[b]	6.3[b]	7.4[b]	6.6[de]	4.8[c]
OD	0.9[de]	9.1[a]	1.7[d]	8.0[cd]	6.4[c]	4.9[cd]	9.3[a]	7.9[bc]	5.2[c]
E15	8.0[b]	6.0[b]	7.4[b]	7.8[cd]	6.7[c]	4.1[d]	7.4[b]	7.0[cd]	5.2[c]
IPM	0.4[c]	7.4[b]	0.9[de]	7.4[d]	3.7[d]	2.1[e]	7.6[c]	5.9[e]	2.5[d]
O15	1.1[d]	9.1[a]	1.6[d]	8.9[ab]	8.0[ab]	5.9[bc]	8.7[a]	7.7[bc]	6.4[b]
EO	2.1[c]	9.3[a]	3.2[c]	9.2[ab]	8.4[ab]	7.4[a]	8.8[a]	8.5[ab]	8.2[a]

Source: Parente, M. E., et al., 2008, *Journal of Sensory Studies*, 23: 149–161.

Note: Means within a column with different superscripts are significantly different ($P \leq 0.05$).

Abbreviations: t0, average score immediately after application; t5, average score after 5 min of application; t10, average score after 10 min of application; ÑO, ñandú oil; CM, cyclomethicone; DM, dimethicone; OD, octyldodecanol; E15, estearin; IPM, isopropyl mirystate; O15, olein; EO, emú oil.

TABLE 24.2

Alkyl Groups in Glycols and Sensory Sensations Provided

Alkyl Group	Fatty Acid Higher Alcohol	Higher Alcohol	Feel of Use
n-12	Laurate	Lauryl alcohol	Coarse
n-14	Myristate	Myristyl alcohol	Silky, lubricious
n-16	Palmitate	Cetyl alcohol	Moisturizing
n-18	Stearate	Stearyl alcohol	Moisturizing
n-22	Behenic acid	Behenyl alcohol	Soft
e-18	Oleic acid	Oleyl alcohol	Oily, moisturizing
iso-16		Hexyldecanol	Light, lubricious
iso-18	Isostearic acid	Isostearyl alcohol	Lubricious
iso-20		Octyldodecanol	Lubricious

Source: Iwata, H. and Shimada, K., 2013, Sensory properties of cosmetics, in *Formulas, Ingredients and Production of Cosmetics: Technology of Skin- and Hair-Care Products in Japan*, p. 103, Tokyo: Springer.

significantly affects its sensory profile. Increasing the solid content leads to higher spreading difficulty, stickiness and lower gloss.

Oils have an alkyl group as the basic structure and include carbohydrates, esters and waxes. Iwata and Shimada (2013) stated that the properties of alkyl groups in oils and surfactants are the key factor in determining the sensory properties of cosmetics. These groups determine the melting point (molecular weight increase corresponds to higher melting point) and polarity. Short alkyl chains generate a 'coarse' feel, those with intermediate length feel smoother and more lubricating, and long alkyl chains provide soft and moisturizing properties for skin care products. Double bonds in alkyl chains increase the moisturizing and oily sensations, whilst branched chains give a silky and light touch (Iwata and Shimada, 2013). Complicated molecular structures can increase the moisturizing and oily sensation, however, oils with larger molecular weights have higher melting points and are solids at higher temperatures providing a less silky and smooth feel. Waxes, which have higher melting points than oils, are adhesive and give a coated feeling on the skin. The functional groups on oils determine the compatibility and interaction with other ingredients and also affect the viscosity, consistency, appearance, hardness, spread and feel on the skin. Ester bonds, hydroxyl and carboxyl groups do not directly affect the touch sensation but affect the viscosity and consistency of the product. Higher viscosities provide smoother and softer sensations, and low viscosities result in lightness (Iwata and Shimada, 2013). Table 24.2 contains examples of the sensations provided by alkyl groups in glycols.

Sensory Analysis Applications in the Cosmetic Industry

Sensory analysis techniques are used in the cosmetic and personal care industries in a range of applications, especially to support product development and marketing. Examples of applications followed by suggestions of sensory tests that can be used are described next:

- Verifying the changes observed in sensory characteristics resulting from variations in the production process, raw material replacement and change in supply of raw ingredients, cost reduction studies, etc. (*Discrimination methods*)
- Determining the choice of appropriate ingredients for formulations (a common example is the choice of fragrance) (*Discrimination, descriptive and/or affective methods*)
- Screening for prototypes resulting from development projects (*Discrimination, descriptive and/or affective methods*)
- Defining consumer's preference – Collecting useful information to optimize the sensory acceptance of a cosmetic formula, according to the expectations of the consumer market (*Affective methods*)

- Evaluating the sensory changes that occur during the storage of a cosmetic product in relation to time, storage conditions and type of packaging; collecting information that is useful to determine the formulation stability and expiry date (*Discrimination+Descriptive+Affective* methods)
- Making a sensory map of the target market, establishing the sensory profile of the products with largest share in that market, with the purpose of determining the attributes that guide consumer acceptance and preference, that is, the *drivers* of consumer preference (*Descriptive+Affective methods*)

The preceding examples serve to illustrate the importance of sensory analysis in the cosmetic and personal care industries. Clearly, having the means to assess man's natural ability to compare, distinguish and quantify the sensory attributes by employing adequate methodology and statistical analysis is a vital element in product development.

Practical Aspects of Sensory Evaluation of Cosmetic Products

Variables That Should Be Controlled and Planned

Testing Rooms

Sensory testing rooms must provide an environment of physical comfort and encourage effective concentration by the assessors. The planning should take into consideration the following topics (ISO 8589-07, 2007).

Odour Control

The test room must be capable of allowing the removal of odours, therefore an efficient exhaust system separated into the testing booths and another one into the sample preparation room is essential. The sensory analysts and assessors should be instructed not to use perfume, moisturizers, hair care products or scented deodorant, and should not chew gum, mints or smoke cigarettes. The cleaning material used in the testing room should be odourless.

Location and Layout

- Far from sources of noise and strong odours (other laboratories, manufacturing lines, restaurants, etc.)
- Easy access (for collaborators and assessors)

Access to the test room should be carefully planned. Individual test areas (booths) should be separated from access to the preparation area of the samples to be evaluated, preventing assessors from receiving undesirable stimulus from brands, preparation tools, packages, odours, and so forth. Exchange of information between assessors arriving to the test room and assessors leaving the test room must be avoided, so a separate entrance and exit is useful. Thus the layout should comprise two separate areas: the test area with individual sensory evaluation booths and the sample preparation area. To obtain a versatile and useful layout for many activities, two additional areas are recommended: a discussion and training room for the descriptive sensory tests, and a reception room for consumer tests. The facilities offered by each of these areas are described next.

Testing Area

The main characteristic of a sensory analysis testing area is the presence of booths for assessors' individual evaluations. The booths must have all necessary features to allow evaluation of the product categories of interest to the company, such as sinks, showers and olfactory booths. Figure 24.1 shows six examples of sensory booths used for cosmetic product evaluation. In some pictures, the presence of an assessor/sensory analyst communication door is observed. This is extremely useful for the delivery of the samples to be evaluated or for any other type of communication.

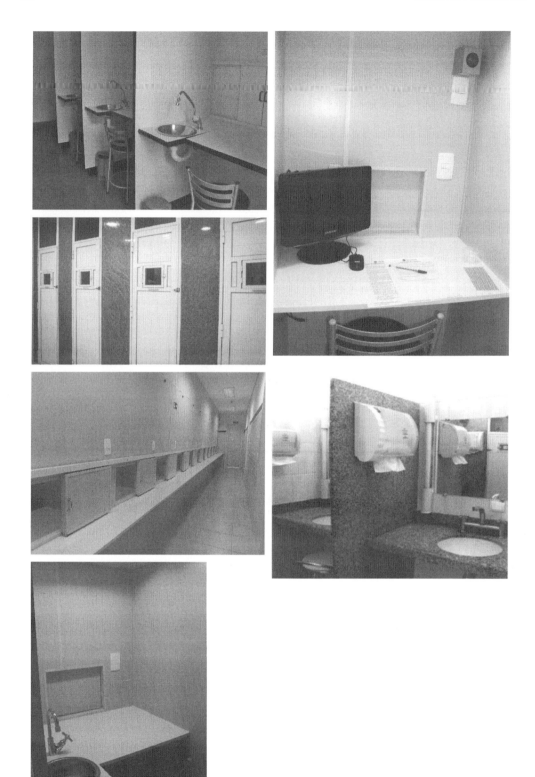

FIGURE 24.1 Examples of sensory booths used for cosmetic products evaluation. (Photographs provided by Perception Sensory and Marketing Research, Brazil.)

The individual booth design provides a quiet and comfortable environment, with control of conditions that might interfere with the assessor's evaluation, such as distraction, peer pressure and exchange of information between assessors. Thus each assessor can concentrate on his or her evaluation, and the analyst is confident of the validity of the results. White lighting (fluorescent daylight) and colour lighting (red and green) in the booths are recommended. Colourful lighting should be used when it is necessary to mask the variation in appearance of samples.

Sample Preparation Area

The sample preparation area should be adjacent to the evaluation areas (booths and training rooms) and should comprise a laboratory with benches, sink, cabinets, semi-analytical scale and other utensils, with the purpose of providing an adequate environment for preparing the samples to be evaluated.

Discussion and Training Room for Descriptive Sensory Analysis

The discussion and training room should accommodate 6–12 assessors and, generally, 2 sensory analysts, and should be adjacent to the sample preparation room. The discussions and training take place at a round table (approximately 1.4 m in diameter) with a rotating center (to facilitate sample evaluation during the discussion). A blackboard, whiteboard or flip-chart (caution: avoid pens and erasers with odour) are also useful. An example of this type of room is shown in Figure 24.2. A 'spy' mirror and an adjoining observation room allow the conduction of some qualitative studies in the same room (focus group, interviews, etc.). The assessor's entrance into the training room should be independent from the other areas.

Reception Area for Testing Consumers

A reception area should be located next to the booths and training room. The purpose of this area is to provide a quiet and relaxing environment for assessors to relax prior to the test, and an environment in which the analyst can provide introductory information and explanations of the test procedures. The

FIGURE 24.2   Training room with a 'spy' mirror. (Photograph provided by Perception Sensory and Marketing Research, Brazil.)

room should have comfortable chairs and all necessary equipment for providing test information/training such as whiteboard/blackboard and computer.

The Product, Its Use and Application Site

The sensory analysis of cosmetic products can be performed in more than one phase:

- *Phase 1* – The sensory characteristics of the product itself: feel, fragrance, consistency, etc.
- *Phase 2* – During skin or hair application (performance attributes): fragrance during application, spreadability, absorption point, oiliness, foam forming rate, amount of foam, etc.
- *Phase 3* – After use (benefits to skin and hair): gloss on skin, fragrance substantivity, velvety film, moisturization, softness, compatibility, clean feel, etc.

Therefore, before performing a cosmetic product sensory test, it is essential to establish the directions of product use, the appropriate application sites (arms, legs, hands, face, hair, feet, etc.) for evaluation and the necessary duration of use in order to assess the practical benefits relative to how it will be used by future consumers.

Phase 1 (the product's sensory characteristics) can always be evaluated in the sensory booths (under controlled conditions of exhaustion, temperature, light). Phases 2 and 3 may or may not require different evaluation conditions such as hair washing chair, sinks with water hardness control, temperature and flow rate, climate control rooms, etc. The assessor often needs to use the product at home under real conditions of use (affective tests and efficacy tests).

Sample Preparation and Presentation

Sample preparation for sensory evaluation must be as uniform and simple as possible, including the packaging and tools used in the test. Most evaluations are blind (assessor and analyst are unaware of product identification) with codes used for the vessels/bottles (e.g. three-digit random numbers or a mix of letters and numbers: 238, 615, J28, M35, etc.). Test products for home use (affective, perceived efficacy) should have labels with the code and directions of use, as well as a contact telephone number of the person in charge of the research to allow the assessor to immediately report any side effects or significant experiences. Sensory analysis studies can be pure monadic (each assessor evaluates one product only; often used in efficacy studies) or sequential monadic (each assessor evaluates a number of products, one at a time; typically used in affective tests). There are two types of psychological errors that can be related to the order in which the products are evaluated:

1. *Position error* – Caused by the position (order) in which each sample is evaluated
2. *Carryover effect* – Effect of the previous sample on the evaluation of the next sample

To minimize these errors, in sequential monadic studies the order of product presentation should be balanced across the products. They can be complete (all products are evaluated by all assessors) or incomplete (some or all of the products are evaluated by all or some of the assessors). All products should be evaluated in the same order position for the same number of times, and preceding all other products the same number of times. Many options of balanced designs are shown in MacFie and Bratchell (1989). Note that to ensure the order of presentation is appropriately balanced, the assessors must be instructed to evaluate the samples from the left to the right in tests where the samples are presented at the same time (e.g. in discrimination tests).

The Assessors

Individuals recruited to perform sensory analysis of cosmetic products can have different characteristics, depending on the objective/type of test, as stated next:

Selected assessors – They have their acuity evaluated by a sensory acuity test (such as odour and taste recognition tests). They can be selected to participate in discrimination tests or to participate in the training of a sensory panel.

Selected and trained assessors – After they are approved in the selection tests, assessors are trained and 'calibrated', with the purpose of creating a sensory panel of trained assessors to perform the descriptive tests. They should have good discrimination ability, reproducibility and concordance with the group of trained assessors.

Consumers – They are recruited to participate in affective tests. They should represent the target audience of the study; in most cases, they are regular consumers of the category, but they can also be recruited following other inclusion criteria (for example, as users of competing products). It is important to point out that trained assessors should not be recruited to evaluate product acceptance and preference, because after they are trained they no longer behave as consumers.

Volunteers – For cosmetic efficacy tests, volunteers should have the condition that the product is intended to treat (such as wrinkles, cellulite, dry skin, cracked feet). As volunteer do not necessarily represent the target audience of potential product buyers, care should be taken to avoid mixing questions related to product attribute acceptance in an efficacy test. Volunteers' answers will not be indicative of answers provided by the target audience but rather they will represent the research study group (and in extreme cases they can be completely opposite to potential buyers of the product evaluated by the efficacy test) (Colipa, 2008).

Assessors' Motivation

- *Trained sensory panelists* – Explaining the importance of the project, valuing participation, if possible, showing results, complimenting, offering gifts at every test, organizing drafts. If the assessors are collaborators from the company, the support of management and the board of directors is crucial.

- *Consumers* – Offering gifts at every test, appreciating their honest evaluation and explaining this will contribute to improving the quality of the cosmetic products offered in the market.

Statistical Analysis

Statistical analysis used to evaluate the sensory test results should be planned as part of the study design. Suitable analysis can include descriptive statistics such as the Friedman test, ANOVA followed by a means test (least significant difference [LSD], Tukey or Dunnett), and principal component analysis (Hinton, 1995).

Sensory analysis data handling requires expert knowledge of the methods (sensory and statistical analysis), and may require the input of a professional statistician to ensure appropriate data analysis and interpretation. The statistical analysis applicable to each test is presented in the chapter, and the table references (for analyzing the discrimination test results) provide further reading.

Factors Influencing Sensory Evaluation

Sensory evaluation relies on the subjective perceptions of human assessors and volunteers, therefore a number of factors that can influence the test results and possibly lead to experimental errors must be recognized.

Psychological errors always occur. That's why we should be aware of their existence to be able to minimize their influence on the study outcomes.

- *Logical error* – Association of characteristics, for example, associating green colour with freshness, blue colour with seaweed.

- *Leniency error* – Influence of the relationship between the assessor and the sensory analyst, or between the consumer and the company. It might have a positive or negative influence on the affective evaluation. This is the main reason why performing affective tests with the company's research collaborators and/or their families and friends is not recommended.

- *Expectation error* – Influenced by previous knowledge of the product. Any information on the product to be evaluated leads to some expectation that will influence the assessor's decision. Therefore, collaborators involved in product development and/or marketing projects should not be asked to participate in sensory tests

- *Stimulus error* – Influences resulting from the way the samples are presented (vessels, packaging, amount).

- *Halo error* – Effect of one answer on the subsequent answers. This often occurs in affective tests when the questionnaire includes questions related to the product attributes (PA) and product liking (PL). The consumer unconsciously tends to be consistent with the former answers. For example, if the first question in the questionnaire is PL and the product is generally well liked, all of its various aspects (PA) – fragrance, consistency, absorption time, and so forth – tend to be rated favourably as well. Conversely, if the product is not well liked, most of the attributes would be rated unfavourably.

Sensory Analysis Methods

Prior to performing a sensory test, it is essential to clarify the desired objectives and precise information to be obtained so that appropriate questions are incorporated. Having established the desired objective(s) and question(s), the appropriate test(s) can be selected and study protocol developed, taking into consideration that more than one test can often be performed to reach the objective and deliver supplementary answers.

A precise evaluation methodology (such as evaluation in the bottle, application to forearms, evaluation after rinsing off), assessor characteristics, experimental conditions (such as test site, room temperature) and statistical analysis must be determined.

Sensory tests, their applications, statistical analysis procedures and recommendations, and examples are presented next. References are provided for further reading to provide more detail than can be discussed in this overview.

Discrimination Methods

Discrimination or 'difference tests' screen the different evaluated products for an evident sensory difference, which can be general to the product overall or specific to a particular attribute of the test product. It is important to accurately determine the minimum number of assessors to be used in difference tests in order to be confident that a null result is due to a lack of perceived difference in the products rather than due to an insufficient number of answers. Meilgaard et al. (2007) dealt with the determination of the minimum number of answers in discrimination tests. Assessors can be selected based on their ability or training to distinguish the specific characteristics of the product that are of interest. Given the nature of the test it is reasonable that the selected assessors may not distinguish differences in the product, and indeed that neither will the consumer. Where the selected assessors perceive a difference, the only way to know if consumers would also perceive the difference is by performing the same test with the heavy users of the product under investigation (that is, the target audience).

This chapter provides an overview of the tests most widely used in cosmetic product sensory analysis to screen for differences between two samples (paired comparison, duo-trio and triangle tests), and three or more samples (ranking test). For detecting and quantifying the degree of difference, the difference from control test is recommended (Meilgaard et al., 2007). The reader is referred to the text by Meilgaard et al. (2007) for detailed information to assist with designing study protocols and data analysis.

Paired Comparison Test

The paired comparison test (or directional difference test; ASTM E2164-08) (ASTM, 2008) is a forced choice test (the assessor must choose one answer) that tests for a specific sensory characteristic between two samples (product gloss, intensity of fragrance, etc.). The samples should be in rank order of presentation (order AB and BA repeated the same number of times) and the probability of random success is 50%.

```
Name: _____ Date: _____


Please, smell the coded soap bar samples, from left to right, and circle the sample code

below with the strongest smell.

                    365                    719

Comment (if any):_____

_____
```

FIGURE 24.3 Example of score sheet for a paired comparison test.

Figure 24.3 shows a score sheet for a paired comparison test of soap fragrance intensity. A significant difference between the samples is deemed to occur at 5% significance.

Example: In a cost-reduction project, sample B was designed with 10% reduction in fragrance concentration. To evaluate if the difference in concentration was sensorially perceived, a paired comparison test was performed, in which 10 selected assessors indicated, in two repetitions, the soap bar samples (A or B) with stronger fragrance in two phases: evaluation in the soap bar and after washing the arms (each arm with a different soap). Sample A was indicated 16 times as having stronger fragrance in the soap bar and 17 times after washing the arms. With reference to tables by Meilgaard et al. (2007) for 20 total answers, the minimum number of correct answers at 5% level of significance is 15, therefore it is concluded that soap A was significantly ($p \leq 0.05$) perceived to have a stronger fragrance than soap B, both in the bar and on the skin (after rinsing off). Based on this conclusion, a number of decisions can be made:

1. Test a new formulation with a smaller percent reduction in fragrance concentration.
2. Investigate an alternative to reduce formulation cost.
3. Perform a paired comparison test of formulations A and B in a sensory booth (evaluating the bar and after washing the arms) with regular consumers of that brand of soap. If regular consumers do not notice a difference in both phases, a 10% reduction in fragrance concentration can be considered.

Duo-Trio Test

The duo-trio test is also a forced choice test (the assessor must choose one answer) that tests for the difference in general characteristics between two samples. The samples should be in rank order of presentation (order AB and BA repeated the same number of times) and the probability of random success is 50%. Figure 24.4 shows the score sheet for a duo-trio test with two shampoo samples.

```
Name: _____ Date: _____


You are receiving a shampoo reference sample (R) and two coded shampoo samples.

Evaluate the consistency of the coded samples, from left to right, and circle the sample

with the same consistency of the reference sample below.

                    734                    198

Comment (if any):_____

_____
```

FIGURE 24.4 Example of score sheet for a duo-trio test.

Example: In order to evaluate an alternative thickener supplier, a duo-trio test was performed with 13 selected assessors comparing the consistency of two shampoo samples (A and B) with a reference sample (A or B) in two repetitions. Sample B was formulated with the thickener from the alternative supplier and A was the actual formula. Fourteen assessors correctly indicated the sample as similar to the reference sample. According to Meilgaard et al. (2007), the minimum number of correct answers at 5% significance is 18, therefore it was concluded that there was no significant difference between the consistency of the evaluated shampoo samples A and B so that the alternative supplier was approved to supply the thickener for the formulation.

Triangle Test (ISO 4120:2004)

The purpose of a triangle test is to test for a significant difference between two samples (A and B) in general, with two equal samples and one different sample, in which the assessor is instructed to indicate the different sample. The samples should be presented in rank order (orders AAB, ABA, BAA, BBA, BAB and ABB repeated the same number of times). This is a forced choice test (the assessor must indicate the different sample), and the random probability of success is 33.33%, which is an advantage of the triangle test in relation to the paired comparison and duo-trio tests. Figure 24.5 shows the score sheet for a triangle test performed with two moisturizing lotion samples.

Example: A company had to replace mineral oil with vegetable oil in the formula of a moisturizing lotion. The formulator developed a number of prototypes with the desired changes and selected the most similar prototype to the line product to be evaluated for skin oiliness by the trained sensory panel. The panel identified a significant difference between the prototype and the line product. The company asked: The trained panel identified the difference in one particular attribute, but will the consumers notice any difference in the product formulation (prototype × line of product) in general? A triangle test was therefore conducted with regular consumers of the product, instructing them to evaluate moisturizing lotions in one specific area of the arms. Of the 60 consumers who performed the triangle test, 23 could tell which was the different sample. According to Meilgaard et al. (2007), the minimum number of correct answers for 5% significance is 27. The company concluded that there was no significant difference and that the prototype was suitable for replacing the line product in the market.

Ranking Test for Difference

The ranking test for difference will detect a significant difference in the perception of a specific sensory characteristic between three or more samples. The samples should be presented in rank order. In the case of three samples, orders ABC, ACB, BAC, BCA, CBA and CAB should be presented the same number of times (or the best possible distribution allowing, for example, for 15 assessors to repeat the orders twice and choose three samples to be presented to the three remaining assessors). Figure 24.6 shows a score sheet for a ranking test for difference of aftershave gel colour intensity.

Example: In a cost-reduction project, two new suppliers of dyes were evaluated by a ranking test for difference with prototypes X and Y (new suppliers) and the line product. Fifteen assessors performed the test using the score sheet shown in Figure 24.6. Data was analyzed by converting the orders into numbers

---

Name: _____ Date: _____

Two samples are equal and one is different. Please, evaluate the coded samples, from left to right, and circle the code of the different sample below.

365   719   852

Comment (if any):_____

_____

FIGURE 24.5   Example of score sheet for a triangle test.

```
┌─────────────────────────────────────────────────────────────────────┐
│                                                                       │
│  Name: ___  _____ Date: _____  │
│                                                                       │
│                                                                       │
│  Please, observe the after-shave gel samples and sort them in         │
│  ascending order of green                                             │
│  color intensity.                                                     │
│                                                                       │
│                        _____    _____    _____                     │
│                        lighter              darker                    │
│                                                                       │
│  Comment (if any):_____    │
│                                                                       │
│  _____  │
│                                                                       │
└─────────────────────────────────────────────────────────────────────┘
```

FIGURE 24.6   Example of score sheet for a ranking test for difference.

TABLE 24.3

Ranking Test for Difference: Ranking Totals

	Line Product	Prototype X	Prototype Y
Ranking totals	30	24	38

with a value of 3 assigned to the darkest sample, a value of 2 assigned to the middle sample and a value of 1 assigned to the lightest sample. Ranking totals were obtained as shown in Table 24.3.

According to Newell and Macfarlane (1987), for 15 answers, a critical difference observed of 13 between the ranking totals indicates a significant difference at 5% between the samples evaluated. In this example, the difference between the ranking totals from prototypes X and Y and the line product total was 6 and 8, respectively, indicating that there was no significant difference in green colour intensity between prototypes X and Y and the line product. The company concluded that both suppliers were approved and chose the supplier offering the best value for money.

Descriptive Methods

Descriptive sensory methods describe and evaluate the intensity of the sensory attributes of the product. Descriptive methods have two aspects: the qualitative aspect related to the description of the evaluated product, and the quantitative aspect that evaluates the intensity of each sensory characteristic present in the product. Quantitative Descriptive Analysis (QDA®), developed by Stone et al. (1974), represents one of the most complete and sophisticated methods for characterizing the sensory attributes of a product. QDA analysis involves trained assessors identifying and quantifying the sensory properties of the product, thus the terminology that describes the attributes is developed through a consensus between assessors, suggesting more agreement among them. In this section, we address the descriptive sensory analysis (DSA) methodology based on the QDA guidelines, adapted to the cosmetic industry with the purpose of quantifying sensory attributes perceived in the product (product sensory characteristics), and the sensory attributes perceived during and after application on the skin as described by Meilgaard et al. (2007) and ASTM E1490-03 (2003).

DSA is frequently used as a technical measure support tool to inform the cosmetic product formulator about the sensory profile of the evaluated products, including the performance sensory attributes, during and after application (amount of foam, spreadability, gloss on skin, etc.). It is also an invaluable tool in Marketing, as it provides sensory characterisation of a cosmetic category in an existing or potential market, thus identifying the attributes driving consumer preference of this product category.

DSA is performed by a panel of trained assessors, who are subject to a number of steps prior to product evaluation:

1. Recruiting candidates – An average of 30 candidates, aged between 18 and 55 years who have graduated from high school, are healthy, and are available to attend all training sessions. Recommended exclusion criteria are skin problems; previous reaction to the tested product

category; other diseases or medications that could interfere directly with the study or pose a risk to the analyst's health; pregnant or lactating women; and criteria specific to each study.

2. Candidates for the trained panel are selected based on their descriptive and discrimination capabilities, communication skills, understanding, interest and availability to participate in activities related to composing the panel, sample evaluation sessions and after the panel validation. Group dynamics can also be assessed to evaluate the candidate's behaviour in the group.

3. The sensory attributes of the products are determined by requesting assessors to describe individually (in the sensory booth) the differences and similarities of two samples. The assessors then participate in a joint session in which the panel leader uses the individual descriptors to stimulate a discussion in order to gather synonyms and find a consensus list of terms that describe the perceived attributes, as described by the assessors. This activity can be repeated with other pair(s) of samples to obtain a better representation of the attributes in the samples of the category evaluated. These terms will then be utilized to inform the DSA.

4. Training the selected candidates – Training duration will depend on the number of attributes and their level of complexity. Two or three weekly sessions are recommended. In the sessions, the samples should be presented with different attribute intensities allowing the assessors to recognize and distinguish the samples in a consensus, according to the intensity of the evaluated attributes.

5. Calibration or pilot study – Evaluation of three representative samples of the sensory universe of the category, with marked differences between them, to be evaluated in three repetitions. The results should be subjected to ANOVA followed by Tukey's test, by attribute, for each assessor (sources of variation: sample and repetition), and will evaluate the panel performance after the training; $p_{sample} < 0.30$ or 0.50 and $p_{repetition} > 0.05$ is desired.

6. Retraining and feedback of the calibration results to assessors – Individually and/or in group with the purpose of showing assessors the calibration results, identifying the opportunities for improvement and reinforcing the training to help improve the assessors' performance.

7. Validating the trained sensory panel – Performed through the same calibration methodology. The assessor will be validated after reaching the desired p_{values} ($p_{sample} < 0.30$ or 0.50 and $p_{repetition} > 0.05$) in most attributes evaluated by the panel.

After validation, the panel is ready to quantify the sensory attributes in a precise fashion on linear scales of 9 to 15 cm, anchored to the far left with absent/low/few terms and to the far right with high/many terms. Figure 24.7 shows an illustration of the evaluation scale of the attribute 'stickiness' in a descriptive sensory analysis of cosmetic emulsions.

The recommended statistical analysis to evaluate the DSA results is the estimate mean and standard deviation for each attribute by evaluated sample and a new ANOVA (causes of variation: sample, assessor and sample*assessor), followed by the mean test to identify significant differences at 5% or 10%. Figure 24.8 shows an example of a chart expressed in polar coordinates in which the mean of each product is plotted against the axis representing the evaluation scale for each attribute, in this case, the DSA of cosmetic emulsions. This chart allows a quick view of the different sensory profiles of cosmetic emulsions evaluated by a trained sensory panel.

In this example, the products showed a marked difference in gloss, stickiness and oiliness, in which product 253 was the stickiest; products 714, 936 and 319 were the glossiest; and products 851 and 714 were the oiliest.

Noronha et al. (2008) developed a DSA study with the purpose of comparing the sensory profile of 12 prototypes of body moisturizers with the marketed benchmark product. The evaluated attributes (see Table 24.4) were evaluated on 10 cm linear scales with quantitative references of slight/none, medium

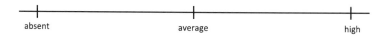

FIGURE 24.7 Example of a linear scale for the evaluation of stickiness in cosmetic emulsions.

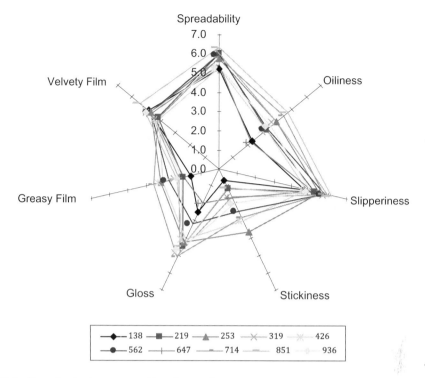

FIGURE 24.8 Example of a chart expressed in polar coordinates, resulting from a DSA of cosmetic emulsions.

and high intensity for each attribute. Table 24.5 shows the means and the results of Tukey's test at 5% significance for all attributes evaluated.

The authors concluded that the prototypes reached the formulators' goals. All prototypes evaluated showed less residual oiliness compared to the benchmark. Prototypes 5, 7, 9 and 10 weren't different from the benchmark in relation to the residual greasy film attribute, but the other prototypes showed less intensity than the benchmark for this attribute. This positive result was also observed for the velvety film attribute, since the prototypes didn't differ significantly from the benchmark.

TABLE 24.4

Sensory Attributes of Body Moisturizers Evaluated in the Descriptive Sensory Analysis

Attribute	Definition
Absorption point	Number of rotations that the product needs to be absorbed by the skin
Spreadability	Ease of moving/spreading the product over the skin
Slipperiness	Ease of moving/slipping the finger over the skin
Immediate skin radiance	Intensity of light reflected on the skin after spreading the product
Residual skin radiance	Intensity of light reflected on the skin 2 minutes after spreading the product
Stickiness	Intensity with which the finger adheres to the skin
Immediate oiliness	Feel of oil on the skin during and after spreading the product
Oiliness	Feel of oil on the skin 2 minutes after spreading the product
Immediate greasy film	Feel of grease, forming a film on the skin, immediately after spreading the product
Residual greasy film	Feel of grease, forming a film on the skin, 2 minutes after spreading the product
Velvety film	'Peach skin' feel
White residue	White film forming on the skin

Source: Noronha, R. L. F., et al., 2008, Perfil sensorial de emulsões para tratamento de pele: avaliação de protótipos e produtos de mercado, https://www.academia.edu/s/30b479e6d9/perfil-sensorial-de-emulsoes-para-tratamento-de-pele-avaliacao-de-prototipos-e-produtos-de-mercado?source=link.

TABLE 24.5A

Means and Results of the Tukey's Test (at Significance Level 5%) of the Evaluated Products

Products	Absorption Point	Spreadability	Slipperiness	Stickiness	Immediate Gloss	Residual Gloss
Prototype 1	1.64 C	5.72 C	5.72 C	0.42 B	2.84 D	0.64 E
Prototype 2	1.91 AB	6.26 B	6.40 AB	1.10 A	4.02 BC	1.55 D
Prototype 3	1.86 AB	6.38 AB	6.43 AB	0.57 AB	4.65 A	2.90 ABC
Prototype 4	1.85 AB	6.38 AB	6.47 AB	1.06 A	4.49 AB	3.07 ABC
Prototype 5	1.82 B	6.31 AB	6.44 AB	0.89 AB	4.88 A	3.39 A
Prototype 6	1.88 AB	6.23 B	6.33 B	0.82 AB	3.72 C	1.23 DE
Prototype 7	1.91 AB	6.49 AB	6.53 AB	0.69 AB	4.74 A	3.06 ABC
Prototype 8	1.92 AB	6.39 AB	6.54 AB	0.53 AB	4.79 A	3.19 AB
Prototype 9	1.88 AB	6.39 AB	6.48 AB	0.69 AB	4.65 A	3.08 ABC
Prototype 10	1.91 AB	6.46 AB	6.56 AB	0.42 B	5.03 A	2.84 ABC
Prototype 11	1.91 AB	6.43 AB	6.54 AB	0.97 AB	4.69 A	3.19 AB
Prototype 12	1.99 A	6.62 A	6.75 A	0.92 AB	4.84 A	2.43 C
Benchmark	1.62 C	6.50 AB	6.47 AB	0.68 AB	4.64 A	2.57 BC

Source: Noronha, R. L. F., et al., 2008, Perfil sensorial de emulsões para tratamento de pele: avaliação de protótipos e produtos de mercado, https://www.academia.edu/s/30b479e6d9/perfil-sensorial-de-emulsoes-para-tratamento-de-pele-avaliacao-de-prototipos-e-produtos-de-mercado?source=link.

Means followed by the same letter, in each row, do not significantly differ from each other at 5% of significance (Tukey's test).

TABLE 24.5B (CONTINUED)

Means and Results of the Tukey's Test (at Significance Level 5%) of the Evaluated Products

Products	Velvety Film	White Residue	Immediate Oiliness	Residual Oiliness	Immediate Greasy Film	Residual Greasy Film
Prototype 1	4.36 A	0.00 A	1.61 E	0.21 BCD	0.99 C	0.08 BC
Prototype 2	4.47 A	0.00 A	2.17 CDE	0.16 CD	1.29 ABC	0.09 BC
Prototype 3	4.21 A	0.00 A	2.67 BC	0.18 CD	1.45 AB	0.13 BC
Prototype 4	4.58 A	0.00 A	2.57 BCD	0.33 BCD	1.31 ABC	0.10 BC
Prototype 5	4.62 A	0.00 A	2.57 BCD	0.42 BC	1.31 ABC	0.19 ABC
Prototype 6	4.55 A	0.00 A	1.92 DE	0.18 CD	1.11 BC	0.07 BC
Prototype 7	4.59 A	0.00 A	3.04 AB	0.46 BC	1.64 A	0.22 ABC
Prototype 8	4.49 A	0.00 A	2.75 ABC	0.27 BCD	1.42 ABC	0.09 BC
Prototype 9	4.54 A	0.00 A	2.55 BCD	0.34 BCD	1.52 AB	0.26 AB
Prototype 10	4.47 A	0.00 A	3.14 AB	0.32 BCD	1.60 A	0.26 AB
Prototype 11	4.39 A	0.00 A	3.01 AB	0.54 B	1.34 ABC	0.17 BC
Prototype 12	4.53 A	0.00 A	2.20 CDE	0.06 D	1.32 ABC	0.02 C
Benchmark	4.06 A	0.00 A	3.33 A	1.06 A	1.44 ABC	0.40 A

Source: Noronha, R. L. F., et al., 2008, Perfil sensorial de emulsões para tratamento de pele: avaliação de protótipos e produtos de mercado, https://www.academia.edu/s/30b479e6d9/perfil-sensorial-de-emulsoes-para-tratamento-de-pele-avaliacao-de-prototipos-e-produtos-de-mercado?source=link.

Means followed by the same letter, in each row, do not significantly differ from each other at 5% of significance (Tukey's test).

Prototypes 1, 2 and 6 were observed to be less effective in increasing residual gloss compared to the benchmark product showing significantly higher intensity for this attribute in comparison to the prototypes. Prototype 5 showed significantly higher residual gloss on the skin than the benchmark, indicating that this prototype can be used as reference to rework the formulations of prototypes 1, 2 and 6 (Noronha et al., 2008).

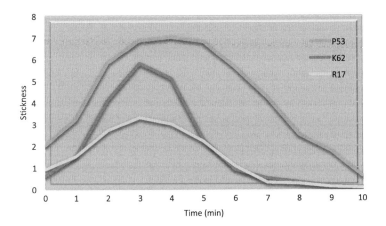

FIGURE 24.9 Representative curves of the average stickiness evaluated in three roll-on deodorants in a 10-minute period of time after applying to the skin.

Another example of high applicability of a trained sensory panel is measuring an attribute over time, as in the case of stickiness in roll-on deodorants. Figure 24.9 shows the results of the average stickiness of three different roll-on deodorants evaluated during a 10-minute period of time.

Stickiness was measured by members of trained sensory panel who applied the deodorants to the their inner elbow area. Evaluations were made after three repeated 'opening–closing' movements of the forearm.

This study was conducted because line product P53 had already received customer complaints, such as high stickiness for a long period of time. The manufacturer was testing two prototypes to improve the product and noted from the results obtained by the trained sensory panel that it could replace the line product with prototype R17, offering a less sticky product to the consumer.

For panels trained for continuous use, weekly sessions of sample evaluation, feedback, and so on, and the implementation of a panel quality monitoring program, are strongly recommended. PanelCheck software (PanelCheck, 2010) has useful graph tools to reach this objective.

One of the significant advantages of DSA is the development of descriptive terminology within the group of assessors, which implies a higher agreement and therefore more likely applicability for the consumer market. This is important, as the sensory characteristics of cosmetic products are commonly used for attracting consumers' attention in advertising campaigns and packaging (Parente et al., 2010). That is why the DSA technique has recently been applied to consumers in order to understand their perceptions, both quantitatively and qualitatively (Bleibaum and Robichaud, 2007). This methodology is described in ASTM E1490-11 (2011) as consumer behaviour approach. This DSA methodology uses, as assessors, consumers who are current users and likers of the product category of interest, and describes their sensory perceptions using a common everyday descriptive language. Products are evaluated following typical usage behaviour expected for that product, for example, hand lotion would be applied to the hands, body lotions applied more broadly, face creams on the face and so forth. ASTM E1490-11 (2011) describes and standardizes how to conduct a DSA with two different approaches – technical expert and consumer behaviour – and presents a complete example of a hand lotion DSA conducted by these two approaches.

Affective Methods: Studies with Consumers

During the cosmetic product development process, consumer feedback is required to inform the product development. This will include which sensory characteristics they expect the product to provide, their emotional associations and how the prototypes are fulfilling their expectations (Parente et al., 2011). Thus if used effectively, these will inform appropriate product development and increase the likelihood of success in the market. Effective testing of the concept, development products and final product will significantly reduce risk of product failure in the market (Moskowitz, 1996), however, they do not guarantee success since the consumer's purchase and repurchase is influenced not only by the sensory characteristics of a product but also by the packaging, price, advertisement and so on.

Affective methods evaluate the extent to which consumers like or dislike a product (acceptance) and consumer preference (direct or inferred through the acceptance averages). They are frequently used to (a) choose prototypes in the preliminary phase of market research, (b) select fragrances (the so-called sniff tests), (c) identify the attributes directing consumer preference, and (d) for sensory claim substantiation studies.

In affective tests the assessors are consumers representing the target audience, that is, with no technical training. Company employees are not recommended since the expectation and leniency errors could lead to biased data. The target audience is usually qualified in relation to consumption frequency and habits, age, sex, social class, attitude, nationality, education, cultural and religious background, and so on.

Affective tests can be performed in sensory booths, in hotel conference rooms, churches and shopping malls (known as central location tests or CLTs), or at home (usually called HUT, home-use test). The HUT represents the most faithful consumer evaluation, since the product is evaluated under real-use conditions, in the consumer's own time and home environment. An example is the affective evaluation of a fragrance in the sensory booth, which can represent the moment of purchase, but when the affective evaluation is performed at home, allowing more contact time with the product, the consumer can actually evaluate the fragrance in continuous use and the affective response can be different from that of the initial impact. Indeed sometimes the consumer may find a fragrance sickening over time, which would clearly decrease the chances of product repurchase. Collecting this data is very important to inform the product development and marketing personnel.

The number of consumers required for an affective test will be dependent on the test variables. Stone and Sidel (2004) recommend a minimum of 30 to 50 consumers for affective tests performed in sensory booths, and 100 consumers for CLTs and HUTs. Hough et al. (2006) identified a minimum number of 112 consumers for acceptance tests, based on desired significance of 5% difference and other statistical parameters to ensure test validity.

The most commonly affective tests used in sensory analysis, described in this chapter, are paired preference test, ranking preference test and acceptance test. Performing the investigation of acceptance and preference together (in the case of two to four samples) allows for a more complete response.

Paired Preference Test

The paired preference test detects a significant difference between two samples according to consumer's preference. This is a forced choice test (the consumer must select one preferred sample). Samples should be presented simultaneously, in rank order (orders AB and BA repeated the same number of times). A sample application form of a paired preference test for intimate hygiene soap after a 3-day use of each sample is shown in Figure 24.10.

The results are analyzed by comparing the largest number of answers for each sample with the corresponding value in the appropriate table (Meilgaard et al., 2007). If the number of answers found is larger than the corresponding number in the table, there is a significant consumer preference for the sample with the largest number of answers (at 5% significance).

Name: _____ Date: _____

After using two intimate hygiene soaps for three days at home, answer, which of the two samples did you like better?

               876   459

Why?_____

_____

FIGURE 24.10   Example of score sheet for a paired preference test.

Ranking Test for Preference

The ranking test for preference detects a significant preference between three or more samples according to the consumer's preference (Greenhoff and MacFie, 1999). The samples' presentation order should follow a balanced block design (MacFie and Bratchell, 1989). In the case of three samples, the orders ABC, ACB, BAC, BCA, CBA and CAB should be presented the same number of times (or the best possible distribution, for example, where 65 consumers can repeat those orders 10 times and choose 5 to present to the 5 remaining assessors). An example of a score sheet for a ranking preference test of shampoo fragrances in the bottle (sniff test) is shown in Figure 24.11. The result should be analyzed following the same procedure as described for the ranking test for difference.

Acceptance Tests

Acceptance tests evaluate global acceptance or acceptance of specific characteristics of one or more products by consumers. They use questionnaires with 10 to 20 questions in which the following scales are used:

Hedonic scale – The most commonly used scale in sensory analysis was developed in 1995 at the Quartermaster Food and Container Institute of the U.S. Armed Forces (Jones et al., 1955) to assess the food preferences of American soldiers. It has since been translated into multiple languages and widely adopted for the evaluation of a wide range of products in the food, personal care, household goods and cosmetic industries. It is used to evaluate both global acceptance (of a product, in general) and acceptance of specific attributes of a product (appearance, colour, fragrance, consistency, etc.). The original version had a 9-point scale but a 7-point scale is commonly used (Figure 24.12).

Just about right (JAR) scale – This scale is of great use for the product formulator and is generally used immediately after the acceptance question. It investigates intensity of attributes in terms of whether or not they are more or less what the consumer likes (Figure 24.13). For example, Question 1: How much did you like the product fragrance after application to your skin? (Answer in hedonic scale). Question 2: Would you say that this fragrance is … (Answer in JAR scale).

Name: _____ Date: _____
Please, evaluate the fragrance of 3 samples of shampoo and sort them in ascending order of your preference, indicating the samples code below.
_____   _____   _____
I like the least   I like the most
Comment, if any:_____

FIGURE 24.11   Example of score sheet for a ranking test for preference.

( ) Like extremely	( ) Like extremely
( ) Like very much	( ) Like very much
( ) Like moderately	( ) Like slightly
( ) Like slightly	( ) Neither like nor dislike
( ) Neither like nor dislike	( ) Dislike slightly
( ) Dislike slightly	( ) Dislike very much
( ) Dislike moderately	( ) Dislike extremely
( ) Dislike very much	
( ) Dislike extremely	

FIGURE 24.12   (Left) Nine-point and and (right) 7-point hedonic scale.

( ) Much stronger than what I like
( ) A little stronger than what I like
( ) Just about right
( ) A little weaker than what I like
( ) Much weaker than what I like

FIGURE 24.13   Just about tight (JAR) scale.

If this product were sold for the normal market price,
you would...

( ) Certainly buy it
( ) Probably buy it
( ) Maybe buy it/ maybe not
( ) Probably not buy it
( ) Certainly not buy it

FIGURE 24.14   Purchase intent scale.

Purchase intent scale – This scale investigates the consumer's intention to buy the evaluated product (Figure 24.14). The use of the purchase intent scale is justified as an independent acceptance answer because a product with high acceptance does not necessarily have high intention to purchase.

Acceptance studies can be pure monadic (only one product is evaluated by the consumers) or sequential monadic (more than one product is evaluated by consumers, one at a time, in rank order between consumers). Results obtained from the various scales can be analyzed by descriptive analysis, means, standard deviations, ANOVA and Tukey's test.

During the tests the consumers are also asked about their 'likes' (What do you like the most about this product?) and 'dislikes' (What do you like the least about this product?), thus adding information about attributes that may positively or negatively influence the consumer's acceptance of the product. The information is pooled (with synonyms) and listed, with the calculation of the number of occurrences. A useful way of presenting the data is through the Pareto chart (Figure 24.15) developed by Speranza et al. (2008) based on a study conducted to evaluate the acceptance of a roll-on deodorant. Duration of

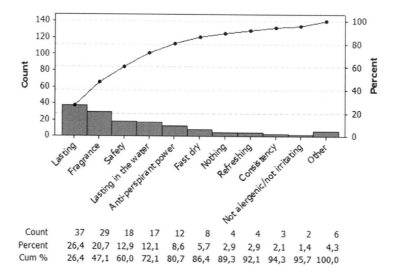

Count	37	29	18	17	12	8	4	4	3	2	6
Percent	26,4	20,7	12,9	12,1	8,6	5,7	2,9	2,9	2,1	1,4	4,3
Cum %	26,4	47,1	60,0	72,1	80,7	86,4	89,3	92,1	94,3	95,7	100,0

FIGURE 24.15   Pareto chart ('What do you like the most about the roll-on deodorant?'). (From Speranza, P., et al., 2008, Teste com consumidores: Uma poderosa ferramenta na comprovação de claims sensoriais, www.academia.edu/s/fb50c97968/teste-com-consumidores-uma-poderosa-ferramenta-na-comprovacao-de-claims-sensoriais?source=link.)

fragrance, fragrance, safety, duration in water and antiperspirant power represented 80.7% of the consumers' spontaneous responses to describe what they liked about the product.

Claim Support Testing

Innovative and efficient methodologies have been developed to substantiate the sensory benefits promised by formulations and thus support the claims made for the product. (Noronha et al., 2011). Steinberg (2009) stated that the claim-use regulation in advertising cosmetic products is still not sufficiently clear for professionals involved in the marketing and development of cosmetic formulations. ASTM (American Society for Testing and Materials) recently edited the 'Standard Guide for Sensory Claim Substantiation' (E1958-12; ASTM, 2012) in an attempt to support the regulation of the sensory claims perceived for all categories of products. It states that there are a number of ways to support sensory claims and that selection of the appropriate methodology will depend on the characteristics of the claim and its classification: comparative (parity or superiority) and non-comparative. Comparative claims are further subdivided into hedonic, which evaluates the acceptance of a product (e.g. as good as brand X), and perception, which evaluates the intensity of a product (e.g. no smell 2 hours after application) (Speranza et al., 2008).

Most claims can be substantiated through efficacy tests (instrumental, clinical and perceived efficacy), although some claims related to consumer acceptance can only be substantiated through consumer studies. Examples of such claims are 'Easier to spread than the usual' and 'Gives a nicer tan than the usual' (Speranza et al., 2008).

ASTM E1958-12 (2012) states that claims substantiation of the attributes and performance of a product can be based on data from consumers who are asked about the performance and/or specific attribute, or sensory test data in which the experimental design must focus on measuring specific attributes. In some cases, both types of tests (consumer and sensory) can be used together to substantiate the same claim(s). When a claim substantiation study is designed, first the claim itself should be known (that is, the sentence with which the product will be advertised), the type of claim (the classification), the product target audience, and the product aspects related to the claim to be tested. The appropriate methodology can then be chosen to generate the correct data that will support the studied claim(s).

Speranza et al. (2008) performed a consumer study with the purpose of substantiating the sensory claims developed to advertise the benefits of a roll-on deodorant for the summer. The evaluated claims were: 'Water resistant'; 'Especially developed for summer'; 'Especially developed for beach use'; 'Feeling of comfort and safety'; 'Better protection during summer'. In addition to evaluating the claims, the following product attributes were evaluated: global acceptance, intention to purchase, fragrance in the bottle and during use, perspiration control, and amount of product released in the package. The study was conducted in Rio de Janeiro with 90 women, aged between 18 and 55 years; in social classes A, B and C; and who were usual roll-on deodorant consumers and deodorant users on the beach. The consumers evaluated the product in use for 7–18 days, including at least one trip to the beach when they would use the product and go in the water. The assertions corresponding to the claims were evaluated on a 7-point Likert scale (agree/disagree scale) and the results are shown on Table 24.6. The authors concluded that the claims 'Offer a feeling of comfort and safety', 'Water resistant' and 'Better protection in the summer' were substantiated because the top two results were above 70% and the top three were above 84.4%.

Sensory Drivers of Consumer Liking

In an increasingly crowded and highly dynamic cosmetic product market it is vital that product developers can identify the sensory drivers of consumers' liking a product.

The techniques for determining the drivers of liking consist basically in the analysis of two groups of data obtained by (1) the descriptive sensory evaluation conducted by trained assessors and (2) an acceptance study conducted with consumers who are regular users of the study product category.

TABLE 24.6

Averages, Standard Deviations (SD), Top Two and Top Three Resulting from Consumer Evaluation of Claims after Using the Product on the Beach and Swimming in the Water

Claim	Top Two		Top Three		Average	SD
	n	%	*n*	%		
Water resistant	68	75.6	76	84.4	6.0	1.4
Specially developed for summer	60	66.7	69	76.7	5.7	1.5
Specially developed for beach use	56	62.2	66	73.3	5.5	1.6
Feeling of comfort and safety	75	83.3	77	85.6	6.2	1.3
Better protection during summer	64	71.1	76	84.4	5.8	1.5

Source: Speranza, P., et al., 2008, Teste com consumidores: Uma poderosa ferramenta na comprovação de claims sensoriais, www.academia.edu/s/fb50c97968/teste-com-consumidores-uma-poderosa-ferramenta-na-comprovacao-de-claims-sensoriais?source=link.

Note: The claims were evaluated on a 7-point Likert scale (1=completely disagree and 7=completely agree). Top Two=completely/moderately agree; Top Three=completely/moderately/slightly agree.

The most commonly statistical analysis used in this analysis is external preference mapping and partial least square regression (PLSR). Figure 24.16 shows an example of an external preference map in which six attributes composing the sensory profile are represented by the vectors; the nine evaluated products are represented by three-digit code numbers (156, 817, 363, 277, 752, 119, 581, 534 and 253) associated to a bullet; and each consumer is represented by a two- or three-digit number, located closer to the sample(s) of their preference.

The preference map provides clear segmentation of consumers in relation to their sample preference. Most consumers preferred samples 119, 752 and 277, which are characterized more by attributes 1 and 6 and less characterized by attributes 2 and 3. Therefore, attributes 1 and 6 can be identified as the sensory drivers of consumer liking.

Parente et al. (2011) studied the drivers of liking of anti-aging creams of Uruguayan consumers through the analysis of two types of data generated in one consumer study: (1) using the check-all-that-apply question to find the descriptors based on consumers' perceptions and (2) asking their overall liking

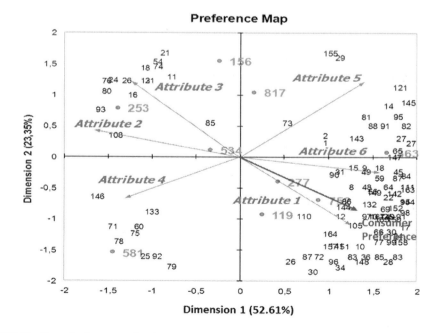

FIGURE 24.16 Example of an external preference map.

using the 9-point hedonic scale. The areas of maximum liking were correlated to products characterized by consumers as having soft and nice perfume, being easily applied and absorbed, and hydrating to the skin. Thus these authors refer to hydration, texture and perfume as key drivers of liking for anti-aging creams. Interestingly, one of the areas of maximum liking corresponded to samples perceived as luxurious and expensive, and associated with positive emotional terms.

Acknowledgements

The authors would like to express their thanks to Perception Sensory and Marketing Research, Brazil (https://perception.net.br) for the pictures shown in Figures 24.1 and 24.2.

REFERENCES

ASTM. 2003. Standard Practice for descriptive skinfeel analysis of creams and lotions. ASTM E1490–03. West Conshohocken, PA: ASTM International.

ASTM. 2008. Standard test for Method for sensory analysis – Duo-trio test. ASTM E2164–08. West Conshohocken, PA: ASTM International.

ASTM. 2011. Standard guide for two sensory descriptive analysis approaches for skin creams and lotions. ASTM E1490–11. West Conshohocken, PA: ASTM International.

ASTM. 2012. Standard guide for sensory claim substantiation. ASTM E1958–12. West Conshohocken, PA: ASTM International.

Bleibaum, R. N. and Robichaud, E. J. 2007. Using consumer's sensory experience to achieve strategic market segmentation. *Cosmetics & Toiletries*, 122 (11): 75–80.

Churchill, A. and Behan, J. 2010. Comparison of methods used to study consumer emotions associated with fragrance. *Food Quality and Preference*, 21 (8): 1108–1113.

COLIPA. Guidelines for the Evaluation of the Efficacy of Cosmetic Products, revised version, May 2008. http://www.colipa.eu/publications-colipa-the-european-cosmetic-cosmetics-association/guidelines.html (accessed June 5, 2011).

Frings, S. 2012. Sensory cells and sensory organs. In: *Sensory Perception. Mind and Matter*, Barth, F. G., Giampieri-Deutsch, P. and Klein, H. D. (eds.), pp. 5–21. New York: Springer Wien.

Gibson, J. J. 1966. Observation on active touch. *Psychological Review*, 69: 477–490.

Gilbert, L., Picard, C., Savary, G. and Grisel, M. 2012. Impact of polymers on texture properties of cosmetic emulsions: A methodological approach. *Journal of Sensory Studies*, 27: 392–402.

Greenhoff, K. and MacFie, H. J. H. 1999. Preference mapping in practice. In: *Measurement of Food Preferences*, MacFie, H. J. H. and Thomson, D. M. H. (eds.), pp. 137–166. Gaithersburg, MD: Aspen Publishers.

Hinton, P. 1995. *Statistics Explained: A Guide For Social Science Students*. London: Routledge.

Hough, G.; Wakeling, I.; Mucci, A.; Chambers IV, E.; Gallardo, I. M. and Alves, L. R. 2006. Number of consumers necessary for sensory acceptability tests. *Food Quality and Preference*, 17: 522–526.

ISO. 2004. Sensory analysis – Methodology – Triangle test. ISO 4120:2004. Switzerland: International Organization for Standardization.

ISO. 2007. Sensory analysis – General guidance for the design of test rooms. ISO 8589:2007. 2nd ed. Switzerland: International Organization for Standardization.

Iwata, H. and Shimada, K. 2013. Sensory properties of cosmetics. In: *Formulas, Ingredients and Production of Cosmetics: Technology of Skin- and Hair-Care Products in Japan*, p. 103. Tokyo: Springer.

Jaeger, S. R., Cardello, A. V. and Schutz, H. G. 2013. Emotion questionnaires: A consumer-centric perspective. *Food Quality and Preference*, 30 (2): 229–241.

Jones, L. V., Peryam, D. R. and Thurstone, L. L. 1955. Development of a scale for measuring soldiers' food preferences. *Food Research*, 20 (5): 512–520.

Kamershwarl, V. and Mistry, N. 2001. Propriedades Sensoriais dos emolientes. *Cosmetics & Toiletries*, 13: 52.

Lukic, M., Jaksic, I., Krstonosic, V., Cekic, S. and Savic, S. 2012. A combined approach in characterization of an effective w/o hand cream: The influence of emollient on textural, sensorial and in vivo skin performance. *International Journal of Cosmetic Science*, 34: 140–149.

MacFie, H. J. and Bratchell, N. 1989. Designs to balance the effect of order of presentation and first-order carry-over effects in hall tests. *Journal of Sensory Studies*, 4: 129–148.

Meilgaard, M., Civille, G. V. and Carr, B. T. 2007. *Sensory Evaluation Techniques*. 4th ed. Boca Raton, FL: CRC Press.

Moskowitz, H. R. 1996. *Consumer Testing and Evaluation of Personal Care Products*. New York: Marcel Dekker.

Newell, G. and MacFarlane, J. 1987. Expanded tables for multiple comparison procedures in the analysis of ranked data. *Journal of Food Science*, 52 (6): 1721–1725.

Noronha, R. L. F., Guerra, E. C., Sewaybricker, M. V. and Alves, L. R. 2008. Perfil sensorial de emulsões para tratamento de pele: avaliação de protótipos e produtos de mercado. https://www.academia.edu/s/30b479e6d9/perfil-sensorial-de-emulsoes-para-tratamento-de-pele-avaliacao-de-prototipos-e-produtos-de-mercado?source=link (accessed January 31, 2017).

Noronha, R. L. F., Braghetto, C. P.; Ferreira, L. D. and Passos, J. L. 2011. Claim substantiation of sensory benefits provided by a make-up product. *Brazilian Journal of Food Technology*, 6: 49–54.

Ozkan, S., Gillece, T. W. and Moore, D. J. 2012. Characterization of yield stress and slip behaviour of skin/hair care gels using steady flow and LAOS measurements and their correlation with sensorial attributes. *International Journal of Cosmetic Science*, 34: 193–201.

PanelCheck 1.3.2. 2010. www.panelcheck.com (accessed August 08, 2010).

Parente, M. E., Gámbaro, A. and Ares, G. 2008. Sensory characterization of emollients. *Journal of Sensory Studies*, 23: 149–161.

Parente, M. E., Ares, G. and Manzoni, A. V. 2010. Application of two consumer profiling techniques to cosmetic emulsions. *Journal of Sensory Studies*, 25: 685–705.

Parente, M. E., Manzoni, A. V. and Ares, G. 2011. External preference mapping of commercial antiaging creams based on consumers' responses to a check-all-that-apply question. *Journal of Sensory Studies*, 26: 158–166.

Peck, J. 2010. Does touch matter? Insights from haptic research in marketing. In: *Sensory Marketing: Research on the Sensuality of Products*, Krishna, A. (ed.), pp. 17–31. New York: Taylor & Francis Group.

Retiveau, A. N., Chambers IV, E. and Milliken, G. A. 2004. Common and specific effects of fine fragrances on the mood of women. *Journal of Sensory Studies*, 19: 373–394.

Salka, B. 1997. Choosing emollients: Four factors will help you decide. *Cosmetics & Toiletries*, 112: 101.

Schifferstein, H. N. J. and Desmet, P. M. A. 2010. Hedonic asymmetry in emotional responses to consumers products. *Food Quality and Preference*, 21: 1100–1104.

Speranza, P., Noronha, R. L. F., Alves, L. R., Vendramini, J. M. M. and Passos, J. L. 2008. Teste com consumidores: Uma poderosa ferramenta na comprovação de claims sensoriais. www.academia.edu/s/fb50c97968/teste-com-consumidores-uma-poderosa-ferramenta-na-comprovacao-de-claims-sensoriais?source=link (accessed January 31, 2017).

Steinberg, D. 2009. Labeling claims. *Cosmetics & Toiletries*. www.cosmeticsandtoiletries.com/ (accessed January 10, 2010).

Stone, H. and Sidel, J. L. 2004. *Sensory Evaluation Practices*. 3rd ed. Redwood City, CA: Elsevier Academic Press.

Stone, H., Sidel, J., Oliver, S., Woolsey, A. and Singleton, R. C. 1974. Sensory evaluation by quantitative descriptive analysis. *Food Technology*, 28(11): 24–34.

Wortel, V. A. L. and Wiechers, J. W. 2000. Skin sensory performance of individual personal care ingredients and marketed personal care products. *Food Quality and Preference*, 11: 121–127.

25

Alternative Methods: New Rules

Jadir Nunes, Chantra Eskes and Lorena Rigo Gaspar Cordeiro

CONTENTS

In the 1930s, tinting eyelashes with dyes containing para-phenylenediamine was common. However, these products were associated with serious adverse reactions such as allergic blepharitis, and were a major concern of regulatory authorities after a woman went blind after 3 months of progressive pain. The scientist and pharmacologist John Draize, head of the ocular toxicity division of the U.S. Food and Drug Administration (FDA), developed a test on rabbits (the Draize test) where the goal was to determine the ocular irritation induced by drugs, cosmetics and other chemicals. In addition, the Food, Drug, and Cosmetic Act of 1938 extended regulatory control over cosmetics for the first time. The first action implemented under this new law was to remove the para-phenylenediamine containing Lash Lure from the market.

Conceptually any animal can be used for testing drugs and cosmetics. The priority is to use a model that presents the best response to a given stimulus due to its sensitivity, for its ease of handling, and the disclosure of the effect or due to anatomical similarities (physiologic or metabolic) with humans. Laboratory animals should be used when there is no validated alternative methods that replace or, in specific cases, after screenings *in vitro* or *in silico*. What is not acceptable is when in doubt using a human being as a guinea pig for carrying out any kind of test.

3Rs Principles of Russell and Burch

In the 1960s, after the publication of the book *The Principles of Humane Experimental Technique* by William Russell and Rex Burch, the principles of the '3Rs' –replacement, reduction and refinement – was created. This concept was adopted by the Ethics Committee of the United Kingdom and by the U.S. government to release stimulus funds to research projects in biomedicine areas.

In August 1999 at the University of Bologna during the Third World Congress of Alternative Methods and Use of Animals in the Life Sciences, the 'Declaration of Bologna' was drafted and all congress attendees strongly endorsed and reaffirmed the principles of Russell and Burch. Humanized science is the basic prerequisite of good science, and consequently the laboratories that use animals based on the 3Rs will get better results. In addition, better and more consistent application of the 3Rs is considered a major opportunity for scientific, economic, and humanitarian cross-benefit, and consequently it is essential to engage scientists in a more meaningful way with the 3Rs in practice.

When we talk about 'replacement', we mean techniques that will replace methods using animals, both in the scientific and educational arena. The progress in the search for methods that will contribute to the replacement could be classified in two ways. First, we can consider the techniques that have been replaced and are valid or validated. Second, we can make our division in terms of the techniques that are in the process of being replaced.

It is important to distinguish between the relative and absolute replacement. The substitution of animal experimentation is still required. On the other hand, no discomfort should be caused to animals exposed to experiments. In absolute replacement, the animals are no longer needed at any stage; however, the proposed test or tests must accurately predict human effect and be validated before being considered an 'alternative method'.

Since 1959, there has been substantial development of techniques for the replacement of animals in education and training mainly due to the development of computer simulation programs, videos, mannequins, models and *in vitro* techniques.

Replacement techniques are especially desirable for humanitarian reasons, but there are also great savings of cost and time, and the discovery of new vitamins and new viruses, to name a few advantages.

When we talk about 'reduction', we mean that the techniques can maintain the same reproducibility and accuracy in their results using a smaller number of animals than would be recommended by the prior art. In this context, we know that the statistical method used has a key role and will specify the minimum number of animals that can be used.

We know that the replacement is the ideal, but we also know that the process to get there is slow and gradual. In the fields where replacement is not possible, reducing the number of animals in the experiments is already a first and important step in the field of alternative methods.

Reduction proposals remain of great importance for their timely implementation and obviously the universality of their acceptance and the advantage in terms of efficiency. Affording better control, the physiological variability among the animals used for experimentation combined with the use of more robust statistical methods will dramatically reduce the number of animals for experimentation.

When we talk about 'refinement', we mean that their use will minimize the conditions that can cause some damage, stress or discomfort to the animal during the experimental period. Environmental conditions, handling, anesthesia and analgesia are some examples of approaches related to this process. When reduction and replacement are not possible, refinement remains as the way to fulfill with the three pillars of the 3Rs.

Regulatory Structure and Adoption

The European Centre for the Validation of Alternative Methods (ECVAM) was established in 1991 pursuant to a requirement in Directive 86/609/EEC that the European Commission (EC) and its member states actively support the development, validation and acceptance of methods to replace, reduce or refine the use of animals in laboratories. ECVAM was the first center for the validation of alternative methods created, and it fostered validation of methodologies that are now used worldwide. The banning of animal testing for cosmetics was implemented gradually and completed only in March 2013 in Europe. Other centres had been created worldwide, including the ICCVAM (Interagency Coordinating Committee on the Validation of Alternative Methods) in the United States, in 2005 with the JaCVAM (Japanese Centre of Alternative Methods Validation) in Japan and in 2011 BraCVAM (Brazilian Centre of Alternative Methods Validation) in Brazil.

Alternative Tests

In Europe, many resources have been invested to reach this level of total replacement of animals for cosmetic testing. For example, a large cosmetic French multinational stated that they invested more than $800 million since 1979 on alternative methods and the biggest reason was not the will to sell more, but the principle that beauty cannot come from animal suffering.

According to this company, animals were used for five-axis tests: skin and eye irritation, skin corrosion, sun effects and genotoxicity. For systemic effects like allergies, a predictive evaluation strategy was created, collecting information and data from studies using multiple databases (*in silico*) that warn of possible reaction risks together with other *in chemico* and *in vitro* methods.

To get an idea of the complexity of the substitution, the case of the Draize test on rabbits to assess the potential for eye irritation will be used as an example. As there is more than one mechanism of action for this type of irritation, an *in vitro* assay is not enough to make a full assessment. The ideal is to obtain data related to these three aspects:

- Vascularization (HET-CAM, hen's egg test on chorioallantoic membrane)
- Opacity/permeability (BCOP, bovine corneal opacity and permeability)
- Cytotoxicity in human reconstructed models

Another advantage is that *in vitro* alternative methods are less variable than animal tests, ensuring the reproducibility of the results. A good example is again the Draize assay, which questioned the validity of extrapolation to humans. The Draize eye test is based more on accumulated experience than on ample comparisons between rabbits and humans. The Draize rabbit eye test is estimated to correctly predict human ocular irritancy for about 85% of toxic materials, but overestimates human ocular irritancy 10% of the time and underpredicts 5% of the time.

Worldwide Harmonization

There is now a worldwide harmonization of the use of animals for cosmetics safety testing. Europe, Israel and India have already banned the use of animals. The United States and Japan recommend alternative tests should be used if possible and are on the way to ban animal tests as soon as possible. China is still an incognita, since in some cases there is a requirement for animal tests, but it is trending towards an animal ban as well.

In Brazil Arouca's Law, passed in 2008, regulates the use of animals in experiments. This law created the Brazilian National Council for the Control of Animal Experimentation (CONCEA) that fostered many advances of the 3Rs in Brazil. CONCEA started the process of reducing animal experimentation in Brazil in July 2014. CONCEA established that Brazil has 5 years to implement 17 alternative methods validated internationally and recognized by the Organisation for Economic Co-operation and Development (OECD). Among these methods are skin sensitization, skin and eye irritation and corrosion potential, acute toxicity, skin absorption, phototoxicity potential and genotoxicity described by OECD test guidelines. Thus after this period testing on animals should not be carried out if these 17 endpoints are addressed and there will be mandatory replacement of the original method by an alternative.

In conceptual terms and based on advances in cosmetic science related to research and development of alternative methods, today we could live without animal testing in cosmetics. However, looking specifically Brazil and the reality of infrastructure, we believe that we still need some more time in order to achieve the same level in Europe and to be able to consistently replace 100% animal testing in cosmetics. New legislation is definitely promoting the development of Brazil in this area and will foster considerable progress on many fronts that will reduce the current reliance on animal testing. New research centers have started their activities focused on the use of alternative methods either *in vitro* or *in silico*.

Brazil still does not have a robust laboratory network throughout the country to cover the more than 2000 existing cosmetic companies registered by the Brazilian Health Surveillance Agency (ANVISA).

There are some isolated groups in Brazil researching alternative methods both for experimentation and for teaching. These groups include official laboratories, private industries and universities. In 2012, the Brazilian Ministry of Science, Technology and Innovation, created the Brazilian Network for Alternative Methods aimed at the adoption, development and validation of alterative methods, including technology catching up and supplying services for both public and private sectors.

The creation of entities such as BraCVAM (Brazilian Centre of Alternative Methods Validation) and SBMAlt (Brazilian Society of Alternative Methods to Animal Experimentation) is encouraging and contributing to the research, development and implementation of alternative methods in Brazil in order to eliminate the use animals in preclinical studies of cosmetics, contributing to the training, updating and interface between professionals, companies and academia. The participation of all institutions that have expertise in alternative methods will help Brazil make a huge leap in these areas.

In this chapter, we intend to point out some considerations in order to foster debate and to place the issue of animal research in a context of the growing social, scientific and ethical debate. We believe that the timing is appropriate before the widespread introduction of a context that emphasized the use of animals for dealing with the protection of human health and where alternative methods are brought up as real possibilities of scientific research.

There is a consensus in the scientific and academic communities that there is no substitute for all research methods that currently use animals, although for cosmetics animal use has been banned in Europe. However, it is not reasonable to infer from that that animal research is the only possible and acceptable form of the scientific method, nor does it invalidate the development of non-animal research methods. The animal's establishment as a scientific method needs to be constantly re-evaluated for its efficacy on the experimental question, and the development of alternative methods is therefore important to obtain important data for scientific research.

We understand that, like the animal, alternative methods have limitations of various natures. However, the contribution of technology has been improving the replacement methods in order to make them more predictive, efficient, safe, accessible and relevant, including helping to reduce the use of animals.

It is essential to emphasize that the validation of an alternative method, using an internationally accepted process, can lead to its regulatory acceptance. The ECVAM, in cooperation with international experts, sets the guidelines for validation (OECD 34 Test Guideline).

The validation of alternative methods has been defined as the process by which the reliability and relevance of the tests are established for a definite purpose and independently. In fact, the process of validation of alternative methods is aimed at assessing the optimization, transfer potential, reproducibility and relevance of the method in order to be submitted to the regulatory agency and, once approved, become officially available for toxicological purposes.

The global availability of validated methods occurs through the OECD and pharmacopoeia. It is essential that regulatory agencies in the country follow these guidelines so that all involved are aligned with the advancement of the process. We understand that part of the research is to meet regulatory requirements and to ensure the health and human security.

When available for use, the national authorities must accept alternative methods in the light of reduction, refinement and/or animal replacement, and adapt existing laws or adopt specific legislation in order to adopted methods efficiently.

Current International Practices Related to the 3Rs

During the last 20 years, extensive developments have been carried out to develop, validate and implement reduction, refinement and replacement alternative methods to animal testing. The present chapter provides a description of the current international practices regarding the acceptance of alternative methods to animal testing and their situation.

The regulatory acceptance of alternative methods includes all procedures that can replace the need for animal experiments, reduce the number of animals required, or diminish the amount of distress or pain experienced by animals. This definition embodies the 3Rs principles proposed by Russell and Burch in *The Principles of Humane Experimental Technique*, and the regulatory requirements of many countries related to the protection of animals used for scientific purposes.

An alternative method for the replacement (or partial replacement) of an animal test is generally defined as the combination of a 'test system', which provides a means of generating physicochemical or *in vitro* data for the chemicals of interest, and a 'prediction model (PM)' or 'data interpretation procedure'.

The prediction model or data interpretation procedure plays an important role in the acceptance process, as it allows converting the obtained data (e.g. *in vitro* or physicochemical) into predictions of toxicological endpoints in animals or humans.

The acceptance process, or the acceptability of a test method, may depend on various factors such as national regulatory requirements, the test method purposes, uses and applicability. In Europe, the original directive on the protection of laboratory animals for experimental and other scientific purposes stated that 'An (animal) experiment shall not be performed if another scientifically satisfactory method of obtaining the result sought, not entailing the use of an animal, is reasonably and practicably available' (Directive 86/609; EC).

As such, for an alternative method to be regulatory accepted, it is critical to demonstrate that the method is scientifically satisfactory, that is, valid for the purpose sought. This is generally carried out through a validation process by where the scientific validity of a test method can be demonstrated.

If initially implemented in Europe, the concept of scientific validation for hazard assessment has since gained international acceptance, and represents nowadays a generally recognized requirement for the acceptance of a test method for safety assessment purposes.

Test method validation is the process whereby the relevance and reliability of the method are characterized for a particular purpose. In the context of a replacement test method, relevance refers to the scientific basis of the test system and to the predictive capacity of an associated PM, whereas reliability refers to the reproducibility of test results, both within and between laboratories, and over time.

The 'purpose' of an alternative method refers to its intended application, such as the regulatory testing of chemicals for a specific toxicological endpoint (e.g. eye irritation). Adequate validation (i.e. to establish scientific validity) of an alternative test requires demonstration that for its stated purpose

- The test system has a sound scientific basis.
- The predictions made by the PM are sufficiently accurate.
- The results generated by the test system are sufficiently reproducible within and between laboratories, and over time.

In the process going from basic research to the regulatory acceptance of alternative test methods, the demonstration of the scientific validity of an alternative method is key for its acceptance within the regulatory framework. Furthermore, depending upon the geographical regulatory framework and the stage of advancement, the data generated by non-validated alternative methods may also be acceptable.

Geographical regulatory acceptance of alternative methods can be viewed as the acceptance for safety assessment of a specific method in a given regulatory context. The current mechanisms and procedures for regulatory acceptance across the world may differ depending on the geographical regions but also depending on the uses and purposes of the test methods (cosmetics, chemicals, pesticides, drugs), including:

- The recognition/tolerance by (control) authorities that manufacturers routinely use alternative approaches in their in-house safety assessments.
- The acceptance of scientifically valid safety alternative approaches as part of safety reviews by authoritative review bodies.
- The formal recommendation/obligation to use certain validated alternative methods in the registration of chemicals.

Validation is an important step within the regulatory acceptance of alternative methods, so that several international efforts have been carried out to favour the harmonization of its processes and peer-reviews with the ultimate aim of promoting harmonization of international acceptance and recognition of alternative methods. Some of the major efforts carried out are described in this chapter.

International acceptance is made by the OECD, comprised of 34 member countries, which has engaged working relationships with most countries worldwide. For over 25 years, the OECD has recognized the need to protect animals in general and in particular those used in experimental work. The Second High-Level Meeting of the Chemicals Group addressed this ethical issue and adopted the following statement:

- The welfare of laboratory animals is important; it will continue to be an important factor influencing the work in the OECD Chemicals Program. The progress in OECD on the harmonization of chemicals control, in particular the agreement on Mutual Acceptance of Data, by reducing duplicative testing, will do much to reduce the number of animals used in testing. Such testing cannot be eliminated at present, but every effort should be made to discover, develop and validate alternative testing systems.

The Mutual Acceptance of Data (MAD) states that data generated in an OECD member country in accordance with OECD Test Guidelines for chemicals and the OECD Principles of Good Laboratory Practice (GLP) shall be accepted in other member and adhering countries for purposes of assessment and other uses relating to the protection of human health and the environment. Accordingly, the MAD allows avoiding duplicative testing whilst at the same time promoting harmonized acceptance of test methods within the OECD member countries.

Since the adoption in 1981 of the first set of test guidelines, many of the short- and long-term toxicity tests, as well as the genetic toxicity tests, have been developed or revised to introduce aspects of the 3Rs principles. Some of the most noteworthy achievements include:

- The deletion in 2002 of Test Guideline 401 (Acute Oral Toxicity), and its replacement Test Guidelines 420 (Acute Oral Toxicity – Fixed Dose Procedure), 423 (Acute Oral Toxicity – Acute Toxic Class) and 425 (Acute Oral Toxicity – Up and Down Procedure) introducing reduction and/or refinement for this type of testing.
- Test Guideline 428 (Skin Absorption: *In Vitro* Method) adopted in 2004 offering a replacement alternative method to original Test Guideline 427 (Skin Absorption: *In Vivo* Method).
- Test Guideline 432 (*In vitro* 3T3 NRU Phototoxicity Test) adopted in 2004 offering a full replacement alternative test method.
- Test Guidelines 430 (*In Vitro* Skin Corrosion: Transcutaneous Electrical Resistance Test), 431 (*In Vitro* Skin Corrosion: Reconstructed Human Epidermis Test Method) and 435 (*In Vitro* Membrane Barrier Test Method for Skin Corrosion) adopted in 2004 and 2006 that offer a partial replacement of Test Guideline 404 (Acute Dermal Irritation/Corrosion).
- Test Guideline 439 (*In Vitro* Skin Irritation) adopted in 2010 which complements Test Guidelines 430, 431 and 435, and offers a further partial replacement (or full replacement depending upon geographical regulatory requirements) of Test Guideline 404.
- Test Guidelines 437 (Bovine Corneal Opacity and Permeability Test Method), 438 (Isolated Chicken Eye Test Method) and 460 (Fluorescein Leakage Test Method) adopted in 2009 and 2012 offer a partial replacement of Test Guideline 405 (Acute Eye Irritation/Corrosion).
- Test Guideline 487 (*In Vitro* Mammalian Cell Micronucleus Test) adopted in 2010 to be used as a part of a test battery for genetic toxicology testing.
- Test Guideline 429 (Skin Sensitization: Local Lymph Node Assay) adopted in 2002 and revised in 2010 introducing refinement and reduction compared to Test Guideline 406 (Skin Sensitization).
- Test Guidelines 442A and 442B nonradioactive versions of the Local Lymph Node Assay adopted in 2010, which provide modifications to Test Guideline 429.
- Test Guidelines 442C and 442D on *in chemico* and *in vitro* test methods for Skin Sensitization adopted in 2015.

Finally, another commitment of the OECD to implement harmonized principles of the 3Rs into regulatory toxicity testing was the adoption in 2000 of the Guidance Document No. 19 'The Recognition, Assessment and Use of Clinical Signs as Humane Endpoints for Experimental Animals Used in Safety Evaluations'. This document gives practical guidance on how to apply the 3Rs principles, with an emphasis on refinement when performing OECD Test Guidelines.

The criteria and processes for the validation of a test method, as previously described, were originally developed and implemented in the European Union in 1991 by the ECVAM, followed by the United States with the creation of the ICCVAM in 1997, and Japan in 2005 through the JaCVAM.

In 2005, the OECD finally adopted internationally agreed validation principles and criteria for the regulatory acceptance of alternative test methods as described in the OECD Guidance Document No. 34 'The Validation and International Acceptance of New or Updated Test Methods for Hazard Assessment'. This guidance document (GD) was the result of an OECD workshop organized in 1996 in Solna, Sweden, and the several ensuing expert meetings involving major stakeholders and the existing validation bodies at that time.

OECD GD 34 details internationally agreed principles and criteria for how validation studies of new or updated test methods should be performed. It represents a document of central importance for promoting harmonized approaches and procedures for the validation and regulatory acceptance of alternative methods within OECD member countries.

New Concepts, New Rules

Omics Technologies

Optimization of existing systems is an important part of the strategy to accelerate the implementation of a mostly animal-free safety science, in addition to the more time-demanding development of entirely new methods. One specific way to improve available tests is the incorporation of highly information-rich endpoints provided by omics technologies. Where classical methods measure only one, or few, endpoints (e.g. metabolites or gene expression levels), the new approaches can yield thousands of data points simultaneously and provide information on a genome-wide scale and, thus, allow insights into the reaction of a network.

Omics technologies provide data-rich endpoints. The biological information flow in the cell leads from gene sequences (the code) via RNA (the message) to enzymes and other functional proteins (the tools). Within this infrastructure, small molecule metabolites may be regarded as the goods that are produced and traded. They comprise energy substrates, building blocks and signaling messengers. As there are feedback loops between all levels, the different omics technologies address these four organization levels. The disturbance of a cell by chemicals may be measured by any single technique. Combinations of more than one approach lead to a better prediction of the true human situation.

Following is an overview of different omics technologies that can inform on chemicals' adverse outcome pathways and underlying modes of action:

- *Genomics* – The biological information flow comes from gene sequences (the code)
- *Transcriptomics* – The biological information flow comes from RNA (the messages)
- *Proteomics* – The biological information flow comes from enzymes and transcription factors (the tools)
- *Metabolomics* – The biological information flow comes from small molecule metabolites (the goods)

Adverse Outcome Pathways (AOPs)

The large number of chemicals requiring assessment for potential human health and environmental impacts challenges regulation. One way to realize this goal is by being strategic in directing testing resources. The hypothesis-driven Integrated Approaches to Testing and Assessment (IATA) proposes practical solutions to such strategic testing.

In parallel, the development of an adverse outcome pathway (AOP) framework, which provides information on the causal links between a molecular initiating event (MIE), intermediate key events (KEs) and an adverse outcome (AO) of regulatory concern, offers the biological context to facilitate development of IATA for regulatory decision making.

Despite some regional differences, regulatory management in general comprises hazard identification, characterisation, and an exposure assessment and risk assessment as its main steps. In some cases the identification of hazards is done prior to market approval, and certain hazards (e.g. carcinogenicity, mutagenicity or reproductive effects) may lead to restrictions on use irrespective of any subsequent risk assessment.

Development and application of AOP-informed IATA represents a new way to evaluate and generate information to meet different regulatory purposes. A conceptual framework for applying IATA considers existing information in the context of an AOP to make an informed decision based on the regulatory context. Frameworks to characterize the scientific confidence of an AOP that are required to meet different regulatory needs are in development.

Consideration needs to be given to the analytical validation of testing and non-testing approaches in order to better characterize their applicability domain (i.e. the types of chemicals that can be reliably assessed). A detailed description of AOPs of regulatory relevance and the establishment of qualitative and quantitative links between MIEs, KEs and AOs will additionally help foster application for different regulatory decisions.

Toxicity Testing in the 21st Century: The Future of Animal-Free Systemic Toxicity Testing

Since March 2013, animal use for cosmetics testing in the European market has been banned. This requires a renewed view on risk assessment in this field. Traditional animal experimentation does not always satisfy requirements in safety testing, as the need for human-relevant information is ever increasing. A general strategy for animal-free test approaches was outlined by the U.S. National Research Council's vision document 'Toxicity Testing in the 21st Century' in 2007. It is now possible to provide a more defined roadmap to implement this vision for the four principal areas of systemic toxicity evaluation: repeat dose organ toxicity, carcinogenicity, reproductive toxicity and allergy induction (skin sensitization), as well as for the evaluation of toxicant metabolism (toxicokinetics). The key recommendations include focusing on improving existing methods rather than favouring the new design, combining hazard testing with toxicokinetics predictions, developing integrated test strategies and incorporating high-content endpoints to classical assays.

The future of toxicity testing has two key features that will distinguish it from the present. First, the present animal-based testing, sometimes followed by *in vitro* tests to supply mechanistic information, will be substituted by ITS (integrated strategy) using *in vitro* and *in silico* approaches, sometime followed by animal tests where further data are needed. Second, according to the vision for a new toxicology, data will be generated for every chemical and possibly also for important mixtures.

This radical new approach to risk assessment has large economic and scientific advantages over the present animal-based system. Considerable work is still needed, but there is also strong consensus that already impressive advances have been achieved and that the goal is well worth the required efforts.

Conclusion

We need to reflect that animal research is a widespread method in scientific activity. However, the 21st century concept of animal testing no longer corresponds to the concept of animal testing upon its establishment in the history of science. The science of animal behaviour recently and repeatedly has demonstrated that animals are beings with consciousness and a complex psychological and cognitive point of view. This causes major ethical reflections on the place of animals in scientific research, motivating a demand from society a resolution that certain sectors of scientific research to review their procedures. We understand that such demand is legitimate, and causes important questions and is salutary to the progress of science as a whole.

As much as advances in human health may have occurred because of the use of animals in the past, we need to consider new ways of doing research, especially in a time where technology is evolving at an accelerated rate. Science should always follow the times, critically review traditional methods of research, and be aware of society's demands in general. Therefore alternative methods to animal research could provide greater mechanistic and scientific data. Besides that, a process more humane, efficient and modern will motivate more scientists to increase the number of screenings of new substances and finished products.

The discussion of alternative methods, as well as the ethical and scientific criticisms that currently are launched against the method of research on animals, must occupy spaces in science education. Especially in undergraduate and postgraduate study, it is essential to have public policies, and specific promotion of research and development of scientific methods without animal use. Thus, engaging in discussion and offering specific academic disciplines that address this issue could stimulate critical thinking in order to create a new generation of students with more receptive minds to new ideas and perspectives that are gradually being established in the global scientific community environment.

The formation of this new scientist must be integrated into the discussions currently established by the social and scientific context, and less tied to old and conservative science.

It is important to emphasize that today there are social, scientific, regulatory and ethic demands that must be considered. Thus, the process must be feasible for technological advance and an increase in humanization and modernization of scientific approaches. We should encourage and contribute to the development of alternative methods, and collaborate with the training, update and professional healthy interface between industries, academia and government.

The challenge to keep to a minimum or even eliminate the use of animals in some branches of scientific research will increase the debate and force deep reflection.

The year 2009 marked the 50th anniversary of the publication of *The Principles of Humane Experimental Technique* by William Russell and Rex Burch. Therefore, we would like to conclude this chapter with the same sentence of the conclusion of Michael Balls in his book *The Three Rs and Humanity Criterion* (2009) that was an abridged version of Russell and Burch's book:

We hope the book may stimulate some experimentalism to devote special attention to the subject, and many others to work in full awareness of its existence and possibilities. Above all, we hope it will serve to present to those beginning work, a unified image of some of the most important aspects of their studies. If it does any of these things, this book will have amply served its purpose.

Acknowledgements

This work was possible by support from Vanessa Sá Rocha, Marize Campos Valadares and Dermeval de Carvalho. We dedicate this chapter to members of the ABC (Brazilian Association of Cosmetology) where we have been working since 2009 in order to disseminate the 3Rs principles and support the development of alternative methods.

BIBLIOGRAPHY

Archer G., Balls M., Bruner L.H., Curren R.D., Fentem J.H., Holzhütter H.-G., Liebsch M. et al. The validation of toxicological prediction models. *ATLA*, 1997, **25**, 505.

Balls M., Blaauboer B., Brusick D., Frazier J., Lamb D., Pemberton M., Reinhardt C. et al. Report and recommendations of the CAAT/ERGATT workshop on the validation of toxicity test procedures. *ATLA*, 1990, **18**, 313.

Council Directive. 86/609/EEC of 24 November 1986 on the approximation of laws, regulations and administrative provisions of the Member States regarding the protection of animals used for experimental and other scientific purposes. *Official Journal of the European Communities*, 1986, **L358**, 1–28.

Guidance Document No. 19 on the Recognition, Assessment and Use of Clinical Signs as Humane Endpoints for Experimental Animals Used in Safety Evaluations. OECD Series on Testing and Assessment. Organisation for Economic Co-operation and Development, Paris, France, 2000, 39 pp.

ICCVAM guidelines for the nomination and submission of new and revised alternative test methods. NIH publication no. 03-4508. National Institute of Environmental Health Sciences, Research Triangle Park, North Carolina, 2003, 50 pp.

Leist M., Hasiwa N., Rovida C., Danashian M., Basketter D., Kimber I., Clewell H. et al. Consensus report on the future of animal-free systemic toxicity testing. *ALTEX*, 2014, **31**(3), 341–56.

Manou I., Eskes C., de Silva O., Renner G., Zuang V. Safety data requirements for the purpose of the Cosmetics Directive. *ATLA*, 2005, **33**(S1), 35–46.

OECD Guidance Document No. 34 on the Validation and International Acceptance of New or Updated Test Methods for Hazard Assessment. OECD Series on Testing and Assessment. Organisation for Economic Co-operation and Development, Paris, France, 2005.

OECD Guidelines for Chemical Testing 429. Skin sensitization: Local lymph node assay. Organisation for Economic Co-operation and Development, Paris, France, 2010, 20 pp.

OECD Guidelines for Chemical Testing 442A. Skin sensitization: Local lymph node assay: DA. Organisation for Economic Co-operation and Development, Paris, France, 2010, 16 pp.

OECD Guidelines for Chemical Testing 442B. Skin sensitization: Local lymph node assay: BrdU-ELISA. Organisation for Economic Co-operation and Development, Paris, France, 2010, 15 pp.

OECD Guidelines for Chemical Testing 460. Fluorescein leakage test method for identifying ocular corrosives and severe irritants. Organisation for Economic Co-operation and Development, Paris, France, 2012, 16 pp.

OECD Guidelines for the Testing of Chemicals 420. Acute oral toxicity – Fixed dose procedure. Organisation for Economic Co-operation and Development, Paris, France, 2001, 14 pp.

OECD Guidelines for the Testing of Chemicals 423. Acute oral toxicity – Acute toxic class method. Organisation for Economic Co-operation and Development, Paris, France, 2001, 14 pp.

OECD Guidelines for the Testing of Chemicals 425. Acute oral toxicity: Up-and-down procedure. Organisation for Economic Co-operation and Development, Paris, France, 2008, 27 pp.

OECD Guidelines for the Testing of Chemicals 428. Skin absorption: *In vitro* method. Organisation for Economic Co-operation and Development, Paris, France, 2004, 8 pp.

OECD Guidelines for the Testing of Chemicals 430. *In vitro* skin corrosion: Transcutaneous Electrical Resistance Test (TER). Organisation for Economic Co-operation and Development, Paris, France, 2004, 12 pp.

OECD Guidelines for the Testing of Chemicals 431. *In vitro* skin corrosion: Human skin model test. Organisation for Economic Co-operation and Development, Paris, France, 2004, 8 pp.

OECD Guidelines for the Testing of Chemicals 432. *In vitro* 3T3 NRU phototoxicity test. Organisation for Economic Co-operation and Development, Paris, France, 2004, 15 pp.

OECD Guidelines for the Testing of Chemicals 435. *In vitro* membrane barrier test method for skin corrosion. Organisation for Economic Co-operation and Development, Paris, France, 2006, 15 pp.

OECD Guidelines for the Testing of Chemicals 437. Bovine corneal opacity and permeability test method for identifying ocular corrosives and severe irritants. Organisation for Economic Co-operation and Development, Paris, France, 2009, 18 pp.

OECD Guidelines for the Testing of Chemicals 438. Isolated chicken eye test method for identifying ocular corrosives and severe irritants. Organisation for Economic Co-operation and Development, Paris, France, 2009, 18 pp.

OECD Guidelines for the Testing of Chemicals 439. *In vitro* skin irritation: Reconstructed human *Epidermis* test method. Organisation for Economic Co-operation and Development, Paris, France, 2010, 18 pp.

OECD Guidelines for the Testing of Chemicals 487. *In vitro* mammalian cell micronucleus test. Organisation for Economic Co-operation and Development, Paris, France, 2010, 18 pp.

Russell W.M.S., Burch R.L. *The Principles of Humane Experimental Technique*. Methuen, London, UK, 1959.

Smyth D.H. *Alternatives to Animal Experiments*. Scolar Press-Royal Defence Society, London, UK, 1978.

Tollefsen KE., Scholz S., Cronin MT., Edwards SW., de Knecht J., Crofton K., Garcia-Reyero N. et al. Applying adverse outcome pathways (AOPs) to support integrated approaches to testing and assessment (IATA). *Regulatory Toxicology and Pharmacology*, 2014, **70**(3), 629–40.

26

In Vitro *Methods: Alternatives to Animal Testing*

Dayane Pifer Luco, Vânia Rodrigues Leite-Silva,
Heather A.E. Benson and Patricia Santos Lopes

CONTENTS

In accordance with the principles of the 3Rs – replacement, reduction and refinement – an alternative method can be used to replace an animal testing method thereby eliminating the need to use any animals in the test; or to refine an animal testing method to minimize the number of animals and/or their pain and suffering. The platforms used to create alternative methods that reduce or replace animal experiments are based on *in vitro* models and computer-based systems (Roi and Grune, 2013).

Establishing the safety of a new cosmetic ingredient involves many steps. First, identification of the chemical identity and composition, with regard to the chemical structure, functional groups and impurities. Second, determination of the physicochemical properties and other molecular descriptors. This includes solubility, stability, form definition, solid-state properties, partition coefficient and ionization constant(s). Third, determination of the physiological kinetics when applied to the body, namely the absorption, distribution, metabolism and excretion. In addition, the mode and/or mechanism of action, dose–response, influence of the route of administration, and adverse outcome pathways must be determined. It is this third evaluation step that historically utilized animal testing and for which now the responses found in alternative assays can lead to a safe cosmetic without the use of animal tests (OECD 194, 2014). This chapter focuses on alternative assays that can replace animal testing for new cosmetic ingredients or products.

The full ban on animal testing for cosmetic products within the European Union (EU) came into force in March 2013. From this date, animal testing for marketing of new cosmetic products in the EU is prohibited. The implementation of the marketing and testing ban follows the Seventh Amendment of the Council Directive on the approximation of the laws of the Member States relating to cosmetic products (76/768/EEC, Cosmetics Directive), which defined the stepwise phase-out of animal testing for cosmetic products and cosmetic ingredients initiated in 2004. In 2009 an EU testing ban for cosmetic ingredients became effective with an extension of three specific areas: repeated dose toxicity (includes skin

TABLE 26.1

Available *In Vitro* Validated Replacement Tests Recognized by the OECD (Organisation for Economic Co-operation and Development)

Endpoints	Situation
Cell transformation assay (carcinogenicity) – limited	No replacement
Cytotoxicity – IC50	OECD (no. 129)
Dermal absorption	Full replacement (TG 428)
Embryotoxicity (reproduction toxicity) – limited	No replacement
Eye irritants	Partial replacement (TG 437, 438, 460)
Mutagenicity/genotoxicity	Partial replacement (TG 487)
Phototoxicity	Full replacement (TG 432)
Skin corrosivity	Full replacement (TG 430, 431)
Skin irritation	Full replacement (TG 439)

Source: Adler, S., et al., 2011, *Arch. Toxicol.* 85:367–485.

sensitization, carcinogenicity and sub-acute/sub-chronic toxicity), reproductive toxicity (includes teratogenicity) and toxicokinetics (Gocht and Schwarz, 2014) (Table 26.1). Although this ban is restricted to the EU, other countries, including the United States and Japan are considering the adoption of alternative methods to test cosmetic ingredients and products. Brazil, China, South Korea, Australia and India have also presented new perspectives in these topics, although the use of animal tests is still recommended in these countries.

In this chapter, the main *in vitro* tests used for risk assessment have been divided into those used for cosmetics ingredients, and tests used for cosmetics products or mixtures.

Cosmetic Ingredients

Cosmetic ingredients can come from a range of sources and chemical classes. Their potential to exert toxicity of the organs and their functions can manifest in many different ways. Thus a series of toxicity tests, including eye irritation, skin irritation, acute toxicity, genotoxicity, teratogenicity must be considered (SCCS, 2012). Assessment of the toxicity potential of cosmetic ingredients should be performed in a logical sequence that is best described by the schematic diagram in Figure 26.1.

Following is a brief description of the principal and most useful tests used to assess the potential *in vitro* toxicity of cosmetic ingredients.

Cytotoxicity

The Interagency Coordinating Committee on the Validation of Alternative Methods (ICCVAM 2006a,b,c) proposed a cytotoxicity test in February 2008. An IC_{50} value (half maximal inhibitory

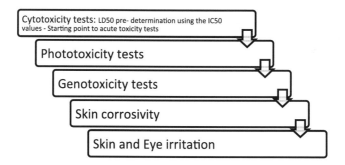

FIGURE 26.1   Schematic diagram of the test sequence to be performed for cosmetics ingredients.

concentration, that is, the concentration of a substance needed to inhibit a given biological process [or component of a process, such as an enzyme, cell, cell receptor or microorganism] by half) from an *in vitro* basal cytotoxicity test is used with the linear regression equation (Registry of Cytotoxicity [RC] regression) to predict the oral LD_{50} value (half-lethal dose, or the individual dose required to kill 50% of a test population). The predicted LD_{50} is used as the starting dose for the acute toxic class (ATC) method or the up-and-down procedure (UDP) test method (Spielmann et al., 1999), and as a first test for all the other cosmetic ingredient *in vitro* tests. Thus the oral LD_{50} for acute oral systemic toxicity is determined.

The test uses BALB/c 3T3 mouse fibroblasts (3T3) or normal human epidermal keratinocytes (NHK) and neutral red uptake (NRU) as the cytotoxicity endpoint. The NRU *in vitro* basal cytotoxicity assay procedure is based on the ability of viable cells to incorporate and bind neutral red (NR), a supravital dye (Borenfreund and Puerner, 1985). NR readily diffuses through the plasma membrane and concentrates in lysosomes. Toxic compounds can alter the cell surface or the lysosomal membrane to cause cell death and/or inhibition of cell growth, which then decreases the amount of NR retained by the culture. The concentration of NR dye desorbed from the cultured cells is directly proportional to the number of living cells, therefore cytotoxicity is expressed as a concentration-dependent reduction of the uptake of NR after chemical exposure (Figure 26.2) (OECD 129, 2010).

Phototoxicity

Phototoxicity is defined as an acute toxic response from a substance applied to the body, which is either elicited or increased after subsequent exposure to light, or that is induced by skin irradiation after systemic administration of a substance (OECD 432, 2004). The test was accepted in 2004 and is applied to identify the phototoxic potential of a test substance induced by the excited chemical after exposure to light.

The test evaluates photocytotoxicity by the relative reduction in viability of cells exposed to the chemical in the presence versus absence of light. The substance is applied to Balb/c 3T3 cells that are exposed to a dose of 5 J/cm^2 (as measured in the UVA range) and then washed to remove the test substance. After 18–22 hours of incubation, the cell viability neutral red uptake assay is performed to evaluate the chemical phototoxicity response (OECD 432, 2004; ECVAM, 2008). Considerations in the design of the test performance are prior validation of the phototoxicity chamber, preparation and concentration of test substances, and the incorporation of appropriate positive and negative controls.

A photoirritation factor (PIF) or mean photo effect (MPE) is determined by processing the data using PHOTOTOX® 2.0 freeware available from ZEBET. A PIF = 1 predicts no phototoxic potential and a MPE < 0.1 is non-phototoxic, but a MPE ≥ 0.1 is phototoxic. This test can identify substances that are likely to be phototoxic *in vivo* following systemic application and distribution to the skin, or after topical application (OECD 432, 2004).

(a)   (b)

FIGURE 26.2   (a) Cytotoxicity NR assay using a 96 microplate and (b) evaluation at 540 nm in a Multiskan EX 355. (Thermo Electron Corporation.)

Genotoxicity

The *in vitro* micronucleus test (MNvit) is a test method that measures the potential for causing damage to chromosomes. In this method, cultured human or rodent (Chinese hamster ovary [CHO]) cells are treated with a test substance and then examined for the presence of chromosome fragments known as micronuclei (Figure 26.3). The test is intended to reduce the number of animals used to identify substances that can lead to cancer and other adverse health effects (OECD 487, 2014; ICCVAM, 2014a). Test Guideline (TG) 487 for the *in vitro* micronucleus test was formally adopted by the OECD in 2010 and with the updated Test Guideline released in September 2014.

At least three test concentrations that meet the acceptability criteria (appropriate cytotoxicity, number of cells, etc.) should be evaluated. A cytokinesis blocker (cytochalasin B [cytoB]) may also be included. S9 fraction (an exogenous metabolic activation system containing cytosol and microsomes) is added to activate metabolism in cells, although the exogenous metabolic activation system does not entirely mimic *in vivo* conditions (OECD 487, 2014). It is mandatory to include clastogenic and aneugenic positive controls, with and without metabolic activation, as well as negative controls, to demonstrate both the ability of the laboratory to identify clastogens and aneugens under the conditions of the test protocol used, and the effectiveness of the exogenous metabolic activation system. Micronuclei in interphase cells can be assessed objectively (Figure 26.3), therefore the slides can be scored relatively quickly and analysis can be automated. This makes it practical to score thousands of cells per treatment, increasing the power of the test.

Concurrent measures of cytotoxicity and/or cytostasis for all treated, negative and positive control cultures should be determined (Zijno et al., 1996). The Cytokinesis-Block Proliferation Index (CBPI) or the Replication Index (RI) should be calculated for all treated and control cultures as measurements of cell cycle delay when the cytokinesis-block method is used.

Providing that all acceptability criteria are fulfilled, a cosmetic ingredient is considered to be clearly positive if at least one test concentration exhibits a statistically significant increase compared with the concurrent negative control, and the increase is dose-related in at least one experimental condition, and finally if any of the results are outside the distribution of the historical negative control data. When all of these criteria are met, the test chemical is considered able to induce chromosome breaks and/or gain or loss in this test system (OECD 487, 2014). A test chemical is considered clearly negative, unable to induce chromosome breaks, if none of the test concentrations exhibit a statistically significant increase compared with the concurrent negative control, there is no concentration-related increase when evaluated with an appropriate trend test, and all results are inside the distribution of the historical negative control data (OECD 487, 2014).

Skin Corrosivity

The *in vitro* skin corrosivity method is based on the production of irreversible damage to the skin defined as visible necrosis, both on the epidermis and dermis, following the application of a test chemical (United Nations, 2010). The method described by OECD TG 430 is based on rat skin transcutaneous electrical resistance (TER). The chemical is applied to three skin discs for a maximum period of 24 hours, and corrosive substances are identified by their ability to produce a loss of normal stratum corneum integrity and barrier function, which is measured as a reduction in the TER below a threshold level (5 kΩ for rat).

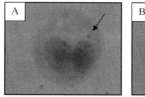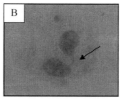

FIGURE 26.3 Micronucleus test performed using CHO cells. (A) Micronucleus (arrow) in a binucleated cell; (B) a bridge (arrow) linking the two nuclei.

Dye binding identifies if the increase in ionic permeability is due to physical destruction of the stratum corneum. This test is most applicable to chemicals but can also be applied to mixtures (OEDC 430, 2013).

For a full evaluation of local skin effects after a single dermal exposure, it is recommended to follow the sequential testing strategy described by OECD TG 404 (2002). This testing strategy describes the conduct of *in vitro* tests for skin corrosion and skin irritation prior to considering testing in live animals (Grindon et al., 2007).

Although the test is used on animal skin, it is considered an alternative test under the refinement principle (Barlow et al., 1991). The *in vitro* skin corrosion test using the reconstructed human epidermis (RHE) test method is described later in the chapter.

Eye Irritation

Evaluation of ocular irritancy potential is essential for the safety of individuals in contact with an increasing number of cosmetics and cosmetic ingredients (Gautheron et al., 1992). The original test that is still used in some countries is the *in vivo* Draize eye irritancy test conducted on rabbits (Draize et al., 1944). This test has raised ethical concerns of animal well-being and been used as a rallying point for concerned individuals and activist organizations regarding how animals are treated in science and industry (Wilhelmus, 2001). Whilst the Draize test protocol was significantly modified in 1981 to reduce suffering, it remains highly controversial. Three alternative methods are described for *in vitro* assessment of eye irritation: bovine corneal opacity (OEDC 437, 2013), isolated chicken eye test (OECD 438, 2013), and fluorescein leakage test method for identifying ocular corrosives and severe irritants (OECD 460, 2012).

Muir (1984, 1985) reported a test that used isolated bovine corneas to assess opacity and permeability after exposure to a substance. The initial protocol was modified to generate the bovine corneal opacity (BCOP) test (OEDC 437, 2013), which assesses toxic effects to the cornea by measurement of the decrease in light transmission (opacity) and increase in the passage of sodium fluorescein dye (permeability). The opacity and permeability assessments are combined to provide an *In Vitro* Irritancy Score (IVIS), which is used to classify the irritancy level of the test chemical (INVITTOX, 1999).

The isolated chicken eye (ICE) test (OECD 438, 2013) follows the same principle as the BCOP test, that is, it utilizes an animal part that is a by-product of human consumption. The ICE test is an organotypic model that permits assessment of the damage caused by the test chemical by determination of corneal swelling (quantitative), opacity and fluorescein retention (qualitative).

The fluorescein leakage test (OECD 460, 2012) involves culturing Madin-Darby canine kidney (MDCK) cells on permeable inserts to generate an epithelial monolayer. Increase in permeability of sodium fluorescein is measured following exposure to the test substance for a short time. The amount of fluorescein leakage is proportional to the chemical-induced damage to the tight junctions, desmosomal junctions and cell membranes, and can be used to estimate the ocular toxicity potential of a test substance.

It is important to note that no single *in vitro* eye irritation test can replace the *in vivo* Draize eye test to predict toxicity across the full range of irritation for different chemical classes. However, strategic combinations of several alternative test methods may be able to replace the animal test (Moldenhauer, 2003; OECD 405, 2012).

Cosmetic Products: Mixtures

Article 2 of Council Directive 76/768/EEC states that a cosmetic product must not cause damage to human health when it is applied under normal and reasonably foreseeable conditions of use. Whilst there are many thousands of different cosmetic products on the market, they are all derived from a far smaller number of ingredients (Rinaldi, 2008; SCCS, 2012). Thus each cosmetic product can be considered as an individual combination of cosmetic substances or ingredients.

The physicochemical and microbiological specifications of raw ingredients are routinely accessed. However, additional information on the finished product could be needed to ensure a thorough safety assessment. There are a number of examples: cosmetics for specific target consumer groups; when the vehicle used increases skin penetration thereby presenting a higher concentration to the tissues than that observed in the toxicity studies on the individual ingredients; a chemical reaction occurs between ingredients rendering the formation of a new, potentially toxic substance; and when there is a claim of reduced skin penetration. Suppliers of new cosmetic products for marketing to the community must ensure that adequate information is provided for a safety analysis of the finished product (SCCS, 2012).

The assessment and evaluation of the safety of a cosmetic product depends on two considerations: how it will be used and its intrinsic toxicological properties. There is a very diverse range of cosmetic product types available to consumers. The variety of application methods (ingestion, inhalation or topical application to skin or mucous membranes), amount and frequency of use of these cosmetics, and the amount of a particular ingredient used in the manufacture of different products is connected to the occurrence of toxicological effects (Salvador and Chisvert, 2007).

The implementation of 3Rs-alternative testing methods as a research tool in toxicity analysis has achieved favorable levels of development during the last few decades, although it is insufficient to overcome the need for animal models in all cases. A combination of many different *in vitro* tests can predict the potential toxic effects of distinct chemicals on human health thus minimizing the need for animal testing (Salvador and Chisvert, 2007; SCCS, 2012). Refinement and reduction improvements have been made to existing *in vivo* guidelines, and partial replacement by *in vitro* test methods has been explained and reinforced in the *in vitro* guidelines. Pressure from the public, policymakers and within the industry has led to this reduction in animal experimentation, and brought us closer to achieving the 3Rs strategy of refinement, reduction and replacement and the ultimate goal of completely eliminating the need for animal testing.

In vitro test methods use whole cells, parts of cells or reconstructed tissues to perform a diversified range of tests and assays. The development of tissue bioengineering related to molecular biology has given rise to the *in vitro* cell culture models to mimic conditions found *in vivo*. Rheinwald and Green's (1975) development of culturing keratinocytes in multiple layers with the discrete presence of keratinization in the surface layers lead to the formation of the first rudimentary epidermal equivalent. Tissue culture models now range from a single cell type monolayer culture, to much more elaborate epithelial and mesenchymal cell co-cultures, and organotypic cultures of three-dimensional skin equivalents that form a highly organized epithelium that closely mimics the cells and tissues in the human body (Fusenig, 1994).

There are a number of validated and recognized skin models that are commercially available. Three common models are based on reconstructed human epidermis (RhE), a three-dimensional epidermal model cultured from human keratinocytes:

- RhE (reconstructed human epidermis) by SkinEthic (Lyon, France) – Used for assessment of skin corrosion, skin irritation, UVB photoprotection, phototoxicity, proof-of-concept of the OECD reference chemicals and percutaneous absorption of chemicals.
- EpiDerm™ tissue model (MatTek Corp., Ashland, Massachusetts) – Used for assessment of acute dermal irritation/corrosion and percutaneous absorption of chemicals.
- EpiSkin (SkinEthic, Lyon, France) – Used for assessment of skin corrosion, irritation, UVB photoprotection, phototoxicity, classification, packaging and labelling of dangerous substances and percutaneous absorption of chemicals.

Development and validation of a new *in vitro* test involves many steps, including the appropriate choice of the assay components, their reproducibility, and accuracy in detecting or grading the response to a raw material or final product. Despite the time and complexity in creating and validating alternative *in vitro* and *in silico* methods, many have been developed and approved by the relevant authorities, thus providing important tools for the replacement of animal models. The European Union Reference Laboratory for alternatives to animal testing (EURL-ECVAM), established in 2011, has validated full replacement RhE methods (as described earlier) in the toxicological areas of skin irritation, skin corrosion, skin absorption/penetration and phototoxicity.

Skin Irritation

The *in vitro* skin irritation test described by OECD TG 439 involves the application of a chemical or mixture for up to 4 hours on the RhE test system and monitoring for reversible damage to the skin caused by a local inflammatory reaction (the innate, non-specific immune system). The three-dimensional RhE model is comprised of non-transformed human-derived epidermal keratinocytes, which have been cultured to form a multilayered, highly differentiated model of the human epidermis. Cell viability is determined by quantitatively measuring the enzymatic conversion of the yellow tetrazolium dye MTT (3-(4,5-dimethylthiazol-2-yl)-2,5-diphenyltetrazolium bromide) into a purple/blue formazan salt. Damage to the cells caused by the application of a test substance can be determined by a change in the amount of dye conversion in the tissues (United Nations, 2010; OECD 439, 2013).

Irritant chemicals or mixtures are identified by their ability to decrease cell viability below defined threshold levels (i.e. $\leq 50\%$, category 2 and 3), whilst chemicals that maintained cell viability above the defined threshold level are considered non-irritants (i.e. $>50\%$, no category) according to the Globally Harmonized System of Classification and Labeling of Chemicals (GHS) (United Nations, 2010; OECD 439, 2013).

Skin Corrosion

The GHS defines 'skin corrosion' as the production of irreversible tissue damage in the skin following the application of a test material. This is manifested as visible necrosis through the epidermis and into the dermis (United Nations, 2010).

The *in vitro* skin corrosion test (OECD TG 431) is based on the topical application of chemicals and mixtures to a three-dimensional RhE model (as performed in OECD TG 439), allowing the identification of non-corrosive and corrosive substances in accordance with GHS, allowing a partial sub-categorization of corrosives (OECD 439, 2013). Identification is based on the ability of the substance to induce a decrease in cell viability due to its penetration into the stratum corneum by diffusion or erosion, and consequent cytotoxicity to the keratinocytes in the underlying layers. Again, cell viability is measured by the MTT assay. Corrosive chemicals are identified by their ability to decrease cell viability below defined threshold levels, labeled category 1; and three further optional subcategories, according to the United Nations Packing Groups I, II and III for the transport of goods (United Nations, 2010; OECD 431, 2013).

Skin Absorption/Penetration

Skin absorption describes the passage of material from topical application on the stratum corneum surface into the skin, whilst skin penetration involves passage through the skin to the systemic circulation (Russell and Guy, 2009). The procedures and parameters for the *in vitro* testing of percutaneous absorption/penetration are well documented and described in the COLIPA Guidelines for Percutaneous Absorption/Penetration, and OECD TG 428. These guidelines describe a general process for measuring the absorption/penetration of test substances through excised fresh or previously frozen mammalian skin (human, pig and rat).

Diffusion of applied chemicals into and across skin disks that separate the donor and receptor chambers of a diffusion cell is monitored over time by removal of the samples and the use of appropriate assays to measure the chemical. Non-viable skin is used to measure diffusion only, whilst freshly excised, metabolically active skin permits simultaneous measurement of diffusion and skin metabolism. The test sample remains in contact with the skin on the donor side for a defined period of time and the receptor fluid is sampled once at the end of the experiment and/or at settled time points during the application period. At the conclusion of the experiment, the amount of chemical in the skin can be measured following extraction from either the whole skin or skin layers obtained by skin stripping. The samples are analyzed by an appropriate method that provides both sensitivity and selectivity (e.g. scintillation counting, high-performance liquid chromatography [HPLC] with various detection methods) (COLIPA, 1997; OECD 428, 2004).

These methods are widely used as a screen for comparing delivery of chemicals into and through skin from different formulations, and can also provide useful models to estimate *in vivo* absorption by extrapolating from suitable *in vitro* data.

Phototoxicity

The 3T3-NRU neutral red uptake (3T3-NRU PT) involving murine cells is an example of a test that is well documented (OECD TG 432, 2004) but has not been certified by regulatory authorities due to reliability issues. This *in vitro* phototoxicity test has some drawbacks as a model for human skin: monolayer cell culture, no stratum corneum and barrier function, requires aqueous solubility of the test substances, mixtures cannot be tested, and so on.

In 1999 EURL ECVAM funded a multilaboratory prevalidation study to develop a phototoxicity test based on the EpiDerm RhE model (Sohn et al., 2009). The eight standard test chemicals used for validation of the 3T3-NRU PT were applied topically to the RhE and UV/vis radiation applied using a solar-simulated light (SSL) source. Cell viability was measured by MTT assay. The EpiDerm tests showed good levels of accuracy and the potential to be an improvement of the current phototoxicity test. In a Draft Guidance Document on Photosafety Testing, the European Medicines Agency (EMA) suggested that confirmatory testing can be performed with this type of skin model (EMEA, 2002; Sohn et al., 2009).

Skin Sensitization

Skin sensitization is described as the toxicological endpoint associated with chemicals that have the intrinsic ability to cause skin allergy (Adler et al., 2011).

Allergic contact dermatitis (ACD) is characterized by localized redness (red rash), swelling, blistering, or itching and burning after direct contact with a skin allergen. This adverse effect results from an overreaction of the adaptive immune system. Induction of sensitization on first contact with allergy symptoms resulted from further contact. Cosmetic ingredients, and other chemicals such as pesticides, are tested for their potential to cause ACD to ensure products are appropriately labeled for safe use and handling (Adler et al., 2011; Kleinstreuer et al., 2014; ICCVAM, 2014b).

Currently, skin sensitization is tested using an animal method such as the mouse local lymph node assay (LLNA) (OECD 429, 2010). Because an alternative *in vitro* test for ACD is not yet available, cosmetic ingredients are evaluated using the adverse outcome pathway (AOP) for skin sensitization. An AOP is a model that links exposure to a substance with a toxic effect, by identifying the sequence of biochemical events required to produce the toxic effect (Kleinstreuer et al., 2014). The AOP for skin sensitization provides a framework for the development of an alternative *in vitro* toxicity test (Adler et al., 2011; ICCVAM, 2014b).

There are several considerations that should be observed in the elaboration of an *in vitro* sensitization model. However, in all cases, the sensitization model substitute should be able to distinguish a sensitizer or a non-sensitizer at the same level as animal models. In this way, the most important aspect in *in vitro* test development is reaching the maximum correlation level of these models with human experimentation. As important as the choice of cell type used to the development of *in vitro* sensitization tests is the kind of biomarker used to evaluate the reactions exhibited by them (Dos Santos et al., 2009).

Conclusion

There has been considerable effort devoted to the development of alternative testing methods for cosmetics and other products. Whilst much progress has been made, many of the currently available methods do not satisfy all the criteria required by the international validation centers and therefore cannot completely remove the need for animal models. However, partial replacement strategies have achieved a substantial reduction in animal testing.

Where the toxicological effects are due to complex multimodal biological processes or 'toxicological endpoints' for which the scientific basis is not fully elucidated, the full replacement of animal testing is not yet achieved (Roi and Grune, 2013). Examples for which standard animal models are still required are repeated-dose toxicity, skin sensitization, carcinogenicity, reproductive toxicity and toxicokinetics. Multiple strategies are being employed to minimize the use of animals where alternative tests are not yet validated. These include the use of existing risk assessment tools (eye irritation, skin corrosion/irritation; phototoxicity, genotoxicity, skin penetration) to provide baseline data so that animal testing is better targeted. New tools are also being developed based on predictive biology, computational/mathematical modelling and informatics, high content screening (HCS) or high content analysis (HCA), and mechanistic chemistry (Reynolds, et al., 2014).

Computer-based approaches (often termed *in silico* or non-testing methods) are becoming increasingly powerful and can be used effectively to predict the toxicity of a chemical from its basic properties. Computer models can also be used to integrate toxicological information derived from complementary *in vitro* and *in silico* methods. An example is the non-testing approach called the 'read-across' technique that is frequently used in the safety assessment of industrial chemicals. Toxicological effects for one chemical are predicted using toxicological data from another chemical that is considered similar in terms of chemical structural, physicochemical properties or bioactivity (De Wever et al., 2012; Anadón et al., 2014).

Finally a different type of alternative method that should be highlighted here is the growing use of 'omics' technologies (e.g. transcriptomics, proteomics and metabolomics) in combination with *in vitro* test systems. These new technologies provide collective characterisation and quantification of pools of biological molecules that translate into the structure, function and dynamics. This allows a comprehensive analysis of the impact of a chemical at the molecular level and can indicate potential toxicity pathways that may lead to adverse health effects (Roi and Grune, 2013; Anadón et al., 2014; Simó et al., 2014).

REFERENCES

Adler, S.; Basketter, D.; Creton, S.; Pelkonen, O.; van Benthem, J.; Zuang, V.; et al. 2011. Alternative (non-animal) methods for cosmetics testing: Current status and future prospects. *Arch. Toxicol.* 85:367–485.

Anadón, A.; Martínez, M.A.; Castellano, V.; Martínez-Larrañaga, M.R. 2014. The role of in vitro methods as alternatives to animals in toxicity testing. *Expert Opin. Drug Metab. Toxicol.* 10:67–79.

Barlow, A.; Hirst, R.A.; Pemberton, M.A.; Rigden, A.; Hall, T.J.; Botham, P.A.; et al. 1991. Refinement of an in vitro test for the identification of skin corrosive chemicals. *Toxicol. Methods.* 1:106–115.

Borenfreund, E.; Puerner, J.A. 1985. Toxicity determination in vitro by morphological alterations and neutral red absorption. *Toxicol. Lett.* 24:119–124.

COLIPA. 1997. Cosmetic ingredients: Guidelines for percutaneous absorption/penetration.

De Wever, B.; Fuchs, H.W.; Gaca, M.; Krul, C.; Mikulowski, S.; Poth, A.; et al. 2012. Implementation challenges for designing integrated in vitro testing strategies (ITS) aiming at reducing and replacing animal experimentation. *Toxicol. In Vitro.* 26:526–534.

Dos Santos, G.G.; Reinders, J.; Ouwehand, K.; Rustemeyer, T.; Scheper, R.J.; Gibbs, S. 2009. Progress on the development of human in vitro dendritic cell based assays for assessment of the sensitizing potential of a compound. *Toxicol. Appl. Pharmacol.* 236(3):372–382.

Draize, J.H.; Woodard, G.; Calvery, H.O. 1944. Methods for the study of irritation and toxicity of substances applied topically to the skin and mucous membranes. *J. Pharmacol. Exp. Ther.* 82:377–390.

ECVAM. 2008. DB-ALM: INVITTOX protocol. 3T3 NRU Phototoxicity Assay. INVITTOX no. 78, p. 21.

EMEA. 2002. Note for guidance on photo safety testing (CPMP/SWP/398/01). London: European Agency for the Evaluation of Medicinal Products, Committee for Proprietary Medicinal Products. http://www.ema.europa.eu/docs/en_GB/document_library/Scientific_guideline/2009/09/WC500003353.pdf.

Fusenig, N.E. 1994. *Epithelial-Mesenchymal Interactions Regulate Keratinocyte Growth and Differentiation In Vitro.* Cambridge, UK: Cambridge University Press.

Gautheron, P.; Dukic, M.; Alix, D.; Sina, J.F. 1992. Bovine corneal opacity and permeability test: An in vitro assay of ocular irritancy. *Fundam. Appl. Toxicol.* 18:442–449.

Gocht, T.; Schwarz, M. 2014. *SEURAT-1 – Towards the Replacement of In Vivo Repeated Dose Systemic Toxicity Testing*, Vol. 4. France: Mouzet Imprimerie.

Grindon, C.; Combes, R.; Cronin, M.T.D.; Roberts, D.W.; Garrod, J.F. 2007. Integrated decision-tree testing strategies for skin corrosion and irritation with respect to the requirements of the EU REACH legislation. *ATLA.* 35:673–682.

ICCVAM – NTP. 2014a. National Toxicology Program, U.S. Department of Health and Human Services. http://ntp.niehs.nih.gov/pubhealth/evalatm/test-method-evaluations/genetic-toxicity/index.html.

ICCVAM – NTP. 2014b. National Toxicology Program, U.S. Department of Health and Human Services. Evaluations of Non-Animal Skin Sensitization Test Methods and Testing Strategies. http://ntp.niehs.nih.gov/pubhealth/evalatm/test-method-evaluations/immunotoxicity/nonanimal/index.html.

ICCVAM. 2006a. Background Review Document: In Vitro Basal Cytotoxicity Test Methods for Estimating Acute Oral Systemic Toxicity. Research Triangle Park, NC: National Institute for Environmental Health Sciences. http://iccvam.niehs.nih.gov/methods/acutetox/inv_nru_brd.htm.

ICCVAM. 2006b. Peer Review Panel Report: The Use of In Vitro Basal Cytotoxicity Test Methods for Estimating Starting Doses for Acute Oral Systemic Toxicity Testing. Research Triangle Park, NC: National Institute for Environmental Health Sciences. http://iccvam.niehs.nih.gov/methods/acutetox/inv_nru_scpeerrev.htm.

ICCVAM. 2006c. ICCVAM Test Method Evaluation Report: In Vitro Cytotoxicity Test Methods for Estimating Starting Doses for Acute Oral Systemic Toxicity Tests. Research Triangle Park, NC: National Institute for Environmental Health Sciences. http://iccvam.niehs.nih.gov/methods/acutetox/inv_nru_tmer.htm.

INVITTOX. 1999. Protocol 124: Bovine Corneal Opacity and Permeability Assay – SOP of Microbiological Associates Ltd. Ispra, Italy: European Centre for the Validation of Alternative Methods (ECVAM).

Kleinstreuer, N.; Strickland, J.; Allen, D.; Casey, W. 2014. Predicting Skin Sensitization Using ToxCast Assays. ToxCast Data Summit Poster. http://ntp.niehs.nih.gov/iccvam/meetings/9wc/posters/kleinstreuer-tox21ss-wc9.pdf.

Moldenhauer, F. 2003. Using *in vitro* prediction models instead of the rabbit eye irritation test to classify and label new chemicals: A post hoc data analysis of the international EC/HO validation study. *Altern. Lab. Anim.* 31(1):31–46.

Muir, C.K. 1984. A simple method to assess surfactant-induced bovine corneal opacity in vitro: Preliminary findings. *Toxicol. Lett.* 22:199–203.

Muir, C.K. 1985. Opacity of bovine cornea in vitro induced by surfactants and industrial chemicals compared with ocular irritancy *in vivo. Toxicol. Lett.* 24:157–162.

OECD. 2002. Test No. 404: Acute Dermal Irritation/Corrosion, OECD Guidelines for the Testing of Chemicals, Section 4, OECD Publishing.

OECD. 2004. Test No. 428: Skin Absorption: In Vitro Method, OECD Guidelines for the Testing of Chemicals, Section 4, OECD Publishing.

OECD. 2004. Test No. 432: In Vitro 3T3 NRU Phototoxicity Test, OECD Guidelines for the Testing of Chemicals, Section 4, OECD Publishing.

OECD. 2010. Test No. 129. Guidance Document on Using Cytotoxicity Tests to Estimate Starting Doses for Acute Oral Systemic Toxicity Tests. OECD Publishing.

OECD. 2010. Test No. 429: Skin Sensitisation: Local Lymph Node Assay, OECD Guidelines for the Testing of Chemicals, Section 4, OECD Publishing. DOI: 10.1787/9789264071100-en

OECD. 2012. Test No. 405: Acute Eye Irritation/Corrosion, OECD Guidelines for Testing of Chemicals, Section 4, OECD Publishing.

OECD. 2012. Test No. 460: Fluorescein Leakage Test Method for Identifying Ocular Corrosives and Severe Irritants, OECD Guidelines for the Testing of Chemicals, Section 4, OECD Publishing.

OECD. 2013. Test No. 430: In Vitro Skin Corrosion: Transcutaneous Electrical Resistance Test Method (TER), OECD Guidelines for the Testing of Chemicals, Section 4, OECD Publishing.

OECD. 2013. Test No. 431: In Vitro Skin Corrosion: Reconstructed Human Epidermis (Rhe) Test Method, OECD Guidelines for the Testing of Chemicals, Section 4, OECD Publishing.

OECD. 2013. Test No. 437: Bovine Corneal Opacity and Permeability Test Method for Identifying i) Chemicals Inducing Serious Eye Damage and ii) Chemicals Not Requiring Classification for Eye Irritation or Serious Eye Damage, OECD Guidelines for the Testing of Chemicals, Section 4, OECD Publishing.

OECD. 2013. Test No. 438: Isolated Chicken Eye Test Method for Identifying i) Chemicals Inducing Serious Eye Damage and ii) Chemicals Not Requiring Classification for Eye Irritation or Serious Eye Damage, OECD Guidelines for the Testing of Chemicals, Section 4, OECD Publishing.

OECD. 2013. Test No. 439: In Vitro Skin Irritation – Reconstructed Human Epidermis Test Method, OECD Guidelines for the Testing of Chemicals, Section 4, OECD Publishing.

OECD. 2014. Test No. 194: OECD Guidance on Grouping of Chemicals, 2nd edition. Series on Testing & Assessment. OECD Publishing.

OECD. 2014. Test No. 487: In Vitro Mammalian Cell Micronucleus Test, OECD Guidelines for the Testing of Chemicals, Section 4, OECD Publishing.

Reynolds, F.; Westmoreland, C.; Fentem, J. 2014. Non-animal approaches and safety science – Toxicity testing. *Biochemist.* 36:19–25.

Rheinwald, J.G.; Green, H. 1975. Formation of keratinizing epithelium in culture by a cloned cell line derived from teratoma. *Cell.* 6(3):317–330.

Rinaldi, A. 2008. Healing beauty? More biotechnology cosmetic products that claim drug-like properties reach the market. *EMBO Rep.* 9:1073–1077.

Roi, A.J.; Grune, B. 2013. *The ECVAM Search Guide – Good Search Practice on Animal Alternatives.* Luxembourg: Publications Office of the European Union.

Russell, L.M.; Guy, R.H. 2009. Measurement and prediction of the rate and extent of drug delivery into and through the skin. *Expert Opin. Drug Deliv.* 6:355–369.

Salvador, A.; Chisvert, A. 2007. *Analysis of Cosmetic Products.* Amsterdam: Elsevier.

SCCS. 2012. The SCCS's Notes of Guidance for the Testing of Cosmetic Substances and their Safety Evaluation, 8th revision. Scientific Committee on Consumer Safety.

Simó, C.; Cifuentes, A.; García-Cañas, V. 2014. *Fundamentals of Advanced Omics Technologies: From Genes to Metabolites.* Madrid: Elsevier.

Sohn, S.; Ju, J.H.; Son, K.H.; Lee, J.P.; Kim, J.; Lim, C.H.; et al. 2009. Alternative methods for phototoxicity test using reconstructed human skin model. *Altex.* 26:136.

Spielmann, H.; Genschow, E.; Liebsch, M.; Halle, W. 1999. Determination of the starting dose for acute oral toxicity (LD50) testing in the up and down procedure (UDP) from cytotoxicity data. *Altern. Lab. Anim.* 27:957–966.

United Nations. 2010. Globally Harmonized System of Classification and Labelling of Chemicals (GHS). 2nd revised edition. http://www.unece.org/trans/danger/publi/ghs/ghs_rev02/02files_e.html.

Wilhelmus, K.R. 2001. The Draize eye test. *Surv. Ophthalmol.* 45,6:493–515.

Zijno, P.; Leopardi, P.; Marcon, F.; Crebelli, R. 1996. Analysis of chromosome segregation by means of fluorescence in situ hybridization: Application to cytokinesis-blocked human lymphocytes. *Mutat. Res.* 372(2):211–219.

Index

Printed in the United States
by Baker & Taylor Publisher Services